METHODS IN MOLECU

Series Editor
John M. Walker
School of Life and Medical Sciences
University of Hertfordshire
Hatfield, Hertfordshire, AL10 9AB, UK

For further volumes:
http://www.springer.com/series/7651

Mitochondrial Bioenergetics

Methods and Protocols

Second Edition

Edited by

Carlos M. Palmeira and António J. Moreno

Department of Life Sciences, University of Coimbra, Coimbra, Portugal

Editors
Carlos M. Palmeira
Department of Life Sciences
University of Coimbra
Coimbra, Portugal

António J. Moreno
Department of Life Sciences
University of Coimbra
Coimbra, Portugal

ISSN 1064-3745 ISSN 1940-6029 (electronic)
Methods in Molecular Biology
ISBN 978-1-4939-7830-4 ISBN 978-1-4939-7831-1 (eBook)
https://doi.org/10.1007/978-1-4939-7831-1

Library of Congress Control Number: 2018941105

Cover caption: Cover image authored by Madalena Palmeira

Printed on acid-free paper

This Humana Press imprint is published by the registered company Springer Science+Business Media, LLC part of Springer Nature.
The registered company address is: 233 Spring Street, New York, NY 10013, U.S.A.

Preface

The purpose of this second edition is to provide a definitive treatise on the conceptual principles and practical considerations regarding research initiatives in mitochondrial function, by adding new chapters related to new approaches that have been developed in this area in recent years. There has been a robust emergence of interest in mitochondrial bioenergetics of late, much of it driven by realization of the impact of drug and environmental chemical-induced disturbances of mitochondrial function as well as hereditary deficiencies and the progressive deterioration of bioenergetic performance with age. These initiatives have fostered the investment of enormous resources into the investigation of genetic and environmental influences on bioenergetics, demanding a certain degree of understanding of the fundamental principles and a level of proficiency in the practice of mitochondrial bioenergetic research.

This book is intended as a bench reference for a broad scope of readers, spanning students of mitochondrial bioenergetics to practitioners in the industries of pharmaceutical and environmental sciences as well as mitochondrial genetics. Each section is prefaced with a short treatise introducing the fundamental principles for that section followed by chapters describing the practical principles and assays designed to derive quantitative assessment of each set of parameters that reflect different aspects of mitochondrial bioenergetics. The hope is that this text will help elevate the quality and rate of investigative discoveries regarding disease states associated with environmental or genetic influences on mitochondrial bioenergetics.

Coimbra, Portugal

Carlos M. Palmeira
António J. Moreno

Contents

Contributors

CHARLES AFFOURTIT • *School of Biomedical and Healthcare Sciences, University of Plymouth, Plymouth, UK*
SALVATORE ANTONUCCI • *Department of Biomedical Sciences, University of Padova, Padova, Italy*
MIGUEL A. AON • *Laboratory of Cardiovascular Science, National Institute on Aging, National Institutes of Health, Baltimore, MD, USA*
ANTONIS A. ARMOUNDAS • *Cardiovascular Research Center, Massachusetts General Hospital, Harvard Medical School, Charlestown, MA, USA*
RUI M. BARBOSA • *Center for Neurosciences and Cell Biology, University of Coimbra, Coimbra, Portugal; Faculty of Pharmacy, University of Coimbra, Coimbra, Portugal*
NICOLE BEZUIDENHOUT • *ESSM UCT Department of Human Biology Sports Science, Institute of South Africa Newlands, University of Cape Town, Cape Town, South Africa*
MASSIMO BONORA • *Department of Experimental and Diagnostic Medicine, Section of General Pathology, Interdisciplinary Center for the Study of Inflammation (ICSI), BioPharmaNet, University of Ferrara, Ferrara, Italy*
MARTIN D. BRAND • *Buck Institute for Research on Aging, Novato, CA, USA*
SILVIA CAMPELLO • *Department of Biology, University of Rome Tor Vergata, Rome, Italy; IRCCS, Fondazione Santa Lucia, Rome, Italy*
RUI A. CARVALHO • *Department of Life Sciences, Faculty of Sciences and Technology, University of Coimbra, Coimbra, Portugal; Centre for Functional Ecology, University of Coimbra, Coimbra, Portugal*
KIERAN J. CLARKE • *Trinity Biomedical Science Institute (TBSI), School of Biochemistry and Immunology, Trinity College Dublin, Dublin 2, Ireland*
SONIA CORTASSA • *National Institute on Aging, National Institutes of Health, Baltimore, MD, USA*
LOUISE T. DALGAARD • *Department of Science and Environment, Roskilde University, Roskilde, Denmark*
SEAN M. DAVIDSON • *The Hatter Cardiovascular Institute, Institute of Cardiovascular Science, University College London, London, UK*
SONI DESHWAL • *Department of Biomedical Sciences, University of Padova, Padova, Italy*
PRATIKSHA DIGHE • *The Buck Institute for Research on Aging, Novato, CA, USA*
CAROLINA DOERRIER • *Oroboros Instruments, Innsbruck, Austria*
MICHAEL R. DUCHEN • *Mitochondrial Biology Group, Cell and Developmental Biology, University College London, London, UK*
JERZY DUSZYNSKI • *Laboratory of Bioenergetics and Biomembranes, Department of Biochemistry, Nencki Institute of Experimental Biology, Warsaw, Poland*
LUIZ F. GARCIA-SOUZA • *Daniel Swarovski Research Laboratory, Mitochondrial Physiology, Department of Visceral, Transplant and Thoracic Surgery, Medical University of Innsbruck, Innsbruck, Austria*
JULIAN GEIGER • *Department of Science and Environment, Roskilde University, Roskilde, Denmark*

ERICH GNAIGER • *Oroboros Instruments, Innsbruck, Austria; Daniel Swarovski Research Laboratory, Mitochondrial Physiology, Department of Visceral, Transplant and Thoracic Surgery, Medical University of Innsbruck, Innsbruck, Austria*
YOUNG-MI GO • *Division of Pulmonary, Allergy and Critical Care Medicine, Department of Medicine, Emory University, Atlanta, GA, USA*
ELISABETH HILLER • *Oroboros Instruments, Innsbruck, Austria*
JAMES HYNES • *Luxcel Biosciences, Cork, Ireland*
IVANA JARAK • *Centre for Functional Ecology, University of Coimbra, Coimbra, Portugal*
DEAN P. JONES • *Division of Pulmonary, Allergy and Critical Care Medicine, Department of Medicine, Emory University, Atlanta, GA, USA*
NINA KALUDERCIC • *Neuroscience Institute, National Research Council of Italy (CNR), Padova, Italy*
TOMOTAKE KANKI • *Department of Cellular Physiology, Niigata University Graduate School of Medical and Dental Sciences, Niigata, Japan*
TIMEA KOMLÓDI • *Oroboros Instruments, Innsbruck, Austria*
GERHARD KRUMSCHNABEL • *Oroboros Instruments, Innsbruck, Austria*
FELIX T. KURZ • *Cardiovascular Research Center, Massachusetts General Hospital, Harvard Medical School, Charlestown, MA, USA; Department of Neuroradiology, Heidelberg University Hospital, Heidelberg, Germany*
JOÃO LARANJINHA • *Center for Neurosciences and Cell Biology, University of Coimbra, Coimbra, Portugal; Faculty of Pharmacy, University of Coimbra, Coimbra, Portugal*
MAGDALENA LEBIEDZINSKA • *Laboratory of Bioenergetics and Biomembranes, Department of Biochemistry, Nencki Institute of Experimental Biology, Warsaw, Poland*
ANA LEDO • *Center for Neurosciences and Cell Biology, University of Coimbra, Coimbra, Portugal*
JOHN J. LEMASTERS • *Center for Cell Death, Injury & Regeneration, Department of Drug Discovery & Biomedical Sciences, and Hollings Cancer Center, Medical University of South Carolina, Charleston, SC, USA; Center for Cell Death, Injury & Regeneration, Department of Biochemistry & Molecular Biology, and Hollings Cancer Center, Medical University of South Carolina, Charleston, SC, USA*
FABIO DI LISA • *Department of Biomedical Sciences, University of Padova, Padova, Italy; Neuroscience Institute, National Research Council of Italy (CNR), Padova, Italy*
JULIA C. LIU • *Systems Biology Center, NHLBI, NIH, Bethesda, MD, USA*
VITOR M. C. MADEIRA • *Department of Life Sciences, Largo Marques de Pombal, University of Coimbra, Coimbra, Portugal*
ANDRÁS T. MÉSZÁROS • *Oroboros Instruments, Innsbruck, Austria*
PIERRE-AXEL MONTERNIER • *Buck Institute for Research on Aging, Novato, CA, USA*
SHONA A. MOOKERJEE • *Touro University California College of Pharmacy, Vallejo, CA, USA; The Buck Institute for Research on Aging, Novato, CA, USA*
ELIZABETH MURPHY • *Systems Biology Center, NHLBI, NIH, Bethesda, MD, USA*
DAVID G. NICHOLLS • *Buck Institute for Research on Aging, Novato, CA, USA*
BRIAN O'ROURKE • *Division of Cardiology, Department of Medicine, Johns Hopkins University, Baltimore, MD, USA*
PAULO J. OLIVEIRA • *MitoXT—Mitochondrial Toxicology and Experimental Therapeutics Laboratory, CNC—Center for Neuroscience and Cell Biology, University of Coimbra, Cantanhede, Portugal*
ADAM L. ORR • *Helen and Robert Appel Alzheimer's Disease Research Institute, Feil Family Brain and Mind Research Institute, Weill Cornell Medicine, New York, NY, USA*

CARLOS MARQUES PALMEIRA • *Department of Life Sciences, Largo Marquês de Pombal, University of Coimbra, Coimbra, Portugal; Center for Neurosciences and Cell Biology, University of Coimbra, Coimbra, Portugal*

RANDI J. PARKS • *Systems Biology Center, NHLBI, NIH, Bethesda, MD, USA*

PAOLO PINTON • *Department of Experimental and Diagnostic Medicine, Section of General Pathology, Interdisciplinary Center for the Study of Inflammation (ICSI), BioPharmaNet, University of Ferrara, Ferrara, Italy*

RICHARD K. PORTER • *Trinity Biomedical Science Institute (TBSI), School of Biochemistry and Immunology, Trinity College Dublin, Dublin 2, Ireland*

CASEY L. QUINLAN • *The Buck Institute for Research on Aging, Novato, CA, USA*

VENKAT K. RAMSHESH • *Center for Cell Death, Injury & Regeneration, Department of Drug Discovery & Biomedical Sciences, and Hollings Cancer Center, Medical University of South Carolina, Charleston, SC, USA; GE Healthcare Life Sciences, Marlborough, MA, USA*

JAMES R. ROEDE • *Skaggs School of Pharmacy and Pharmaceutical Sciences, University of Colorado, Aurora, CO, USA*

ANABELA PINTO ROLO • *Department of Life Sciences, Largo Marquês de Pombal, University of Coimbra, Coimbra, Portugal; Center for Neurosciences and Cell Biology, University of Coimbra, Coimbra, Portugal*

ANA M. SILVA • *MitoXT—Mitochondrial Toxicology and Experimental Therapeutics Laboratory, CNC—Center for Neuroscience and Cell Biology, University of Coimbra, Cantanhede, Portugal; Department of Life Sciences, University of Coimbra, Cantanhede, Portugal*

LUCA SIMULA • *Department of Biology, University of Rome Tor Vergata, Rome, Italy; Department of Pediatric Hematology and Oncology, IRCCS Bambino Gesù Children's Hospital, Rome, Italy*

ONDREJ SOBOTKA • *Department of Physiology, Faculty of Medicine in Hradec Kralove, Charles University, Prague, Czech Republic*

STEVEN J. SOLLOTT • *National Institute on Aging, National Institutes of Health, Baltimore, MD, USA*

JAN SUSKI • *Laboratory of Bioenergetics and Biomembranes, Department of Biochemistry, Nencki Institute of Experimental Biology, Warsaw, Poland; Department of Experimental and Diagnostic Medicine, Section of General Pathology, Interdisciplinary Center for the Study of Inflammation (ICSI), BioPharmaNet, University of Ferrara, Ferrara, Italy*

RACHEL L. SWISS • *Pfizer Global R&D, Compound Safety Prediction-WWMC, Cell Based Assays and Mitochondrial Biology, Groton, CT, USA*

JOÃO SOEIRO TEODORO • *Department of Life Sciences, Largo Marquês de Pombal, University of Coimbra, Coimbra, Portugal; Center for Neurosciences and Cell Biology, University of Coimbra, Coimbra, Portugal*

MARIUSZ R. WIECKOWSKI • *Laboratory of Bioenergetics and Biomembranes, Department of Biochemistry, Nencki Institute of Experimental Biology, Warsaw, Poland*

YVONNE WILL • *Pfizer Global R&D, Compound Safety Prediction-WWMC, Cell Based Assays and Mitochondrial Biology, Groton, CT, USA*

YVONNE WOHLFARTER • *Oroboros Instruments, Innsbruck, Austria*

HOI-SHAN WONG • *Buck Institute for Research on Aging, Novato, CA, USA*

SHUN-ICHI YAMASHITA • *Department of Cellular Physiology, Niigata University Graduate School of Medical and Dental Sciences, Niigata, Japan*

Chapter 1

Overview of Mitochondrial Bioenergetics

Vitor M. C. Madeira

Abstract

Bioenergetic science started in the eighteenth century with the pioneer works by Joseph Priestley and Antoine de Lavoisier on photosynthesis and respiration, respectively. New developments were implemented by Pasteur in the 1860s with the description of fermentations associated with microorganisms, further documented by Buchner brothers who discovered that fermentations also occurred in cell extracts in the absence of living cells. In the beginning of the twentieth century, Harden and Young demonstrated that orthophosphate and other heat-resistant compounds (*cozymase*), later identified as NAD, ADP, and metal ions, were mandatory in the fermentation of glucose. The full glycolysis pathway has been detailed in the 1940s with the contributions of Embden, Meyeroff, Parnas, and Warburg, among others.

Studies on the citric acid cycle started in 1910 (Thunberg) and were elucidated by Krebs et al. in the 1940s.

Mitochondrial bioenergetics gained emphasis in the late 1940s and 1950s with the works of Lehninger, Racker, Chance, Boyer, Ernster, and Slater, among others. The prevalent "chemical coupling hypothesis" of energy conservation in oxidative phosphorylation was challenged and replaced by the "chemiosmotic hypothesis" originally formulated in the 1960s by Mitchell and later substantiated and extended to energy conservation in bacteria and chloroplasts, besides mitochondria, with clear-cut identification of molecular proton pumps.

After identification of most reactive mechanisms, emphasis has been directed to structure resolution of molecular complex clusters, e. g., cytochrome c oxidase, complex III, complex II, ATP synthase, photosystem I, photosynthetic water-splitting center, and energy collecting antennae of several photosynthetic systems.

Modern trends concern to the reactivity of radical and other active species in association with bioenergetic activities. A promising trend concentrates on the cell redox status quantified in terms of redox potentials.

In spite of significant development and advances of bioenergetic knowledge, major issues remain mainly related with poor experimental designs not representative of the real native cell conditions. Therefore, a major effort has to be implemented regarding direct observations in situ.

Key words ATP synthase, Bioenergetics, Chloroplast, Chemical coupling hypothesis, Chemiosmotic hypothesis, Citric acid cycle, Complexes II and III, Cytochrome c oxidase, Energy collecting antennae, Fermentation, Glycolysis, Mitochondria, Photosystem, Radical, Redox potential, Water-splitting center

Carlos M. Palmeira and António J. Moreno (eds.), *Mitochondrial Bioenergetics: Methods and Protocols*,
Methods in Molecular Biology, vol. 1782, https://doi.org/10.1007/978-1-4939-7831-1_1,

1 Overview

Very often, bioenergetics is wrongly considered a modern science of the twentieth century. Rather, the pioneers started already in the middle of the eighteenth century, with the works of Joseph Priestley, in London, and Antoine de Lavoisier, in Paris, in parallel with the development of chemistry.

Joseph Priestley [1, 2] described several important gases, including oxygen, named *dephlogisticated* air. He discovered that plants, in the presence of daylight, could purify the bad air (unsuitable for respiration) which resulted from combustions and metal calcinations (oxidations).

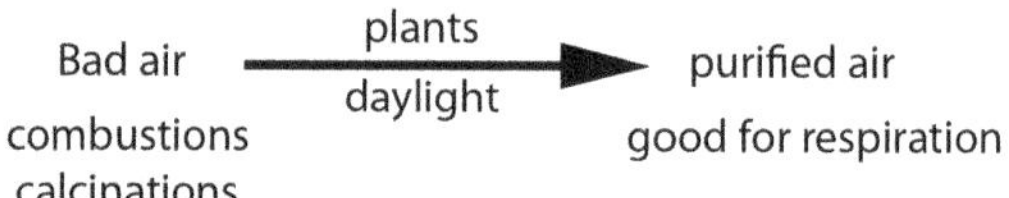

In Priestley's terminology, the bad or *phlogisticated* air (loaded with *phlogist*, the principle of fire and calcinations) (cf. Phlogiston Theory) turned into *dephlogisticated* air. Later, Priestley concluded that *dephlogisticated* air was in fact a fraction of the total air.

Antoine de Lavoisier clearly established that Priestley's *dephlogisticated* air was an independent gas, initially named *vital air* and later *oxygine* (acid generator, in Greek), the acidifying principle of common acids [3, 4]. Additionally, Lavoisier carried out precise experiments on animal and human respiration, showing that pure air turned into *acide crayeux* (lime acid, carbon dioxide) in lungs during respiration. He further demonstrated that there is a constant relationship between the heat released and the amount of air that entered the lungs [4]. Therefore, the first measurement of a metabolic rate has been reported by Lavoisier ca. 1780.

In 1860, Louis Pasteur described fermentation of sugar into alcohol strictly linked to living yeast cells [5]. By the time, it was believed that the ferments (enzymes) could only be active inside the living cell "organized" by the "vital principle" (Vitalism Theory). However, by the end of the nineteenth century, chemists Hans Buchner and Eduard Buchner [6, 7] clearly showed that filtered yeast juice (cells absent) was very active in the fermentation of sugar into alcohol, meaning that the ferments are active when "disorganized", i. e., without the need of any "vital principle." This basic idea opened the way of modern biochemistry, allowing detailed studies of metabolic pathways carried out later in the twentieth century.

Details on fermentation were provided in 1905 by Harden and Young who demonstrated that yeast extracts rapidly fermented glucose into alcohol and that orthophosphate is required and consumed in the phosphorylation of hexose [8, 9]. They separated the yeast extract in two types of compounds: *zymase*, nondialyzable and easily inactivated by heat, and *cozymase*, a dialyzable fraction resistant to heat. The *zymase* fraction contained the glycolytic enzymes, and *cozymase*, besides orthophosphate, contained NAD^+, ADP, and metal ions.

Full glycolysis pathway has been detailed in the 1940s with the major contributions of Embden, Meyeroff, Parnas, Warburg, and other researchers [10].

Studies on citric acid cycle started in 1910 with the work of Thunberg, followed by important achievements of Szent-Györgyi in 1935 [5]. However, full elucidation has been provided by Krebs [11] who established the condensation reaction as the cycle closing step.

Mitochondrial bioenergetics gained emphasis on the late 1940s and 1950s with the relevant work of eminent scientists, e. g., Lehninger, Racker, Chance, Boyer, Ernster, and Slater [12]. It has been shown that oxidations of NADH and $CoQH_2$, formed at the expense of oxidation of citric acid cycle intermediates, resulted in energy yielding effective in ATP synthesis. The coupling of oxidations and ATP synthesis has been explained on the basis of the "chemical coupling hypothesis" [5] proposing that energy of oxidations could be conserved in "high energy compounds," e. g., phosphate esters as in the case of "substrate-level phosphorylations" occurring in glycolysis and the citric acid cycle [5]. However, these compounds were never demonstrated or isolated.

In the 1960s, Mitchell in his "chemiosmotic hypothesis" proposed that the energy of oxidations is transduced into physicochemical states through the establishment of a transmembrane proton gradient generating a transmembrane electric potential and/or a pH gradient [13] described by the famous equation of the "protonmotive force":

$$\Delta p = \Delta\Psi - Z\Delta pH \ (Z = 59 \text{ at } 25^{\circ}\text{C})$$

where the transmembrane potential $\Delta\Psi = \Psi_{in} - \Psi_{out}$ and the transmembrane $\Delta pH = pH_{in} - pH_{out}$.

The protonmotive force (Δp) is expressed in mV and represents the electrochemical potential (free energy change) of the transmembrane proton electrochemical gradient divided by the Faraday's constant. It is analogous to the electromotive force in electricity.

This hypothesis is generally accepted and explains conveniently the energy transduction in mitochondria, bacteria, and chloroplasts [14].

Δp is provided by protonmotive force generators at the expense of transmembrane proton transport: from matrix to cytoplasm in mitochondria, from cytoplasm to the intermembrane space in bacteria, and from stroma to thylakoid space in chloroplasts [14].

Protonmotive force development has been detected in all known membrane redox systems [13], viz., mitochondria, bacteria, chloroplasts, and Archaea [15].

Protonmotive force generators function in the basis of molecular proton pumps and the ubiquinone pool associated with the activity of mitochondrial complex III or equivalent in bacteria and chloroplasts as proposed originally by Mitchell in his looping mechanisms [14].

Molecular proton pumps have been identified in mitochondria [16] and bacterial cytochrome c oxidases [17]. Proton pumps have been also putatively assigned to complex I [18].

The protonmotive force energy may be used in several activities: ATP synthesis, ion transport, orthophosphate transport, nucleotide exchange, transhydrogenase, heat generation, and flagellar motion [10, 14].

Redox processes are carried out in complex protein clusters: I (NADH-CoQ oxidoreductase), II (succinate-CoQ oxidoreductase), III ($CoQH_2$-cytochrome c oxidoreductase), and IV (cytochrome c-O_2 oxidoreductase) or the equivalents in bacteria. In chloroplasts, complexes I, II, and IV are absent, and photosynthetic PSI and PSII clusters are present.

Complexes diffuse laterally in the lipid membrane [19] with relatively low diffusion constants for the big complexes (I, II, III, and IV), and the transfer of reducing equivalents is achieved randomly during effective encounters. These are significantly facilitated and accelerated by the ubiquinone pool (associating complexes I and II with III) and cytochrome c (associating complexes III and IV), owing to the fast diffusion constants [19].

This strategy involving big complex clusters and small fast components (ubiquinone and cytochrome c or equivalents) is a common motif in all known redox systems: mitochondria, bacteria, and chloroplasts. In several complexes (I and II), redox sequences occur in two distinct segments: a two-electron event (e. g., oxidation of NADH to NAD^+ and succinate to fumarate) is followed by a one-electron event (iron-sulfur centers of complexes I and II). The two events are coupled by a flavin center able to process either two or one electron at a time, undergoing intermediate semiquinone species [10]. This strategy is a common motif in all redox systems where a redox segment of two-electrons is followed by an one-electron segment, e.g., bacteria [20], photosynthetic bacteria [20] (synthesis of NADH), chloroplasts [10] (synthesis of NADPH), methanogens [20], and nitrogenase systems in *Rhizobium* [21].

After identification of basic reactive mechanisms, emphasis has been directed to the structure resolution of complex clusters. Detailed structures of cytochrome c oxidases [22, 23], complex III [24], and complex II [25] have been described. Complex I has been also partially resolved [18]. Structure of ATP synthase [26] and its clustering with the orthophosphate carrier and the adenine nucleotide carrier [27] (ATP synthasome) have been established. Available also are the structures of photosystem I [28, 29], water-splitting center [30], and energy collecting antennae [31–33] of several photosynthetic systems. The structural trend is still in progress for other bioenergetic assemblies.

Modern research trends are concerned with oxidative processes related with the formation and reactivity of radical species (oxygen related and other) in association with mitochondrial and other cell activities. A promising trend, attempting a quantitative description, regards the integration of the radical chemistry with the cell redox status in terms of $NAD^+/NADH$, $NADP^+/NADPH$, GSSG/GSH, and other redox balances in relation to redox potentials [34]. These efforts may effectively contribute for a clear quantitative appraisal how the redox balances affect the radical chemistry and its involvement in major cell functions, viz., mitochondrial transition permeability, apoptosis, necrosis, enzyme, and gene activities.

In spite of the relevant progress in bioenergetic knowledge, there are still major issues and challenges to be accomplished. Most issues relate to poor experimental conditions that roughly deviate from the native cell conditions. Most available data has been collected with isolated preparations which contain fragments of the original chondriom framework. Therefore, it should be not surprising that relevant functions got lost and other severely modified during the crude isolation procedures. Hence, drawn conclusions must be evaluated with caution. Furthermore, most experimental setups and reaction media are diverse from in situ conditions, e. g., temperature is generally set at 25 °C (for technical reasons), instead the actual cell temperature (37 °C), in experiments of oxygen tracing by electrometry. The chemical composition of media (sucrose, salt concentrations, buffers, pH) often does not minimally relate with cytoplasm condition. Furthermore, oxygen experiments are carried out at saturation (240 μM, at 25 °C), a situation far from the expected low oxygen activity in the living cell. Therefore, data on oxygen radicals may be severely questioned.

It is of general concern that a significant effort has to be implemented regarding direct observations in situ, looking at bioenergetic activities directly in living cells. This is a tremendous and difficult challenge for the near future. If not, we will never be certain if the observations are facts or artifacts.

References

1. Priestley J (1775) An account of further discoveries in air. Philosoph Trasact 65:384–394
2. Priestley J (1775) Experiments and observations on different kinds of air, 2nd edn. J. Johnson, London
3. Lavoisier AL (1789) Traité elementaire de chimie. Cuchet, Paris
4. Lavoisier AL (1864) In oeuvres de Lavoisier, Tome II. memoires de chimie et de physique. Imprimerie Imperiale, Paris

5. Lehninger AL (1975) Biochemistry. Worth Publishers, Inc., New York
6. Buchner E (1897) Alkoholische gärung ohne hefezellen. Berichte der Deutschen Chemischen Gesellshaft 30:117–124
7. Buchner E, Rapp R (1899) Alkoholische gärung ohne hefezellen. Berichte der Deutschen Chemischen Gesellshaft 32:2086–2094
8. Mahler HR, Cordes EH (1971) Biological chemistry, 2nd edn. Harper and Row, New York, p 495
9. Harden A, Young JW (1905) Proc Chem Soc 21:189–195
10. Stryer L (1995) Biochemistry, 5th edn. W. H. Freeman and Co., New York, pp 483–484
11. Krebs HA (1970) The history of the tricarboxylic acid cycle. Prespect Biol Med 14:154–170
12. Lehninger A (1965) The mitochondrion: molecular basis of structure and function. Benjamin, Menlo Park, CA
13. Mitchell P, Moyle J (1969) Estimation of membrane potential and pH difference across the crystal membranes of rat liver mitochondria. Eur J Biochem 7:471–478
14. Nicholls DG, Ferguson SJ (1992) Bioenergetics 2. Academic Press, London
15. Bott M, Thauer RK (1989) Proton translocation coupled to oxidation of carbon monoxide to CO_2 and H_2 in Methanosarcina barkeri. Eur J Biochem 179:469–472
16. Wikström MKF (1977) Proton pump coupled to cytochrome c oxidase in mitochondria. Nature 266:271–273
17. Solioz M, Carafoli E, Ludwig B (1982) The cytochrome c oxidase of Paracoccus denitrificans pumps protons in a reconstituted system. J Biol Chem 257:1579–1582
18. Yagi T, Matsuno-Yagi (2003) The proton-translocating NADH-quinone oxidoreductase in respiratory chain: the secret unlocked. Biochemistry 42:2266–2274
19. Hackenbrock CR (1981) Lateral diffusion and electron transfer in mitochondrial inner membrane. Trends Biochem Sci 6:151–154
20. Madigan MT, Martinko JM, Parker J (1997) Brock biology of microorganisms. Prentice Hall, London
21. Nelson DL, Cox MM (2000) Lehninger principles of biochemistry, 3rd edn. Worth Publishers, New York
22. Tsukihara T, Aoyama H, Yamashita E et al (1996) The whole structure of 13-subunit oxidized cytochrome c oxidase at 2.8 Å. Science 272:1136–1144
23. Iwata S, Osteimer C, Ludwig B, Michel H (1995) Structure at 2.8 Å resolution of cytochrome c oxidase from Paracoccus denitrificans. Nature 376:660–669
24. Zang Z, Huang L, Shulmeister VM et al (1998) Electron transfer by domain movement in cytochrome bc1. Natura 392:677–684
25. Yankovskaya V, Horsefield R, Törnroth S et al (2003) Architecture of succinate dehydrogenase and reactive oxygen species generation. Science 299:700–704
26. Abrahams JP, Leslie AGW, Luter R, Walker JE (1994) Structure at 2.8 Å resolution of F1-ATPase from bovine heart mitochondria. Nature 370:621–628
27. Chen C, Ko Y, Delannoy M, Ludtke J, Chiu W, Pedersen PL (2004) Mitochondrial ATP synthasome. J Biol Chem 23:31761–31768
28. Deisenhofer J, Epp O, Sinning I, Michel H (1995) Crystallographic refinement at 2.3 Å resolution and refined model of the photosynthetic reaction centre from Rhodopseudomonas viridis. J Mol Biol 246:429–457
29. Jordan P, Fromme P, Witt HT, Klukas O, Saenger W, Krauss N (2001) Three-dimensional structure of cyanobacterial photosystem I at 2.5 Å resolution. Nature 411:909–917
30. Ferreira KN, Iverson TM, Maghlaoui K, Barber J, Iwata S (2004) Architecture of the photosynthetic oxygen-evolving center. Science 303:1831–1838
31. McDermott G, Prince SM, Freer AA, Hawthornthwaite-Lawless AM, Papiz MZ, Cogdell RJ, Isaacs NW (1995) Crystal structure of an integral membrane light-harvesting complex from photosynthetic bacteria. Nature 374:517–521
32. Karrasch S, Bullough PA, Gosh R (1995) The 8.5 Å projection map of the light-harvesting complex I from Rhodospirillum rubrum reveals a ring composed of 16 subunits. EMBO J 14:631–638
33. Liu Z, Yan H, Wang K, Zhang J, Gui L, An X, Chang W (2004) Crystal structure of spinach major light-harvesting complex at 2.72 Å resolution. Nature 428:287–292
34. Jones DP (2006) Disruption of mitochondrial redox circuitry in oxidative stress. Chem Biol Interact 163:38–53

Chapter 2

Evaluation of Respiration with Clark-Type Electrode in Isolated Mitochondria and Permeabilized Animal Cells

Ana M. Silva and Paulo J. Oliveira

Abstract

In many studies, the evaluation of mitochondrial function is critical to understand how disease conditions or xenobiotics alter mitochondrial function. One of the classic end points that can be assessed is oxygen consumption, which can be performed under controlled yet artificial conditions. Oxygen is the terminal acceptor in the mitochondrial respiratory chain, namely, at an enzyme called cytochrome oxidase, which produces water in the process and pumps protons from the matrix to the intermembrane space. Several techniques are available to measure oxygen consumption, including polarography with oxygen electrodes or fluorescent/luminescent probes. The present chapter will deal with the determination of mitochondrial oxygen consumption by means of the Clark-type electrode, which has been widely used in the literature and still remains to be a reliable technique. We focus our technical description in the measurement of oxygen consumption by isolated mitochondrial fractions and by permeabilized cells.

Key words Mitochondria, Permeabilized cells, Clark-type electrode, Cellular respiration, Mitochondrial respiratory chain, Respirosome, Oxygen consumption rate, ADP/O ratio, Respiratory control ratio

1 Introduction

Aerobic eukaryotic cells present very developed intracellular machinery aimed at energy production, the mitochondrion, where molecular oxygen, O_2, is the terminal electron acceptor of a series of electron transfers through proteins comprising the so-called respiratory chain [1].

The diagram in Fig. 1 (adapted from refs. 2–4) represents the classical view of the mitochondrial respiratory chain, where electrons flow from NADH (oxidized in complex I) or $FADH_2$ (via succinate conversion to fumarate in complex II, coupling the Krebs cycle with the mitochondrial respiratory chain) to the final electron acceptor, molecular oxygen (O_2). The different complexes contain several cofactors and prosthetic groups. Complex I contains FMN and 22–24 iron-sulfur (Fe-S) proteins in 5–7 clusters. Complex II

Carlos M. Palmeira and António J. Moreno (eds.), *Mitochondrial Bioenergetics: Methods and Protocols*, Methods in Molecular Biology, vol. 1782, https://doi.org/10.1007/978-1-4939-7831-1_2,

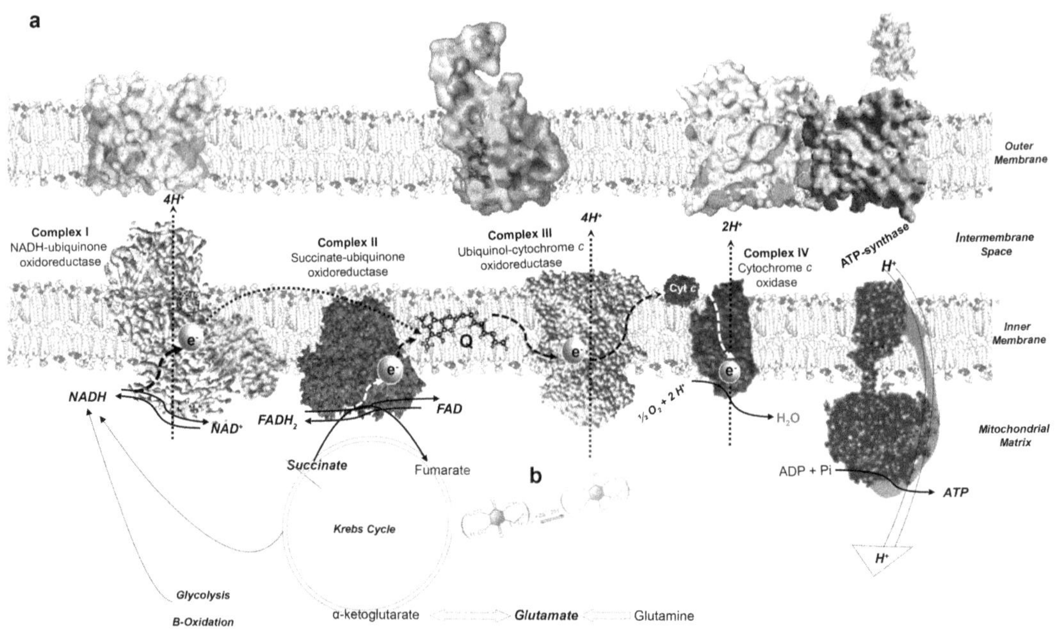

Fig. 1 (**a**) Representation of the mitochondrial respiratory chain and ATP synthase, which mediate the process of oxidative phosphorylation, integrated with the possible sources of reducing equivalents. The generic linear architecture of the five mitochondrial complexes in the inner mitochondrial membrane is shown; the outer membrane is featuring some typical proteins (porins, Tom20, Tom7, Mdm10, Mdm34, representative of some N-anchored, tail-anchored, and with intramembrane domain proteins), just to avoid the classical erroneous visual concept of a nude outer mitochondrial membrane. (**b**) Inset presenting reduced/oxidized forms of ubiquinone. **H⁺**, protons; **e⁻**, free electrons; ***Q***, ubiquinone (coenzyme Q_{10}); ***Cyt C***, cytochrome c

contains FAD, 7–8 Fe-S proteins in 3 clusters, and cytochrome b_{560}. Complex III contains cytochrome *b* (the ultimate electron acceptor in this complex), cytochrome c_1, and one Fe-S protein. Complex IV contains cytochrome *a*, cytochrome a_3, and two copper ions. It is proposed that when two electrons pass through complex I, four protons (H^+) are pumped into the intramembrane space of the mitochondrion. In the same manner, as each pair of electrons flows through complex III, four protons are pumped into the intramembrane space. Electrons are then used to reduce O_2 to H_2O, on cytochrome *c* oxidase. The protons in intramembrane space return to the mitochondrial matrix down their concentration gradient by passing through the ATP synthase, also called complex V, which synthesizes ATP at the expense of the transmembrane electric potential generated by the proton imbalance between the two sides of the membrane [2]. The process of ATP synthesis by mitochondria is termed oxidative phosphorylation (OXPHOS), which describes the coupling between substrate oxidation and ADP phosphorylation.

When OXPHOS and electron transfer system (ETS) in particular need to be evaluated, an approach to monitor oxygen dynamics is a possibility, especially when coupled to proper substrates and inhibitors of the electron transfer and ADP phosphorylation process.

Recently, the structure of the ETS reached a new level of understanding based on the cryo-electron microscopy imaging, after the solubilization of mitochondrial membranes and blue native polyacrylamide gel electrophoretic separation of their resident proteins. The observations suggest that respiratory complexes (I, III, and IV), based in a fluid model of membrane diffusion, physically interact with each other, forming a functional supercomplex [3, 4]. This supercomplex is named the respirosome, committed with oxygen consumption [4]. In this model, ubiquinone must diffuse and find a proper position, being surrounded by the proteins inside a hydrophobic layer provided by their 3D arrangement (Fig. 2a, adapted from 3, 4).

Complex II is a distinct entity in the respiratory chain, reinforcing its different origin and metabolic position over other complexes; the complex II membrane-embedded portion anchors it to the lipid bilayer, constituting the electron transfer path to ubiquinone, not interacting with other complexes' proteins, and transferring electrons to ubiquinone, and from that to supercomplexes III–IV, and later on to oxygen (Fig. 3a, adapted from 3, 4). One of the reasons of the tight interactions among complexes' proteins is the need to avoid electron leakage and therefore prevent reactive oxygen species production [4]. In order to be fully functional, complex V (ATP synthase) is apparently present as a dimer, as supported by cryo-electron microscopy imaging.

Several systems and technologies have been developed to follow oxygen consumption in different biological preparations. Historically, the electrochemical reduction of oxygen was discovered by Heinrich Danneel and Walter Nernst in 1897 (revised by Severinghaus and Astrup [5]). In the 1940s, the technology was tested in biological tissues to follow oxygen supply or availability by the implantation of platinum electrodes, in an amperometric approach. The major problem of the direct contact of electrodes with biological preparations was (and is) contamination, diminishing the electrochemical response, as well as the possible contamination of platinum electrodes by compounds such as cyanide and iodide [6]. The fundamental innovation introduced by Clark [5] was the presence of a permeable membrane, which allows the transport of ions and gases such as oxygen but restricts the contact of higher molecular weight substances (e.g., proteins) with the electrode surface [7]. Nevertheless, it is important to know the properties of the chosen oxygen permeable membrane, paying special attention to the electrochemical delays and the possibility of organic degradation or physicochemical interactions with some inhibitors such as antimycin A or oligomycin, among others [6], that could still be present from one experiment to the next.

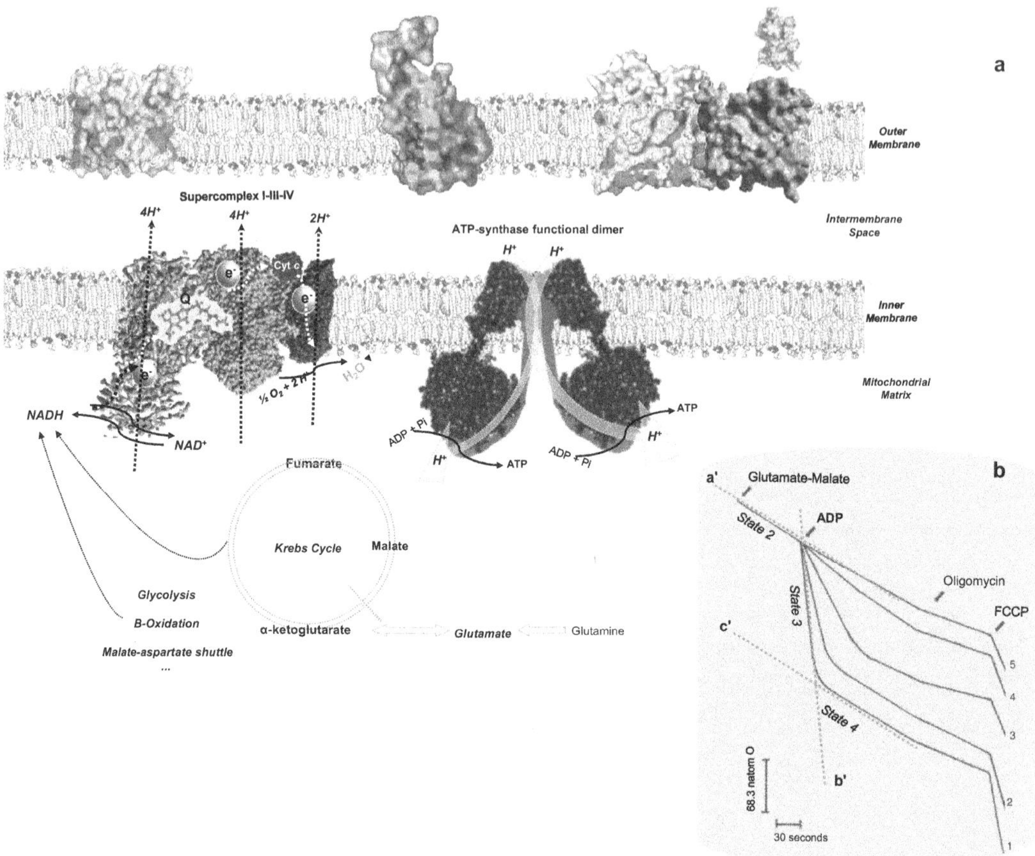

Fig. 2 Oxygen consumption measurement in a flatbed recorder in the presence of the substrate glutamate/malate, which enables the feeding of NADH:ubiquinone oxidoreductase unit of complex I with NADH. (**a**) The supramolecular organization of the respiratory chain in terms of supercomplexes is shown here. Eukaryotic cells show different complexes I, III, and IV macromolecular assembling, but the most common is the one that comprises the direct interaction of complexes I, III, and IV (representing the respirosome). In this assembling, electrophilic transfers from and to ubiquinone imply a specific hydrophobic pocket interaction between this molecule and proteins, as depicted in the figure. Part of the phosphorylative system relies on a dimerized complex V (ATP synthase) to be fully functional. (**b**) Representative polarograph tracing of oxygen consumption by isolated mitochondria (in a flatbed recorder, numbered lines represent different experimental conditions), using a Clark-type electrode. Dashed lines (**a'**, **b'**, and **c'**) were drawn (for plot 1, as an example) for ADP/O and RCR calculations (see text body). $\mathbf{H^+}$, protons; $\mathbf{e^-}$, free electrons; ***Q***, ubiquinone (coenzyme Q_{10}); ***Cyt C***, cytochrome c

It is thus necessary to obtain information from the current-concentration relationship, thus maintaining a desired selectivity. The chemical transformations that take place at the electrodes after the passage of electrical current are ruled by Faraday's law and by the current-voltage equation. The Clark-type electrode-associated oxygen consumption apparatus may be assembled as described in Fig. 4 (adapted from 8–10). The figure represents a classical and easy manner to set up equipment. The temperature-controlled

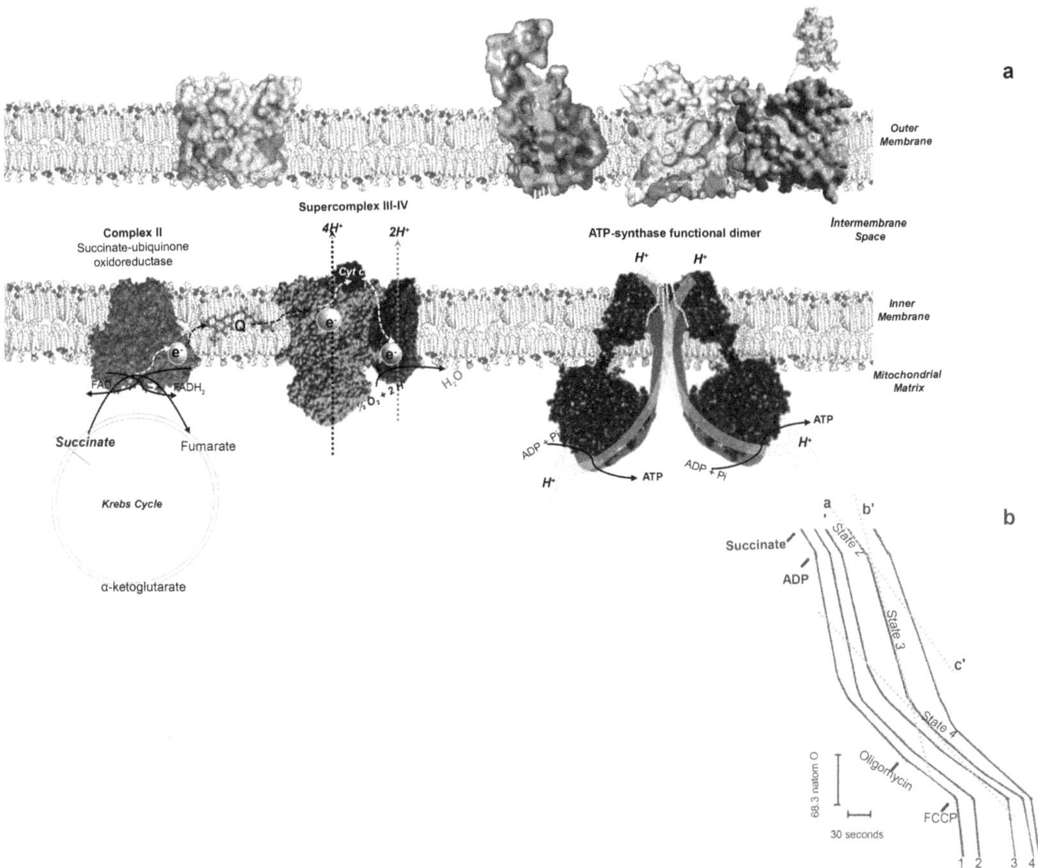

Fig. 3 Oxygen consumption measurement in a flatbed recorder in the presence of the substrate succinate, which is processed in the succinate dehydrogenase/succinate-coenzyme Q reductase unit of complex II, reducing FAD cofactor, and ultimately transfers electrons to the mitochondrial chain until O_2 is reduced in complex IV. (**a**) Similarly to Fig. 2, we show here the supercomplex assembling as a functional setting. Complex II remains an individualized structure inserted in the inner membrane, whereas complexes III and IV are also part of a supercomplex, considering a tight interaction between their proteins. Complex V (ATP synthase) functional structure is considered to be in the dimer form. (**b**) Representative polarograph tracing of oxygen consumption by isolated mitochondria (in a flatbed recorder, numbered lines represent different experimental conditions), using a Clark-type electrode. Dashed lines (**a'**, **b'**, and **c'**) were drawn (for plot 3, as an example) for ADP/O and RCR calculations (see text body). Respiratory plots in the presence of succinate. **H^+**, protons; **e^-**, free electrons; ***Q***, ubiquinone (coenzyme Q_{10}); ***Cyt C***, cytochrome c

incubation chamber allows the addition of cell or isolated mitochondrial fractions. One advantage of this setup, comparatively with some other oxygen consumption methods, is the possibility to simultaneously measure mitochondrial membrane potential using a tetraphenylphosphonium cation (TPP^+) electrode, substituting the stopper by a well-fitted and insulated electrode, sensitive to TTP^+.

In vivo cell respiration, considering organ and whole-body organization, is regulated by intracellular non-saturating ADP levels during normal states of activity [11] and increases when

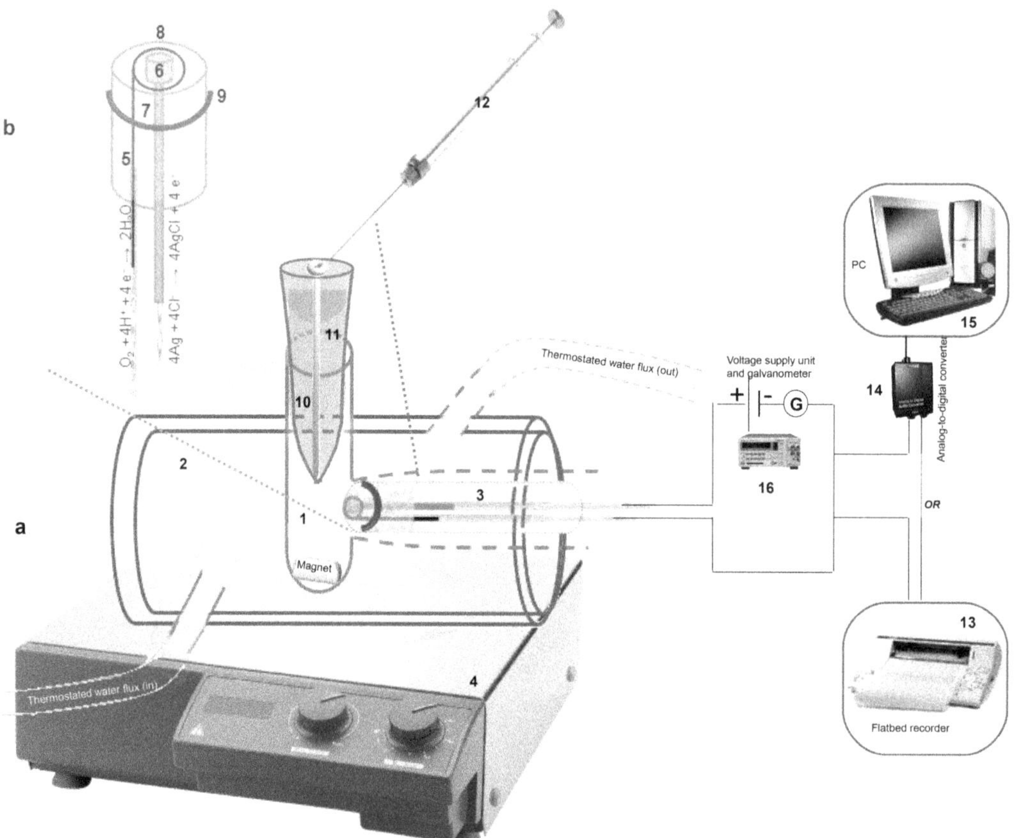

Fig. 4 General representation of the Clark-type electrode and experimental setup. (**a**) The biological preparations (mitochondrial or cell suspensions) are introduced in a well-defined volume of media (depending on the preparations), inside a temperature-controlled incubation chamber (1) (with water jacket, 2) with an oxygen electrode inserted (3) and magnetic stirrer coupled (4). The electrode determines 0_2 concentration in aqueous solutions over a period of time. (**b**) Oxygen electrode inset (adapted from ref. 5). The electrode itself is located inside the chamber, in horizontal position (or depending on the general apparatus design). The platinum cathode (5) is located surrounding a rod-like center anode (6) made by silver (reference Ag/AgCl electrode). These two electrodes connect with each other by a thin layer of electrolyte (50% saturated KCl solution, 7). Directly placed on top of the rod, there is an oxygen permeable Teflon membrane (8), which is tight fitted by a rubber ring (9). The 0_2 recordings can be done in open or closed chamber mode. In the last case, the reciprocal air solution 0_2 diffusion is avoided, which allows for better determination of the respiratory rates. Experimental additions of solutions, substrates, mitochondria, and substrates/inhibitors are done through the top of the chamber (10) (open mode) or through the small hole (11) inside the stopper, respectively, using a glass syringe (12). The electric signal may be directly measured in a flatbed recorder (13) or digitally converted (14) for later processing in a personal computer (15). The electrodes (*see* inset **b**) are polarized (16) by a constant voltage between 0.4 and 1.2 V, generating a current which reduces 0_2

submitted to a specific stimulus. One reason to use mitochondria in intact or permeabilized cells is that mitochondria interact with the cytoskeleton network, being grouped in functional clusters in close contact with other cell organelles and structures, probably an essential feature for the correct function of the whole mitochondrial bioenergetic apparatus [12]. Cell permeabilization may be achieved by different methodologies, but the most popular is the one that considers the presence of amphipathic substances. Detergents (such as digitonin and saponin) have been used for selective membrane permeabilization, but their usage conditions must be carefully controlled. Both detergents have a high affinity for cholesterol, allowing the selective permeabilization of cytoplasmic membrane, while leaving the outer and inner mitochondrial membranes intact [13]. By treating cells with specific detergents, it is possible to induce membrane permeabilization by promoting loss of lipid content, although without losing important intracellular proteins. When preparing a permeabilization protocol, it is important to consider the preliminary titration of the detergent, aiming to reach the best concentration. Mitochondrial substrates, such as pyruvate, succinate, glutamate, and malate that cannot cross the cytoplasmic membrane in intact cells, can now be added to cells after permeabilization, reaching mitochondria; subsequently it will be easier to follow the oxygen consumption in the presence of different substrates, nucleotides, and respiratory inhibitors (*see* **Note 1**).

Cell permeabilization by electroporation, applying an electrical field to increase the permeability of the cell membrane, is also technically possible. Electrically permeabilized cells were used to allow the uptake of exogenously added ATP, which is essential for receptor-mediated activation of the respiratory burst. For example, blood cells, such as neutrophils, are ATP-depleted prior to permeabilization and lately fed with it [14]. The negative side of electroporation is that cells reseal after the electric pulse, not allowing continuous cytoplasmic membrane permeabilization during oxygen consumption assays.

Differences between intact cell respiration and substrate oxidation in digitonin-permeabilized cells can be polarographically measured with an adapted Clark-type oxygen electrode in a micro-jacketed chamber [15, 16], using smaller working volumes when compared with the ones used in the traditional electrode represented in Fig. 4, thus useful for limited quantity of biological samples, although quenched-fluorescence oxygen sensing can be also used in intact cells [17]. Also, more recently, cultured cells or isolated mitochondria oxygen consumption may be analyzed in a high-throughput manner, in smaller volumes and lower amounts of samples, using a Seahorse Extracellular Flux device [18, 19]. However, Clark-type electrodes allowed to achieve more reliable absolute oxygen concentration/tension values. Precision respirometry can also be obtained using the Oxygraph device from Oroboros, which can also be coupled to other methods to measure mitochondrial function [11].

In alternative to the use of intact or permeabilized cells, isolated mitochondrial fractions have been a useful biological tool, being a basic model widely used to investigate the direct effect of modulators on the various components of the mitochondrial electron transport and on the ADP phosphorylation system. This simpler model has been used to investigate the toxicity of clinical relevant molecules [20, 21].

In the next sections of the present book chapter, some representative procedures to prepare permeabilized cells from solid or liquid organs/tissues, such as the muscle, brain, liver, or heart, as well as cells from blood or immune system, using different permeabilization methods, and isolated mitochondria are demonstrated, as well as some hypothetical schemes for the assessment of oxygen consumption and mitochondrial respiration modulation.

2 Materials

2.1 Cell Culture/Preparation and Permeabilization

2.1.1 Electropermeabilized Cells: Neutrophils (Adapted from 14)

1. Ficoll 400 and dextran T-500 (Pharmacia LKB Biotechnology Inc.).
2. RPMI 1640, AMP-PNP, ATP, EGTA, GDP, GTP, NADPH, glucose, 2-deoxy-D-glucose, Coomassie blue, analytical grade salts (Sigma).
3. Bicarbonate-free medium RPMI 1640 (buffered to 7.3, with 25 mM NaHEPES). Sterilize by autoclaving.
4. Permeabilization medium: 140 mM KCl, 1 mM $MgCl_2$, 1 mM EGTA, 10 mM KHEPES (pH 7.0), and $CaCl_2$ enough to reach a final concentration of 100 nM, 1 mM ATP (except for experiments without ATP), and 10 mM glucose (except for ATP-depletion experiments, where it is replaced by 2-deoxy-D-glucose). Store it on ice (±4 ° C).
5. Reaction medium for oxygen consumption assays: 140 mM NaCl, 5 mM KCl, 10 mM HEPES, 1 mM $MgCl_2$, 1 mM $CaCl_2$, 1 mM $MgCl_2$, and 10 mM NaHEPES (pH 7.3).
6. HEPES-RPMI 1640 medium and all cell buffers must be at 37 °C previously to usage.
7. All reagents (salts, sucrose, etc.) with analytical grade, from Sigma (St. Louis, MO).

2.1.2 Saponin Permeabilization: Saponin-Skinned Muscle Fibers (Adapted from 12)

1. Isolation/resuspension medium: 10 mM Ca-EGTA (0.1 μM free Ca^{2+}), 20 mM imidazole, 20 mM taurine, 50 mM K-MES, 0.5 mM DTT (dithiothreitol), 6.56 mM $MgCl_2$, 5.77 mM ATP, and 15 mM phosphocreatine (pH 7.1). Store on ice.
2. Saponin (Sigma S-2149) and other reagents (salts, sucrose, etc.) with analytical grade, from Sigma (St. Louis, MO).

2.1.3 Digitonin Permeabilization: Lymphocytes (Adapted from 16)

1. Histopaque 10771 and other reagents (salts, etc.) with analytical grade, from Sigma (St. Louis, MO).
2. Phosphate buffer: 137 mM NaCl, 2 mM KCl, 1 mM KH_2PO_4, and 10 mM Na_2HPO_4 (pH 7.4).
3. Digitonin (Sigma-Aldrich D-141) and other reagents (salts, sucrose, etc.) with analytical grade, from Sigma (St. Louis, MO).

2.2 Isolation of Mitochondrial Fraction

1. *Brain isolation buffer (BIB)*: 225 mM mannitol, 75 mM sucrose, 5 mM HEPES, 1 mM EGTA, and 1 mg/mL bovine serum albumin (fatty acid-free) (pH 7.4).
2. *Brain resuspension buffer (BRB)*: 225 mM mannitol, 75 mM sucrose, and 5 mM HEPES (pH 7.4).
3. *Liver and heart isolation buffer (LIB or HIB)*: 250 mM sucrose, 10 mM HEPES, 0.5 mM EGTA, and for liver isolation supplement medium with 1 mg/mL bovine serum albumin (fatty acid-free) (pH 7.4).
4. *Liver and heart resuspension buffer (LRB or HRB)*: 250 mM sucrose and 10 mM HEPES (pH 7.4).
5. Protease (subtilisin fraction VIII) and other reagents (salts, sucrose, etc.) with analytical grade, from Sigma (St. Louis, MO).

2.3 Oxygen Electrode

2.3.1 Oxygen Electrodes and Polarographic Systems

Commercially available, being the most common, are the following: Oroboros Oxygraph (Physica respirometer, Paar Physica, Oroboros, Austria); Hansatech Instruments Limited (Norfolk, UK); Yellow Springs Instrument Company (YSI Inc., Ohio, USA, the one we use in our laboratory); Gilson Medical Electronics (Middleton, WI); Mitocell microrespirometer (Strathkelvin Instruments Limited, Scotland). See specifications with manufacturers.

2.3.2 Oxygen Permeable Membranes

We considered preferentially PTFE films [poly(tetrafluoroethylene), e.g., Teflon] with the reference YSI 5776, Oxygen Probe Service Kit (Yellow Springs, Ohio, 45387 USA).

2.4 Oxygen Consumption Assays: Permeabilized Animal Cells

1. *Neutrophil respiration medium* (mM): 140 NaCl, 5 KCl, 10 HEPES, 1 $MgCl_2$, 1 $CaCl_2$, 1 $MgCl_2$, and 10 NaHEPES (pH 7.3).
2. *Saponin-skinned muscle fiber respiration medium* (in mM): 110 mannitol, 60 KCl, 10 KH_2PO_4, 5 $MgCl_2$, 60 Tris–HCl, and 0.5 Na_2EDTA (pH 7.4).
3. *Lymphocyte respiration medium* (in mM): 300 mannitol, 10 KCl, 10 HEPES, 5 $MgCl_2$, and 1 mg/mL bovine serum albumin (fatty acid-free) (pH 7.4).

2.5 Oxygen Consumption Assays: Isolated Mitochondria

1. *Brain mitochondrial respiration medium* (in mM): 100 sucrose, 100 KCl, 2 KH_2PO_4, 5 HEPES, and 10 μM EGTA (pH 7.4).
2. *Heart mitochondrial respiration medium* (in mM): 50 sucrose, 100 KCl, 1 KH_2PO_4, 10 Tris, and 10 μM EGTA (pH 7.4).
3. *Liver mitochondrial respiration medium* (in mM): 135 sucrose, 65 KCl, 5 KH_2PO_4, 5 HEPES, and 2.5 $MgCl_2$ (pH 7.4).

It must be stressed that other reaction media are possible, using different compositions and, if required, different osmolarities, although typically media with 250–300 mOsm are used.

3 Methods

3.1 Cell Culture and Permeabilization

3.1.1 Electro-permeabilized Cells: Neutrophils (Adapted from 14)

1. Isolate neutrophils from fresh heparinized human blood by dextran sedimentation followed by Ficoll-Hypaque gradient centrifugation.
2. Add NH_4Cl for red cell lysis; centrifuge again and wash with HEPES-buffered RPMI.
3. Count cells (using Coulter ZM counter, Neubauer chamber or other similar device), and resuspend them at 10^7 cell/mL in HEPES-buffered RPMI [for ATP-depleted cells, cell (10^7/mL) incubation at 37 °C, during 5 min, in the specific buffer (with 2-deoxy-D-glucose, instead of glucose), prior to permeabilization].
4. Electroporation: sediment cells (with ATP or ATP depleted) by centrifugation and resuspend in ice-cold permeabilization solution at 10^7 cells/mL. Add 800 μL cell suspension/cuvette in a Bio-Rad Pulser (although other similar devices can be used), and apply two electric discharges (pulses – charges of 5 kV/cm from a 25-microfarad capacitor) as described by the manufacturer. Sediment cells by centrifugation in 1.5 mL conic tubes (Eppendorf 5415 microcentrifuge), between pulses, and resuspend in ice-cold permeabilization solution (at 10^7 cells/mL).
5. Electroporated cells are ready to be used by picking cells to a final density of 2×10^6 cell/mL or can be stored on ice up to 15 min.

3.1.2 Saponin Permeabilization: Saponin-Skinned Muscle Fibers (Adapted from 12)

1. Muscle fibers should undergo mechanical separation. Under a microscope dissect small bundle fibers (submerged on isolation/resuspension medium ice-cold) at one end and separate down to the muscle mid-belly (using fine jeweler forceps).
2. Place muscle bundles in ice-cold isolation medium (8–15 wet weight/mL); add 50 μg/mL saponin, and gently mix the suspension during 30 min, at 4 °C.
3. Wash muscle fibers with resuspension medium to remove all saponin.

4. Measure protein by Lowry et al. [17] method, using a 5 μL aliquot. Consider 0.5 mg/mL for oxygen consumption assays.
5. Store on ice (muscle fibers in resuspension medium) until the start of respiration analysis in specific reaction medium.

3.1.3 Digitonin Permeabilization: Lymphocytes (Adapted from 16)

1. Isolate human lymphocytes introducing 5 mL heparinized blood in a tube filled with a solution containing polysucrose and sodium diatrizoate adjusted to a density of 1.077 g/mL (Histopaque® 10771-Sigma) (the proportion between Histopaque and blood should be 1/3 and 2/3, respectively).
2. Centrifuge (Eppendorf 5702 centrifuge) at 500 × *g* for 30 min (20 ° C) until mononuclear cells form a distinct layer at the plasma-Histopaque® interface (the blood must remain on top; do not mix). Wash with phosphate-buffered saline (PBS), centrifuge, and resuspend. Repeat the procedure twice.
3. Resuspend in a final volume of 25 μL of the same buffer. Measure protein by the Lowry method [19], using a 5 μL aliquot. Consider 0.5 mg/mL for oxygen consumption assays.
4. Digitonin (1%) is added to lymphocytes just after introduction in the oxygen consumption chamber.

3.2 Isolation of Mitochondrial Fraction

Independently for cell mitochondrial origin, the first important procedure to achieve good and feasible results on oxygen respiration is the isolation of tightly coupled mitochondrial fractions. Several factors have to be taken into account, including the composition and temperature of all the solutions used. Homogenization followed by centrifugation is crucially dependent of the composition of the isolation medium. The proper gradient build-up and osmolarity stabilization are related with the purity of mitochondrial fraction that will be obtained (*see* **Note 2**).

Mitochondrial isolation from the brain, heart, and liver (from Wistar rats).

3.2.1 Brain Mitochondria (Adapted from 22)

1. After being anesthetized using different possible processes, rats should be sacrificed accordingly with the ethical proceedings established by the *Guide for the Care and Use of Laboratory Animals*, eighth edition, published by the US National Institutes of Health (revised in 2011).
2. Decapitate rat (reserve other body parts for other projects, if possible) and remove rapidly the cerebellum. Wash and mince it (using sharp scissors) and later homogenize (in a small glass Potter-Elvejhem) at 4 °C in 10 mL of brain isolation buffer (BIB) containing 5 mg of bacterial protease.
3. Brain homogenate is brought to 30 mL, and centrifuge 2000 × *g* for 3 min (Sorvall RC6 plus centrifuge; SS-34 rotor).

4. Resuspend pellet, including the fluffy synaptosomal layer, in 10 mL BIB containing 0.02% digitonin and centrifuge at 12,000 × *g* for 8 min.
5. Resuspend brown mitochondrial pellet without the synaptosomal layer in 10 mL BRB and centrifuge at 12,000 × *g* for 10 min.
6. Resuspend mitochondrial pellet in 300 μL BRB. Store mitochondria on ice until assays start.
7. Determine mitochondrial protein by biuret method [23] calibrated with bovine serum albumin (BSA). Consider 0.8 mg/mL for oxygen consumption assays, although other protein concentrations can be used.

3.2.2 Heart Mitochondria (Adapted from 24)

1. Open the rat chest with scissors and remove the heart.
2. Place the heart in isolation buffer (HIB), and cut slowly the organ in little pieces with cold scissors.
3. Wash heart pieces as many times as needed with HIB to remove all blood.
4. Homogenize heart pieces in a Potter-Elvejhem with 20 mL HIB complemented with 35 μL protease (subtilis in fraction VIII or nagarse). Be careful to maintain temperature under 4 °C.
5. The homogenization takes place with three or four homogenizations with a refrigerated piston (preferentially). The suspension is then left under incubation during 1 min on ice. After that period of time, homogenize 2–3 more times.
6. Place heart homogenate in centrifuge tubes and fill up with HIB.
7. The tubes are centrifuged at 12,000 × *g* during 10 min (Sorvall RC6 plus centrifuge; SS-34 rotor).
8. The supernatant is discharged and the pellet is freed after addition of HIB. Transfer to a smaller glass manual Potter-Elvejhem homogenizer, and promote gentle homogenization.
9. Centrifuge at 2000 × *g* during 10 min. Pour supernatant to new refrigerated centrifuge tubes, and fill up with HIB.
10. Centrifuge supernatant at 12,000 × *g* during 10 min.
11. Discharge the new supernatant, and gently homogenize pellet with a smooth paintbrush, adding heart resuspension buffer (HRB). Place homogenate in centrifuge tubes and fill up with HRB.
12. Centrifuge at 12,000 × *g* during 10 min. Discharge supernatant and resuspend isolated mitochondria in 200–300 μL in HRB. Store mitochondria on ice until the start of oxygen respiration assays.
13. Determine mitochondrial protein by biuret method [23] calibrated with BSA. Consider 0.5 mg/mL for oxygen

consumption assays, although different mitochondrial protein concentrations can be used, according to the respiratory activity of the preparation.

3.2.3 Liver Mitochondria (Adapted from 25)

1. Open rat abdominal cavity with a scissor and remove the liver.
2. Place the liver in isolation buffer (LIB) and cut slowly in little pieces with a scissor.
3. Wash liver pieces as many times as needed with LIB to remove all blood.
4. Homogenize liver pieces in a Potter-Elvejhem with 60 mL LIB. Be careful to maintain temperature under 4 °C.
5. Place liver homogenate in centrifuge tubes and fill up with LIB.
6. The tubes are centrifuged at 2000 × *g* during 10 min (Sorvall RC6 plus centrifuge; *SS-34* rotor). Pour supernatant to new refrigerated centrifuge tubes, and fill up with LIB. Avoid contamination with debris from pellet.
7. Centrifuge supernatant at 12,000 × *g* during 10 min.
8. Discharge the new supernatant, and gently homogenize pellet with a smooth paintbrush, adding heart resuspension buffer (LRB). Place homogenate in centrifuge tubes and fill up with LRB.
9. Repeat **steps** 7 and **8**.
10. Centrifuge at 12,000 × *g* during 10 min. Discharge supernatant and resuspend isolated mitochondria in 200–300 μL in LRB. Store mitochondria on ice until the start of oxygen respiration assays.
11. Quantify mitochondrial protein by biuret method [23] calibrated with BSA. Consider 1.5 mg/mL for oxygen consumption assays, although alternatives in mitochondrial protein content can be considered.

3.3 Oxygen Electrode Preparation and Maintenance

The Clark-type electrode is an apparatus that measures oxygen on a catalytic platinum surface. The classical Clark electrode is basically composed of two electrodes (cathode and anode). The electric signal arises from the current developed on the cathode. The voltage supply unit (associated with a galvanometer) feeds the system with electrons, and when oxygen diffuses through an oxygen permeable membrane (commonly Teflon based) to the platinum electrode, O_2 is reduced to water. The current is proportional to the oxygen tension in the solution ([8], *see* **Note 3**) where mitochondria or cells are placed. The silver ions combine with chloride ions in solution, precipitating silver chloride over the silver electrode, according with the overall equation $\mathbf{4Ag(s) + 4Cl^- + O_2 + 4H^+ + 4e^- \rightarrow 4AgCl\ 2H_2O}$ [9, 10]. The electric signal can be analogically recorded using a flatbed recorder, or the electrode output current can pass through an analog-to-digital converter and later analyzed through a signal transducer coupled to a personal computer (as exemplified in ref. 26).

3.4 Modulators of Mitochondrial Complexes

Several mitochondrial substrates and inhibitors can be used during oxygen consumption assays to pinpoint possible alterations in respiratory fluxes or individual respiratory complex assessment (*see* Table 1).

The classical protocols designed to follow mitochondrial chain activity through the means of oxygen consumption assessment in isolated mitochondria often investigate the entire span of the respiratory chain, by feeding the NADH:ubiquinone oxidoreductase

Table 1
Most used substrates and inhibitors during oxygen consumption assays

Name	Classification	Place of action
Antimycin A	Inhibitor	Complex III
Ascorbate	Substrate	Complex IV (by donating electrons to cytochrome *c*) usually through artificial carriers, as TMPD (see below)
Atractyloside	Inhibitor	Adenine nucleotide translocator (no ADP/ATP exchange occurs)
Azide, cyanide	Inhibitor	Complex IV (competitive inhibition, by preventing the binding of oxygen)
Cyanocinnamite derivatives	Inhibitor	Pyruvate transporter antagonist that restricts maximal respiration[a]
FCCP	Protonophore and uncoupler of mitochondrial OXPHOS	Mitochondrial membrane (although proton-shuttling activity can be facilitated by some mitochondrial proteins)
Fumarate	Substrate	Intermediate in Krebs cycle
Glutamate	Substrate	Complex I (converted into NADH in the matrix) and an intermediate in Krebs cycle
KCN	Inhibitor	Complex IV
Malate	Substrate	Complex I (converted into NADH in the matrix) and an intermediate in Krebs cycle
Malonate	Inhibitor	Inhibition of succinate oxidation
Myxothiazol	Inhibitor	Complex III
Oligomycin	Inhibitor	Complex V
Pyruvate	Substrate	Feed Krebs cycle (via acetyl-CoA)
Rotenone	Inhibitor	Complex I
Succinate	Substrate	Complex II (generating $FADH_2$ in complex II) and an intermediate in Krebs cycle
TMPD	Reducing co-substrate	Bypassing electrons to complex IV (via cytochrome *c*), as an artificial electron mediator

FCCP carbonyl cyanide p-trifluoromethoxyphenylhydrazone, *KCN* potassium cyanide, *TMPD* *N,N,N′,N′*-tetramethyl-p-phenylenediamine
[a]Used as potent inhibitors of mitochondrial respiration supported/activated by pyruvate, but not by other substrates

unit with NADH, derived from pairs glutamate/malate or pyruvate/glutamate. In fact, and as depicted in Fig. 2a, mitochondrial NADH matrix enrichment can be achieved from different sources. For example (Fig. 2b, adapted from ref. 25), it is possible to feed the Krebs cycle with glutamate, which is deaminated to α-ketoglutarate, increasing NADH levels through Krebs cycle activation. In order to maintain the catabolic/anabolic intermediates, it is very important to consider the simultaneous addition of malate, maintaining the function of malate-aspartate shuttle, as well as enabling the indirect transfer of reducing equivalents between the two sides of the inner mitochondrial membrane. The experiments designed to follow complex II activation (consider the example depicted in Fig. 3b, adapted from ref. 25) usually feed mitochondria with succinate that is processed in the succinate dehydrogenase/succinate-coenzyme Q reductase unit of complex II, reducing $FADH_2$ cofactor in the complex. The inhibitors used are molecular tools that allow the researcher to selectively manipulate the components of the ETS, as well as find the mitochondrial limits in terms of maximal activity (*see* next section).

3.5 Respiratory Parameters

Two important criteria used to assess the integrity or quality of a determined mitochondrial fraction after isolation or after incubation with test compounds are the respiratory control ratio (RCR) or the ADP/O.

The RCR is a measure of the coupling between substrate oxidation and phosphorylation and basically informs the researcher how intact or how coupled mitochondria are. The initial step to obtain RCR is to independently calculate state 3 and state 4 respiration rates. State 3 respiration is triggered in a suspension of isolated mitochondria by the addition of ADP. The increase in respiration denotes the use of the proton-motive force for the synthesis of ATP, being restored by the augmented proton pumping activity of the respiratory chain. When all ADP is phosphorylated into ATP by the action of the ATP synthase, the respiration returns to or closes to the initial pre-ADP addition values. The new respiratory state is now termed state 4. Some authors describe other respiratory states including state 2, which is the mitochondrial respiration in the presence of substrate only (pre-ADP addition), while others even establish a respiratory state 1, when mitochondria are consuming oxygen by oxidizing internal substrates only.

Regardless of the respiratory state being measured, units are usually expressed as nmol O_2/min/mg. protein or natom O/min/mg protein. There are several important values to account when calculating the rates of oxygen consumption. By the units involved, the exact time in which an *x* amount of oxygen is consumed is critical. Also, the amount of mitochondrial protein used in the experiment must be well known. There are several precise protein quantification methods including Bradford and Biuret methods, among others [23, 27]. For cells, the oxygen consumption may be rationalized for the number of cells considered in oxygen reaction chamber and data present in function of nmol O_2/min/cell number (cell number may be quantified, depending from the biological samples), or when considering the total protein, we can use the method described by Lowry et al. [19].

Some authors may normalize oxygen consumption rates to the activity of citrate synthase to standardize for mitochondrial content [28, 29] (*see* **Note 4**). If oxygen consumption is being measured in a flatbed recorder, one important parameter that should be taken into account is the scale of the paper, which will determine which distance in the paper will correspond to 100% oxygen saturation in water or in saline [several tables can be consulted for the values at the desired temperature (*see* **Notes 5** and **6**)]. Also, paper recording velocity, which is adjustable, must be recorded even before the assays start. In this regard, it is advised to reach a compromise between a slow paper velocity, which would allow for a better measurement of the rates, and a faster paper velocity, which turns the measurement of the faster rates harder but allows to spare the researcher the waste of long stretches of paper. After measuring state 3 and state 4 respiration rates, the RCR is easily calculated by performing the rate between state 3 and state 4. The final value should not present any units. There is a wide variety of tables showing values that are consistent with a good mitochondrial preparation (for example, see ref. 25). As an example, we routinely measure the RCR of isolated mitochondria in our laboratory just after the isolation or when investigating the effects of different xenobiotics. We usually have RCRs in the order of 6–8 for complex I substrates and 2–4 for complex II substrates (we have slightly lower values for the heart vs. the liver, most likely because of a higher intrinsic ATPase activity in our heart mitochondrial fractions, which leads to a higher state 4 respiration). When using ascorbate plus *N*,*N*,*N*′,*N*′-tetramethyl-p-phenylenediamine (TMPD) to directly feed complex IV via cytochrome *c*, the RCR value is around 1.5–2, which occurs due to the very fast respiration during state 4 respiration, since only one proton pump is working (complex IV or cytochrome c oxidase).

Damaged mitochondria usually have low RCRs, being 1 in the theoretical minimum value. One test compound will decrease the RCR as it leads to increased state 4 respiration, decreased state 3 respiration, or both.

The ADP/O has a completely different meaning. This parameter measures the efficiency of the mitochondrial phosphorylative system; calculated as the ADP/O index, which has no units, and it will give us a measure of how much oxygen the respiratory chain reduces to water per ADP phosphorylated. The index is calculated as the number of nmol of ADP added to the system divided by the number of natoms of oxygen consumed during state 3 respiration.

And this last point is critical, since a common mistake is to calculate the ADP/O as the amount of ADP added in nmol divided by the respiration rate during state 3. After measuring the absolute oxygen consumption and converting it into natoms, the calculation is now easy to do. One common strategy to calculate the amount of oxygen during state 3 respiration is to draw lines (as shown in Figs. 2b and 3b) on top of the respiration curves during states 2, 3, and 4 and use the intersection points between state 2/3 and state 3/4 as a measure to calculate the absolute distance in the paper, parallel to the axis corresponding to oxygen concentration in the media.

There is a large controversy on the theoretical values for the ADP/O value [30], since the ADP/O is dependent on the coupling ratios of proton transport. New discoveries on the structure and activity of the ATP synthase justify a new assessment of the established values, also because several other mechanisms account for the different ADP/O values found in the literature [30, 31]. Under our conditions, we routinely have values around 2.4–3.1 for complex I substrates and 1.4–1.9 for complex II substrates. Common during bench work is also the appearance of ADP/O values much higher or lower than the expected. The researcher must be aware that an incorrectly measured ADP stock solution will lead to an incorrect determination of ADP/O values (*see* **Note 7**). Finally, one common source of error is when the oxygen electrode used is not responding fast enough to the alterations in the media of oxygen content. When this happens, the transition between each state is slow and does not have an abrupt profile, which will lead to the underestimation of the ADP/O value. A good and fast responding oxygen electrode is thus essential for correct estimate of the index (*see* **Note 8**). Also, membranes needed to have good O_2 permeability properties (*see* **Note 9**).

During oxygen consumption experiments, we can also consider other mitochondrial respiration states [11], such as "state FCCP" (uncoupled respiration, maximal activity of the respiratory chain) or "state oligomycin," which is usually used to determine oxygen consumption with inhibited ATP synthase.

3.6 Oxygen Consumption Assays: Permeabilized Animal Cells

Oxygen consumption assessment implies the choice of the best-suited reaction chamber, accordingly with the biological preparation volumes.

For all types of presented cells, we need to consider:

1. Check if temperature in reaction/O_2 consumption chamber is 37 ° C.
2. Calibrate O_2 scale using distilled water.
3. Introduce the desired volume of respiration buffer, and just place biological preparations after reaching a steady baseline on the chart.

4. In electropermeabilized, digitonin-permeabilized, or saponin-permeabilized cells, a schematic sequence of additions (using, for example, microsyringes) such as 10 mM pyruvate +5 mM malate, 10 mM glutamate +5 mM, 2 mM ascorbate + TMPD, and 2.5 μM FCCP (*see* **Note 1**) and the addition of ADP (125–250 nM) after a substrate can be used to check OXPHOS function.

3.7 Oxygen Consumption Assays: Isolated Mitochondria

1. Define appropriate temperature, usually between 25 and 30 ° C.
2. Calibrate appropriate O_2 scale using distilled water (for example, a 1.5× scale can be appropriate in most cases to get a better signal resolution for isolated mitochondria).
3. Introduce the desired volume of respiration buffer, and just place biological preparations after reaching a steady baseline on the chart.
4. Add mitochondrial suspension with the desired protein concentration.

Sequential additions of mitochondrial substrates and inhibitors can allow characterizing different segments of the respiratory chain in one experiment only.

We can use a schematic sequence of additions such as 10 mM pyruvate +5 mM malate followed by ADP (125–250 nmol) and later rotenone (2 μM), followed by 5 mM succinate, ADP (same as before), followed by antimycin (2 μM). If still far from total chamber oxygen consumption, a subsequent addition can be performed with 2 mM ascorbate + TMPD, plus ADP. Another possibility is the addition of 1 μM FCCP (*see* **Note 1** and Table 1) to uncouple respiration. The number of additions depends on the oxygen availability in the chamber and enough time should be elapsed between additions to warranty correct measurement of oxygen consumption rates.

4 Notes

1. It is necessary to consider particular substrates and/or inhibitors for individual respiratory complex assessment during oxygen consumption assays, as previously mentioned (Table 1). Several substrates have the same target, with the choice relying on the availability and how affordable are it for the researcher. The biological preparations also dictate the best choice: isolated mitochondria are easily fed with different substrates comparatively to intact cells. The choice of the inhibitors is more delicate, considering the sequence of modulators planned

to be added during the assay. Some inhibitors are very toxic, making it impossible to continue the assay after the addition.

2. It is critical to obtain well-purified mitochondrial fractions. For example, original protocols to isolate brain mitochondria (mainly synaptosomal) considered sucrose gradients that expose mitochondria to undesired osmotic conditions, reducing organelle viability. Other hydrophilic polysaccharides, such as Ficoll, can be considered, although there are disadvantages regarding the need of long ultracentrifugation time. Percoll, another molecule used for gradients and which is constituted by polyvinylpyrrolidone (PVP)-coated colloidal silica particles, creates a mild fractionation gradient easy to reach, with lighter centrifugations, yielding good mitochondrial preparations [32].
3. Electric circuitry controls oxygen electrode. One tension division circuit (inside galvanometer) generates an electrode potential difference of 0.71 V (between anode and cathode) which is proportional to the activity of water-dissolved O_2 solution [8].
4. When performing respiratory assays, it is important to start with a known number of cells or a specified quantity of mitochondria in a specific volume at the Clark-type electrode chamber. Isolated cells are easy to count, but cell aggregates, or fibers imply total protein quantification. By measuring the latent citrate synthase activity (calculating the difference between total citrate synthase activity and free citrate synthase activity [29]) or citrate synthase ratio (latent citrate synthase activity/free citrate synthase activity), it is possible to infer not only the structural integrity of the mitochondrial preparation but also to correlate oxygen consumption with the quantity of the functional mitochondria or cells with functional mitochondria, instead of the total protein. Rationing oxygen consumption by citrate synthase content is time-consuming and thus rarely used. Still, if using citrate synthase activity measurements for normalization, the protocol described by Lanza and Nair [33] could be useful to determine it.
5. Oxygen solubility differs according to the type of media we use. We can find many tabled values for distilled water and different saline buffers [34, 35], but our experience tells us that a good approach is to use standard oxygen-saturated deionized water for calibration.
6. The most correct step to calibrate the O_2 scale is by comparing the signal of air-saturated water with the zero level created by adding dithionite to the water in the reaction chamber (this starts a reaction that quickly consumes the dissolved oxygen) or flushing the cell with nitrogen [9]. But when comparing oxygen consumption from specific conditions to a control, knowing or not the precise values of dissolved O_2 will not have a meaningful impact on data analysis.

7. Spectrophotometric determination of ADP stock solutions should always be conducted before starting experiments with a new prepared batch. ADP concentration can be determined by measuring the absorbance of the batch at 259 nm (molar extinction coefficient = 15.4×10^3 M^{-1} cm^{-1})[36].
8. To maintain the electrode in good conditions, it should be cleaned after use and before prolonged storage. Never allow the electrode to dry without the electrolyte in place, as crystallization of the electrolyte may occur and cause the platinum/ epoxy resin seal to be breached and allow the deposition of electrolyte salt crystals around the cathode, leading to necessary electrode replacement. The silver electrode (anode) must be cleaned to remove the excessive deposits of black oxides, which could cause deterioration. For that, we use a small cotton bud moistened with distilled water. Some commercial products can be used with the purpose to clean silver electrodes, but avoid any harsh abrasive substance.

   Platinum cathode cleaning involves the usage of a non-corrosive polishing agent (as a soft white toothpaste, with small silica spheres in the composition) on top of a cotton bud moistened with few drops of distilled water. Circular movements in the polished platinum electrode surface can be performed to obtain a final shiny effect. Make sure that all electrical connections are preserved and completely dry [37].
9. Several different membrane materials can be used for the oxygen electrode including Teflon, polyethylene, and silicon rubber, among others. Thin hydrophilic polymers are usually not mechanically strong which may cause problems. It is possible to obtain some degree of selectivity by choosing the material of this membrane according to the conditions of application. The diffusion through such a structure is more complicated. For radial geometry, the steady-state current, ***I***, is given by the mathematical expression

$$i = \frac{4\pi F_0 D_s S_s P(r_1)}{D_s(r_1 - r_0)/D_m r_1 + S_m r_0 r_1}$$

D corresponds to the diffusion coefficient (in solution and in the membrane), r_0 is the outer radius of the membrane, and r_1 is the radius of the electrode. *S* represents oxygen solubility in solution and in the membrane (S_s and S_m, respectively). Another important parameter is the oxygen partial pressure on the membrane pressure, $P(r_1)$ [8].

Literature reports a wide range of membranes for Clark-type oxygen sensors. Jobst et al. [38] considered a membrane based on a thin-layer polymer, to prepare a planar oxygen sensor with a three-electrode configuration (platinum working and counter electrodes and Ag/AgCl reference electrode), that allows the regeneration of oxygen consume in the cathode. The electrolyte is based in a hydrogel layer, preventing the typical buffer degradation and self-oxygen-consuming behavior. But for mitochondrial preparations, the system is not the best suited to be coupled to the common oxygen recording chamber, and membrane preparation requires a complex experimental setup.

5 Conclusion

The polarographic evaluation of oxygen consumption has been a feasible and reliable methodology, being continuously technologically updated along the time, in terms of shielding and electrical insulation, chamber volume capacity, as well as signal transduction and software development for data analysis. Still, the technique is always relying on the basic physicochemical principles of oxygen reduction, allied to the fact that it is an inexpensive and easy technique to adapt to different experimental requirements in the context of mitochondrial bioenergetics or in the cell metabolism in general.

Acknowledgments

Work in the author's laboratory was funded by FEDER funds through the Operational Programme Competitiveness Factors – COMPETE – and national funds by FCT (Foundation for Science and Technology) under research grant PTDC/DTP-FTO/2433/2014, POCI-01-0145-FEDER-007440, POCI-01-0145-FEDER-016659. AMS is supported by a Ph.D. fellowship from the Foundation for Science and Technology (SFRH/BD/76086/2011).

References

1. Du G, Mouithys-Mickalad A, Sluse FE (1998) Generation of superoxide anion by mitochondria and impairment of their functions during anoxia and reoxigenation in vitro. Free Radic Biol Med 25(9):1066–1074
2. Berg JM, Tymoczko JL, Stryer L (2012) Biochemistry, 7th edn. W. H. Freeman and Company, San Francisco
3. Chaban Y, Boekema EJ, Dudkina NV (2014) Structures of mitochondrial oxidative phsophorylation supercomplexes and mechanisms for their stabilisation. Biochim Biophys Acta 1837:418–426
4. Letts JA, Fiedorczuk K, Sazanov LA (2016) The architecture of respiratory supercomplexes. Nature 537:644–648
5. Severinghaus JW, Astrup PB (1986) History of blood gas analysis. IV. Leland Clark's oxygen electrode. J Clin Monit 2(2):125–139
6. Estabrook RW (1967) Mitochondrial respiratory control and the polarographic measurement of ADP:O ratios. Methods Enzimol 10:41–47
7. Janata J (2009) Principles of chemical sensors, 2nd edn. Springer, New York
8. Moreno AM (1992) Estudo do efeito de Insecticidas de usos comum na bioenergética mitocondrial. Tese de Doutoramento, Universidade de Coimbra
9. Renger G, Hanssum B (2009) Oxygen detection in biological systems. Photosynth Res 102:487–498

10. Diepart C, Verrax J, Calderon PB, Feron O, Jordan BF (2010) Comparison of methods for measuring oxygen consumption in tumor cells in vitro. Anal Biochem 396:250–256
11. Gnaiger E (ed) (2007) Respiratory states and coupling control ratios. In: Mitochondrial pathways and respiratory control, 2nd edn. Oroboros MiPnet Publications, Innsbruck
12. Wenchich L, Drahota Z, Honzík T, Hansíková H, Tesařová M, Zeman J, Houštěk J (2003) Polarographic evaluation of mitochondrial enzymes activity in isolated mitochondria and in permeabilized human muscle cells with inherited mitochondrial defects. Physiol Res 52:781–788
13. Zhang J, Nuebel E, Wisidagama DR, Setoguchi K, Hong JS, Van Horn CM, Imam SS, Vergnes L, Malone CS, Koehler CM, Teitell MA (2012) Measuring energy metabolism in cultured cells, including human pluripotent stem cells and differentiated cells. Nat Protoc 7(6):1068–1085
14. Grinstein S, Furuya W, Lu DJ, Mills GB (1990) Vanadate stimulates oxygen consumption and tyrosine phosphorilation in electropermeabilized human neutrophils. J Biol Chem 265:318–327
15. Conget I, Barrientos A, Manzanares JM, Casademon J, Viñas O, Barceló J, Nunes V, Gomis R, Cardellach F (1997) Respiratory chain activity and mitochondrial DNA content of nonpurified and purified pancreatic islet cells. Metabolism 46(9):984–987
16. Artuch R, Colomé C, Playán A, Alcaine MJ, Briones P, Montoya J, Vilaseca MA, Pineda M (2000) Oxygen consumption measurement in lymphocytes for the diagnosis of pediatric patients with oxidative phosphorylation diseases. Clin Biochem 33(6):481–485
17. Rolo AP, Palmeira CM, Cortopassi GA (2009) Biosensor plates detect mitochondrial physiological regulators and mutations in vivo. Anal Biochem 385(1):176–178
18. Salabei JK, Gibb AA, Hill BG (2014) Comprehensive measurement of respiratory activity in permeabilized cells using extracellular flux analysis. Nat Protoc 9(2):421–438
19. Lowry OH, Rosenbrough NJ, Far AL, Randall RJ (1951) Protein measurement with the Folin phenol reagent. J Biol Chem 193(1):265–275
20. Pereira CV, Moreira AC, Pereira SP, Machado NG, Carvalho FS, Sardão VA, Oliveira PJ (2009) Investigating drug-induced mitochondrial toxicity: a biosensor to increase drug safety? Curr Drug Saf 4(1):34–54
21. Pereira SP, Pereira GC, Moreno AJ, Oliveira PJ (2009) Can drug safety be predicted and animal experiments reduced by using isolated mitochondrial fractions? Altern Lab Anim 37 (4):355–365
22. Moreira PI, Santos MS, Moreno AM, Seiça R, Oliveira CR (2003) Increased vulnerability of brain mitochondria in diabetic (Goto-Kakizaki) rats with aging and amyloid-β exposure. Diabetes 52:1440–1456
23. Gornall AG, Bardawill CS, David MM (1949) Determination of serum proteins by means of the biuret reaction. J Biol Chem 177:751–766
24. Alves M, Oliveira PJ, Carvalho RA (2009) Mitochondrial preservation in celsior versus histidine buffer solution during cardiac ischemia and reperfusion. Cardiovasc Toxicol 9(4):185–193
25. Pereira GC, Branco AF, Matos JA, Pereira SL, Park D, Perkins EL, Serafim TL, Sardão VA, Santos MS, Moreno AM, Holy J, Oliveira PJ (2007) Mitochondrially targeted effects of Berberine [natural yellow 18, 5,6-dihydro-9,10-dimethoxybenzo(g)-1,3-benzodioxolo(5,6-a) quinolizinium] on K1735-M2 mouse melanoma cells: comparison with direct effects on isolated mitochondrial fractions. J Pharmacol Exp Ther 323(2):636–649
26. Kopustinskas A, Adaskevicius R, Krusinskas A, Kopustinskiene DM, Liobikas J, Toleikis A (2006) A user-friendly PC-based data acquisition and analysis system for respirometric investigations. Comput Methods Programs Biomed 82:231–237
27. Bradford MM (1976) A rapid and sensitive for the quantitation of microgram quantitites of protein utilizing the principle of protein-dye binding. Anal Biochem 72:248–254
28. Bugger H, Chemnitius J, Doenst T (2006) Differential changes in respiratory capacity and ischemia tolerance of isolated mitochondria from atrophied and hypertrophied hearts. Metabolism 55:1097–1106
29. Chemnitius JM, Manglitz T, Kloeppel M, Doenst T, Schwartz P, Kreuzer H, Zech R, (1993) Rapid preparation of subsarcolemmal and interfibrillar mitochondrial subpopulations from cardiac muscle. Int J Biochem 25(4):589–596
30. Hinkle PC (2005) P/O ratios of mitochondrial oxidative phosphorylation. Biochim Biophys Acta 1706(1–2):1–11
31. Lee CP, Gu Q, Xiong Y, Mitchell RA, Ernster L (1996) P/O ratios reassessed: mitochondrial P/O ratios consistently exceed 1.5 with succinate and 2.5 with NAD-linked substrates. FASEB J 10(2):345–350
32. Kiss DS, Toth I, Jocsak G, Sterczer A, Bartha T, Frenyo LV, Zsarnovszky A (2016)

Preparation of purified perikaryal and synaptosomal mitochondrial fractions from relatively small hypothalamic brain samples. MethodsX 3:417–429

33. Lanza IR, Nair KS (2009) Functional assessment of isolated mitochondria in vitro. Methods Enzymol 457:349–372
34. Atkins PW (1998) Physical chemistry, 6th edn. Oxford University Press, Oxford
35. Rasmussen HN, Rasmussen UF (2003) Oxygen solubilities of media used in electrochemical respiration measurements. Anal Biochem 319:105–113
36. Young M (1967) On the interaction of adenosine diphosphate with myosin and its enzymatically active fragments. J Biol Chem 242(11):2790
37. Hansatech Intruments (2006) Operations Manual – Setup, installation & maintenance – Electrode preparation & maintenance. *Version 2.2* Hanstech Instruments Ltd, Norfolk, UK
38. Jobst G, Urban G, Jachimowicz A, Kohl F, Tilado O, Lettenbichler I, Nauer G (1993) Thin-film Clark-type oxygen sensor based on novel polymer membrane systems for in vivo and biosensor applications. Biosens Bioelectron 8(3–4):123–128

Chapter 3

High-Resolution FluoRespirometry and OXPHOS Protocols for Human Cells, Permeabilized Fibers from Small Biopsies of Muscle, and Isolated Mitochondria

Carolina Doerrier, Luiz F. Garcia-Souza, Gerhard Krumschnabel, Yvonne Wohlfarter, András T. Mészáros, and Erich Gnaiger

Abstract

Protocols for High-Resolution FluoRespirometry of intact cells, permeabilized cells, permeabilized muscle fibers, isolated mitochondria, and tissue homogenates offer sensitive diagnostic tests of integrated mitochondrial function using standard cell culture techniques, small needle biopsies of muscle, and mitochondrial preparation methods. Multiple substrate-uncoupler-inhibitor titration (SUIT) protocols for analysis of oxidative phosphorylation (OXPHOS) improve our understanding of mitochondrial respiratory control and the pathophysiology of mitochondrial diseases. Respiratory states are defined in functional terms to account for the network of metabolic interactions in complex SUIT protocols with stepwise modulation of coupling control and electron transfer pathway states. A regulated degree of ***intrinsic uncoupling*** is a hallmark of oxidative phosphorylation, whereas pathological and toxicological ***dyscoupling*** is evaluated as a mitochondrial defect. The ***noncoupled*** state of maximum respiration is experimentally induced by titration of established uncouplers (CCCP, FCCP, DNP) to collapse the protonmotive force across the mitochondrial inner membrane and measure the electron transfer (ET) capacity (open-circuit operation of respiration). Intrinsic uncoupling and dyscoupling are evaluated as the flux control ratio between non-phosphorylating LEAK respiration (electron flow coupled to proton pumping to compensate for proton leaks) and ET capacity. If OXPHOS capacity (maximally ADP-stimulated O_2 flux) is less than ET capacity, the phosphorylation pathway contributes to flux control. Physiological substrate combinations supporting the NADH and succinate pathway are required to reconstitute tricarboxylic acid cycle function. This supports maximum ET and OXPHOS capacities, due to the additive effect of multiple electron supply pathways converging at the Q-junction. ET pathways with electron entry separately through NADH (pyruvate and malate or glutamate and malate) or succinate (succinate and rotenone) restrict ET capacity and artificially enhance flux control upstream of the Q-cycle, providing diagnostic information on specific ET-pathway branches. O_2 concentration is maintained above air saturation in protocols with permeabilized muscle fibers to avoid experimental O_2 limitation of respiration. Standardized two-point calibration of the polarographic oxygen sensor (static sensor calibration), calibration of the sensor response time (dynamic sensor calibration), and evaluation of instrumental background O_2 flux (systemic flux compensation) provide the unique experimental basis for high accuracy of quantitative results and quality control in High-Resolution FluoRespirometry.

Key words Substrate-uncoupler-inhibitor titration, Human vastus lateralis, Needle biopsy, HEK, HPMC, HUVEC, Fibroblasts, PBMCs, Platelets, ROUTINE respiration, Oxidative phosphorylation, Q-junction, Pyruvate, Glutamate, Malate, Succinate, Leak, Coupling control, Uncoupling, O2k-FluoRespirometer, O_2 flux, Residual O_2 consumption, Instrumental background

Carlos M. Palmeira and António J. Moreno (eds.), *Mitochondrial Bioenergetics: Methods and Protocols*, Methods in Molecular Biology, vol. 1782, https://doi.org/10.1007/978-1-4939-7831-1_3,

List of Abbreviations (*see* also Table 1)

CCP	Coupling control protocol
ET	Electron transfer
E	ET capacity
FCR	Flux control ratio
FCF	Flux control factor
HRFR	HRFR High-Resolution FluoRespirometry
L	LEAK respiration
mt	Mitochondrial
O2k	O2k Oxygraph-2k
P	OXPHOS capacity
POS	Polarographic oxygen sensor
R	ROUTINE respiration
Rox	Residual oxygen consumption
SUIT	Substrate-uncoupler-inhibitor titration
W_w	Wet weight

1 Introduction

Mitochondrial respiration is a key element of cell physiology. Cell respiration channels metabolic fuels into the bioenergetic machinery of oxidative phosphorylation, regulating and being regulated by molecular redox states, ion gradients, mitochondrial membrane potential as a part of the protonmotive force, the phosphorylation state of the ATP system, and heat dissipation in response to intrinsic and extrinsic energy demands. Complementary to anaerobic energy conversion, mitochondrial respiration is the aerobic flux of life. It integrates and transmits a wide range of physiological and pathological signals within the dynamic communication network of the cell. This provides the background for an interdisciplinary and clinical interest in measurement of mitochondrial respiratory function in biology and medicine (*see* **Notes 1** and **2**).

Respirometry reflects the function of mitochondria as structurally intact organelles. It provides a dynamic measurement of metabolic flux (rates), in contrast to static determination (states) of molecular components, such as metabolite and enzyme levels, redox states and membrane potential, concentrations of signaling molecules, or RNA and DNA levels. Mitochondrial respiratory function, therefore, cannot be measured on frozen tissue samples but usually requires minimum storage times of biological samples and delicate handling procedures to preserve structure and function or highly specific cryopreservation. Moreover, mitochondrial respiration yields an integrative measure of the dynamics of complex coupled metabolic pathways, in contrast to monitoring activities of isolated enzymes. Measurement of respiratory flux in different metabolic states is required for evaluation of the effect on oxidative phosphorylation of changes in metabolite levels, membrane permeability, or activity of individual enzymes. Small imbalances in metabolic flux can result in large cumulative changes of state of a metabolic system. Vice versa, even a large defect of individual enzymes may result in minor changes of flux, due to threshold effects. Hence high quantitative resolution of respirometry is required for diagnostic applications, particularly when amounts of cells or tissue are limiting. Understanding mitochondrial respiratory control, in turn, requires experimental modulation of metabolite levels, electrochemical potentials, and enzyme activities. High-resolution FluoRespirometry (HRFR) has been developed to meet these demands and to provide the instrumental basis for modular extension with additional electrochemical and optical sensors for investigations of mitochondrial respiratory physiology [2–4].

This chapter describes applications of HRFR with intact and permeabilized cells, permeabilized muscle fibers, and isolated mitochondria for functional mitochondrial diagnosis (*see* **Note 3**). Protocols are presented for plasma membrane permeabilization, biopsy sampling, short-term storage, fiber preparation, and respirometric titration regimes in substrate-uncoupler-inhibitor titration (SUIT) protocols (*see* **Note 4**). The SUIT protocols provide diagnostic tests for evaluation of cell viability; membrane integrity (coupling of oxidative phosphorylation; cytochrome *c* release); respiratory inhibition resulting from defects in the phosphorylation pathway or electron transfer (ET) pathway, including respiratory complexes and activities of dehydrogenases particularly in the tricarboxylic acid (TCA) cycle; and metabolite transporters across the inner mitochondrial membrane [5]. We focus on the measurement of oxygen (O_2) concentration and O_2 flux with HRFR, while applications of the fluorometric channels of the O2k-FluoRespirometer are described elsewhere. Quality control (QC) in HRFR includes traceability of basic sensor calibration, systemic flux compensation, kinetic evaluation of ADP saturation and O_2 dependence of respiration from isolated mitochondria to permeabilized fibers [3, 6], and mitochondrial respiration medium (MiR05-Kit; [7]). QC integrates the general features of the Oroboros O2k and is supported by the newly developed DatLab 7 software, which incorporates user-friendly instrumental and experimental protocols as guides through standardized measurement routines, complete documentation, and traceability of measurements. Terminological standards are adopted according to a position statement in the frame of COST Action CA15203 MitoEAGLE [8].

2 Materials

2.1 Media and Chemicals

1. Mitochondrial respiration medium (Oroboros MiR05-Kit), 110 mM sucrose, 60 mM K^+-lactobionate, 0.5 mM EGTA, 3 mM $MgCl_2$, 20 mM taurine, 10 mM KH_2PO_4, 20 mM

HEPES. The powder is dissolved in distilled water, 1 g L^{-1} BSA essentially fatty acid-free is added, and pH is adjusted to 7.1 with KOH at 37 °C [7]. MiR06 is MiR05 plus 280 units mL^{-1} catalase. 20 mM creatine may be added to MiR05 or MiR06 (MiR05Cr or MiR06Cr) [9]. The solutions can be stored at −20 °C.

2. Relaxing and biopsy preservation solution for muscle fibers and various tissue biopsies (BIOPS), 50 mM K^{+}-MES, 20 mM taurine, 0.5 mM dithiothreitol, 6.56 mM $MgCl_2$, 5.77 mM ATP, 15 mM phosphocreatine, 20 mM imidazole pH 7.1 adjusted with 5 N KOH at 0 °C, 10 mM Ca-EGTA buffer (2.77 mM CaK_2EGTA + 7.23 mM K_2EGTA; 0.1 μM free calcium; ref. 10). ATP is hydrolyzed at least partially during fiber storage, thus generating mM levels of inorganic phosphate. BIOPS can be stored at −20 °C.
3. Selected substrates, uncouplers, and inhibitors are listed in Table 1, with corresponding Hamilton syringes used for manual titrations into a 2 mL O2k-Chamber. The sources of chemicals change according to availability and evaluation of quality and price. Information is updated on http://www.bioblast.at/index.php/MitoPedia

2.2 Preparation of Tissue Homogenate and Permeabilized Muscle Fibers

1. Nonmagnetic forceps are used for tissue preparation: one pair with sharp straight tips, one pair with sharp rounded tips, and two pairs with very sharp angular tips.
2. Microbalance with five digits; 0.01 mg (Mettler Toledo XS105DU or XS205DU) for measurement of tissue wet weight.
3. The PBI Shredder O2k-Set (Oroboros) is an auxiliary tool including forceps and scissors for tissue homogenate preparation using small samples of a few mg wet weight. Simpler homogenization is possible of soft tissues, using a precooled glass potter (tight fit; WiseStir homogenizer HS-30E, witeg Labortechnik GmbH, Wertheim, Germany), e.g., at 1000 rpm with 10 strokes for mouse motor cortex and striatum and 15 strokes for hippocampus and brainstem [11].

2.3 The O2k for High-Resolution FluoRespirometry

New methodological standards have been set by High-Resolution FluoRespirometry (HRFR) with the Oroboros Oxygraph-2k (O2k, Fig. 1; Oroboros Instruments, Austria; www.oroboros.at; refs. 2–4, 12). The principle of closed-chamber HRFR is based on monitoring O_2 concentration, c_{O2}, in the incubation medium over time, and plotting O_2 consumption by the biological sample (J_{O2}) and c_{O2} continuously while performing the various titrations in a respirometric protocol (Figs. 2, 3, 4, 6, 8, and 9). HRFR integrates respirometry and fluorometry for simultaneous measurement in the

Table 1
Selected substrates, uncouplers, and inhibitors used in SUIT protocols with mitochondrial preparations

Substrates	Abbr.	Site of action	Concentration in syringe (solvent)	Storage (°C)	Final conc. in 2 mL 02k-Chamber	Titration (μL) into 2 mL	Syringe (μL)
Pyruvate	P	N	2 M (H_2O)	Fresh	5 mM	5	25
Malate[a]	M	N	0.4 M (H_2O)	−20	2 mM	10	25
Glutamate	G	N	2 M (H_2O)	−20	10 mM	10	25
Succinate[b]	S	S	1 M (H_2O)	−20	50 mM	100	100
Octanoylcarnitine	Oct	F	0.1 M (H_2O)	−20	0.2 mM	4	10
Glycerophosphate	Gp	Gp	1 M (H_2O)	−20	10 mM	20	50
Ascorbate	As	CIV	0.8 M (H_2O)	−20	2 mM	5	25
TMPD[c]	Tm	CIV	0.2 M (H_2O)	−20	0.5 mM	5	25
Cyt. *c*	c	CIV	4 mM (H_2O)	−20	10 μM	5	25
ADP	D	CV, ANT	0.5 M (H_2O)	−80	1–5 mM	4–20	10
ATP	T	CV, ANT	0.5 M (H_2O)	−80	1–5 mM	4–20	10
Uncoupler							
CCCP[d]	U	F_{H+}	1 mM (EtOH)	−20	0.5 μM steps	1 μL steps	10
CCCP	U	F_{H+}	0.1 mM (EtOH)	−20	0.05 μM steps	1 μL steps	10
Inhibitors							
Rotenone	Rot	CI	1.0 mM (EtOH)	−20	0.5 μM	1	10
Malonic acid	Mna	CII	2 M (H_2O)	Fresh	5.0 mM	5	25
Antimycin A	Ama	CIII	5 mM (EtOH)	−20	2.5 μM	1	10
Myxothiazol	Myx	CIII	1 mM (EtOH)	−20	0.5 μM	1	10
Sodium azide	Azd	CIV	4 M (H_2O)	−20	≥100 mM	≥50	50
KCN	Kcn	CIV	1 M (H_2O)	Fresh	1.0 mM	2	10
Oligomycin	Omy	CV	5 mM (EtOH)	−20	2.5 μg mL^{-1}	1	10
Atractyloside	Atr	ANT	50 mM (H_2O)	−20	0.75 mM	30	50

Abbreviations (Abbr.) and site of action are ET-pathways (N, NADH; S, succinate; F, fatty acid oxidation; Gp, glycerophosphate; CIV, cytochrome *c* oxidase), the phosphorylation pathway (CV, ATP synthase; ANT, adenine nucleotide translocase), the protonmotive force (F_{H+}), or inhibition of specific enzymes

Modified from ref. 1

[a]In several mitochondrial preparations, malate at concentrations higher than 0.5 mM exerts an inhibitory effect on succinate-linked (S) and NADH- and succinate-linked (NS) respiration. However, S-pathway is involved when NADH-linked (N) substrates are used. The contribution of S-pathway depends on the substrate combinations employed. Unpublished data suggest the use of 2 mM of malate to minimize S-contribution with a mild inhibitory effect on mitochondrial respiration (Z. Sumbalova et al. in prep.)

[b]The concentration of succinate is increased from 10 to 50 mM to compensate for the inhibitory effect of malate

[c]*N,N,N,N′,N′*-tetramethyl-*p*-phenylenediamine dihydrochloride

[d]Carbonyl cyanide *m*-chlorophenyl hydrazine

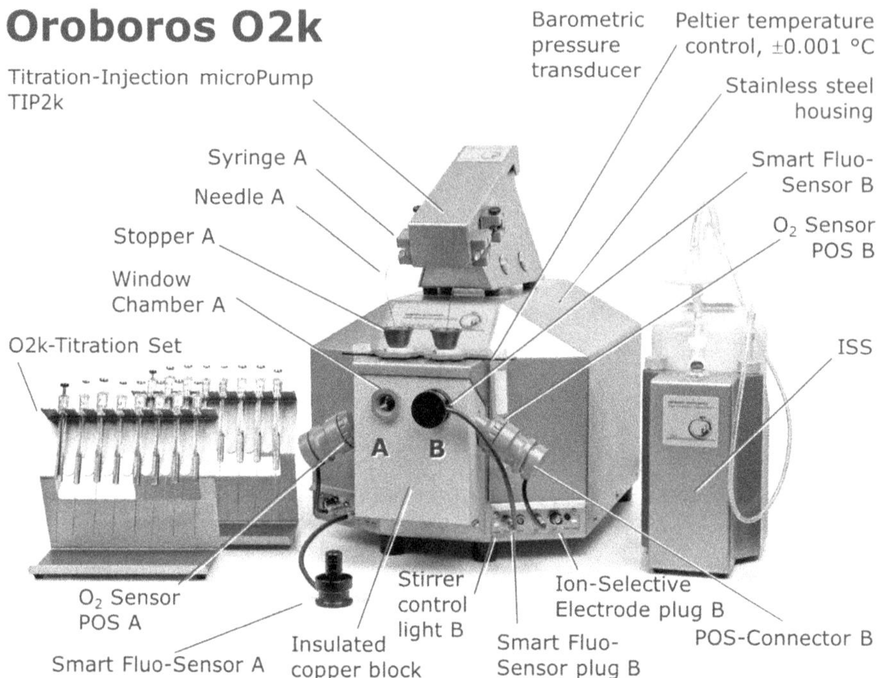

Fig. 1 Oroboros O2k-FluoRespirometer with TIP2k and integrated suction system (ISS), supporting two Smart Fluo-Sensors which can be attached to the windows of chambers A and B. The glass chambers are housed in an insulated copper block with electronic Peltier temperature control. Polarographic oxygen sensors (POS) are sealed by butyl rubber gaskets against the angular plane on the glass chambers. The magnetic stirrer bars are coated by oxygen-impermeable PVDF or PEEK and are powered by electrically pulsed magnets inserted in the copper block. Stoppers contain a capillary for extrusion of gas bubbles and insertion of a needle for manual or automatic titrations with the TIP2k. Additional capillaries through the stopper (PEEK) are drilled for insertion of various electrodes, the signals of which are simultaneously recorded by the DatLab software. Copyright ©2017 by Oroboros Instruments. Reproduced with permission; www.oroboros.at

same chamber. Smart Fluo-Sensors (pre-calibrated LED with specified wavelength, photodiode, filter cap attached with specific optical filter for the LED and photodiode) are incorporated in HRFR.

Without compromise on HRFR features, the O2k provides robustness and reliability of routine instrumental performance. To increase throughput particularly in research with cell cultures and biopsy samples, the user-friendly integrated concept with full software support (DatLab) makes it possible to apply several instruments in parallel, each O2k with two independent chambers (Fig. 1). DatLab supports all measurement channels, quality control, documentation, and traceability of measurements with the Oroboros O2k. Chambers and sensors are thermostated in a Peltier-controlled copper block. The electronics is shielded in a stainless steel housing. Angular insertion of the O_2 sensor into the cylindrical glass chamber places the polarographic oxygen sensor into an optimum position for stirring [4]. Integrated electronic control includes Peltier temperature regulation (2–47 °C, stability ±0.001 °C), stirrer control, an electronic barometric pressure transducer for air calibration, and the optional automatic Titration-Injection microPump TIP2k (Fig. 1).

3 Methods

3.1 Respirometry with Intact Cells: Coupling Control Protocol

Aerobic and anaerobic metabolism is physiologically controlled in the ROUTINE state of cell respiration. Different coupling control states [5, 8] are induced by application of membrane-permeable inhibitors and uncouplers in a coupling control protocol (CCP; Table 2). 0.3–1.0 million fibroblasts or endothelial cells per experiment are sufficient using the Oroboros O2k at 37 °C [2–4]. Cell densities are adjusted to obtain maximum fluxes in the range of 13–300 pmol s^{-1} mL^{-1}. It takes 50–90 min for evaluation of mitochondrial coupling states (Fig. 2b) [9].

Respiratory flow, I_{O2}, of intact cells is expressed per viable cell (amol s^{-1} $cell^{-1}$ = pmol s^{-1} 10^{-6} cells; Table 2). Cell viability should be >0.95 in the control group of various cell cultures (HUVEC, fibroblasts). Mass-specific O_2 flux, J_{O2}, is expressed per mg dry weight [16] or cell protein [20]. Mitochondrial marker-specific respiration is obtained by normalization of flux relative to a mt-marker (mitochondrial element, mte), such as citrate synthase or cytochrome *c* oxidase activity [8, 20]. Internal normalization yields flux control ratios relative to flux in a common reference state (Table 2).

3.1.1 ROUTINE Respiration

Cellular ROUTINE respiration (*R*) and growth is supported by exogenous substrates in culture media [4, 15–23]. In media without energy substrates, respiration is based on endogenous substrates. Physiological energy demand, energy turnover, and the degree of coupling (intrinsic uncoupling and pathological decoupling) control the levels of respiration and phosphorylation in the physiological ROUTINE state of intact cells. The capacity of oxidative phosphorylation (OXPHOS capacity) cannot be studied by external addition of ADP to intact cells, since the plasma membrane is impermeable to ADP and many mitochondrial substrates [16, 19–21, 23]. However, evaluation of the Crabtree effect upon addition of glucose [24] is possible only in intact cells (Fig. 2b). Respiration of intact cells may be measured in MiR05 adding pyruvate (P) as an external energy source. Using a mitochondrial respiration medium allows the extension of the coupling control protocol to include a cell viability test and cytochrome *c* oxidase activity assay (Fig. 2c, d). *R* in human cells (Table 2 [amol O_2 s^{-1} $cell^{-1}$]) ranges from 6 in the small peripheral blood mononuclear cells (PBMCs; [14]), 15 in HEK cells [22], 30–40 in HUVEC and

Table 2
Electron transfer capacity and flux control ratios (*FCR*)[a] in the coupling control protocol with intact human (and 32D mouse) cells (37 °C; mean ± SD)

Cell type	ET, *E* amol s^{-1} $cell^{-1}$	ROUTINE *R/E*	LEAK *L/E*	netROUTINE *(R − L)/E*	Residual *Rox/E*	Ref.
Platelets[b]	0.22 ± 0.03	0.33 ± 0.11	0.07 ± 0.01	0.26 ± 0.11	0.07 ± 0.03	[14]
PBMCs[c]	13.91 ± 1.11	0.27 ± 0.02	0.08 ± 0.02	0.19 ± 0.01	0.05 ± 0.01	[14]
HEK 293[d]	47 ± 7	0.31 ± 0.03	0.09 ± 0.00	0.23 ± 0.02	0.01 ± 0.00	[15]
32D (mouse)[e]	81 ± 11	0.39 ± 0.02	0.10 ± 0.02	0.29 ± 0.02	0.03 ± 0.01	[4]
CEM—control[f]	54 ± 11	0.40 ± 0.03	n.d.	n.d.	0.02 ± 0.03	[16]
CEM—G1-phase, apopt.[f]	31 ± 6	0.41 ± 0.03	n.d.	n.d.	0.03 ± 0.03	[16]
CEM—S-phase, apopt.[f]	85 ± 13	0.38 ± 0.03	n.d.	n.d.	0.01 ± 0.01	[16]
HUVEC[g]	114 ± 18	0.26 ± 0.02	0.13 ± 0.02	0.13 ± 0.00	0.05 ± 0.04	[17]
HPMC control[h]	181 ± 58	0.40 ± 0.09	0.09 ± 0.01	0.31 ± 0.08	0.005 ± 0.01	[18]
HPMC + IL-1β[h]	142 ± 47	0.41 ± 0.09	0.08 ± 0.02	0.32 ± 0.08	0.02 ± 0.01	[18]
Fibroblasts—young[i]	111 ± 24	0.34 ± 0.03	0.14 ± 0.02	0.20 ± 0.02	0.07 ± 0.03	[19]
Fibroblasts—young-arrest[i]	138 ± 22	0.23 ± 0.01	0.05 ± 0.01	0.18 ± 0.02	0.05 ± 0.00	[19]
Fibroblasts—senescent[i]	285 ± 72	0.42 ± 0.05	0.21 ± 0.04	0.21 ± 0.05	0.07 ± 0.03	[19]

[a]Capacity of the electron transfer system (pmol s^{-1} 10^{-6} cells = amol s^{-1} $cell^{-1}$) is the reference for normalization of *FCR*, $E = E' - Rox$, where E' is the apparent (uncorrected) electron transfer capacity. Similarly, $R = R' - Rox$ and $L = L' - Rox$. To calculate total ROUTINE respiration, R' (amol s^{-1} $cell^{-1}$): $R' = \left(R/E + \frac{Rox/E'}{1-Rox/E'}\right) \cdot E$

[b]Human platelet cells (147 × 10^6 mL^{-1}); $N = 3$; in culture medium RPMI. Omy (2.5 μM f.c.) displays a strong inhibitory effect on *E* in human platelets; therefore values of *E* displayed in the table are obtained without Omy (Fig. 3b)

[c]Human peripheral blood mononuclear cells (2.3 × 10^6 mL^{-1}); $N = 3$; in culture medium RPMI. *E* is obtained in the absence of Omy. *E* after Omy (2.5 μM f.c.) showed significant lower values

[d]Transformed human embryonic kidney cells (10 × 10^6 mL^{-1}); $N = 3$ to 8 independent cell cultures in culture medium DMEM

[e]Mouse parental hematopoietic cells (1.1 × 10^6 mL^{-1}); $n = 6$, replicate O2k measurements of a single suspension culture in culture medium RPMI

[f]Human leukemia cells (1.0 × 10^6 mL^{-1} to 10^6 mL^{-1}); controls ($N = 27$), and 30% apoptotic cultures preincubated with dexamethasone, arrested in the G1-phase ($N = 9$) or with gemcitabine, arrested in the S-phase ($N = 12$) in culture medium RPMI; n.d., not determined

[g]Human umbilical vein endothelial cells (0.9 × 10^6 mL^{-1}); $N = 3$; in culture medium EGM

[h]Human peritoneal mesothelial cells (0.6 × 10^6 mL^{-1}); $N = 5$; cultured from five donors, incubated for 48 h without (controls) or with recombinant IL-1β. ROUTINE respiration in MiR05 with succinate and ADP

[i]Human foreskin fibroblasts; young ($n = 12$; 1.0 × 10^6 mL^{-1}), young-cell cycle arrest ($n = 5$; 1.1 × 10^6 mL^{-1}), and senescent ($n = 12$; 0.2 × 10^6 × mL^{-1}), in culture medium DMEM

fibroblasts [16, 17], and 70 in mesothelial cells [23] to 250 in the much larger rat hepatocytes [25].

3.1.2 LEAK Respiration

Following stabilization of *R*, ATP synthesis is inhibited by oligomycin (atractyloside or carboxyatractyloside; Table 1). In this

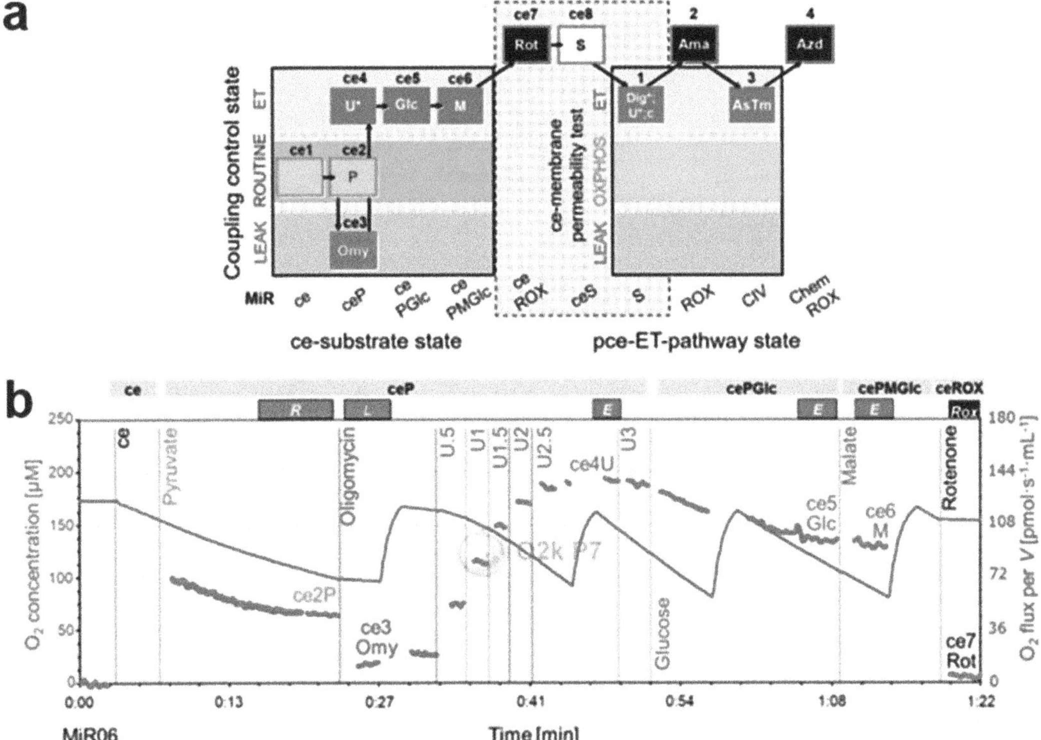

Fig. 2 Coupling control protocol and respirometric cell viability test (CCV protocol). (**a**) Coupling/substrate-pathway control diagram. Titration steps with intact cells, ce-substrate states (left block): ce1: addition of cells, ROUTINE respiration in MiR05; ce2P: 10 mM pyruvate; ce3Omy: 2.5 μM oligomycin, LEAK state; ce4U*: multiple uncoupler titrations to obtain ET capacity; ce5Glc: 25 mM glucose, Crabtree effect; ce6M: 2 mM malate; ce7Rot: inhibition by 0.5 μM rotenone, residual oxygen consumption, ceROX state; ce8S: 10 mM succinate stimulates dead cells with functional mitochondria only. Titration steps with permeabilized cells, pce-ET-pathway states (right block): 1Dig: after digitonin titrations all cells are permeabilized; 1c (c): 10 μM cytochrome *c* for testing the integrity of the outer mitochondrial membrane. 2Ama: 2.5 μM antimycin A for inducing the ROX state. 3AsTm: cytochrome *c* oxidase (CIV) assay with 2 mM ascorbate (As) and 0.5 mM TMPD (Tm). 4Azd: 200 mM azide for inhibition of CIV and measurement of chemical *Rox*. (**b–d**) Respiration of cryopreserved human embryonic kidney 293 cells (HEK 293; 1.5×10^6 cells mL^{-1}); 2 mL O2k-Chambers, 37 °C. 94% viability measured by trypan blue staining (1.4×10^6 viable cells mL^{-1}). O2k traces of oxygen concentration (μM) (blue line) and oxygen flux per volume [pmol s^{-1} mL^{-1}] (red line). (**b**) Section of the CCV protocol with intact cells, ce, uncoupler titration of CCCP at 0.5 μM steps. (**c**) Cell viability test in the CCV protocol (continued from panel **b**); U4: uncoupler titration to check if maximum stimulation of mitochondrial respiration is reached. (**d**) CIV assay (continued from panel **c**). Experiment 2017-02-23 P8-03

non-phosphorylating or resting LEAK state (analogous to State 4; ref. 18), LEAK respiration (*L*) reflects intrinsic uncoupling as (1) compensation for the proton leak at maximum protonmotive force, (2) proton slip (decoupled respiration), (3) electron slip which diverts electrons toward reactive oxygen species (ROS) production, and (4) cation cycling (Ca^{2+}, K^+) [4, 5, 8]. Monitoring *L* should be limited to <5 min (Fig. 2b) to avoid secondary effects on coupling and respiratory capacity.

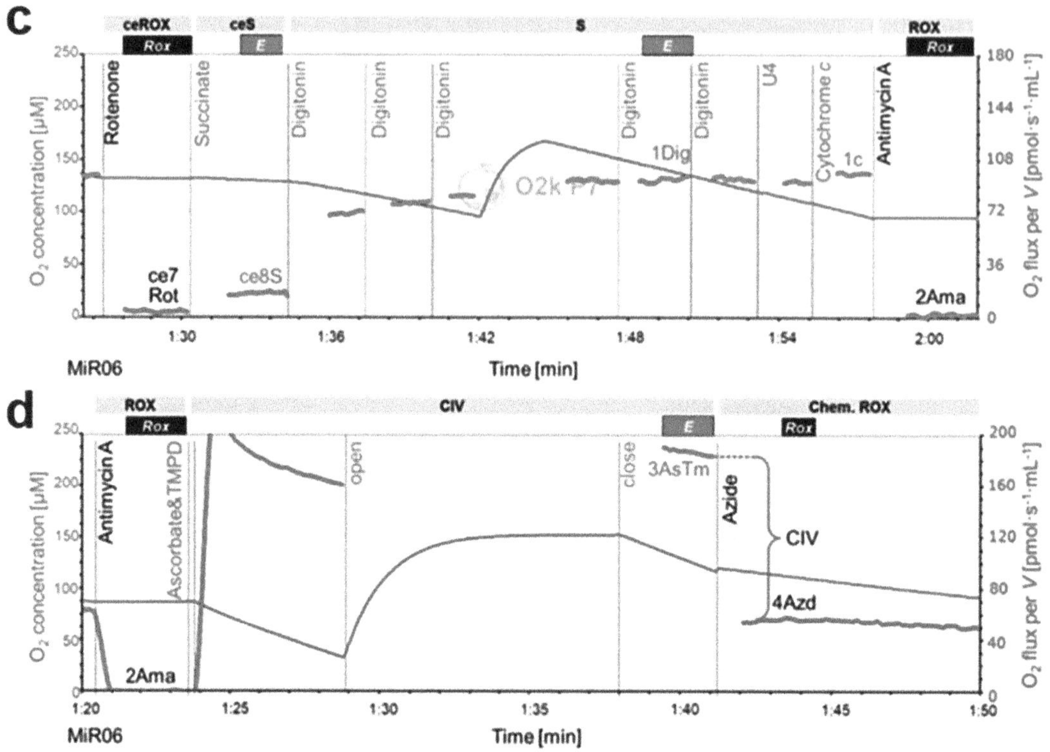

Fig. 2 (continued)

3.1.3 Electron Transfer Capacity

Mitochondrial respiratory control by the phosphorylation pathway is partially or fully released by pathophysiological uncoupling and dyscoupling or experimentally by titration of a protonophore such as CCCP (Table 1). In the noncoupled, open proton circuit state of maximum ET capacity, the electrochemical proton potential across the inner mitochondrial membrane is largely but not fully collapsed [26]. The electrochemical backpressure on the proton pumps (complexes CI, CIII, and CIV) is removed, which stimulates respiration maximally at level flow as a measure of ET capacity, *E*, in the noncoupled ET state (Fig. 2). It is important to titrate an optimum concentration of uncoupler, beyond which respiration is inhibited [4, 16]. Optimum uncoupler concentrations depend on cell type, cell concentration, medium, and permeabilized versus intact cells. Inhibition of *E* by oligomycin should be evaluated by uncoupler titrations in the absence of inhibitor (Fig. 3). *E* ranges from 14 to 180 amol O_2 s^{-1} $cell^{-1}$ in HEK, CEM, HUVEC, fibroblasts, and mesothelial cells, largely depending on cell size. ET capacity per cell doubles with cell size in CEM cells after cell cycle arrest in the G_1- versus S-phase and in senescent fibroblasts (Table 2). Glucose addition in the ET state can be used for the evaluation of the Crabtree effect in the respiratory medium MiR05.

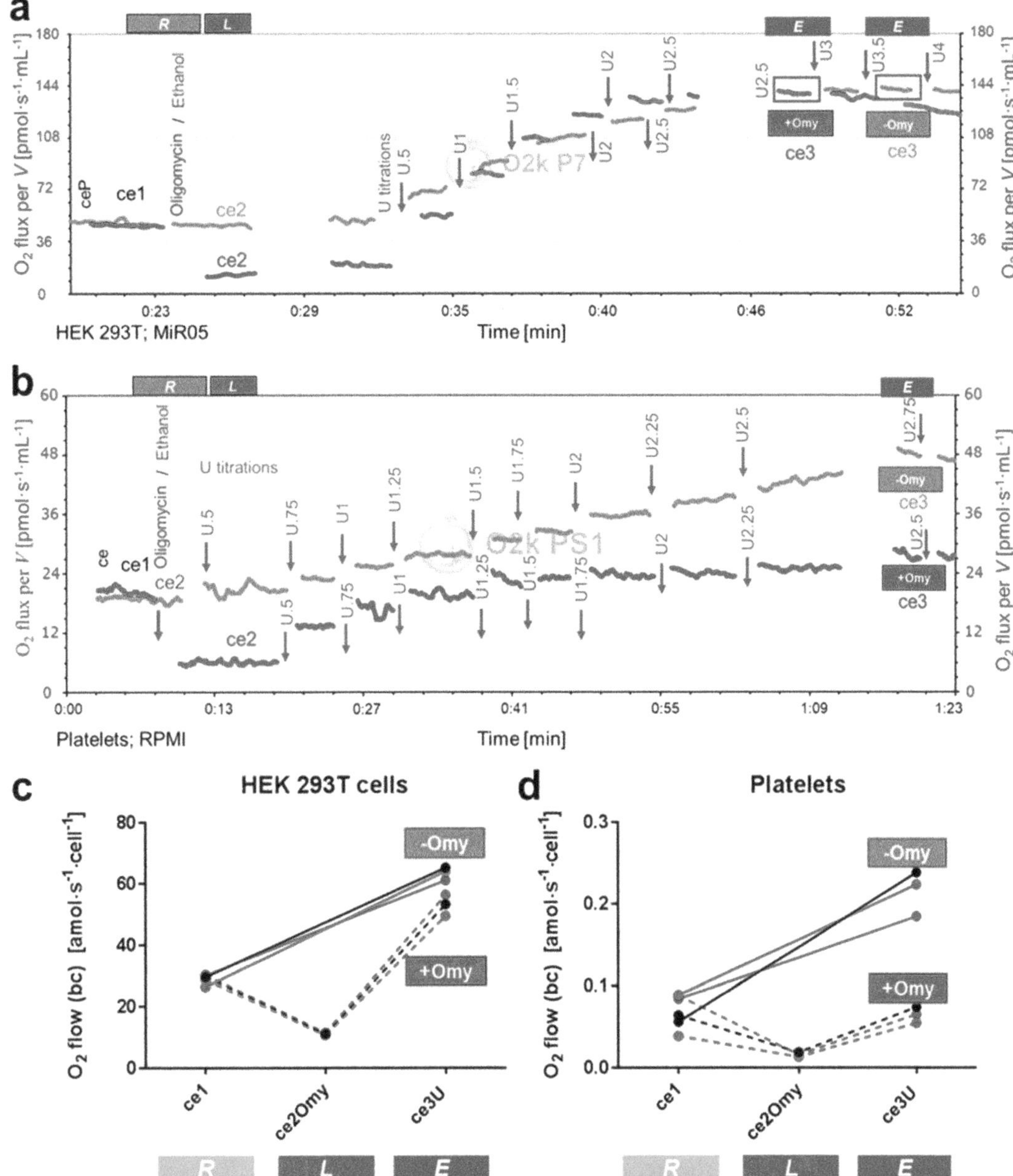

Fig. 3 Coupling control protocol. Superimposed volume-specific oxygen flux [pmol O_2 s^{-1} mL^{-1}] in (**a**) cryopreserved HEK 293T cells (1.5×10^6 cells mL^{-1}) and (**b**) human platelets (147×10^6 cells mL^{-1}). Red lines, +Omy; green lines, controls −Omy. (**c** and **d**) Respiration of HEK cells ($n = 3$ replica) and human platelets ($N = 3$ subjects), I_{O2} [amol O_2 s^{-1} $cell^{-1}$ = pmol O_2 s^{-1} 10^{-6} cells] baseline corrected (bc for *Rox*). Experiments 2017-06-08 P7-02 and 2016-09-26 PS1-02

3.1.4 Residual Oxygen Consumption

Residual oxygen consumption (*Rox*) remains after inhibition of the ET pathway in the ROX state. We distinguish the state, ROX, from the rate, *Rox*. Mitochondrial respirations *R*, *L*, and *E* are corrected for *Rox* (Table 2). Many cellular O_2-consuming enzymes and autoxidation reactions give rise to *Rox*, including peroxidase and oxidase activities which partially contribute to ROS production. It is difficult to evaluate exactly the extent to which inhibitors of the ET pathway (Table 1) exert an influence on *Rox*. Cyanide and azide inhibit CIV and other heme-containing enzymes, such as catalase, and may thus modify *Rox*. Valuable information on *Rox* is obtained by sequential titration of inhibitors (Fig. 2). Rotenone inhibits cell respiration of human fibroblasts and HUVEC, without a further decline of *Rox* after addition of antimycin A [15, 17].

3.1.5 Flux Control Factors from the Coupling Control Protocol

Flux control factors (*FCF*) express the control of respiration in a step by a metabolic control variable (X). X acts upon flux in the background state (Υ) at low flux by stimulating (or activating) it and yielding a higher flux in relation to Υ in the reference state (with flux Z). On the other hand, X can inhibit Z to Υ. $FCF = (Z - \Upsilon)/Z = 1 - \Upsilon/Z$ are independent of mitochondrial content and cell size. *FCF* provide a theoretical lower and upper limit of 0.0 and 1.0 or 0% and 100% [1].

1. Rox/E': The Rox/E' ratio is low (0.01–0.07; Table 2), but *Rox* contributes to a significant extent to LEAK respiration, with corresponding Rox/L' ratios ranging from 0.1 to 0.3 and up to 0.5 in growth-arrested fibroblasts (Table 2).
2. $1 - L/E$: The ET coupling efficiency, $1 - L/E$, is a *FCF* based on measurement of a coupling control ratio (LEAK control ratio, L/E). In the ET coupling efficiency, the background state is the LEAK state (LEAK respiration, L) which is stimulated to the reference ET state (ET capacity, E) by uncoupler titration. $1 - L/E$ is an index of coupling from 0.0 at zero coupling ($L = E$) to 1.0 at the limit of a fully coupled system ($L = 0$). This FCF decreases with the uncoupling or dyscoupling at constant ET capacity. However, this factor may decrease if ET capacity declines. It is, therefore, important to evaluate potential defects of ET capacity per mt-marker, e.g., E per citrate synthase activity [17, 20, 23].
3. $1 - R/E$: The apparent excess $E - R$ capacity factor indicates the scope of stimulation from ROUTINE respiration (well-coupled respiration) to ET capacity (noncoupled respiration). $1 - R/E = 0$ defines the lower limit of zero excess capacity ($R = E$), while $1 - R/E = 1$ defines the upper limit at $R = 0$.
4. $(R - L)/E$: The netROUTINE control ratio, $(R - L)/E$, expresses phosphorylation-related respiration (corrected for LEAK respiration) as a fraction of ET capacity. 0.1–0.3 of ET capacity is used for oxidative phosphorylation under ROUTINE conditions (Table 2). $(R - L)/E$ remains constant, if decoupling is fully compensated by an increase of ROUTINE respiration and a constant rate of oxidative phosphorylation is

maintained (fibroblasts in Table 2). $(R - L)/E$ increases if ROUTINE respiration is stimulated by an increased ATP demand or if the ET capacity declines without effect on R; this indicates that a higher proportion of the ET capacity is activated to drive ATP synthesis. $(R - L)/E$ declines to zero in either fully uncoupled cells ($R = L = E$) or in cells under metabolic arrest ($R = L < E$).

5. *FCFc*: The cytochrome *c* control factor obtained by adding externally cytochrome *c* reflects the integrity of the outer mt-membrane. This test indicates cytochrome *c* release induced by (a) sample preparation or (b) treatment. For calculating the cytochrome *c* control factor, the reference state is the flux after cytochrome *c* addition, and the background state is the flux before cytochrome *c* is added. Cytochrome *c* is not permeable across the plasma membrane; therefore, it cannot be assessed in intact cells. However, the extension added to the CCP module allows the possibility to test the intactness of the outer mt-membrane after the selective cell permeabilization by digitonin (Fig. 2c).
6. If the CCP is extended by measurement of cytochrome *c* oxidase (Fig. 2d), then the ratio of CIV activity and noncoupled respiration is an index of the apparent excess capacity of this enzyme step in the ET pathway. Autoxidation of ascorbate and TMPD (Table 1) is extremely high in culture media; hence, a mitochondrial respiration medium is used [20].

3.1.6 Respirometric Viability Index

A quantitative index of cell viability is derived from the extended coupling control and cell viability protocol (CCV protocol) shown in Fig. 2. Noncoupled cell respiration is inhibited by rotenone (ce7Rot) to the level of *Rox*, since succinate production by the TCA cycle is stopped when NADH cannot be oxidized, and there are no cytosolic sources of succinate. Then any stimulatory effect of externally added succinate (Fig. 2c: ce8S-ce7Rot) depends on plasma membrane permeability in nonviable cells, since succinate cannot penetrate the intact plasma membrane, i.e., the plasma membrane is impermeable to succinate. After digitonin titration (1Dig), S-ET capacity of permeabilized cells is obtained. Subsequent uncoupler and cytochrome *c* titrations (U4 and 1c) indicate that the maximum flux in the ET pathway is reached and the outer mitochondrial membrane remained impermeable for cytochrome *c* (Fig. 2c). Cell viability is the ratio of viable cells in the total cell count, N_{vce}/N_{ce}. The respirometric viability index is calculated as 1 − (ce8S-ce7Rot)/(1Dig-ce7Rot). Previously, less quantitative indices for plasma membrane permeability have been reported [9, 16, 21, 23, 27].

The respiratory viability index is based on preserved respiratory function in mt-respiration medium after plasma membrane injury, whereas respiration of dead cells is fully inhibited by high Ca^{2+} in culture media [27]. The respirometric approach was confirmed by agreement between respirometric viability (0.87 ± 0.03; mean ± SD) and cell viability (0.92 ± 0.05; mean ± SD) in cryopreserved HEK cells obtained with a Countess II Automated Cell Counter (ThermoFisher Scientific, USA).

3.2 Preparation of Permeabilized Cells and Muscle Fibers

Extended functional OXPHOS analysis requires isolation of mitochondria or controlled plasma membrane permeabilization, with effective washout of free cytosolic molecules including adenylates, substrates, and cytosolic enzymes, making externally added compounds accessible to the mitochondria [19, 28–32]. Full mechanical permeabilization of plasma membranes is achieved in liver tissue [32]. Biopsy sampling and mechanical fiber preparation lead to partial permeabilization of skeletal muscle. Without homogeneous plasma membrane integrity, respiration cannot be studied in the ROUTINE state. At low concentrations, digitonin or saponin permeabilize the plasma membranes completely and selectively due to their high cholesterol content, whereas mitochondrial membranes with lower cholesterol content are affected only at higher concentrations. Mitochondrial isolation is more time-consuming than plasma membrane permeabilization. Merely 1 or 2 mg wet weight of cardiac or skeletal muscle fibers is sufficient for individual assays with the Oroboros O2k, but >70 mg is required even for micro-preparations of isolated mitochondria [33]. The homogeneous suspension of isolated mitochondria yields a representative average for large tissue samples, whereas tissue heterogeneity contributes to the variability of results with small samples of permeabilized fibers. Mitochondria can be isolated to separate different mitochondrial subpopulations [34]. Isolated mitochondria and small cultured cells are the appropriate models for the study of mitochondrial oxygen kinetics [2, 3, 6, 12, 16, 22, 24, 35–37].

3.2.1 Permeabilization of Cells

1. After air calibration of the oxygen sensors in the O2k-Chamber with MiR05 or MiR06 (or with Cr added), restart data acquisition with a new file, and select a DL-Protocol as a guide through standardized SUIT steps.
2. Suspended cells are titrated into the closed chambers with a Hamilton microsyringe. Siphon off any excess cell suspension from the receptacle of the stoppers.
3. Alternatively, cells are pipetted into the open chambers with a volume up to 3 mL. While rotation of the stirrers is maintained, subsamples can be collected from the homogeneous cell suspension, for analysis of cell count, cell volume, and cell viability (Countess II Automated Cell Counter, ThermoFisher Scientific, USA), protein concentration, and enzyme assays. A minimum of 2.1 mL cell suspension must remain in the chamber.

Close the chambers by fully inserting the stoppers into the volume-calibrated position, thereby extruding all gas bubbles. Siphon off any excess cell suspension from the receptacle of the stoppers.

4. The final cell concentration is optimized such that ROUTINE respiration yields a volume-specific O_2 flux of about 20 pmol s^{-1} mL^{-1} or higher (Figs. 2, 3, and 4). Special quality control is required if ROUTINE respiration is as low as 4 pmol s^{-1} mL^{-1}. Allow endogenous R to stabilize for 15–20 min (Fig. 2).
5. Add digitonin at optimum concentration, e.g., 10 μg × 10^{-6} cells. Observe a gradual decline of respiration due to plasma membrane permeabilization and loss of adenylates from the cytosol.
6. Optimum digitonin concentrations for complete plasma membrane permeabilization of cultured cells can be determined separately in a respirometric protocol (Fig. 4), which may be used simultaneously for selecting optimum experimental cell concentrations. After inhibition of endogenous R by rotenone, respiration of intact cells (viability >0.95) is not stimulated by the addition of succinate and ADP (compare Fig. 4 with cell viability >0.95 to Fig. 2 in ref. 9 with cell viability of 0.8). Subsequent stepwise digitonin titration yields gradual permeabilization of plasma membranes, indicated by the increase of respiration up to full permeabilization (Fig. 4a, at 10 μg/mL in 1×10^6 cells; Fig. 4b, at 25 μg/mL in 0.3×10^6 cells). Respiration is constant over a range of optimum digitonin concentrations but is inhibited at higher concentrations when the outer mitochondrial membrane becomes affected and cytochrome *c* is released (cytochrome *c* test; Subheading 3.3.2). Therefore, after the titration of digitonin to obtain the optimal concentration for the plasma membrane permeabilization, it is recommended to add 10 μM cytochrome *c*. Respiration of permeabilized cells is stable in SUIT protocols in the presence of an optimum digitonin concentration.

3.2.2 Muscle Biopsy

The human muscle most extensively studied for functional diagnosis of mitochondrial diseases is the quadriceps (m. vastus lateralis; ref. 5) as it is easily accessible and major nerves and blood vessels lie close to the femur and are unlikely to be injured during biopsy sampling [38]. For details of biopsy sampling, *see* ref. 9.

3.2.3 Mechanical Preparation of Permeabilized Fibers

1. The tissue sample with BIOPS solution is transferred onto a small Petri dish on ice.
2. Connective tissue is removed using two pairs of very sharp angular forceps (Fig. 5a).

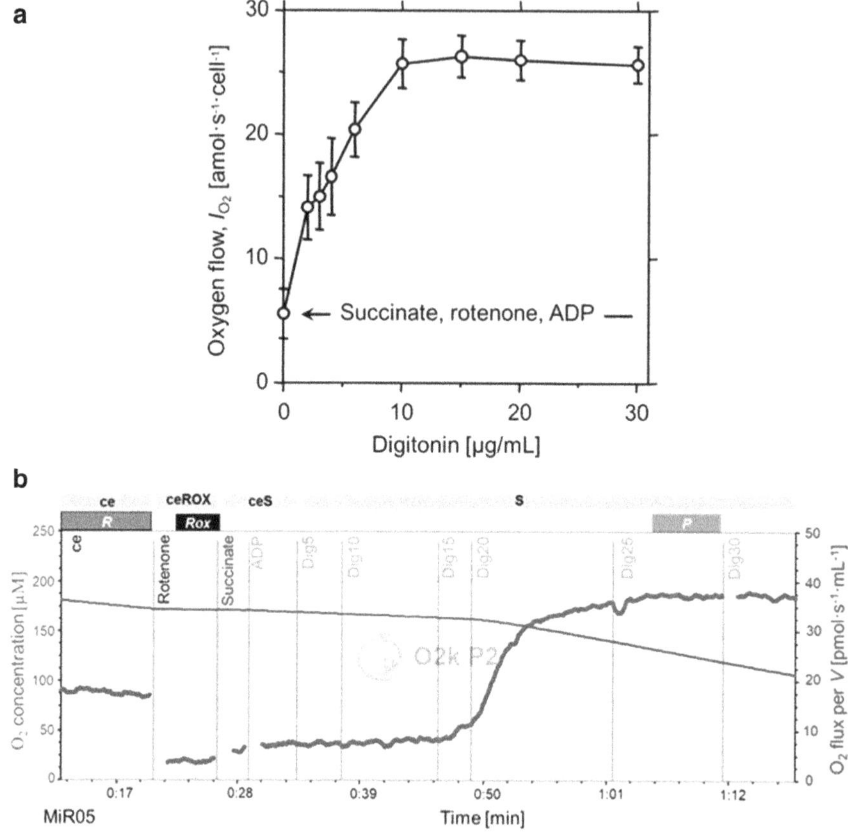

Fig. 4 Respirometric determination of optimum digitonin concentration for selective cell membrane permeabilization in MiR05. (**a**) Human umbilical vein endothelial cells transformed by lung carcinoma ($1.02 \pm 0.16 \times 10^6$ cells mL^{-1}; $N = 6$; ±SD), 10 mM succinate, 0.5 μM rotenone, and 1 mM ADP, titration of digitonin. 12–14 min time intervals between titrations up to 3 μg mL^{-1}, 4–5 min at higher digitonin concentrations. Permeabilization at a digitonin concentration of 10 μg 10^{-6} cells is optimum for ADP-stimulated respiration. Experiments 2010-06-24 P4-01, 2010-07-12 P4-01. From ref. 13. (**b**) MG-63 cells (0.3×10^6 cells mL^{-1}), O2k representative trace: 0.5 μM rotenone, 10 mM succinate, 2.5 mM ADP, digitonin titrations up to 30 μg mL^{-1}. 25 μg mL^{-1} was the optimum concentration for selective plasma membrane permeabilization. Experiment 2014-05-05 P2-01

3. Fiber bundles are separated mechanically with these forceps over a standardized period of 4 min for preparation of a 2 mg sample of human v. lateralis. Fibers are partially teased apart and stretched out, remaining connected in a mesh-like framework (Fig. 5c). Proper separation and a change from red to pale color is best observed against a dark background (Fig. 5a–c). At least during a start-up period, it is recommended to use a dissecting scope for effective removal of connective tissue and observation of the mechanical separation. Initially, difficulties arise frequently from application of excess tissue, which makes mechanical separation tedious.

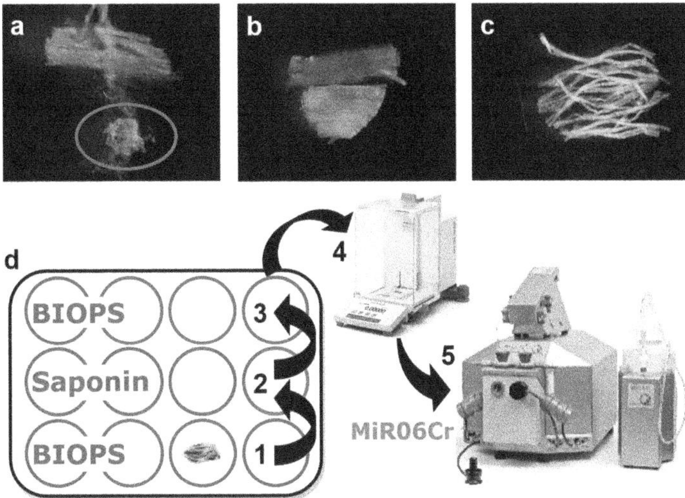

Fig. 5 Preparation of permeabilized muscle fibers from a small biopsy of human vastus lateralis. (**a**) Connective tissue (circle) is removed. (**b**) Muscle fiber bundle. (**c**) Fiber bundles after mechanical separation using a pair of forceps with very sharp angular tips over a standardized period of 4 min. Mesh-like structure and change in color from reddish to pale due to loss of myoglobin and removal of remaining vessels. (**d**) After mechanical separation, the fibers are placed into wells of a 12-well plate (Falcon 35/3043) on ice: (*1*) with BIOPS; (*2*) transfer to BIOPS with saponin for chemical permeabilization; (*3*) washing step in BIOPS; (*4*) wet weight determination; (*5*) transfer of fibers into the O2k-Chambers for respirometric measurements. Modified from ref. 9

4. Fiber bundles of similar mass are placed sequentially into 2 mL ice-cold BIOPS in individual wells (Fig. 5d).

3.2.4 Chemical Permeabilization and Wet Weight of Muscle Fibers

1. After fibers for all simultaneously operated O2k-Chambers are mechanically prepared and placed into the wells with ice-cold BIOPS, the fiber bundles are transferred quickly into 2 mL freshly prepared saponin solution (50 µg mL^{-1} BIOPS; add 20 µL saponin stock solution of 5 mg saponin mL^{-1} BIOPS into 2 mL BIOPS; Fig. 5d).
2. Shake by gentle agitation on ice for 20 min.
3. Transfer all samples from the saponin solution to 2 mL of BIOPS (Fig. 5d). Continue shaking by gentle agitation for 10 min on ice.
4. Wet weight measurements are made after permeabilization, which reduces osmotic variations in water contents. Loosely connected fiber bundles (skeletal muscle 1–3 mg W_w; heart 0.5–2 mg W_w) are taken with the pair of sharp forceps (rounded tip) and placed for 40 s onto blotting paper (GE Healthcare). During this time, wipe off any liquid from the tip of the forceps with another filter paper. Take the sample from the blotting paper, touch it once more shortly onto a dry area of filter paper while holding it with the forceps, and place

the sample onto aluminum paper (or small cup) on the table of the tared balance.

5. Immediately after reading the W_w, the sample is transferred into an individual well with ice-cold MiR06 (each well contains a sample for a respirometric experiment). Alternatively, the sample may be transferred directly into the O2k-Chamber.
6. A pair of forceps with straight tips is used to fully immerse the fibers into the medium in the O2k-Chamber. Check if the entire tissue sample has been retrieved from the well and no parts of the tissue remain on the tip of the forceps.
7. Full permeabilization is validated by a decline of LDH activity to 1% of intact tissue [10] or more quickly by respirometry. Respiration of fully permeabilized tissue is not increased by titration of saponin or digitonin in the presence of substrate and ADP. A stimulatory effect of saponin, however, indicates incomplete permeabilization of muscle fibers that were not incubated in saponin solution prior to the experimental run. Saponin permeabilization in the respiration chamber does not yield maximum respiratory capacity of muscle fibers. The larger saponin stimulation of respiration indicates a lower degree of permeabilization, which correlated with lower mass-specific O_2 flux even after addition of saponin [9].

3.3 High-Resolution FluoRespirometry with Permeabilized Muscle Fibers, Permeabilized Cells, and Isolated Mitochondria

3.3.1 Temperature

Temperature exerts a strong influence on mitochondrial respiration. Respiratory capacity should be evaluated at physiological temperature [5, 39]. The further the experimental conditions differ from the physiological reference state, the larger the error becomes which may result from adjustment to 37 °C of respiratory fluxes in various metabolic states, applying a commonly assumed constant temperature coefficient. Assuming a Q_{10} of 2 (multiplication factor for flux at a 10 °C difference), the temperature coefficients for rates measured at 22 °C, 25 °C, or 30 °C are 2.83, 2.30, and 1.62, respectively, to convert to respiration at 37 °C. Some fundamental functional properties of mitochondria change at 25 °C, for instance, there is a shift from proton leak at 37 °C (proton flux through the membrane) to proton slip at 25 °C (protons pulled back into the matrix phase within a proton pump; ref. 40). In mitochondrial physiology, therefore, experimental temperature close to body temperature has become a standard for quantitative evaluation of mitochondrial respiratory function in mammalian cells (Table 2) and tissues [5, 39].

3.3.2 Substrates: Electron Donors

Mitochondrial respiration depends on a continuous flow of electron-supplying substrates across the mitochondrial membranes into the matrix space. Many substrates are strong anions that cannot permeate lipid membranes and hence require carriers. Various anion carriers in the inner mitochondrial membrane are involved in the transport of mitochondrial metabolites. Their distribution across the mitochondrial membrane varies mainly with ΔpH and not $\Delta\psi$, since most carriers (but not the glutamate-aspartate carrier) operate non-electrogenically by anion exchange or co-transport of protons. Depending on the concentration gradients, these carriers also allow for the transport of mitochondrial metabolites from the matrix into the cytosol and for the loss of intermediary metabolites into the incubation medium. Export of intermediates of the TCA cycle plays an important metabolic role in the intact cell. This must be considered when interpreting the effect on respiration of specific substrates used in studies of permeabilized cells and isolated mitochondria [1, 5]. Some typical saturating substrate concentrations used in respiratory studies are listed in Table 1.

1. Electron transfer in the NADH pathway (N): NADH-linked substrate combinations such as pyruvate and malate (PM) and glutamate and malate (GM) stimulate dehydrogenases yielding reduced nicotinamide adenine dinucleotide (NADH), which feeds electrons into CI (NADH-ubiquinone oxidoreductase) and hence down the thermodynamic cascade through the Q-cycle, CIII, cytochrome *c*, CIV, and ultimately O_2. Electrons flow from NADH to O_2 with three proton pumps (CI, CIII, CIV) in series.
2. Complex II is the only membrane-bound enzyme in the TCA cycle. The flavoprotein succinate dehydrogenase is the largest polypeptide of CII. In succinate oxidation (S-pathway), CII transfers electrons to the quinone pool [41]. Whereas CI is NADH-linked *upstream* to the dehydrogenases of the TCA cycle, CII is $FADH_2$-linked *downstream* with subsequent electron flow to the Q-junction [1, 5]. Electrons flow from succinate to O_2 with two proton pumps (CIII, CIV) in series (*see* **Note 5**).
3. Studies of fatty acid oxidation (F-pathway) involve a large variety of substrates, such as palmitic acid, palmitoylcarnitine, or palmitoyl-CoA with carnitine. Like CII, electron-transferring flavoprotein complex (CETF) is located on the matrix face of the inner mitochondrial membrane. It supplies electrons from fatty acid β-oxidation to coenzyme Q (CoQ). For β-oxidation to proceed, convergent electron flow into the Q-junction is obligatory from both CI and CETF. Malate is provided, therefore, simultaneously with fatty acid substrates. Fatty acid oxidation is blocked by inhibition of CI. Concentrations of fatty acid substrates must be optimized carefully to reach substrate saturation without inducing inhibitory and uncoupling effects. The octanoylcarnitine (Oct) concentration was 0.5 mM in the present protocol. In previous

experiments performed in human muscle fibers, 0.2 mM octanoylcarnitine was employed [9, 29, 42]. Higher concentrations did not yield higher flux.

4. Glycerophosphate dehydrogenase complex (CGpDH) is a complex of the electron transfer system localized at the outer face of the inner mt-membrane. CGpDH oxidizes glycerophosphate (Gp) to dihydroxyacetone phosphate and feeds two electrons into the Q-junction (Gp pathway).

5. Ascorbate and TMPD (Table 1) are artificial electron donors reducing cytochrome *c* for measurement of cytochrome *c* oxidase (CIV) as an isolated step. CIII is inhibited by antimycin A or myxothiazol. Ascorbate (As) is added to maintain the subsequently added TMPD (Tm) in a reduced state. Autoxidation of ascorbate and TMPD depends on (a) their concentrations, (b) O_2 concentration, (c) concentration of added cytochrome *c*, and (d) the medium. Histidine stimulates autoxidation of ascorbate and is therefore omitted from MiR05. Chemical background O_2 flux plus *Rox* is determined at the end of an experimental run after inhibition of CIV by cyanide or azide at low O_2 concentration, continued after reoxygenation at high O_2 concentration. The keto acid pyruvate and α-ketoglutarate remove cyanide from CIV, forming the respective cyanohydrins. Reversibility of cyanide inhibition is particularly effective at high O_2 concentrations [43]. Therefore, cyanide cannot be used as a CIV inhibitor in the presence of pyruvate. High azide concentrations must be applied for full inhibition of CIV (Table 1). To separate autoxidation (chemical background) from *Rox*, O_2 consumption is determined in the absence of biological material under experimental conditions as a function of O_2 concentration. The chemical and instrumental background is subtracted from total measured O_2 flux to obtain CIV activity. Current findings in different mitochondrial preparations (e.g., cardiac isolated mitochondria, HEK permeabilized cells) showed that a simplified CIV assay yields reliable results, when subtracting O_2 consumption after azide inhibition from O_2 flux obtained with ascorbate and TMPD (Figs. 2d and 6c, d).

6. Cytochrome *c* (c) does not pass the intact outer mt-membrane. Comparable to the succinate test for plasma membrane permeability, a cytochrome *c* test can be applied to evaluate the intactness of the outer mt-membrane in mitochondrial preparations [31]. Permeabilized fibers of human v. lateralis and cardiac biopsies (healthy controls) [13, 28, 29, 44], human prostate tissue homogenate [45], or permeabilized cardiac fibers and isolated mitochondria of mouse (Figs. 6 and 8) do not show a cytochrome *c* effect when 10 μM cytochrome *c* is added. Cytochrome *c* is added early in the protocol (after ADP; refs. 13, 28) to obtain all active fluxes in a comparable *c*-activated state, in a

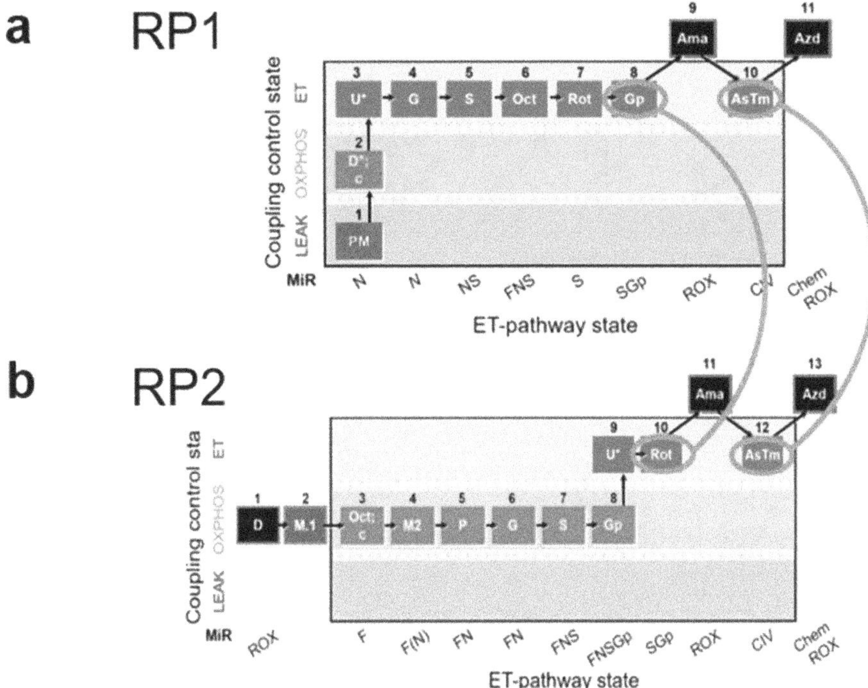

Fig. 6 Substrate-uncoupler-inhibitor titration (SUIT) reference protocols RP1 and RP2. (**a** and **b**): coupling/ET-pathway control diagrams. (**c** and **d**): oxygen concentration (μM) and volume-specific oxygen flux [pmol O_2 s^{-1} mL^{-1}] as a function of time, isolated cardiac mitochondria from mouse. (**a** and **c**): RP1, linear coupling control from 1 PM: 5 mM pyruvate and 2 mM malate as NADH-linked (N) substrates, LEAK; to 2D: 2.5 mM ADP, OXPHOS (2c: 10 μM cytochrome *c* for assessment of outer mitochondrial membrane integrity) and 3U*: 0.5 μM stepwise titrations of CCCP, ET state. ET-pathway control: 4G: 10 mM glutamate; 5S: 50 mM succinate; 6Oct: 0.5 mM octanoylcarnitine; 7Rot: 0.5 μM rotenone, 8Gp: 10 mM glycerophosphate. 9Ama: 2.5 μM antimycin A for measurement of residual oxygen consumption (*Rox*). Cytochrome *c* oxidase (CIV) activity assay: 10AsTm: 2 mM ascorbate and 0.5 mM TMPD; 11Azd: 200 mM azide. (**b** and **d**): RP2 with ET-pathway control in the OXPHOS state. 1D: 2.5 mM ADP; 2M.1: malate at low concentration (0.1 mM); 3Oct: 0.5 mM octanoylcarnitine for evaluation of fatty acid oxidation capacity, F, corrected for 2M.1 (3c: 10 μM cytochrome *c*); 4M2: high concentration of malate (2 mM) did not stimulate the anaplerotic pathway; 5P: 5 mM pyruvate; 6G: 10 mM glutamate; 7S: 50 mM succinate; 8Gp: 10 mM glycerophosphate. 9U*: CCCP titration (0.5 μM steps), ET capacity in the FNSGp pathway. 10Rot: 0.5 μM rotenone, SGp-pathway. 11Ama: 2.5 μM antimycin A, ROX. CIV activity assay as in RP1. MiR06Cr, 37 °C, 2 mL chamber. Experiments 2017-02-08 P2-02.DLD and 2017-02-08 P2-02.DLD

series of OXPHOS states [31], or at a late state in the protocol, e.g., after 100 min incubation [29], indicating stability of the outer mt-membrane during prolonged incubation. The kinetic response to external oxidized cytochrome *c* is monophasic hyperbolic and identical in cytochrome *c*-depleted permeabilized fibers and isolated mitochondria of rat heart (treated by hypoosmotic shock), with a c_{50} of 0.4 μM cytochrome *c* supporting half-maximum flux with succinate and rotenone and saturating ADP [31, 36]. 10 μM cytochrome *c*, therefore, is sufficient

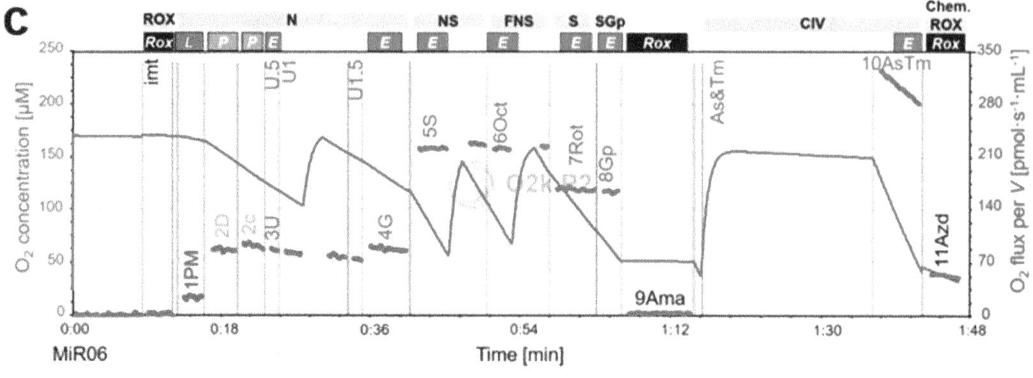

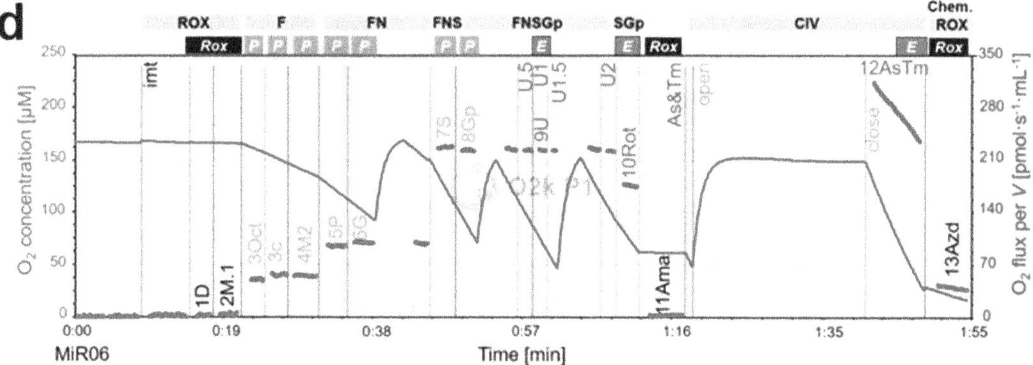

Fig. 6 (continued)

to saturate electron transfer in the S-pathway. In the presence of 0.5 mM TMPD and 2 mM ascorbate, the kinetic response to cytochrome *c* is biphasic, with a high-affinity K_m' of 0.5 and 0.9 µM in isolated mitochondria versus permeabilized fibers and a low-affinity K_m' of 12 µM in both preparations. Then 10 µM cytochrome *c* saturates the velocity of CIV to only 63% and 75% in fibers and mitochondria, respectively [31, 36].

3.3.3 ADP and Inorganic Phosphate

ADP and inorganic phosphate are added to permeabilized cells and fibers at high concentrations to kinetically saturate OXPHOS capacity. The transmembrane proton pumps drive H^+ out of the matrix phase against an electrochemical backpressure, which is used in turn to fuel phosphorylation of ADP and release of ATP at the ATP synthase. The proton circuit is well coupled in the OXPHOS state, but a fraction of the electrochemical gradient is dissipated through proton leaks. Diffusion restriction as shown by oxygen kinetics (Subheading 3.3.6) and the outer mt-membrane generate barriers for inorganic phosphate and ADP different from isolated mitochondria [3, 10, 30, 46]. MiR05 contains 10 mM phosphate. Kinetic saturation by ADP requires testing by titrations. At a high apparent K_m for ADP of 0.5 mM [30], flux at 2.5 and 5 mM ADP is ADP-limited by

13% and 7% (assuming $L/P = 0.2$). 2.5 mM ADP is saturating in many cases, yet a further increase of ADP concentration provides a test for saturating [ADP]. This is particularly important for evaluation of OXPHOS versus ET capacity (*P* versus *E*; Fig. 6 in [9]).

3.3.4 Substrate-Uncoupler-Inhibitor Titration (SUIT) Protocols

Sequential titrations of substrates, inhibitors, and uncouplers are applied in SUIT protocols with mitochondrial preparations, for evaluation of mitochondrial pathway capacities in coupling control states of LEAK, OXPHOS, and ET. A new perspective of mitochondrial physiology and respiratory control emerged from a series of studies based on High-Resolution FluoRespirometry with novel SUIT protocols [1, 29, 39, 44–48]. SUIT protocols are designed to test specific hypotheses in mitochondrial research (*see* **Notes 4** and **6**). To cover a wide range of respiratory states for comparison of mitochondrial respiratory control in different species, tissues, and cells, we developed the SUIT reference protocols (SUIT-RP; Fig. 6). The SUIT-RP includes RP1 and RP2, which are harmonized at two common states: SGp_E and CIV_E (Fig. 6a, b) [48]. Abbreviations and sites of action are listed in Table 1. Coupling control in ET-pathway states is shown by subscripts. To provide a unique code for each SUIT protocols, SUIT steps are indicated in numerical sequence of titrations which induce a new respiratory ,state (*see* **Note 6**).

RP1 (Fig. 6a, c): Coupling control in the N-pathway state (PM); variation of ET-pathway states in the noncoupled ET state.

$$1PM; 2D; 2c; 3U; 4G; 5S; 6Oct; 7Rot; 8Gp; 9Ama; 10AsTm; 11Azd \quad (1)$$

1PM: N_L, pyruvate (P) and malate (M) titrated in immediate sequence to induce the N-LEAK state, in the absence of ADP (no adenylates).

2D: N_P, OXPHOS capacity (*P*) at kinetically saturating [ADP] (D), flux increases to active respiration, limited by substrate supply to the Q-junction through the N-pathway.

2c: Nc_P, cytochrome c to evaluate the integrity of the outer mitochondrial membrane.

3U: N_E, ET capacity after step titration of the uncoupler (U) CCCP to obtain the optimal concentration (noncoupled state). Activation by uncoupling is expected if the phosphorylation pathway (ANT, ATP synthase, phosphate transporter) limits OXPHOS capacity [1, 5], but also if ADP is not kinetically saturating. From 1PM to 3U, the linear coupling control, *L*–*P*–*E*, is evaluated in the N-pathway.

4G: N_E, stimulation by glutamate, an additional N-linked substrate.

5S: NS_E, respiration is further stimulated by adding succinate, activating convergent electron flow into the Q-cycle (NS pathway; CI- and CII-linked) [1, 5].

6Oct: FNS_E, octanoylcarnitine addition inducing the FNS-ET capacity.

7Rot: S_E, by inhibiting the CI with rotenone, only S-ET capacity is evaluated.

8Gp: SGp_E, with glycerophosphate NFSGp pathway, flux converges at the Q-junction.

9Ama: ROX is determined after the inhibition of the ET pathway by antimycin A (inhibitor of CIII).

10AsTm: CIV_E, CIV activity in the ET coupling state is obtained using ascorbate (As) and TMPD (Tm).

11Azd: Azide inhibits CIV.

RP2 (Fig. 6b, d): F-pathway versus N-pathway capacity in the OXPHOS state; variation of ET-pathway states in the OXPHOS state.

1D;2M.1;3Oct;3c;4M2;5P;6G;7S;8Gp;9U;10Rot;11Ama;12AsTm;13Azd (2)

1D: ADP for the depletion of endogenous substrates.

2M.1: Malate at low concentration (0.1 mM) is sufficient to kinetically saturate the F-pathway but is sufficiently low to prevent stimulation of the anaplerotic pathway, thus preventing overestimation of F-pathway capacity.

3Oct: F_P, octanoylcarnitine addition for stimulating F-OXPHOS with an obligatory contribution of the N-pathway, but corrected for 2M.1.

3c: $F(N)_P$, cytochrome *c* test.

4M2: $F(N)_P$, a second titration of 2 mM malate is performed. At this concentration, malate alone supports N-OXPHOS if the anaplerotic pathway (e.g., mitochondrial malic enzyme) is highly active.

5P: FN_P, pyruvate in the presence of 2 mM malate and octanoylcarnitine for FN-OXPHOS capacity.

6G: FN_P, glutamate as an additional N-linked substrate.

7S: FNS_P, succinate for obtaining FNS-OXPHOS.

8Gp: $FNSGp_P$, with glycerophosphate additivity of FNSGp is evaluated in the OXPHOS state.

9U: $FNSGp_E$, FNSGp-ET capacity is determined by titrating CCCP, thus obtaining the *E–P* excess capacity in the state of multiple convergent electron input into Q.

10Rot: SGp_E, SGp-ET capacity after inhibition of CI by rotenone.

11Ama: Antimycin A for ROX.

12AsTm: CIV_E, CIV activity in the ET coupling state with ascorbate and TMPD.

13Azd: Azide inhibits CIV.

In the design of multiple, complementary protocols, it is important to include one or several overlapping coupling/pathway states, providing a quantitative link between the separate experimental incubations (harmonization) for statistical analysis of reproducibility of the assay with different subsamples (Fig. 6a, b). On the other hand, the different harmonized protocols yield additional information on mitochondrial respiratory control patterns when compared with a strictly repetitive approach.

A simplified SUIT protocol is shown in Fig. 7 with examples of differential diagnosis of severe CI and CII injuries (*see* **Note 6**). Intact cell respiration is equally inhibited when TCA cycle function is interrupted at the level of CI or CII. OXPHOS analysis in mitochondrial preparations is required to localize bioenergetic deficiencies or adaptations in the complex framework of mitochondrial respiratory control [13, 28, 29, 31, 39, 45].

3.3.5 Oxygen Flux, Normalization of Flux, Flux Control Ratios, and Flux Control Factors

Mass-specific flux of permeabilized tissues is expressed per mg wet weight (Fig. 7), integrating mitochondrial quality and quantity (mt-density). Rather than tabulating mitochondrial respiration in an unnecessary variety of units, SI units provide a standard for expressing O_2 flow, I_{O2} [mol s^{-1}; pmol s^{-1}] (Table 2), and mass-specific O_2 flux, $J_{m,O2}$ [pmol s^{-1} mg^{-1} = nmol s^{-1} g^{-1}]. Multiply "bioenergetic" units [ng atom O min^{-1}] by 8.33 to convert to SI units [pmol O_2 s^{-1}].

To separate the effects of mt-quality from mt-density, a common mt-marker is used for normalization, such as mtDNA [44], citrate synthase activity or CIV activity [20], or cytochrome aa_3 content [10]. Subsamples or the entire contents can be collected from the O2k-Chamber for analysis of CS activity [28, 32]. Respiratory flux control ratios, *FCR*, however, are internal ratios within an experimental run and thus minimize several experimental errors, providing the most powerful normalization of flux [5, 13, 28, 29]. ET-pathway control ratios are *FCR* at constant coupling state, whereas coupling control ratios are *FCR* at constant ET-pathway state, relating *L* and *P* to *E*. Complementary to the *FCR*, flux control factors, *FCF*, express the control of respiration in a step by a metabolic control variable (Subheading 3.1.5).

1. The LEAK control ratio, L/E, expresses uncoupling or decoupling, provided that specific limitations of flux by *E* are considered. L/E increases with uncoupling from a theoretical minimum of 0.0 for a fully coupled system to 1.0 for a fully dyscoupled state.
2. The phosphorylation pathway control ratio, P/E, increases from a minimum of L/E if the capacity of the phosphorylation pathway is zero to the maximum of 1.0 if the capacity of the phosphorylation pathway fully matches the ET capacity (or in fully dyscoupled mitochondria), when there is no limitation of *P* by the phosphorylation pathway or the proton backpressure. It is important to separate the effect of ADP limitation [9] from

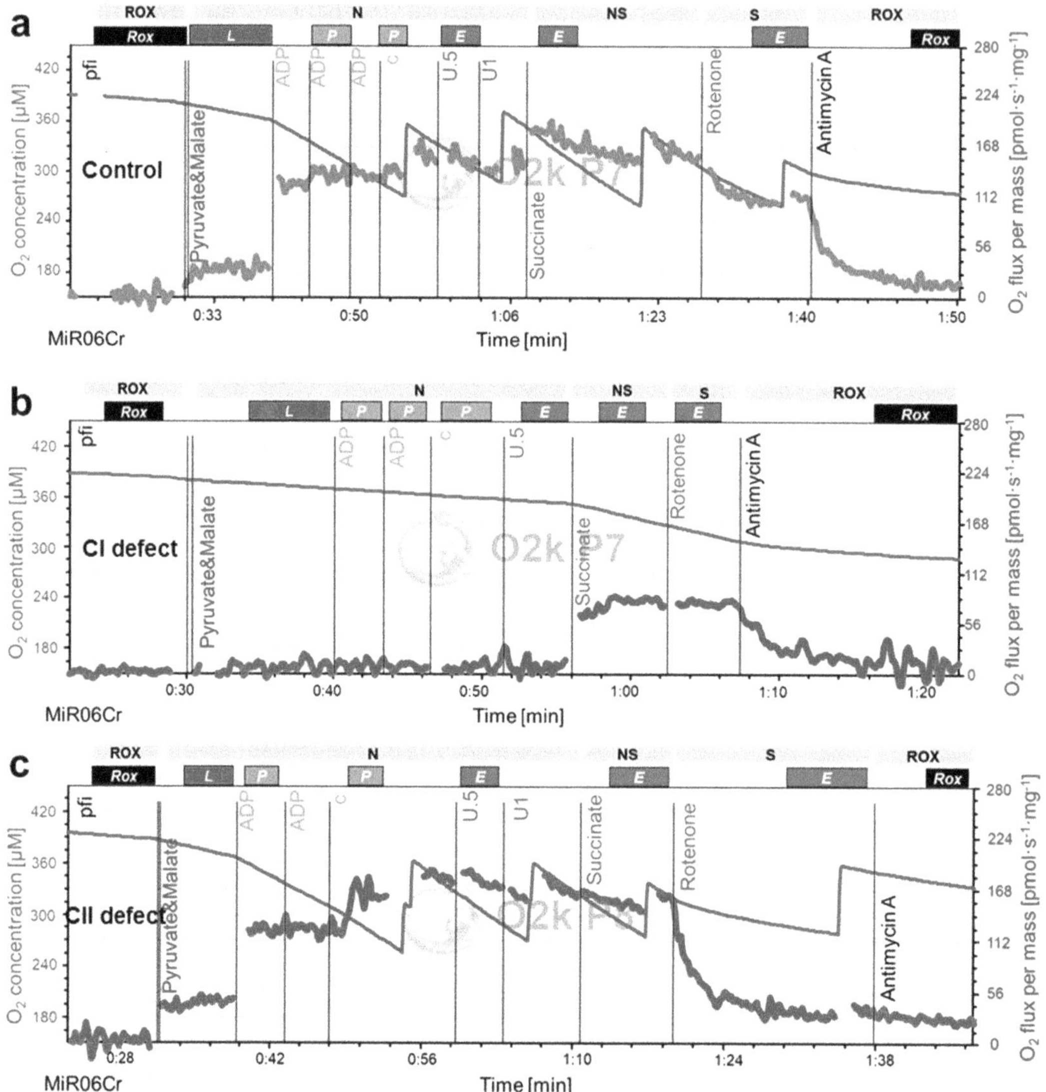

Fig. 7 Shortened SUIT protocol linked to RP1 with diagnostic examples. N-linked LEAK, OXPHOS, and ET states with pyruvate (5 mM) and malate (2 mM). Sequential addition of succinate (10 mM) and rotenone (0.5 μM) yields NS-ET and S-ET capacity. ROX after inhibition of CIII with antimycin A (2.5 μM). Oxygen concentration (μM) and tissue mass-specific oxygen flux [pmol O_2 s^{-1} mg^{-1}] of permeabilized mouse cardiac fibers as a function of time. Reoxygenations in MiR06 with H2O2 titrations. (**a**) Control sample. (**b**) CI defect induced by 0.5 μM of rotenone. (**c**) CII defect induced by 5 mM malonic acid. Experiments 2017-06-14 P7-02.DLD and 2017-06-14 P8-02.DLD

limitation by enzymatic capacity at kinetically saturating ADP concentration (Fig. 6).

3. The biochemical coupling efficiency may be expressed as the OXPHOS coupling efficiency or ET coupling efficiency. OXPHOS coupling efficiency, $1 - L/P$, is the free OXPHOS capacity (total OXPHOS capacity corrected for LEAK

respiration) which reaches a maximum of 1.0 for a fully coupled system ($L = 0$) and 0.0 in a system with zero OXPHOS capacity [29]. ET coupling efficiency, $1 - L/E$, displays the free ET capacity where $1 - L/E$ is 0.0 at zero coupling ($L = E$) and 1.0 at the limit for a fully coupled system ($L = 0$). Respiration may be stimulated first by saturating ADP (OXPHOS capacity, P) with subsequent uncoupler titration to ET capacity.

4. The conventional respiratory control ratio, RCR (State 3/State 4 or P/L), increases from 1.0 to infinity from fully dyscoupled to fully coupled mitochondria. But the RCR declines with increasing levels of coupling as a function of the phosphorylation pathway control ratio, P/E. For mathematical reasons, it is more appropriate to use $1 - L/P$ with the theoretical boundaries of 0.0 at zero coupling to the maximum of 1.0 in fully coupled mitochondria [1].
5. The apparent excess *E-P* capacity factor, $1 - P/E$, shows a minimum value of 0.0 when OXPHOS is not limited by the phosphorylation pathway at zero ET excess capacity ($P = E$). The apparent excess capacity increases with increasing control of the phosphorylation pathway over OXPHOS capacity, exhibiting a maximum value of 1.0 at the limit of zero phosphorylation capacity.
6. The cytochrome *c* control factor, *FCFc* (Subheading 3.1.5), is a test for the evaluation of the outer mt-membrane integrity.

3.3.6 Oxygen Dependence of Respiration of Permeabilized Fibers

High O_2 concentrations in respirometry with permeabilized muscle fibers are necessary to avoid an artificial oxygen limitation of respiration (*see* **Note 7**). The high degree of oxyconformance in permeabilized fibers is not a kinetic property of the mitochondria but is largely determined by the geometry of the fiber bundle, with diffusion distances increased from 5 to 10 μm in cells to >150 μm in the intertwined bundle [6]. As a consequence, a compromise is suggested to maintain O_2 concentrations in the range of 500–250 μM (37 °C) to minimize O_2 limitation of respiration but avoid extremes of hyperoxia experienced by the peripheral or partially separated mitochondria in the O2k-Chamber [9].

1. Add 2.2 mL medium into the O2k-Chamber and insert the stopper incompletely, leaving an air space above the stirred medium. A Stopper-Spacer is used for optimal and reproducible positioning. In this state, the oxygen sensors are air-calibrated (Subheading 3.4.1), while fibers are prepared for respirometry. Remove the stopper, insert a permeabilized fiber bundle into the medium, and replace the stopper incompletely. Inject a few mL of O_2 from a gas injection syringe through a needle inserted into the injection port of the stopper and extending into the gas phase, but not into the aqueous

phase. Thereby O_2 pressure is increased in the gas phase above the stirred aqueous medium in which the partial O_2 pressure, p_{O2}, starts to equilibrate rapidly. When the targeted O_2 concentration above 400 μM is nearly reached, close the chamber, thereby removing the gas phase and stopping the equilibration process.

2. Small reoxygenation steps during the experiment can be performed by titrating a few μL of H_2O_2 (200 mM in H_2O, adjust to pH 6 and keep on ice to minimize autoxidation) into MiR06 (containing catalase). This is sufficient to raise O_2 concentrations from 250 μM again to 400 μM. After injecting H_2O_2, the time required for the flux to stabilize depends on the step change of O_2 concentration; smaller steps require less time for stabilization (Fig. 7). Using H_2O_2 increases total gas pressure with O_2 pressure. This can generate bubbles in steps from air saturation to >400 μM. Therefore, O_2 gas is applied initially.

3.3.7 Oxygen Kinetics of Mitochondrial Respiration

In closed-chamber HRFR, O_2 flux of mitochondrial preparations is assessed in pseudo steady states [8]. During such measurements, fuel substrates of the ET pathway and O_2 are available at saturating concentrations to avoid progressive kinetic limitation of flux (*see* **Note 7**). O_2 concentrations below 30–50 μM (p_{O2} 3–5 kPa) are usually avoided. However, extracellular O_2 concentrations in most mammalian tissues are as low as 40–10 μM [6, 49], rendering most protocols in respirometry strongly hyperoxic for mitochondria. Myoglobin saturation levels in heart and skeletal muscle indicate intracellular p_{O2} as low as 0.3 kPa (2 Torr; 3 μM; 1.5% air saturation) [6, 50]. Evaluation of oxygen kinetic parameters is, therefore, important in basic physiological research [12, 35, 51]. In addition, oxygen kinetics provides diagnostic information relevant for evaluation of pathophysiological conditions.

In isolated mitochondria and small cells, respiration is a near hyperbolic function of extracellular (extramitochondrial) p_{O2}, which resembles classic Michaelis-Menten enzyme kinetics [2, 3, 12]. In isolated enzymes, V_{max} is the maximum reaction velocity at saturating substrate concentrations, and K_m' is the substrate concentration at half-maximum reaction velocity. By comparison, J_{max} denotes maximum O_2 flux at saturating O_2 concentrations, and p_{50} is the p_{O2} at half-maximum J_{max}, when kinetics deals with pathways in cells and mitochondrial preparations rather than single enzymes (Fig. 8). In isolated mitochondria and small cells, O_2 does not exert a significant control over mitochondrial respiration down to 0.2–0.05 kPa (ca. 20–5 μM O_2 at 37 °C), due to the high affinity of CIV for O_2 [36, 52]. In contrast, there is a pronounced diffusion limitation in tissue slices, large cells, and muscle fibers, which leads to high extracellular p_{50} values [6, 9].

Importantly, alterations in O_2 dependence of mitochondrial respiration are not only a direct consequence of changes in respiratory state and enzyme turnover rate [36, 52], but the presence of several endogenous molecules (nitric oxide, hydrogen sulfide, carbon monoxide) also strongly influences the affinity of mitochondria for O_2 [22]. Moreover, deep tissue hypoxia (e.g., occlusive and nonocclusive ischemia), chemical hypoxia (poisoning), and other pathophysiological states may lead to altered O_2 affinity. p_{50} values, therefore, integrate the effects of the O_2 affinity of CIV (linearly dependent on enzyme turnover), O_2 diffusion limitation, and specific inhibitors.

Oxygen kinetics is assessed in closed-chamber respirometers during aerobic-anaerobic transitions when O_2 is consumed by mitochondria until the O_2 concentration declines practically to zero [2]. Kinetic parameters are calculated from a hyperbolic fit of volume-specific O_2 flux, $J_{V,O2}$, plotted as a function of p_{O2}. Using polarographic oxygen sensors with a high signal-to-noise ratio eliminates the necessity of smoothing, which otherwise may severely distort the kinetic analysis. Secondary time effects on the analysis of oxygen kinetics are avoided at high $J_{V,O2}$ with a concomitant rapid aerobic-anaerobic transition. On the other hand, limitations of time resolution induce kinetic artefacts during excessively rapid transitions [37]. Reducing the data recording interval to 0.2 s is recommended (Fig. 8). If p_{50} is relatively high (>0.1 kPa) or $J_{V,O2}$ is low (<100 pmol s^{-1} mL^{-1}], a data recording interval of 1 s can be used. A first-order exponential time constant (τ) is determined for deconvolution of the O_2 signal [3]. Accurate correction for instrumental background O_2 flux is critically important for oxygen kinetics. The effective zero O_2 concentration is calculated by iterative hyperbolic fitting, correcting for drift of the zero current of the oxygen sensor (internal zero O_2 calibration; Fig. 8).

3.4 High-Resolution FluoRespirometry

3.4.1 Calibration of the Polarographic Oxygen Sensor

Dissolved O_2 concentration is measured amperometrically by Clark-type polarographic oxygen sensors (POS), containing a gold cathode and Ag/AgCl anode connected electrically by a KCl electrolyte and separated from the sample by an O_2-permeant FEP membrane (0.25 µm). A polarization voltage of 0.8 V is applied to reduce O_2 that diffuses from the incubation medium to the cathode through the membrane. O_2 is reduced to water, generating a current (hence amperometric) that is linearly proportional to p_{O2}, in the stirred experimental solution [53]. After current-to-voltage conversion and amplification, the raw signal is obtained (1 V μA^{-1}). A data sampling interval of 2 s is sufficient for routine applications (Figs. 2, 3, 4, 6, and 7); 1–0.2 s is recommended for kinetic experiments (Fig. 8). One hundred data points are averaged at each data sampling interval and represented as a data point of the signal. The limit of detection of O_2 concentration extends to 0.005 µM (5 nM) O_2 with internal zero calibration (Fig. 8). The

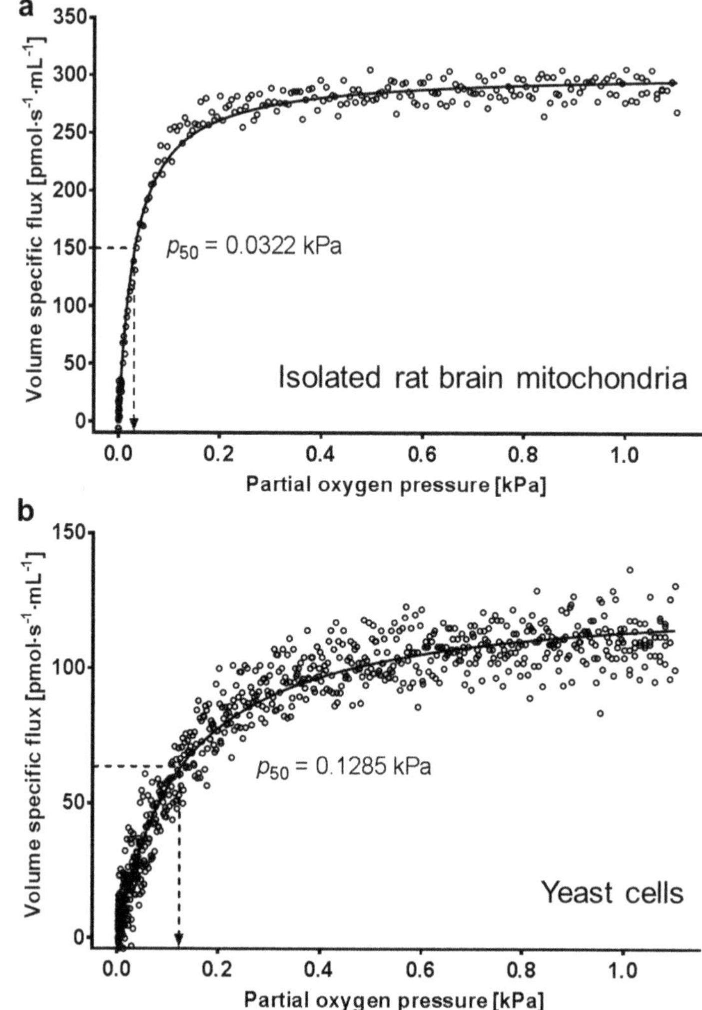

Fig. 8 Oxygen kinetics of respiration in the closed 2 mL chamber of the O2k-FluoRespirometer. Volume-specific flux, $J_{V,O2}$ [pmol O_2 s^{-1} mL^{-1}] (corrected for instrumental background O_2 flux), as a function of oxygen partial pressure, p_{O2} (kPa), measured in a continuous aerobic-anaerobic transition. The data recording interval was set to 0.2 s to ensure high time resolution without smoothing of the signal. The exponential time constant of the polarographic oxygen sensor ($\tau = 2.3$ s) for signal deconvolution [3] was assessed in a separate calibration run (not shown) by switching on the stirrer bar (stirrer test) and exponential fit of the signal over the step change. Circles represent individual data points. Solid lines are hyperbolic fits over the low oxygen range (<1.1 kPa, ca. 10 μM), yielding the p_{O2} at half-maximum $J_{V,O2}$ (p_{50}) as a parameter. Zero oxygen signal calibrated internally, for cutoff of data points when the compensation point of mitochondrial respiration and O_2 back diffusion is reached [2]. Calculations performed automatically by a p_{50} software module (Python). (**a**) Rat brain mitochondria (0.25 mg protein mL^{-1}) in MiR06 (oxygen solubility 9.72 μM kPa^{-1}) at 37 °C in the presence of 10 mM glutamate, 2 mM malate, 50 mM succinate, and 2.5 mM ADP. (**b**) Baker's yeast (freeze-dried, 0.2 mg dry weight mL^{-1}) in Na-phosphate buffer (50 mM, pH 7.1, oxygen solubility 10.05 μM kPa^{-1}) at 37 °C in the ROUTINE state of respiration with endogenous substrates

digital resolution is 2 nM, yielding a 500,000-fold dynamic range up to O_2 saturation. The polarographic oxygen sensors (Orobo-POS) are stable for several months without exchange of membranes or electrolyte [4]. A standardized calibration of the linear oxygen sensor includes quality control of signal stability (noise and drift; static two-point sensor calibration) and dynamic calibration of the sensor response time [3]. The DatLab 7 software (DL) developed for high-resolution FluoRespirometry (HRFR) incorporates function for quality control, documentation, and traceability of measurements with the Oroboros O2k (*see* **Note 8**).

1. For storage, fill up the clean O2k-Chambers completely with 70% ethanol, with the POS and stirrers remaining in the assembled chamber, and the stopper loosely inserted and covered by a lid. Storage in 70% ethanol between experiments can be extended over periods of months, keeping the chamber sterile and the POS immediately ready to use [4].
2. After switching on the instrument, connect to DL to set the experimental temperature, wash with distilled or deionized water while the stirrer is on (750 rpm or 12.5 Hz is optimal), and add 2.2 mL experimental medium (MiR05 or MiR06). Insert the stoppers slowly to their volume-calibrated position (2 mL effective volume, plus 0.08 mL to fill the stopper capillary). Siphon off excess medium ejected through the stopper capillary with the integrated suction system (ISS; Fig. 1).
3. Lift the stoppers slightly to introduce an air space above the stirred aqueous medium (open position with Stopper-Spacer), and allow for sufficient time to obtain temperature stability and O_2 equilibration between the gas and aqueous phases. The gas volume has to be exchanged for air, if the medium has not been near air saturation initially, to ensure a well-defined p_{O2} in the gas phase during air equilibration. Equilibration is a slow process, but stability should be reached within 30–60 min (Fig. 9).
4. Select the instrumental DL-Protocol for air calibration identical for O2k-Chambers A and B, and follow the step-by-step instructions for events, i.e., actions at a time point during the experiment for a chamber, and marks to be set over periods of time on a plot (Fig. 9a).
5. During the equilibration time, a quick stirrer test is performed for dynamic sensor calibration, switching off the stirrer with the consequence of a sharp drop of the POS signal and observing the monoexponential increase of the signal after the stirrer is switched on automatically after 30 s (Fig. 9a, quality control n°1). The corresponding response time is a sensitive indicator of dynamic sensor performance, and deconvolution of the signal is possible for high time resolution in kinetic studies [2, 3, 12].

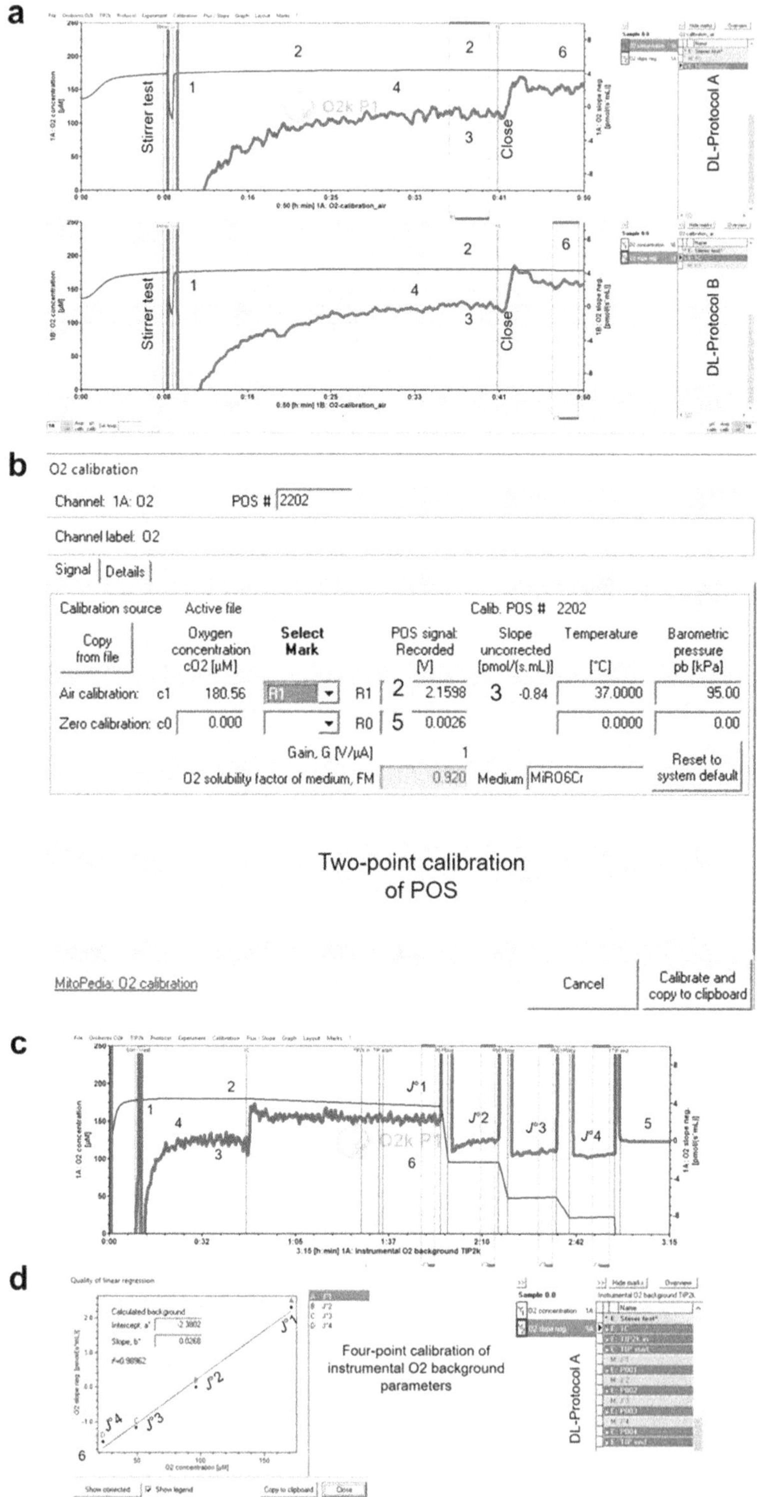

Fig. 9 MitoFit quality control in High-Resolution FluoRespirometry with software DatLab 7: (**a**) air calibration and quality control criteria 1–4 and 6. O_2 concentration and negative O_2 slope. (**b**) DatLab calibration window

6. The stable signal at air saturation provides the first calibration point, with raw signal R1 in the range from 1 to 3 V (Fig. 9a, blue line; quality control n°2). Set a mark on the plot for O_2 concentration, open the calibration window (Fig. 9b), and select mark R1.
7. During the marked period, the slope of O_2 over time after calibration must be within ±1.0 pmol s^{-1} mL^{-1} (Fig. 9a, b, quality control n°3), indicating proper signal stability of the POS. Air calibrations are performed daily before starting an experiment [54].
8. Noise of the time derivative (O_2 slope neg.) should be less than ±2 or ±4 pmol s^{-1} mL^{-1} (Fig. 9a, quality control n°4).
9. Close the glass chambers to evaluate if there is any biological contamination of the medium and/or the chamber. After the stabilization of the signal, the O_2 slope neg [pmol s^{-1} mL^{-1}] should be in the range of 2–4 (Fig. 9a, quality control n°6).
10. Titrate 100 μL freshly prepared 10 mM solution of sodium dithionite to fully exhaust the dissolved O_2 concentration to zero. The zero signal, R0, should be <3% of R_1, but most importantly, R0 must be stable (higher stability and lower noise than at air saturation) (Fig. 9b, quality control n°5). Occasional checks over a period of months are sufficient [4], except in studies of oxygen kinetics, when short-term zero drift must be accounted for by internal zero calibration after O_2 depletion by mitochondrial respiration, for resolution in the nM O_2 range [2, 3, 12, 37].

3.4.2 Oxygen Solubility and Concentration

To convert p_{O2} (kPa) to O_2 concentration (c_{O2} [μM]), the O_2 solubility of the medium is calculated as a function of temperature and salt concentration. O_2 calibration is fully supported by the Oroboros DatLab software and combines the following information [54]:

1. Raw signal, R1 (V), obtained at air saturation of the medium (Fig. 9b; n°2).
2. Raw signal, R0 (V), obtained at zero O_2 concentration (Fig. 9b; n°5).
3. Experimental temperature, T [°C], measured in the thermoregulated copper block (Figs. 1 and 9b).

Fig. 9 (continued) and quality control criteria 2, 3, and 5. Experiment 2017-02-08 P1-01.DLD. (**c**) Instrumental O_2 background test with four steps of oxygen concentration and background flux, $J^{\circ}1$ to $J^{\circ}4$, automatically controlled by the Titration-Injection microPump, as quality control no°6. (**d**) DatLab window plotting $J^{\circ}1$ to $J^{\circ}4$ as a function of O_2 concentration, and display of linear background parameters, $a^{\circ} = -2.4$ pmol O_2 s^{-1} mL^{-1} (back diffusion at zero O_2), and slope, $b^{\circ} = 0.0268$. The instrumental DL-Protocol is shown on the right. Experiment 2017-01-02 P1-03.DLD

4. Barometric pressure, p_b (kPa), measured by an electronic pressure transducer (Fig. 9b).
5. The O_2 partial pressure, p_{O2} (kPa), in air saturated with water vapor, as a function of barometric pressure and temperature, calculated by the DatLab software [53, 54].
6. The O_2 solubility, S_{O2} (μM kPa^{-1}), in pure water as a function of temperature, calculated by the DatLab software [53, 54].
7. The O_2 solubility factor of the incubation medium (F_M) which expresses the effects of salt concentration on O_2 solubility relative to pure water. In MiR05, F_M is 0.92 determined at 30 and 37 °C, and F_M is 0.89 in serum at 37 °C [53]. The same factor of 0.89 can be used for various culture media such as RPMI. At standard barometric pressure (100 kPa), the O_2 concentration at air saturation is 207.3 μM at 37 °C (19.6 kPa partial O_2 pressure; [53]). In MiR05 and serum, the corresponding saturation concentrations are 191 and 184 μM.

3.4.3 Oxygen Flux and Instrumental Background

Long-term signal stability and low noise of the O_2 signal are a basis for real-time display of O_2 flux calculated as the negative time derivative of O_2 concentration. The limit of detection in HRFR of O_2 flux is 1 pmol s^{-1} mL^{-1} (0.001 μM s^{-1}). With small amounts of sample and correspondingly low respiratory flux per volume, the oxygen capacity in the chamber provides for sufficient time to evaluate the stability of respiratory activity in each metabolic state and to permit complex titration regimes (Fig. 6). At a constant volume-specific flux of 100 pmol s^{-1} mL^{-1}, 180 μM O_2 is exhausted within 30 min.

O_2 consumption by the POS and O_2 diffusion below or above air saturation contributes to instrumental background O_2 flux, which is minimized in HRFR and corrected for [2–4]. At air saturation, the POS generates a current of about 2 μA at a stoichiometry of four electrons/O_2. The O_2/electron ratio divided by the Avogadro constant ($F = 96{,}485.53$ C mol^{-1}) yields the amount of O_2 per coulomb [$1/(4 \cdot F) = 2.591$ μmol O_2 C^{-1}] or O_2 flow per current (2.591 pmol s^{-1} μA^{-1}). At 2 μA per 2 mL, therefore, volume-specific O_2 flux or O_2 consumption by the POS ($J_{O2,POS}$) is 2.6 pmol s^{-1} mL^{-1}. $J_{O2,POS}$ declines to zero as a strictly linear function of p_{O2} or c_{O2} under constant experimental conditions. Hence correction for $J_{O2,POS}$ is simple and accurate and does not influence the limit of detection of biological flux. In contrast, the contribution of O_2 diffusion to instrumental background is unpredictable in various respirometric systems and needs to be determined empirically in the closed chamber in the absence of biological material, as a function of O_2 concentration in the experimental range. Effects of O_2 back diffusion are minimized in the O2k by the large volume (2 mL) and selection of diffusion-tight materials in contact with the respiration medium: glass chambers and PVDF or titanium stoppers (not Perspex), magnetic stirrer bars coated by PVDF or PEEK (not Teflon), and Viton O-rings (not silicon). Compared to aqueous media, plastic materials such as Teflon have a >10-fold higher O_2 solubility. Plastic is not feasible for respirometry, since uncontrolled O_2 back diffusion distorts the respiratory decline of O_2 concentration in a closed chamber.

Instrumental background tests were designed to detect and eliminate possible O_2 leaks, introducing this integrated systemic calibration as a key component of quality control in HRFR [2–4, 55] (Fig. 9c). The use of an instrumental DL-Protocol for the O_2 background test and application of the Titration-Injection microPump (TIP2k; Fig. 1) provides a guide and automatic instrumental background test (Fig. 9c).

1. Close the chamber by fully inserting the stoppers after stabilization at air saturation, excluding any gas bubbles. After 10–15 min, observe instrumental background flux ($J°1$) which is due to $J°_{O2,POS}$ only (Fig. 9c, quality control n°6). $J°_{O2,POS}$, is 2.5–3.5 pmol s^{-1} mL^{-1} at air saturation in the 2 mL O2k-Chamber at 37 °C. Agreement with the predicted flux validates the instrumental limit of detection of flux. Higher $J°1$ is due to microbial contamination of the medium and chamber. $J°1$ would increase to 25 pmol s^{-1} mL^{-1} in a 200 μL chamber. Whereas $J°_{O2,POS}$ decreases linearly to zero at anoxia, it increases to 8 pmol s^{-1} mL^{-1} at 500 μM O_2, but instrumental background flux $J°_{O2}$ does not simply conform to $J°_{O2,POS}$ at these O_2 concentrations.
2. (a) Lower experimental O_2 concentrations are obtained by stepwise titration of small volumes of freshly prepared 10 mM solution of sodium dithionite ($Na_2S_2O_4$; 1.7 mg mL^{-1} phosphate buffer, pH 8) into MiR06. Standardized four-step background tests ($J°_{O2}$ at air saturation, 100, 50, and 20 μM O_2) can be performed automatically using the programmed Oroboros Titration-Injection microPump (TIP2k; [55]). Alternatively, the chamber is opened intermittently for flushing the gas phase with nitrogen or argon and closed at reduced O_2 concentration. The near-linear dependence of $J°_{O2}$ on O_2 concentration extrapolates to zero O_2 concentration with an intercept $a°$ of −1.5 to −2.5 pmol s^{-1} mL^{-1}, which is the O_2 back diffusion per volume of the chamber at zero O_2 concentration (Fig. 9d). A typical value of the slope $b°$ is 0.025. More negative values of $a°$ indicate an O_2 leak in the system.

   (b) In experiments with permeabilized fibers at elevated O_2 concentrations, instrumental background is measured in the experimental range after increasing the O_2

concentration in MiR06 (see above) and stepwise measurement of $J°_{O2}$ at four O_2 concentrations matching the experimental O_2 regime. The recommended range is 400–250 μM.

3. In a plot of $J°_{O2}$ as a function of c_{O2}, $a°$ and $b°$ are calculated by linear regression, $J°_{O2} = a° + b° \cdot c_{O2}$ (Fig. 9d, quality control n°6).

4. Background-corrected volume-specific O_2 flux, $J_{V,O2}$ [pmol s^{-1} mL^{-1}], is calculated real time in DatLab over the entire experimental O_2 range [2–4]:

$$J_{V,O2} = -dc_{O2}/dt \cdot 1000 - (a^{\circ} + b^{\circ} \cdot c_{O2})$$

where cO_2 [μM or nmol cm^{-3}] is O_2 concentration measured at time t, dcO_2/dt is the time derivative of O_2 concentration, and the expression in parentheses is instrumental background O_2 flux.

4 Notes

1. OXPHOS protocols presented in this chapter and instrumental standards in HRFR address new challenges for quality control (QC) and data reporting for clinical applications of mitochondrial respiratory physiology and pathology.
2. Demands are increasing for quality control, quality assurance, traceability of calibrations, and standardization of protocols for functional mitochondrial diagnosis in biomedical research and clinical applications. A requirement for consistency of nomenclature in mitochondrial respiratory physiology and bioenergetics has become increasingly apparent, aiming at the development of databases of mitochondrial respiratory function in species, tissues, and cells [8].
3. Emphasis is placed on intact cells, permeabilized cells, permeabilized muscle fibers, and isolated mitochondria. It has not been shown if isolation of mitochondria involves the selective loss of damaged mitochondria, but in any case, all types of mitochondria are accessible experimentally in permeabilized cells and tissues. Respiration of permeabilized skeletal muscle fibers and isolated mitochondria yields comparable results on OXPHOS capacity [5].
4. Mitochondrial preparations from tissues and cells can be studied with identical SUIT protocols and methodological consistency in the O2k, for direct comparison [45] and transfer of results based on standardized SUIT protocols (reference protocols) into a rigorously monitored database [8].

5. OXPHOS and ET capacity are measured in SUIT protocols with various substrate-uncoupler-inhibitor combinations. Partial reconstitution of TCA cycle function requires physiological NS-substrate combinations and yields higher fluxes compared to separate N- and S-pathway capacities [1, 5, 29, 33, 44–48].
6. The basic coupling control protocol with intact cells can be extended to include a respirometric cell viability test and cytochrome *c* oxidase assay. Short SUIT protocols for mitochondrial preparations are suitable for cohort studies with increased throughput. Complex SUIT protocols, such as RP1 and RP2, are considered for comprehensive OXPHOS analysis with multiple electron entries into the NADH- and Q-junctions, including the N- and S-pathways but also fatty acid and glycerophosphate oxidation.
7. Kinetically saturating concentrations of fuel substrates, ADP, inorganic phosphate, and oxygen are used to obtain reference values on OXPHOS and ET capacities. Kinetic studies provide a closer link to cell physiology. In particular, oxygen kinetics of mitochondrial respiration addresses physiological intracellular oxygen pressures in contrast to effective hyperoxia at air saturation, represents a standardized and simple measurement for cells and isolated mitochondria, and is of high diagnostic value. Permeabilized fibers and to a lesser extent tissue homogenates are characterized by oxygen diffusion limitation relative to isolated cells and mitochondria and thus require elevated oxygen concentrations to obtain kinetically saturated respiratory rates.
8. Standardization of quality control of High-Resolution FluoRespirometry as presented in the present compendium provides the basis for extending respiratory measurements by HRFR to simultaneous monitoring of respiration and fluorometric and potentiometric signals. It is recommended to apply comparatively short SUIT protocols to such multisensory measurements to optimize specific outputs for mitochondrial membrane potential [26], hydrogen peroxide production [56, 57], proton flux, Ca^{2+} handling, ADP → ATP phosphorylation, and spectrophotometric measurement of cytochrome redox states [58].

Acknowledgments

We thank Philipp Gradl and his team (WGT-Elektronik GmbH & Co KG) for O2k hardware and electronics development, Lukas Gradl for software development (DatLab 3 to 7), and Markus Haider for software development for p_{50} analysis. Marielle Hansl and Stephanie Droescher performed some experiments with intact cells. This work is an extension of the original presentation by Pesta and Gnaiger [9], was supported by K-Regio project MitoFit, and is a contribution to COST Action CA15203 MitoEAGLE.

Competing Financial Interests

E.G. is the founder and CEO of Oroboros Instruments, Innsbruck, Austria.

References

1. Gnaiger E (2014) Mitochondrial pathways and respiratory control. Oroboros MiPNet Publications, Innsbruck. http:/www.oroboros.at
2. Gnaiger E, Steinlechner-Maran R, Méndez G, Eberl T, Margreiter R (1995) Control of mitochondrial and cellular respiration by oxygen. J Bioenerg Biomembr 27:583–596
3. Gnaiger E (2001) Bioenergetics at low oxygen: dependence of respiration and phosphorylation on oxygen and adenosine diphosphate supply. Respir Physiol 128:277–297
4. Gnaiger E (2008) Polarographic oxygen sensors, the oxygraph and high-resolution respirometry to assess mitochondrial function. In: Dykens JA, Will Y (eds) Mitochondrial dysfunction in drug-induced toxicity. Wiley, New York, pp 327–352
5. Gnaiger E (2009) Capacity of oxidative phosphorylation in human skeletal muscle. New perspectives of mitochondrial physiology. Int J Biochem Cell Biol 41:1837–1845
6. Gnaiger E (2003) Oxygen conformance of cellular respiration: a perspective of mitochondrial physiology. Adv Exp Med Biol 543:39–56
7. Gnaiger E, Kuznetsov AV, Schneeberger S, Seiler R, Brandacher G, Steurer W, Margreiter R (2000) Mitochondria in the cold. In: Heldmaier G, Klingenspor M (eds) Life in the cold. Springer, New York, pp 431–442
8. MitoEAGLE preprint 2017-11-11(16). The protonmotive force and respiratory control: building blocks of mitochondrial physiology Part 1. http://www.mitoeagle.org/index.php/MitoEAGLE_preprint_2017-09-21
9. Pesta D, Gnaiger E (2012) High-resolution respirometry. OXPHOS protocols for human cells and permeabilized fibres from small biopsies of human muscle. Methods Mol Biol 810:25–58
10. Veksler VI, Kuznetsov AV, Sharov VG, Kapelko VI, Saks VA (1987) Mitochondrial respiratory parameters in cardiac tissue: a novel method of assessment by using saponin-skinned fibres. Biochim Biophys Acta 892:191–196
11. Burtscher J, Zangrandi L, Schwarzer C, Gnaiger E (2015) Differences in mitochondrial function in homogenated samples from healthy and epileptic specific brain tissues revealed by high-resolution respirometry. Mitochondrion 25:104–112
12. Gnaiger E, Méndez G, Hand SC (2000) High phosphorylation efficiency and depression of uncoupled respiration in mitochondria under hypoxia. Proc Natl Acad Sci U S A 97:11080–11085
13. Lemieux H, Semsroth S, Antretter H, Hoefer D, Gnaiger E (2011) Mitochondrial respiratory control and early defects of oxidative phosphorylation in the failing human heart. Int J Biochem Cell Biol 43:1729–1738
14. Sumbalová Z, Garcia-Souza LF, Veliká B, Volani C, Gnaiger E (2017) Analysis of mitochondrial function in human blood cells. In: Gvozdjáková A (ed) Recent advances in mitochondrial medicine and Coenzyme Q10. NOVA Sciences, New York
15. Hütter E, Unterluggauer H, Garedew A, Jansen-Dürr P, Gnaiger E (2006) High-resolution respirometry - a modern tool in aging research. Exp Gerontol 41:103–109
16. Steinlechner-Maran R, Eberl T, Kunc M, Margreiter R, Gnaiger E (1996) Oxygen dependence of respiration in coupled and uncoupled endothelial cells. Am J Physiol 271:C2053–C2061
17. Hütter E, Renner K, Pfister G, Stöckl P, Jansen-Dürr P, Gnaiger E (2004) Senescence-associated changes in respiration and oxidative phosphorylation in primary human fibroblasts. Biochem J 380:919–928
18. Chance B, Williams GR (1955) Respiratory enzymes in oxidative phosphorylation. I. Kinetics of oxygen utilization. J Biol Chem 217:383–393

19. Stadlmann S, Rieger G, Amberger A, Kuznetsov AV, Margreiter R, Gnaiger E (2002) H_2O_2-mediated oxidative stress versus cold ischemia-reperfusion: mitochondrial respiratory defects in cultured human endothelial cells. Transplantation 74:1800–1803
20. Renner K, Amberger A, Konwalinka G, Kofler R, Gnaiger E (2003) Changes of mitochondrial respiration, mitochondrial content and cell size after induction of apoptosis in leukemia cells. Biochim Biophys Acta 1642:115–123
21. Steinlechner-Maran R, Eberl T, Kunc M, Schröcksnadel H, Margreiter R, Gnaiger E (1997) Respiratory defect as an early event in preservation/reoxygenation injury in endothelial cells. Transplantation 63:136–142
22. Aguirre E, Rodríguez-Juárez F, Bellelli A, Gnaiger E, Cadenas S (2010) Kinetic model of the inhibition of respiration by endogenous nitric oxide in intact cells. Biochim Biophys Acta. https://doi.org/10.1016/j.bbabio.2010.01.033
23. Stadlmann S, Renner K, Pollheimer J, Moser PL, Zeimet AG, Offner FA, Gnaiger E (2006) Preserved coupling of oxidative phosphorylation but decreased mitochondrial respiratory capacity in IL-1β treated human peritoneal mesothelial cells. Cell Biochem Biophys 44:179–186
24. Smolková K, Bellance N, Scandurra F, Génot E, Gnaiger E, Plecitá-Hlavatá L, Ježek P, Rossignol R (2010) Mitochondrial bioenergetic adaptations of breast cancer cells to aglycemia and hypoxia. J Bioenerg Biomembr. https://doi.org/10.1007/s10863-009-9267-x
25. Jones DP (1986) Intracellular diffusion gradients of O_2 and ATP. Am J Physiol 250: C663–C675
26. Krumschnabel G, Eigentler A, Fasching M, Gnaiger E (2014) Use of safranin for the assessment of mitochondrial membrane potential by high-resolution respirometry and fluorometry. Methods Enzymol 542:163–181
27. Gnaiger E, Kuznetsov AV, Rieger G, Amberger A, Fuchs A, Stadlmann S, Eberl T, Margreiter R (2000) Mitochondrial defects by intracellular calcium overload versus endothelial cold ischemia/reperfusion injury. Transpl Int 13:555–557
28. Pesta D, Hoppel F, Macek C, Messner H, Faulhaber M, Kobel C, Parson W, Burtscher M, Schocke M, Gnaiger E (2011) Similar qualitative and quantitative changes of mitochondrial respiration following strength and endurance training in normoxia and hypoxia in sedentary humans. Am J Physiol Regul Integr Comp Physiol 301: R1078–R1087
29. Gnaiger E, Boushel R, Søndergaard H, Munch-Andersen T, Damsgaard R, Hagen C, Díez-Sánchez C, Ara I, Wright-Paradis C, Schrauwen P, Hesselink M, Calbet JAL, Christiansen M, Helge JW, Saltin B (2015) Mitochondrial coupling and capacity of oxidative phosphorylation in skeletal muscle of Inuit and Caucasians in the arctic winter. Scand J Med Sci Sports 25(Suppl 4):126–134
30. Saks VA, Veksler VI, Kuznetsov AV, Kay L, Sikk P, Tiivel T, Tranqui L, Olivares J, Winkler K, Wiedemann F, Kunz WS (1998) Permeabilized cell and skinned fibre techniques in studies of mitochondrial function in vivo. Mol Cell Biochem 184:81–100
31. Kuznetsov AV, Schneeberger S, Seiler R, Brandacher G, Mark W, Steurer W, Saks V, Usson Y, Margreiter R, Gnaiger E (2004) Mitochondrial defects and heterogeneous cytochrome c release after cardiac cold ischemia and reperfusion. Am J Physiol Heart Circ Physiol 286:H1633–H1641
32. Kuznetsov AV, Strobl D, Ruttmann E, Königsrainer A, Margreiter R, Gnaiger E (2002) Evaluation of mitochondrial respiratory function in small biopsies of liver. Anal Biochem 305:186–194
33. Rasmussen UF, Rasmussen HN (2000) Human quadriceps muscle mitochondria: a functional characterization. Mol Cell Biochem 208:37–44
34. Palmer JW, Tandler B, Hoppel CL (1977) Biochemical properties of subsarcolemmal and interfibrillar mitochondria isolated from rat cardiac muscle. J Biol Chem 252:8731–8739
35. Gnaiger E, Lassnig B, Kuznetsov AV, Margreiter R (1998) Mitochondrial respiration in the low oxygen environment of the cell: effect of ADP on oxygen kinetics. Biochim Biophys Acta 1365:249–254
36. Gnaiger E, Kuznetsov AV (2002) Mitochondrial respiration at low levels of oxygen and cytochrome *c*. Biochem Soc Trans 30:252–258
37. Scandurra FM, Gnaiger E (2010) Cell respiration under hypoxia: facts and artefacts in mitochondrial oxygen kinetics. Adv Exp Med Biol 662:7–25
38. Dubowitz V, Sewry CA (2006) Muscle biopsy: a practical approach. Saunders Elsevier, Philadelphia
39. Lemieux H, Blier PU, Gnaiger E (2017) Remodeling pathway control of mitochondrial respiratory capacity by temperature in mouse

heart: electron flow through the Q-junction in permeabilized fibers. Sci Rep 7:2840

40. Dufour S, Rousse N, Canioni P, Diolez P (1996) Top-down control analysis of temperature effect on oxidative phosphorylation. Biochem J 314:743–751
41. Sun F, Huo X, Zhai Y, Wang A, Xu J, Su D, Bartlam M, Rao Z (2005) Crystal structure of mitochondrial respiratory membrane protein Complex II. Cell 121:1043–1057
42. Puchowicz MA, Varnes ME, Cohen BH, Friedman NR, Kerr DS, Hoppel CL (2004) Oxidative phosphorylation analysis: assessing the integrated functional activity of human skeletal muscle mitochondria – case studies. Mitochondrion 4:377–385
43. Delhumeau G, Cruz-Mendoza AM, Lojero CG (1994) Protection of cytochrome *c* oxidase against cyanide inhibition by pyruvate and α-ketoglutarate: effect of aeration in vitro. Toxicol Appl Pharmacol 126:345–351
44. Boushel R, Gnaiger E, Schjerling P, Skovbro M, Kraunsøe R, Dela F (2007) Patients with Type 2 Diabetes have normal mitochondrial function in skeletal muscle. Diabetologia 50:790–796
45. Schöpf B, Schäfer G, Weber A, Talasz H, Eder IE, Klocker H, Gnaiger E (2016) Oxidative phosphorylation and mitochondrial function differ between human prostate tissue and cultured cells. FEBS J 283:2181–2196
46. Scheibye-Knudsen M, Quistorff B (2009) Regulation of mitochondrial respiration by inorganic phosphate; comparing permeabilized muscle fibres and isolated mitochondria prepared from type-1 and type-2 rat skeletal muscle. Eur J Appl Physiol 105:279–287
47. Aragonés J, Schneider M, Van Geyte K et al (2008) Deficiency or inhibition of oxygen sensor Phd1 induces hypoxia tolerance by reprogramming basal metabolism. Nat Genet 40:170–180
48. Doerrier C, Sumbalova Z, Krumschnabel G, Hiller E, Gnaiger E (2016) SUIT reference protocol for OXPHOS analysis by high-resolution respirometry. Mitochondr Physiol Netw 21(06):1–12
49. Erecinska M, Silver IA (2001) Tissue oxygen tension and brain sensitivity to hypoxia. Respir Physiol 128:263–276
50. Richardson RS, Noyszewski EA, Kendrick KF, Leigh JS, Wagner PD (1995) Myoglobin O_2 desaturation during exercise. Evidence of limited O_2 transport. J Clin Invest 96:1916–1926
51. Larsen FJ, Schiffer TA, Sahlin K, Ekblom B, Weitzberg E, Lundberg JO (2011) Mitochondrial oxygen affinity predicts basal metabolic rate in humans. FASEB J 25:2843–2852
52. Krab K, Kempe H, Wikstrom M (2011) Explaining the enigmatic K(M) for oxygen in cytochrome c oxidase: a kinetic model. Biochim Biophys Acta 1807:348–358
53. Gnaiger E, Forstner H (eds) (1983) Polarographic oxygen sensors. Aquatic and physiological applications. Springer, New York
54. Gnaiger E (2016) O2k Quality Control 1: polarographic oxygen sensors and accuracy of calibration. Mitochondr Physiol Netw 6.3(15). http:/www.oroboros.at
55. Fasching M, Gnaiger E (2016) O2k Quality Control 2: instrumental oxygen background correction and accuracy of oxygen flux. Mitochondr Physiol Netw 14.6(05):1–8. http:/www.oroboros.at
56. Makrecka-Kuka M, Krumschnabel G, Gnaiger E (2015) High-resolution respirometry for simultaneous measurement of oxygen and hydrogen peroxide fluxes in permeabilized cells, tissue homogenate and isolated mitochondria. Biomolecules 5:1319–1338
57. Komlodi T, Sobotka A, Krumschnabel G, Doerrier C, Bezuidenhout N, Hiller E, Gnaiger E (2018) Comparison of mitochondrial incubation media for measurement of respiration and hydrogen peroxide production. In: Palmeira CM, Moreno AJ (eds) Mitochondrial bioenergetics: methods and protocols, Springer, New York
58. Harrison DK, Fasching M, Fontana-Ayoub M, Gnaiger E (2015) Cytochrome redox states and respiratory control in mouse and beef heart mitochondria at steady-state levels of hypoxia. J Appl Physiol 119:1210–1218

Chapter 4

High-Throughput Analysis of Mitochondrial Oxygen Consumption

James Hynes, Rachel L. Swiss, and Yvonne Will

Abstract

Interest in the investigation of mitochondrial dysfunction has seen a resurgence over recent years due to the implication of such dysfunction in both drug-induced toxicity and a variety of disease states. Here we describe a methodology to assist in such investigations whereby the oxygen consumption of isolated mitochondria is assessed in a high-throughput fashion using a phosphorescent oxygen-sensitive probe, standard microtiter plates, and plate reader detection. The protocols provided describe the required isolation procedures, initial assay optimization, and subsequent compound screening. Typical data is also provided illustrating the expected activity levels as well as recommended plate maps and data analysis approaches.

Key words Mitochondria, OXPHOS, Oxygen consumption, Toxicity, Respiration, Polarimetry, Oxygen-sensitive probes, Toxicity, Drug safety testing

1 Introduction

Mitochondrial dysfunction is a common mechanism of drug-induced toxicity and has been implicated with a variety of drug classes [1, 2]. This has led to a requirement for a high-throughput method of assessing the metabolic implications of drug treatment. Oxygen consumption measurements are favored in this regard as they provide direct information on the activity of oxidative phosphorylation and are therefore highly sensitive to perturbations in mitochondrial function. Such measurements allow the identification of compounds that specifically perturb mitochondrial function while also providing information on the mechanisms involved; electron transport chain (ETC), ATPase, and adenine nucleotide translocator (ANT) inhibitors, for example, will cause a decrease in

Electronic supplementary material: The online version of this chapter (https://doi.org/10.1007/978-1-4939-7831-1_4) contains supplementary material, which is available to authorized users.

Carlos M. Palmeira and António J. Moreno (eds.), *Mitochondrial Bioenergetics: Methods and Protocols*, Methods in Molecular Biology, vol. 1782, https://doi.org/10.1007/978-1-4939-7831-1_4,

ADP-activated respiration, while uncouplers cause as an increase in basal respiration. Traditionally, these measurements were performed using standard polarography; however, limited throughput precludes such an approach to this type of application. The necessary throughput is instead achieved using a phosphorescent water-soluble oxygen probe which forms the basis of the MitoXpress® Xtra Oxygen Consumption Assay, which facilitates a microplate-based analysis of mitochondrial oxygen consumption [3, 4]. Probe emission is quenched by molecular oxygen via a physical (collisional) mechanism, whereby depletion of dissolved oxygen causes an increase in probe emission. Measuring this signal therefore allows the quantification of dissolved oxygen in 96- or 384-well plates, with changes in probe signal reflecting changes in oxygen concentration within the sample. For the specific identification of mitochondrial inhibitors, measurements are typically performed in the presence of ADP (State 3) and the required substrate (glutamate/malate, succinate, fatty acids) with decreases in the rates of probe signal change indicative of mitochondrial inhibition. For the identification of mitochondrial uncouplers, measurements are performed in the absence of ADP (State 2), whereby increases in the rate of probe signal change are indicative of uncoupling. Successful analysis requires rates of oxygen consumption which exceed the rates of back diffusion from ambient air. For this reason, a sealing layer of mineral oil is applied to limit such back diffusion, thereby increasing assay sensitivity. Detailed protocols are presented describing mitochondrial isolation from relevant tissues, and typical activity values are provided. Assay optimization is also addressed, and suggestions are provided on how to perform compound screening and on the recommended approach to data interpretation.

2 Materials

2.1 Mitochondrial Isolation

1. Glass tissue homogenizer with Teflon pestle (100 ml).
2. Glass beakers.
3. Glass stirring rods.
4. Plastic funnel.
5. Centrifuge tubes (50 ml).
6. Power drill (handheld or static).
7. Ultra-Turrax tissue homogenizer (IKA, T25).
8. Cheesecloth.
9. Refrigerated high-speed centrifuge.
10. Several ice buckets.
11. BCA kit for protein determination.

12. Eppendorf tubes.
13. Standard clear bottom 96-well plate.
14. Automated pipettes: Gilson P20, P200, P1000.
15. Eppendorf syringe dispenser with 2.5 ml plastic syringes.
16. Absorbance plate reader.
17. Triton X-100.
18. Type XXIV protease.

2.2 Assay Optimization (See Note 1(a))

1. Isolated mitochondria of known concentration.
2. Ice (for storage of isolated mitochondria).
3. Glutamate/malate: 0.5/0.5 M in H_2O, pH 7.4, aliquot and store at −20 °C.
4. Sodium succinate: 1 M in H_2O, pH 7.4, aliquot and store at −20 °C.
5. Palmitoyl CoA/carnitine/malate: 1.6 mM palmitoyl-CoA, 80 mM carnitine, 200 mM L-(−)-malic acid in H_2O. Aliquot and store up to 6 months at −20 °C.
6. Adenosine 5′-diphosphate (ADP): 100 mM in H_2O, aliquot and store at −20 °C.
7. "Respiration buffer": 250 mM sucrose, 15 mM KCl, 1 mM EGTA, 5 mM $MgCl_2$, 30 mM K_2HPO_4, pH 7.4.
8. MitoXpress® Xtra Oxygen Consumption Assay (isolated Mito) (Luxcel Biosciences).
9. Black body clear bottom 96-well plate (Costar 3631 or equivalent).
10. Automated pipettes: Gilson P20, P200, P1000.
11. 8- or 12-channel 100 μl pipette.
12. Gilson Distriman® pipette with 1250 μl cartridges.
13. Multi-Blok® heater.
14. 2 ml clear Eppendorf tubes.
15. Water bath, 30 °C (for warming solutions).
16. Fluorescence plate reader with temperature control and monochromators or suitable filters (*see* Table 1).

2.3 Screening

Materials as listed in Subheading 3.2 plus:

1. PCR plates for compound dilutions.

2.4 Data Analysis

1. Plate reader software [e.g., MARS (BMG Labtech), Magellan (Tecan), Gen 5 (BioTek), SoftMax Pro (Molecular Devices), MyAssays (PerkinElmer)].
2. Data processing software (MS Excel, Microcal Origin, GraphPad Prism).

Table 1
Summary of recommended instrument setting for common fluorescence plate readers

Instrument	Optical configuration	Integ 1 (D1/W1) Integ 2 (D2/W2)	Optimum mode	Ex (nm) Em (nm)
BMG Labtech: CLARIOstar	Filter-based Top or bottom read	30/30 μs 70/30 μs	Dual-read TR-F (lifetime)	Ex 340 ± 50 nm (TR-EX) Em 665 ± 50 nm or Em 645 ± 10 nm *with LP-TR dichroic*
BMG Labtech: FLUOstar Omega POLARstar Omega	Filter-based Top or bottom read	30/30 μs 70/30 μs	Dual-read TR-F (lifetime)	Ex 340 ± 50 nm (TR-EXL) Em 655 ± 25 nm (BP-655)
BioTek: Cytation 3 or 5	Filter-based Top or bottom read	30/30 μs 70/30 μs	Dual-read TR-F (lifetime)	Ex 380 ± 20 nm Em 645 ± 15 nm
BioTek: Synergy H1 or Neo or 2	Filter-based Top or bottom read	30/30 μs 70/30 μs	Dual-read TR-F (lifetime)	Ex 380 ± 20 nm Em 645 ± 15 nm
Tecan: Spark 10M or 20M	Filter-based/fusion optics Top read	30/30 μs 70/30 μs	Dual-read TR-F (lifetime)	Ex 380 ± 20 nm Em 670 ± 40 nm (or Em 650 ± 20 nm fusion optics)
Tecan: Infinite M1000Pro or F200Pro	Monochromator/ filter-based Top or bottom read	30/30 μs 70/30 μs	Dual-read TR-F (Lifetime)	Ex 380 ± 20 nm Em 650 ± 20 nm or Em 670 ± 40 nm
Perkin Elmer: VICTOR series, X4 or X5	Filter-based Top read	30/30 μs 70/30 μs	Dual-read TR-F (Lifetime)	Ex 340 ± 40 nm (D340) Em 642 ± 10 nm (D642)
BMG Labtech: PHERAstar FS	Filter-based Top or bottom read	40/100 μs n/a	TR-F	Ex 337 nm (HTRF module) Em 665 nm (HTRF module)

BMG Labtech: FLUOstar Optima POLARstar Optima	Filter-based Top or bottom read	30/100 μs n/a	TR-F	Ex 340 ± 50 nm (TR-EXL) Em 655 ± 50 nm (BP-655)
Perkin Elmer: EnVision	Filter-based Top read	40/100 μs n/a	TR-F	Ex 340 ± 60 nm (X340) Em 650 ± 8 nm (M650)
Perkin Elmer: EnSpire	Monochromator-based Top read	40/100 μs n/a	TR-F	Ex 380 ± 20 nm Em 650 ± 20 nm
Tecan: Infinite M200Pro, Safire or Genios Pro	Monochromator/ filter-based Top or bottom read	30/100 μs n/a	TR-F	Ex 380 ± 20 nm Em 650 ± 20 nm
Molecular devices: SpectraMax M series, FlexStation, or Gemini, i3x, i3, or Paradigm	Monochromator-based Top or bottom read	n/a n/a	Intensity (prompt)	Ex 380 nm Em 650 nm

3 Methods

The most critical aspect of functional mitochondrial measurements such as those outlined here is the quality of the mitochondria preparation used. Quality control of the mitochondrial preparation is therefore of critical importance such that variations in source tissue or possible batch-to-batch variability can be accounted for prior to the assessment of effector action. It is also critically important that the preparation be well coupled. This is determined from the respiratory control ratio (RCR), the ratio of ADP-stimulated to basal respiration (State 3/State 2).

Screening for compounds with a mitochondrial liability can be carried out on a variety of substrate combinations, the most common of which are glutamate/malate and succinate feeding complex 1 and complex 2 of the electron transport chain (ETC), respectively (*see* Figs. 1–3). Fatty acid can also be used as substrates (*see* **Note 2 (i)**; Fig. 4). Respiration can then be assessed in basal (State 2) or ADP-stimulated (State3) conditions, with inhibitor screening generally performed in State 3 and uncoupler screening performed in State 2. As the resultant oxygen consumption rate is dependent on both substrate and ADP availability, it is necessary to establish the optimum protein concentration for such screening.

3.1 Mitochondrial Isolation

3.1.1 Procedure for Isolation of Liver Mitochondria

1. Prepare the following buffers (*see* **Note 1(a)**):

   Buffer I: 210 mM mannitol, 70 mM sucrose, 5 mM HEPES, 1 mM EGTA, 0.5% BSA, pH 7.4.

   Buffer II: 210 mM mannitol, 70 mM sucrose, 10 mM $MgCl_2$, 5 mM K_2HPO_4, 10 mM MOPS, 1 mM EGTA, pH 7.4.

2. Euthanize animals with an overdose of carbon dioxide (*see* **Note 1(b)**), excise organs rapidly, and place in ice-cold Buffer I.
3. Using a pair of scissors, finely mince approximately 6 g of liver tissue, and then wash repeatedly in Buffer I until the homogenate is blood-free. Then add 5 volumes of Buffer I and homogenize using 6–8 passes of a smooth glass grinder with Teflon pestle driven by a power drill on low speed.
4. Adjust to 8 volumes with Buffer I and centrifuge at 700 × *g* at 4 °C for 10 min, then filter through two layers of cheesecloth, and re-centrifuge for 10 min at 14,000 × *g* to precipitate the mitochondrial fraction.
5. Discard the supernatant, using a glass stirring rod, re-suspend the mitochondrial pellet in 20 ml of isolation Buffer I, and re-centrifuge at 10,000 × *g* for 10 min at 4 °C.
6. Repeat this wash step in Buffer II.

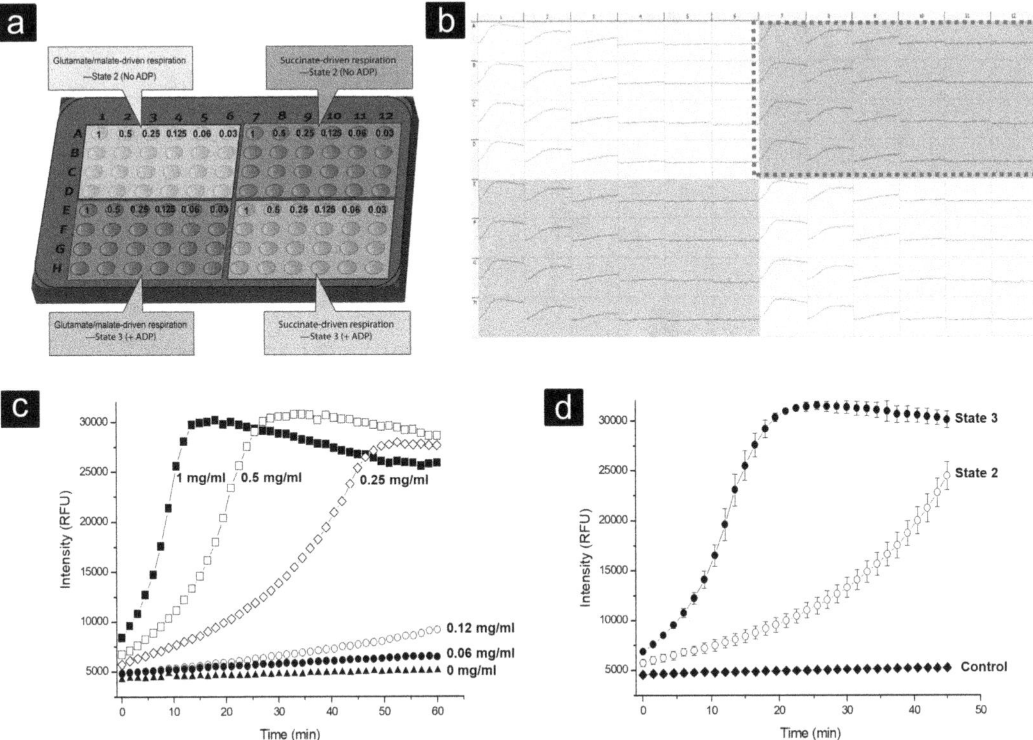

Fig. 1 Assay optimization. (**a**) Typical plate layout for the initial optimization of liver mitochondrial protein concentration needed for screening of NCEs. Typical data output is presented in (**b**) as a MARS plate view (BMG Labtech) showing a serial dilution of mitochondrial protein at the indicated concentrations (mg/ml) measuring both glutamate-/malate- (*left*) and succinate (*right*)-driven respiration in both basal (State 2, *top*) and ADP-stimulated (State 3, *bottom*). State 2 succinate-driven respiration profiles are presented in detail in (**c**), and the effect of ADP addition on succinate-driven respiration is presented in (**d**). If cardiac mitochondria are used, the same plate layout can be used to optimize the protein amount for respiration using fatty acids

7. Re-suspend the resultant mitochondrial pellet in 0.7 ml of Buffer II and store on ice until required (*see* **Note 1(c)**).
8. Determine protein concentration using a BCA kit as per manufacturer's instructions, briefly:
   (a) *Construct a standard curve using albumin stock solutions at 1.5, 1.0, 0.75, 0.5, 0.25, 0.125, 0.025, and 0 mg/ml prepared in 1%* $^{v}/_{v}$ *Triton X* (*see* **Note 1(d)**).
   (b) *Dilute samples of the mitochondrial preparation 1:60, 1:80, and 1:100, using 1%* $^{v}/_{v}$ *Triton X as dilute.*
   (c) *Mix 9.8 ml protein reagent A with 200 μl protein reagent B to produce the developing reagent. Then add 20 μl of each standard, or sample into 96-well plates followed by 200 μl of developing reagent.*
   (d) *Incubate the plate for 30 min at 37 °C, then read absorbance at 520 nm, and calculate mitochondrial protein concentration (see* ***Note 1(c)****).*

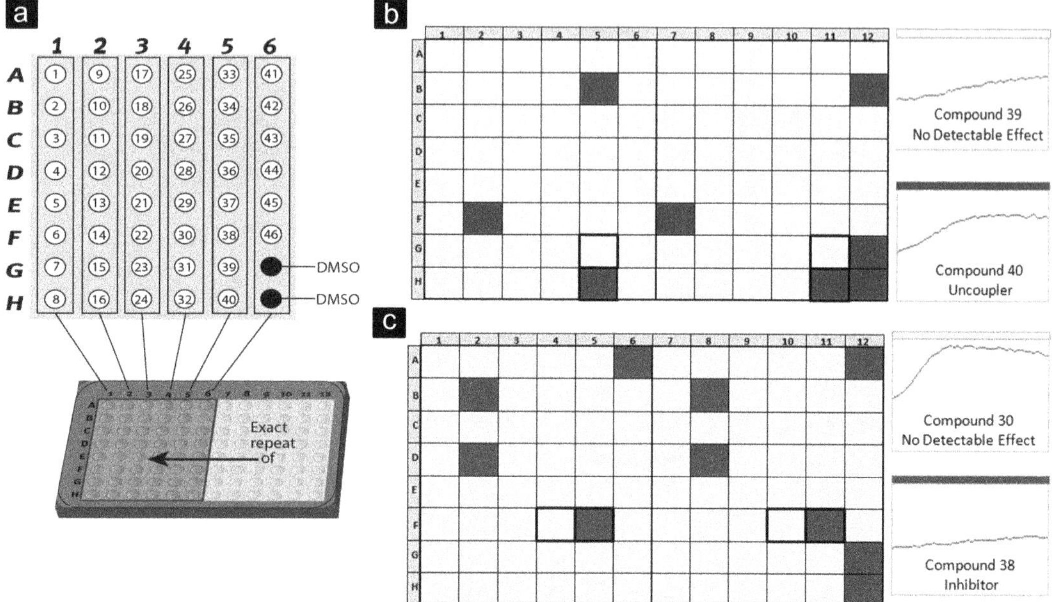

Fig. 2 Single-dose screening. Forty-six NCEs can be screened in duplicate on one plate using the plate layout presented in (**a**). Two vehicle controls (DMSO) are included on the *left* side of the plate, whereas on the *right* side of the plate, one would include FCCP [for State 2, (**b**)] or rotenone [for State 3, (**c**)], as positive controls giving the maximum uncoupling or inhibition. Typically, compounds showing more than 50% change from control values are flagged and taken forward into dose-response analysis

3.1.2 Procedure for Isolation of Cardiac and Skeletal Muscle Mitochondria

1. Prepare the following buffers (*see* **Note 1(a)**):

   Buffer I: 100 mM KCl, 40 mM Tris–HCl, 10 mM Tris base, 5 mM $MgCl_2$, 1 mM EDTA, and 1 mM ATP, pH 7.4.

   Buffer II: 100 mM KCl, 40 mM Tris–HCl, 10 mM Tris-base, 1 mM $MgSO_4$, 0.1 mM EDTA, 0.2 mM ATP, and 2% BSA, pH 7.4.

   Buffer III: 100 mM KCl, 40 mM Tris–HCl, 10 mM Tris-base, 1 mM $MgSO_4$, 0.1 mM EDTA, and 0.2 mM ATP, pH 7.4.

   Buffer IV: 220 mM mannitol, 70 mM sucrose, 10 mM Tris–HCl, and 1 mM EGTA, pH 7.4.

2. Euthanize animals with an overdose of carbon dioxide (*see* **Note 1(b)**), excise either two rat hearts or 10 g of mixed muscle (EDL/gastrocnemius/soleus), and place in ice-cold Buffer I.
3. Add type XXIV protease to a concentration of 5 mg/g of wet tissue and mince finely using a pair of scissors. Incubate for 7 min intermittently mixing and mincing and then add an equal volume of Buffer I to terminate the digestion.
4. Using an Ultra-Turrax tissue homogenizer (IKA, T25) set at setting 1 (11,000 rpm), homogenize the resultant mixture for

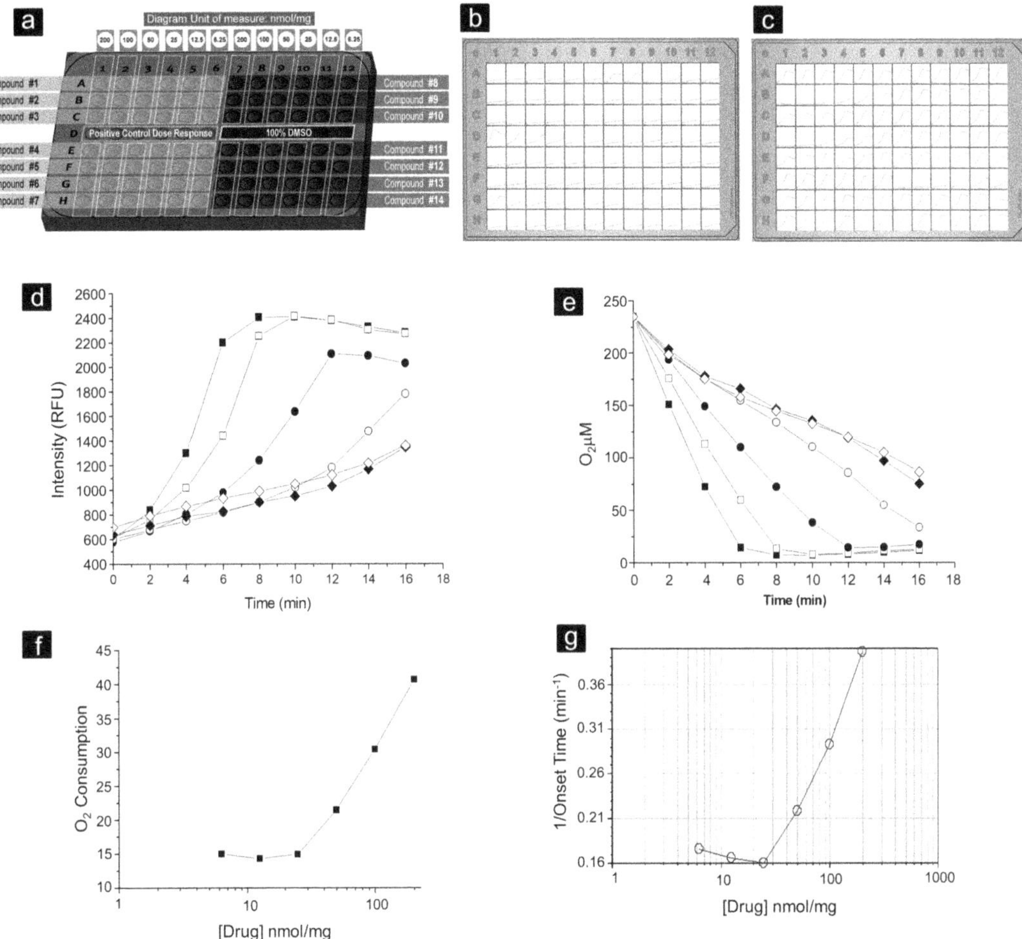

Fig. 3 Dose-response analysis. Typical plate layout for generation for NCE dose-response curves. Fourteen compounds can be screened at six different concentrations (200, 100, 50, 25, 12.5, and 6.25 nmol/mg protein) using the plate map presented (**a**). Typical data output for uncouplers and inhibitors is shown in (**b**) and (**c**), respectively (Tecan Magellan plate views). For State 2 analysis, an FCCP dose response is included in H1–H6, and for State 3 respiration, a nefazodone dose response is included. A vehicle control is included in D7–12. Raw data are analyzed and UC_{50}/IC_{50} values generated. Sample uncoupler raw data (A7–12), transformed data, and the resultant dose-response curves are presented in (**d**), (**e**), and (**f**), respectively. Dose-response curve (SoftMax Pro, Molecular Devices) generated using the more basic time-to-threshold data analysis approach available on most plate readers is shown in (**g**). *See* **Note 2(e)**

30 s and then centrifuge the homogenate at 4 °C for 10 min at 700 × *g*.

5. Filter the supernatant through two layers of cheesecloth and re-centrifuge at 14,000 × *g* for 10 min at 4 °C.
6. Discard the supernatant, and using a glass stirring rod, re-suspend the mitochondrial pellet in Buffer II and then centrifuge at 7000 × *g* for 10 min at 4 °C.

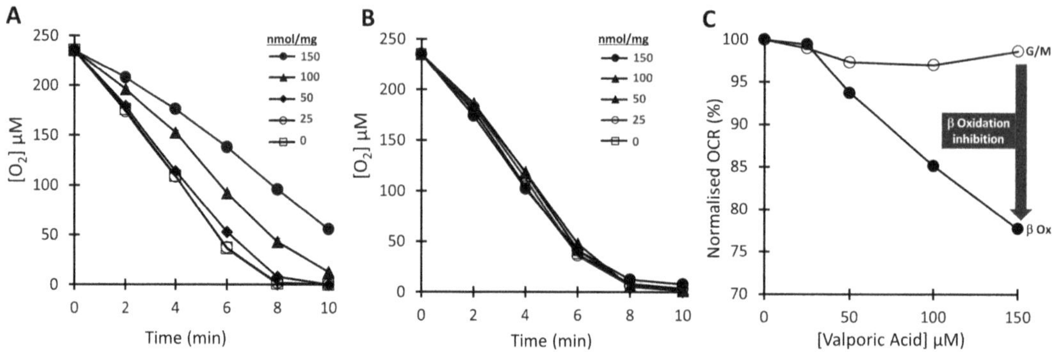

Fig. 4 Differential sensitivity of State 3 (+ADP) palmitoyl CoA/carnitine/malate (**a**) and palmitoyl CoA-/carnitine-/glutamate-/malate (**b**)-driven mitochondrial oxygen consumption to increasing concentration of valproic acid (150 (●), 100 (▲), 50 (◆), 25 (○), 0 (□) nmol/mg protein). Data is presented in oxygen scale. Specific FAO inhibitors impair palmitoyl CoA-/carnitine-/malate-driven respiration (**a**) but have no effect on addition of glutamate/malate (**b**), due the direct supply of NADH to CI of the ETC, thereby bypassing β oxidation inhibition. Inhibition in both conditions is indicative of a specific ETC inhibitor or a multifactorial response where β Oxidation inhibition contributes only partially to the overall metabolic impairment caused by compound treatment. In such instances, differential IC_{50} values can be assessed. This differential sensitivity is summarized in (**c**) and highlights how the assay can identify FAO inhibition

7. Discard the supernatant, and using a glass stirring rod, re-suspend the mitochondrial pellet in 20 ml of Buffer III and then centrifuge at 3500 × *g* for 10 min at 4 °C. Then re-suspend the mitochondrial pellet in a minimal volume of solution IV for further use (*see* **Note 1(c)**).
8. Protein concentrations should be determined as described in **step 8** of the preceding protocol.

3.2 Oxygen Consumption Analysis: Assay Optimization

1. Reconstitute the MitoXpress® Xtra reagent in 1 ml of respiration buffer mixing to insure re-suspension. Then dilute to 10 ml with the same buffer and warm to 30°C (*see* **Note 2(a)**).
2. Set instrument temperature to 30 °C and allow sufficient time for target temperature to be reached. Prepare a kinetic protocol using the recommended optical settings (*see* Table 1) reading test wells at 0.5–1.5 min intervals over 30–60 min (*see* **Note 2 (b)**). Recommended measurement protocols are available from Luxcel Biosciences.
3. Warm mineral oil to 30 °C.
4. For State 2 respiration (basal), add 150 μl of substrate stock solution (succinate or glutamate/malate) to 1.35 ml of incubation buffer and warm to 30 °C.
5. For State 3 respiration (ADP-stimulated), add 150 μl of substrate stock and 100 μl ADP stock to 1.25 ml of incubation buffer and warm to 30 °C.

6. Using respiration buffer, prepare a six-point mitochondrial protein dilution series to a 1.5 ml total volume for each concentration. Recommended concentrations (mg/ml): 1.5, 1.0, 0.5, 0.25, 0.125, and 0.63.
7. Place a clear bottomed, 96-well plate on a plate heater equilibrated to 30 °C and, using automatic or multichannel pipette, dispense solutions as per the recommended plate map (Fig. 1) adding the following to each well:
   (a) *100 μl of assay buffer containing MitoXpress® Xtra reagent.*
   (b) *50 μl of mitochondrial stock dilutions giving the desired mitochondria concentration.*
   (c) *50 μl of substrate solution (final concentration 25 mM for succinate or 12.5/12.5 mM glutamate/malate).*
   (d) *50 μl of substrate/ADP solution (final concentration 25 mM for succinate, 12.5/12.5 for mM glutamate/malate, 1.65 mM for ADP).*
8. Using a syringe dispenser, add 100 μl of prewarmed heavy mineral oil to each well (*see* **Note 2(c)**).
9. Insert the microplate into the fluorescence plate reader and read using the settings outlined above. When measurement is completed, remove the plate and save the data to file.

3.3 Oxygen Consumption Analysis: Screening

1. Reconstitute MitoXpress® Xtra reagent in 1 ml of assay buffer mixing to insure re-suspension. Then dilute to 10 ml with the same buffer and warm to 30 °C.
2. Set instrument temperature to 30 °C and allow sufficient time for target temperature to be reached. Prepare a kinetic protocol using the recommended optical settings (*see* Table 1) reading test wells at 0.5–1.5 min intervals over 30–60 min.
3. Prepare compounds in DMSO to 100× the required concentration (*see* **Note 2(d)**).
   (a) *For initial screening assay at a single concentration (typically 100 nmol/mg) in duplicate as per plate map* (Fig. 2).
   (b) *For subsequent mechanistic elucidations and dose-response analysis, 1:2 dilution seria are prepared for each compound at one data point per individual concentration as per plate map* (Fig. 3).
4. Prewarm mineral oil to 30 °C.
5. Prepare substrate stocks and prewarm reagents to 30 °C.
   (a) *For standard inhibitor/uncoupler screening:*

5.1a: State 2 analysis (uncoupler screening)—Mix 5.4 ml of incubation buffer and 600 μl of either succinate or glutamate/malate stock.

5.1b: State 3 analysis (inhibitor screening)—Mix 5 ml of incubation buffer and 600 μl of either succinate or glutamate/malate stock and 400 μl of ADP stock.

(b) *For FAO-specific screening:*

5.2a: FAO- and G/M-driven respirations—Mix 5 ml of incubation buffer and 600 μl of palmitoyl CoA/carnitine/malate substrate stock and 400 μl of ADP stock.

5.2b: FAO-driven respiration—Mix 5 ml of incubation buffer and 600 μl of palmitoyl CoA/carnitine/malate substrate stock, 600 μl of glutamate/malate stock, and 400 μl of ADP stock.

6. Prepare 6 ml of stock mitochondria at 4× the required protein concentration (optimized above) and store on ice.
7. Place a black 96-well plate on a plate heater equilibrated to 30 °C and dispense the following solutions using either an appropriate automatic or multichannel pipette (*see* **Note 2(c)**):
   (a) *100 μl of prewarmed assay buffer.*
   (b) *1 μl of compound at desired concentration, according to the appropriate plate map.*
   (c) *50 μl of mitochondria at optimal dilution.*
   (d) *50 μl of prewarmed substrate stock solution to each well.*
   (e) *100 μl of prewarmed heavy mineral oil.*
8. Place the microplate in a fluorescence reader preset as described above and commence reading. When completed, save data to file.

3.4 Data Analysis

1. Standard Analysis

3.4.1 General Approach

The standard data analysis approach entails calculating *slope* values from plots of raw intensity or lifetime versus time using standard linear regression (Fig. 1). This can be performed on the software of most readers [MARS (BMG Labtech), Gen 5 (BioTek), MyAssays (PerkinElmer), Magellan (Tecan), SoftMax Pro (Molecular Devices)], or on packages such as MS Excel.

2. Advanced Analysis

For high-throughput screening, where additional standardization is sometimes required, two further approaches can be deployed (*see* **Note 2(e)**): The first is to convert MitoXpress® Xtra signal to an oxygen scale and subsequently apply linear regression analysis to the transformed profiles (Fig. 3d–f). Oxygen conversion can be approximated by normalizing profiles (dividing each value in the profile but the first value) and using the following equation: $[O_2](t) = 235 \times 1 \times (S - X)/(X \times (S - 1))$, where S is the signal increase on deoxygenations. For a more detailed analysis, *see* **Note 2(e)** and Supplemental Data 1. This approach provides the most quantitative data output.

The second involves using a "time-to-threshold" value calculated using plate reader software and relating the reciprocal of this value to drug concentration to generate dose-response information. This can be performed on most plate reader software (Fig. 3g). This approach simplifies data reduction and is useful when processing very large data sets.

3.4.2 Assay Optimization

1. Open the saved file and either analyze on plate reader software (preferable) or export to external software package such as MS Excel. Then plot raw intensity or lifetime versus time (Fig. 1).
2. Calculate an RCR and compare to expected values to insure the mitochondrial preparation is of sufficient quality (*see* **Note 2 (f)**). RCR is calculated by determining the State 3/State 2 ratio using reciprocal slopes and is best performed using data in O_2 scale (*see* **Note 2(e)**).
3. If this value is sufficiently high, examine individual profiles and select enzyme concentrations that produce reliably measurable signal changes to allow analysis of both inhibition and uncoupling. These are then used for subsequent screening (*see* **Notes 2(g)** and **(i)**).

3.4.3 Screening

1. Open the saved file and either analyze on plate reader software (preferable) or export to an external software package such as MS Excel. Then plot raw intensity or lifetime versus time.
2. Calculate "slope" values from plots of raw intensity or lifetime versus time using standard linear regression and compare calculated slopes to those of the untreated sample to determine what compounds, if any, are inhibiting/uncoupling (Fig. 2).

Dose-based compound ranking can also be performed if necessary by calculating an IC_{50} (inhibitory effect in State 3) or UC_{50} (uncoupling effect in State 2 using FCCP as 100% uncoupling) value for each compound. This is best performed using one of the approaches outlined above (*see* **Note 2(e)** and Supplemental Data 1).

4 Notes

1. Mitochondrial Isolation.
   (a) Solutions should be prepared using Millipore grade water and stored in pre-washed glassware. pH should be

adjusted with HCl and KOH. Do *not* use NaOH. All substrate stocks can be prepared in advance and stored in aliquots at −80 °C. Buffers should be freshly prepared on a weekly basis, and any BSA additions should be made on the day of use.

(b) Sprague Dawley rats (Charles River, Wilmington, MA) or equivalent strain. Care and maintenance need to be in accordance with the principles described in the *Guide for the Care and Use of Laboratory Animals* (NIH Publication 85-23, 1985) or equivalent. For best results the animals should weigh between 150 and 180 g. Rats are housed in pairs in a controlled environment with constant temperature (21 ± 2 °C) and a 12-h light/dark cycle. Food and water are provided at ad libitum. Avoid anesthetics as they can have adverse effects on mitochondrial quality.

(c) Mitochondria should be stored on ice at a protein concentration > 30 mg/ml and used within 4–6 h.

(d) Protein standards can be made in batches and stored at 4 °C for several weeks. Albumin is provided as a 2 mg/ml stock solution.

2. Oxygen Consumption Analysis.

(a) Standard reagent package is for one 96-well plate (or ~100 assay points). Reagent diluted in respiration buffer should be used on the same day. Adjust volumes accordingly for smaller numbers of samples. MitoXpress® Xtra reagent reconstituted in 1 ml H_2O can be stored in the dark between +2 and +8 °C for several days or stored as aliquots at −20 °C for use within 1 month (avoid freeze thaw).

(b) Instrument-specific measurement settings are presented in Table 1. A filter-based optical configuration is recommended where available. Time-resolved measurements are also recommended where available. While measurements in fluorescence intensity mode work very well and are widely published, additional robustness can be achieved using dual TRF measurements (lifetime mode). Compatible instruments are listed in Table 1. Using these dual intensity reads as outlined in Table 1, the corresponding lifetime is calculated using the following relationship: $\tau = 40/\ln(R1/R2)$, where $R1$ and $R2$ represent read one and read two, respectively, providing lifetime values with units of μs. This can be conducted on plate reader software or after export to software packages such as MS Excel or Microcal Origin. If lifetime measurements are not possible, data can be generated in standard "intensity" mode. Recommended instrument-specific measurement protocols are available from Luxcel Biosciences.

(c) To minimize oxygen depletion in samples prior to the measurement, plate preparation time should be kept to a minimum.

(d) Concentration range is defined by the user. As drug concentrations are usually expressed in nmol/mg of mitochondrial protein, starting concentration should be altered based on the mitochondrial protein concentration being used. DMSO content should not exceed 0.5% $^{v}/_{v}$, and dilution plates should be prepared no earlier than the 24 h pretreatment.

(e) The simplest approach to data reduction is to apply linear regression analysis to the MitoXpress® Xtra signal profiles and relate this to mitochondrial activity. An additional level of quantification can be achieved by converting these values to oxygen scale. The procedure for this conversion is detailed below and illustrated in Fig. 3. For large datasets where simple data analysis and parameter standardization is required, a time-to-threshold approach can be deployed. The procedure for this is also detailed below.

Oxygen Conversion and Linear Regression.

Converting to an oxygen scale allows the use of simple linear regression for the generation of dose-response data. Oxygen concentrations can be estimated from measured intensities using the following relationship:

$$[O_2](t) = [([O_2]_a \times I_a \times (I_o - I_t)]/[I_t \times (I_o - I_a)].$$

$[O_2]_a$ is oxygen concentration in air-saturated buffer (235 μM at 30 °C), and I_t, I_a and I_o are fluorescent signal at time t, signal in air-saturated buffer, and signal in deoxygenated buffer (maximal signal), respectively.

By normalizing each profile for initial intensity (dividing each profile by the initial read) and measuring the signal increase on complete deoxygenation (S), this can be simplified to:

$$[O_2](t) = 235 \times 1 \times (S - X)/(X \times (S - 1)).$$

This specific transformation is only valid at 30 °C, and the signal increase on deoxygenation is specific to reader type, measurement settings, and buffer composition. Plates should be at measurement temperature prior to the reading. It is also important that no significant deoxygenation has occurred prior to read one of the kinetic analysis. It should be noted that this transformation, in this instance, is not primarily intended as an analytical determination of dissolved oxygen but rather as an estimation of oxygen concentration to facilitate simple linear regression analysis to be performed on large data sets.

Threshold Time.

By assessing the time at which a particular signal threshold is reached and using the inverse of this value (1/threshold time) as a metric of oxygen consumption rate, dose-response data can be quickly generated using standard plate reader software. Oxygen consumption rates and the optimum threshold should however be optimized such that the threshold is not exceeded prior to the beginning of the assay (uncoupler analysis) or that significant activity is not missed (inhibitor analysis).

(f) The respiratory control ratio (RCR) is calculated by assessing the effect of ADP addition on mitochondrial oxygen consumption and is an important metric in assessing the quality of a mitochondria preparation. A lower than expected value indicates that the mitochondria are insufficiently coupled, probably due to membrane damage during preparation. Rat liver mitochondria generally show RCRs of approximately 5.0 for glutamate-/malate-driven respiration or approximately 3.0 for succinate-driven respiration. It should be noted however that RCR values are tissue-dependent. Calculated RCR values should be compared with previous experiments or literature data [5].

(g) A concentration producing large signal changes without inducing rapid signal saturation is deemed "optimal," thereby allowing reliable analysis of both uncoupling and inhibition of mitochondrial. These optimal concentrations will be dependent on the tissue of origin, the substrate used, and the availability of ADP-specific. Typical values for rat liver mitochondria are as follows:

Basal respiration glutamate/malate:	1 mg/ml
ADP-driven respiration with glutamate/malate:	0.25 mg/ml
Basal respiration with succinate:	0.5 mg/ml
ADP-driven respiration with succinate:	0.25 mg/ml

(h) Prior to screening for compounds with a mitochondrial liability, it should be insured that assay performance is consistent. This is determined from CV values for intra- and inter-assay variations. These values should not exceed 15% and normally be at ~10%. Once mitochondrial preparations are seen to be reproducible and the assay is well established ($n = 3$–5), this optimization becomes unnecessary.

(i) Interpreting FAO screening data: If a compound inhibits FAO only, a decrease in oxygen consumption with palmitoyl CoA/carnitine/malate plus ADP will be observed, while oxygen consumption in the presence of palmitoyl CoA/carnitine/malate plus glutamate/malate plus ADP will be unaffected (Fig. 4). In contrast, if a compound inhibits palmitoyl CoA/carnitine/malate plus ADP-driven oxygen consumption *and* palmitoyl CoA/carnitine/malate plus glutamate/malate plus ADP-driven oxygen consumption, the compound could (also) be an inhibitor of Complexes I–V.

References

1. Dykens JA, Marroquin LD, Will Y (2007) Strategies to reduce late-stage drug attrition due to mitochondrial toxicity. Expert Rev Mol Diagn 7:161–175
2. Wallace KB (2008) Mitochondrial off targets of drug therapy. Trends Pharmacol Sci 29:361–366
3. Papkovsky DB, Hynes J, Will Y (2006) Respirometric screening technology for ADME-Tox studies. Expert Opin Drug Metab Toxicol 2:313–323
4. Will Y, Hynes J, Ogurtsov VI, Papkovsky DB (2007) Analysis of mitochondrial function using phosphorescent oxygen-sensitive probes. Nat Protoc 1:2563–2572
5. Hynes J, Hill R, Papkovsky DB (2006) The use of a fluorescence-based oxygen uptake assay in the analysis of cytotoxicity. Toxicol In Vitro 20:785–792

Chapter 5

Modulation of Cellular Respiration by Endogenously Produced Nitric Oxide in Rat Hippocampal Slices

Ana Ledo, Rui M. Barbosa, and João Laranjinha

Abstract

Nitric oxide (•NO) is an ubiquitous signaling molecule that participates in molecular processes associated with several neural phenomena ranging from memory formation to excitotoxicity. In the hippocampus, neuronal •NO production is coupled to the activation of NMDA type glutamate receptors. Cytochrome c oxidase has emerged as a novel target for •NO, which competes with O_2 for binding to this mitochondrial complex. This reaction establishes •NO as a regulator of cellular metabolism and, possibly, mitochondrial production of reactive oxygen species which participate in cellular signaling. A major gap in the understanding of •NO bioactivity, namely, in the hippocampus, has been the lack of knowledge of its concentration dynamics. Here, we present a detailed description of the simultaneous recording of •NO and O_2 concentration dynamics in rat hippocampal slices. Carbon fiber microelectrodes are fabricated and applied for real-time measurements of both gases in a system close to in vivo models. This approach allows for a better understanding of the current paradigm by which an intricate interplay between •NO and O_2 regulates cellular respiration.

Key words Nitric oxide, Oxygen, Hippocampus, Mitochondrial respiration, Neuronal nitric oxide synthase, Glutamate NMDA receptor, Carbon fiber microelectrode, Nafion®, *o*-Phenylenediamine

1 Introduction

Nitric oxide (•NO), like oxygen, is a small hydrophobic molecule that diffuses fast in the cellular environment [1], crossing cell membranes unassisted by receptors or channels [2]. In the late 1980s, •NO was shown to regulate multiple signaling pathways in the nervous, immune, and endothelial systems [3].

The better characterized intracellular molecular target of •NO is soluble guanylate cyclase, an enzyme activated by •NO binding that then converts GTP to cGMP [4] which, in turn, initiates signaling cascades leading, for instance, to vasodilation and neuromodulation [5, 6]. Increasing evidence shows that •NO may act as a master regulator of brain energy metabolism [7] and that these effects of •NO critically depend on its concentrations at the reaction

Carlos M. Palmeira and António J. Moreno (eds.), *Mitochondrial Bioenergetics: Methods and Protocols*, Methods in Molecular Biology, vol. 1782, https://doi.org/10.1007/978-1-4939-7831-1_5,

site and the period of exposure. Nitric oxide can react with several complexes of the mitochondrial respiratory chain at different rates. While the reaction with complex III is sluggish [8], those with complex I and complex IV (cytochrome c oxidase or CcO) are fast and, at least to an extent, reversible but, most importantly, critically dependent on the concentration dynamics of •NO [9]. Cytochrome c oxidase is the terminal complex of the mitochondrial respiratory chain: it accepts electrons from cytochrome c and reduces O_2 to H_2O. Low concentrations of •NO can reversibly bind to CcO and inhibit mitochondrial respiration by competing with O_2 [10–13]. Nitric oxide binds to the active site of CcO with a high affinity ($K_D = 0.2$ nM, [14]), increasing the apparent K_M of the enzyme for O_2 [11]. Several mechanisms have been proposed to explain the molecular details of the inhibition of CcO by •NO. However, due to the difficulties associated with the measurement of •NO binding to CcO in vivo, most of what is known is based on theoretical models [15–17].

The concept of •NO inhibition of CcO has been clearly demonstrated in cellular and subcellular preparations, namely, mitochondria and synaptosomes [10–12]. Here, we present a method that allows for the simultaneous and real-time recording of the profiles of •NO and O_2 in hippocampal slices at physiological O_2 concentrations. This approach uses a complex biological preparation which retains the cytoarchitectural organization found in vivo and allows for a better understanding of the mechanisms of •NO regulation of mitochondrial respiration [18].

Both the preservation of tissue integrity and the partial pressure of O_2 (pO_2) are critical issues when considering the interplay of •NO and O_2 in modulating CcO activity. Firstly, both compounds are diffusible and, therefore, tissue tortuosity might affect diffusion. Secondly, at values of pO_2 (such as atmospheric pO_2 or even at higher values observed in cell culture experiments), •NO autoxidation by O_2 may spuriously occur, yielding reactive species, thus diverting •NO bioactivity.

The hippocampus is a structure of the temporal lobe in the central nervous system involved in memory formation [19]. In this region, •NO is produced upon activation of the NMDA type of glutamate receptor [20, 21] and acts as a neuromodulator in molecular phenomena associated with long-term potentiation and depression of synaptic strength [22–25]. In the hippocampus, the neuronal circuit (input ->dentate gyrus ->CA3 subregion ->CA1 subregion ->output) is organized in a lamellar fashion, so in transversal sections this trisynaptic loop is maintained [26, 27]. Furthermore, such brain slices are electrophysiologically and biochemically operative [28]. Hence, the great interest in the use of acute hippocampal slices.

In the method presented here, both endogenous •NO and O_2 are measured electrochemically using carbon fiber microelectrodes.

This methodology allows real-time and direct measurement of the electroactive species of interest with high temporal and spatial resolution. In the case of •NO, the use of carbon fiber microelectrodes modified with Nafion® and *o*-phenylenediamine has been established by others and us as convenient probes for measurements in vitro [29, 30] in ex vivo preparations [29–33] and in vivo in anesthetized rats [34, 35]. The use of these polymer films covering the carbon fiber surface allows for an increase of sensor selectivity, a relevant concern considering the relatively high oxidation potential of •NO and the presence of various potential electroactive interferents present in tissue preparations from the nervous system, such as ascorbic acid, biogenic amines, and indoles and their metabolites.

In the case of O_2, it is habitual to use platinum microelectrodes due to their well-known catalytic properties for O_2 reduction [36]. However, some authors argue that, contrary to platinum electrodes, carbon-based electrodes consume less O_2 from the medium, and thus there is a lesser risk of the electrode *per se* affecting O_2 distribution in the tissue [37, 38].

In the following sections, we will provide a detailed description of the steps involved in the use of carbon fiber microelectrodes for the evaluation of how endogenously produced •NO can modulate cellular respiration in hippocampal slices, namely, microelectrode fabrication, evaluation and application in simultaneous electrochemical recordings of •NO and O_2, preparation and determination of saturated •NO solutions, and, finally, preparation and use of acute hippocampal slice.

2 Materials and Solutions

Unless otherwise mentioned, all reagents used are analytical grade.

2.1 Saturated Nitric Oxide Stock Solutions

The preparation of saturated •NO stock solutions is laborious, and standard procedures should be followed to guarantee that it is free of N_xO_y impurities that may originate in the compressed gas tank [39]. All handling of •NO gas should be performed in a fume hood. Reaction of gaseous •NO and atmospheric O_2 results in the production of toxic nitrogen oxides such as $•NO_2$ and N_2O_3. The setup, as shown in Fig. 1, should contain only inert materials, such as glass, Teflon, and stainless steel tubing and fittings. Solvents should be carefully removed of O_2 by saturating with high purity argon. Required materials for the preparation of this solution are as follows:

1. A tank of •NO gas, 99.9% (Air Liquide).
2. Saturated solution of •NO is prepared in deoxygenated ultrapure Milli-Q water containing 100 μM DTPA

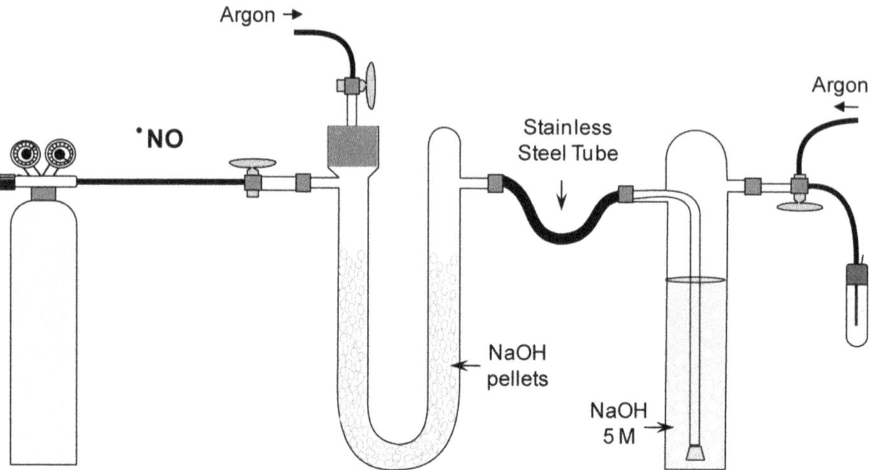

Fig. 1 Schematic representation of the setup used to prepare purified standard solutions of •NO from a commercial gas tank. All tubing and seals are Teflon and stainless steel. Oxygen is removed from the lines and solutions by bubbling purified argon

(diethylenetriaminepentaacetic acid, Sigma) in a glass tube sealed with a Teflon cap (e.g., Vacutainer tube).

3. The •NO gas is washed from higher oxides (such as $•NO_2$, N_2O_3, and N_2O_4) by a first passage through a glass column packed with NaOH pellets followed by a passage through a trap containing 5 M NaOH prior to being bubbled in the solution vial.
4. Determination of •NO concentration in the stock solution is performed on the ISO-NOP 2 mm Pt sensor connected to an ISO-NO Mark II (World Precision Instruments, Inc., USA).
5. The ISO-NOP sensor is calibrated in accordance to the specifications and requires the following solutions: 100 μM $NaNO_2$, 0.2 M KI, and 0.2 M H_2SO_4.

2.2 Microelectrode Fabrication and Evaluation

The fabrication of microelectrodes for the electrochemical measurement of •NO entails the construction, surface modification (for •NO microsensors), evaluation of general recording properties, and calibration of each microelectrode. The required materials for this process are as follows:

2.2.1 Microelectrode Fabrication

1. Carbon fibers. We use different types of carbon fibers:
   (a) For •NO microsensors, we use 30 μm diameter fibers from Specialty Materials Inc., USA.
   (b) For O_2 microelectrodes, we use 10 μm diameter fibers from Amoco Corp., USA.
2. Borosilicate glass capillaries (1.16 mm i.d. × 2.0 o.d.) from Harvard Apparatus Ltd., UK.

3. Acetone for washing carbon fiber and insertion into capillary.
4. Copper wire (we use individualized copper wires from network cables).
5. Silver conductive paint (RS, UK).
6. Cyanoacrylate glue.
7. Vertical puller (Harvard Apparatus Ltd., UK).
8. A pair of small tweezers (0.5 mm tip).
9. Microscope (inverted will work best).

2.2.2 Modification of Carbon Fiber Microelectrodes

To produce •NO microsensors, the carbon fiber microelectrodes are modified with Nafion® and *o*-phenylenediamine (o-PD). For this procedure, required materials and solutions are:

1. Nafion® solution, by Aldrich (5 wt% in a mixture of aliphatic alcohols).
2. A freshly prepared 5 mM *o*-PD solution in deoxygenated 0.05 M PBS lite containing 100 μM ascorbic acid (Sigma). The composition of the PBS lite is (in mM): 10 NaH_2PO_4, 40 Na_2HPO_4, and 100 NaCl (pH 7.4).
3. An oven (must reach 170 °C).

2.2.3 Microelectrode Evaluation and Calibration

1. All microelectrodes are systematically pretested for their general recording properties by fast cyclic voltammetry (FCV) using a potentiostat (Ensman Instruments, USA).
2. The evaluation and calibration of •NO and O_2 microelectrodes are performed in PBS lite.
3. The calibration of O_2 microelectrodes requires argon and Carbox (95% O_2/5% CO_2) in order to vary pO_2 in PBS lite.
4. Calibration of O_2 microelectrodes is performed in the slice recording chamber in a two-electrode circuit with an Ag/AgCl pellet as a reference electrode and the microelectrode as the working electrode. Measurement of O_2 is performed with a CompactStat potentiostat (Ivium Technologies B.V., The Netherlands).
5. For calibration of •NO microsensor, a saturated •NO stock solution prepared as described above is used. During the calibration of the •NO microsensor, selectivity is also evaluated, using fresh solutions of ascorbic acid and NO_2^- (20 mM) in deoxygenated Milli-Q water.
6. Calibration of •NO microsensor is performed in a two-electrode circuit using an Ag/AgCl 3 M reference electrode (RE-5B, BASi) and the microsensor as the working electrode, in a 40 mL beaker. The electrochemical recordings of •NO are performed using a CompactStat potentiostat (Ivium Technologies B.V., The Netherlands).

2.3 Simultaneous Recording of •NO and O_2 in Rat Hippocampal Slices

2.3.1 Preparation of Rat Hippocampal Slices

1. Isolation and recovery of rat hippocampal slices is performed using aCSF with the following composition (in mM): 124 NaCl, 2 KCl, 25 $NaHCO_3$, 1.25 KH_2PO_4, 1.5 $CaCl_2$, 1.4 $MgCl_2$, and 10 D-glucose. The aCSF solution must be saturated with Carbox to assure pH buffering and oxygenation (*see* **Note 1**).
2. Dissection materials required for preparation of hippocampal slices are: two medium (10 cm) and one small (5 cm) petri dishes, one scalpel, two pairs of scissors (one of which should be able to cut scalp bone—if larger animals are used, using serrated scissors may be considered useful), one pair of large forceps (these will be used to hold the decapitated rat head, so the distance of the tips should be close to the distance between animal eyes), two small forceps with curved and serrated tips (tip curvature should be between 45° and 90°; tip diameter is not critical, but around 0,8 mm will be adequate), one rongeur, one spatula, filter paper (to fit the medium petri dish), and clear acetate sheet cut to the size of the tissue chopper stage.
3. Other materials needed are a tissue chopper (McIlwain Tissue Chopper, Campden Instruments, UK) and a tissue slice pre-incubation chamber (BSC-PC model, Harvard Apparatus). In order to transfer slices from one place to another, a 1 or 5 mL automatic pipette can be used with a disposable tip cut to allow aspiration of slices without touching them (approx. 5 mm in diameter).
4. Slice separation: separating slices after sectioning is a critical step in the procedure and care should be taken to avoid lesion of the tissue. To perform this separation, we use Pasteur pipettes in which the tip has been blunted and slightly curved by heating. Alternatively, one can use a very small paintbrush, with only a couple of threads.

2.3.2 Electrochemical Recording of •NO and O_2 in Hippocampal Slices

1. Thermostatic bath is placed above the level of recording chamber to allow solutions to perfuse in a siphon system.
2. Electrochemical recordings in slices are performed in a perfusion recording chamber (BSC-BU with BSC-ZT top, from Harvard Apparatus, USA) coupled to a temperature controller (model TC-202A, Harvard Apparatus, USA).
3. The amperometric recordings of •NO and O_2 are performed using a CompactStat bipotentiostat (Ivium Technologies B.V., The Netherlands).

   To position the microsensors a micromanipulator is required.
4. All work done using this recording system requires the use of a binocular stereomicroscope on a boom stand. In our setup, we use a SZ60 from Olympus.

3 Methods

3.1 Preparation and Determination of Saturated •NO Stock Solution

This solution is prepared freshly on the day of use.

1. Remove O_2 from the Milli-Q water (containing DTPA) by bubbling argon or N_2 for 40 min.
2. To obtain a saturated stock solution, bubble •NO gas for 40 min. After this period, the concentration of •NO has been determined to be 1.98 ± 0.15 mM [39].
3. The concentration of this saturated •NO solution can be confirmed by means of the ISO-NOP 2 mm Pt electrode. This is a platinum electrode covered with a gas permeable membrane that confers selectivity of the sensor to •NO. This sensor is calibrated daily using the recommended calibration procedure, by chemical generation of •NO in the calibration vial from the following reaction:

$$2NaNO_2 + 2KI + 2H_2SO_4 \rightarrow 2 \bullet NO + I_2 + H_2O + 2Na_2SO_4$$

Add $NaNO_2$ aliquots to a mixture of 0.1 M KI and 0.1 M H_2SO_4. In the presence of excess reductant (I^-), all NO_2^- is converted to •NO.

4. After calibrating the ISO-NOP electrode, determine the •NO concentration in deoxygenated PBS lite. Allow baseline current to stabilize and then add an aliquot of •NO solution. Repeat this step three times and extrapolate the •NO concentration of the stock solution.

3.2 Carbon Fiber Microelectrodes

3.2.1 Fabrication of Microelectrodes

1. Place capillaries in a petri dish filled with acetone with one extremity of the capillary immersed in the liquid and the other over the opposite edge of the petri dish protruding out of the dish. The capillaries should be filled with acetone. Using a pair of tweezers, insert an individual fiber into each capillary making sure that the fiber has crossed the whole length. This procedure should be performed near a source of natural light or using a cold light source—convection from lights will make the fibers move in the air, making insertion into the capillary more difficult.
2. Allow the acetone to evaporate (*see* **Note 2**).
3. Mount the capillary with a fiber onto the vertical puller. After pulling, retain the half containing the fiber, and cut the exposed section to a length of approx. 1 cm from the end of the glass (*see* **Note 3**).
4. Mount the microelectrode on a microscope, and cut exposed carbon fiber at desired length using small tweezers. For •NO

microsensors, we cut fibers at 100–150 μm of length, and, for O_2, fibers are cut at 10–20 μm (*see* **Note 4**).

5. Use a syringe with a long/thin tip to deposit a small amount of conductive silver paint inside the capillary, at the stem end.
6. Insert a thin copper wire in the capillary. Be sure that the copper is exposed at both extremities (remove insulating plastic from tips). The exposed tip must reach the conductive silver paint.
7. To finish the construction process, add a small drop of cyanoacrylate glue at the top end of the capillary to fix the copper wire in position.
8. The general recording properties of the microelectrodes are evaluated in PBS lite by FCV performed between −0.4 and +1.6 V vs Ag/AgCl at a scan rate of 200 V/s. This potential range offers an electrical pretreatment of the carbon fiber, enhancing sensitivity. The observation of stable background current and sharp transients at the reversal potentials indicated suitable recording properties.

3.2.2 Nitric Oxide Microsensors

While the electrochemical measurement of O_2 can be performed using bare carbon fiber microelectrodes, for •NO detection, the active carbon surface must be modified in order to achieve sensor selectivity. One of the problematic issues associated with the electrochemical recording of •NO is the high potential (+0.9 V vs Ag/AgCl) required to drive the oxidation reaction at the electrode surface. At this working potential, many other electroactive species present in biological preparation (namely, from the nervous system, biogenic amines and indoles, ascorbic acid, and NO_2^-) can become interferents. One strategy used to increase selectivity is the chemical modification of the active carbon surface of the electrode. Here, the active surface of the carbon fiber is modified by coating with Nafion® and electropolymerization of *o*-PD. Nafion® is a sulfonated tetrafluoroethylene-based fluoropolymer-copolymer acting as an anionic repellant, thus preventing access of molecules such as ascorbic acid to the active surface of the microelectrode. The polymerized *o*-PD creates an exclusion layer at the carbon fiber surface minimizing the access of interfering molecules such as ascorbic acid, dopamine, and 5-hydroxytryptamine to the carbon surface and also preventing fouling of the carbon surface [30, 40].

1. Dry the microelectrodes in an oven at 170 °C for 4 min to remove traces of humidity.
2. Dip the microelectrode tip into the Nafion® solution at room temperature for 1–2 s, and then dry at 170 °C for 4 min. The best results are obtained when two layers of Nafion® are applied (*see* **Note 5**). After coating with Nafion®, the microelectrodes can be stored at 4 °C for 1 week until usage.

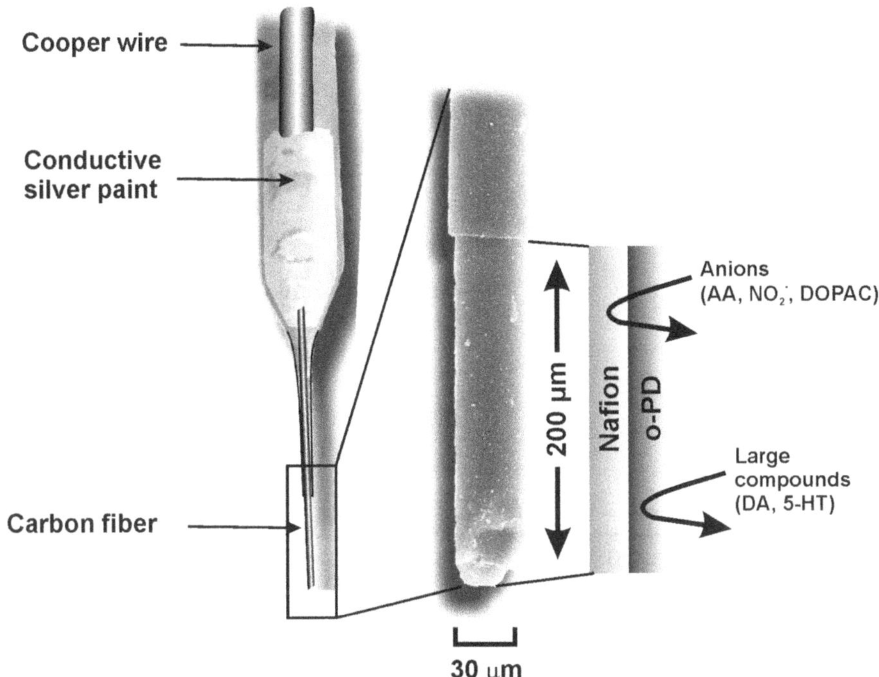

Fig. 2 Schematic representation of a carbon fiber microelectrode for •NO (•NO microsensor) showing tip detail, seal between the carbon fiber and the pulled glass capillary as well as chemical modification of carbon surface with Nafion® and o-PD which increase electrode selectivity toward negatively charged species and large compounds, respectively: AA, ascorbic acid; DA, dopamine; DOPAC, 3,4-dihydroxyphenylacetic acid; 5-HT, 5-hydroxytryptamine

3. Submerge the tip of the microelectrode coated with Nafion® in a freshly made o-PD solution. Electropolymerization is achieved by applying a constant potential +0.7 V vs Ag/AgCl for 30 min, with stirring. The microsensor (coated microelectrode) is then rinsed with deionized water and placed in a 50 mL beaker containing PBS lite. The electrodes are coated and used on the day of use and not reused (*see* **Note 6**).

Figure 2 shows a schematic representation of a carbon fiber microelectrode and surface modification details.

3.2.3 Calibration of the •NO Microsensor

1. Calibration of the •NO microsensor is performed at room temperature in a 40 mL beaker containing 20 mL deoxygenated PBS lite with gentle stirring.
2. Place •NO microsensor in PBS lite (using a micromanipulator), and polarize the electrode at +0.9 V vs Ag/AgCl.
3. Allow the oxidation current to stabilize.
4. In order to evaluate microsensor selectivity, add 250 μM of ascorbic acid prior to addition of •NO.
5. Add •NO from stock solution. We usually perform three additions of 0.5–1 μM •NO each.

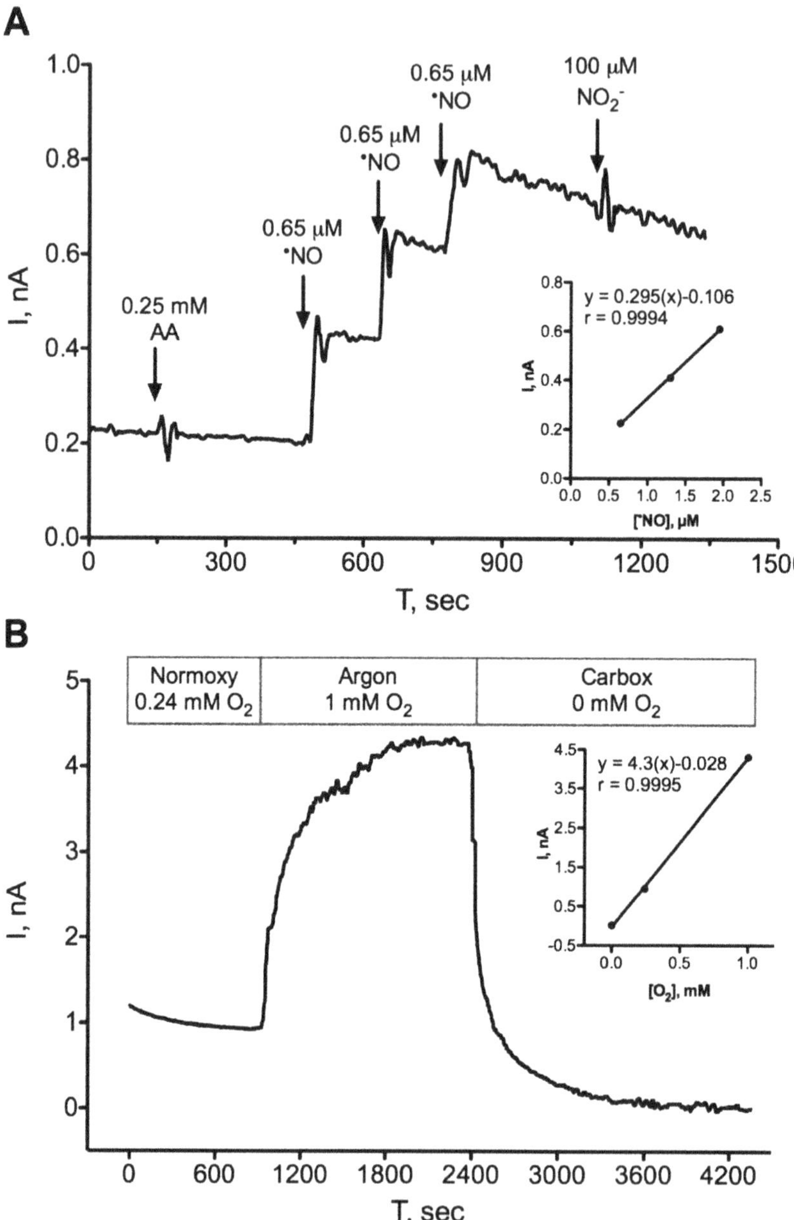

Fig. 3 (**a**) Typical amperometric recording and calibration curve (inset) of the •NO microsensor. The microelectrode selectivity is tested by adding ascorbic acid prior to •NO and NO_2^- afterward. As shown, the microsensor response to these two interferents is negligible. (**b**) Typical amperometric recording and calibration curve (inset) of an O_2 microelectrode

6. To further evaluate microsensor selectivity, add 100 μM NO_2^- after addition of •NO. Testing of other possible electroactive interferents is recommended (*see* **Note 7**).

   A typical calibration recording and calibration curve is presented in Fig. 3a.

3.2.4 Calibration of the O_2 Microelectrode

1. The O_2 microelectrode is calibrated in the perfusion chamber used for recordings in slices. This approach allows the calibrations to be performed at 32 °C, the same temperature used for slice recordings.
2. The calibration of the O_2 microelectrode requires three solutions at different O_2 concentrations. We use PBS lite saturated with argon (0 mM O_2), at atmospheric O_2 concentration (0.24 mM O_2 at 32 °C) and saturated with Carbox (1 mM O_2 at 32 °C) [41]. These solutions are preheated in the bath at 34 °C and perfused to the recording chamber, at 32 °C. The first solution used is the PBS lite at atmospheric [O_2].
3. Using a micromanipulator, submerge the microelectrode tip into the perfusion media in the recording chamber.
4. Polarize the microelectrode at −0.8 V vs Ag/AgCl pellet and allow the current to stabilize.
5. Once a stable current is obtained, switch the perfusing solution to PBS lite 0 mM O_2, and once again allow current to stabilize. Finally, switch to 1 mM O_2.
   A typical calibration curve is shown in Fig. 3b.

3.3 Monitoring of the Interacting Profiles of Endogenous •NO and O_2 in Rat Hippocampal Slices

We typically use male Wistar rats with ages between 5 and 6 weeks (approx. 150 g). However, the procedure is the same for animals of other ages.

3.3.1 Preparation of Rat Hippocampal Slices

1. Prepare ice-cold aCSF (2 × 50 mL) and fill the preincubation chamber with aCSF at room temperature. Allow both solutions to saturate with Carbox by bubbling for 40 min.
2. Sacrifice the animal by cervical displacement and decapitate (*see* **Note 8**).
3. Use a pair of scissors to cut the skin and expose the cranium. Then, cut through the cranial bones along the sagittal suture, and remove the parietal skull bones to each side, exposing the brain. If older animals are used, serrated scissors may be required because cranial bone is very hard. The frontal bone can be removed with the assistance of the rongeur. Make sure the meningeal membranes are fully retracted prior to removing the brain as they can cut through the soft cerebral tissue.
4. With a spatula cut the optical nerves (with one caudal-frontal movement of the spatula inserted under the brain), and then remove the whole brain into a medium petri dish with filter paper. Add ice-cold aCSF saturated with Carbox. One can additionally bubble Carbox in the medium during dissection (*see* **Note 9**).
5. With the help of a scalpel, separate the cerebellum from the encephalon, and then cut the two hemispheres apart.

6. Flip one hemisphere in order for the cortex to be facing down and the ventral side facing up. With one pair of curve forceps, gently hold the encephalon at the frontal end; using the curved section of the second pair of curved forceps, remove the midbrain and expose the hippocampus. Gently flip the hippocampus over and out cut along the subiculum. Repeat this process with the other hemisphere.
7. Place the hippocampi on the stage of the tissue chopper (on the acetate film) with the long axis perpendicular to the blade. The tissue should dry and the blade slightly wet.
8. Cut 400 μm sections (*see* **Note 10**).
9. Transfer sectioned hippocampi into a small petri dish (transfer the acetate with the tissue into the dish), and add ice-cold aCSF saturated with Carbox.
10. Separate the slices using Pasteur pipettes with sealed end or single thread paintbrush. This is a critical step, and care should be taken not to damage the tissue.
13. Using the 1 or 5 mL cut pipette tip, transfer separated slices into the pre-recording chamber containing room temperature aCSF continuously bubbled with Carbox. Discard slices that are not intact (during dissection and separation, the inferior blade of the dentate gyrus may be loss, or cuts to the hippocampus may disrupt the tissue integrity).
14. Maintain slices in the pre-recording chamber at room temperature for at least 1 h prior to use, to allow for good tissue recovery.

3.3.2 Simultaneous Recording of •NO and O_2 Concentration Dynamics in Rat Hippocampal Slices

1. Preheat solutions perfused in the recording chamber at 34 °C to prevent formation of gas bubbles in the tubing and chamber, which cause interference in electrochemical recordings. Before placing a slice in the recording chamber, make sure that the temperature has stabilized—temperature fluctuations alter recorded oxidation and reduction currents.
2. Transfer one hippocampal slice into the recording chamber and fix it to the nylon mesh (*see* **Note 11**).
3. Place each microelectrode on the holder of a micromanipulator, and insert them into the CA1 subregion of the hippocampal slice (or other regions of interest). The microelectrodes are inserted at a 45°–60° angle, and the exposed tip is placed at a depth of 200–300 μm into the tissue (*see* **Note 12**). Figure 4 shows a schematic representation of the setup used to perform these simultaneous recordings.
4. For high temporal and spatial resolution of the simultaneous recording, the distance between the electrodes should be as small as possible (<50 μm).

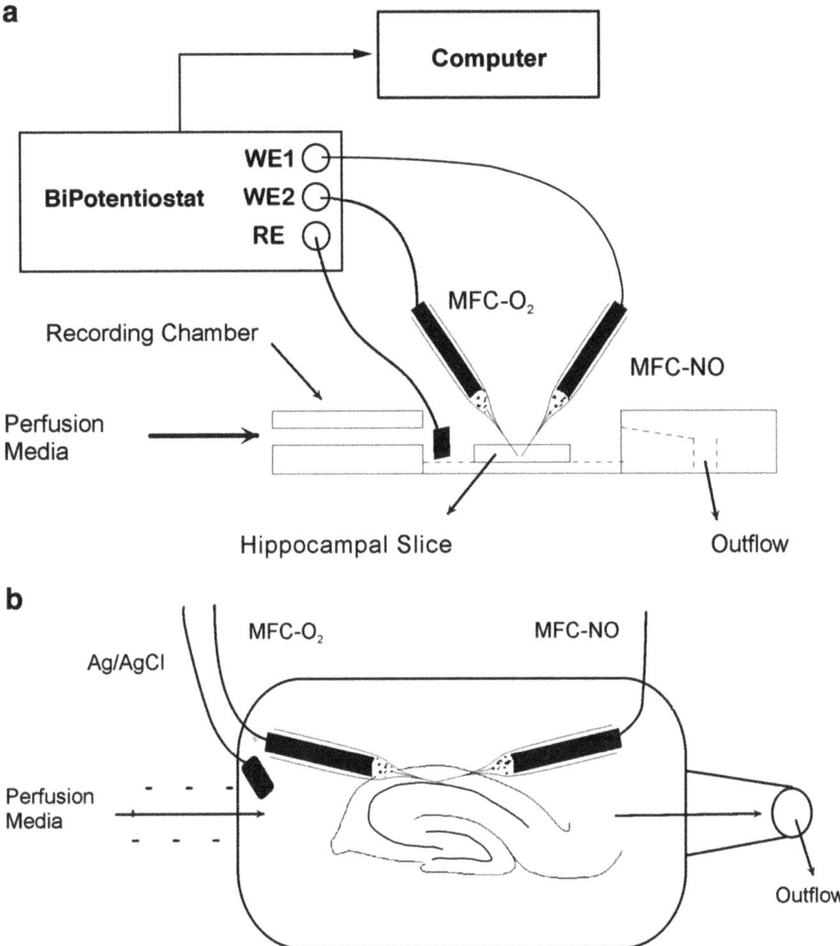

Fig. 4 Schematic representation of the setup used in the simultaneous recording of •NO and O_2 concentration dynamics in the hippocampal slice. In (**a**) a lateral view of the recording chamber top is shown, with the electrodes placed in the tissue and instrumentation used. In (**b**), a top view showing the two electrodes placed in the CA1 subregion of the hippocampal slice

5. Allow both current recordings to stabilize and perform tissue stimulation. Typically, NMDA is perfused at concentrations 10, 50, and 100 μM for 2 min. Add NMDA to Carbox saturated aCSF in a different beaker, and switch the perfusion solution using a stopcock. This stimulation protocol results in a transient increase in •NO concentration in the hippocampal slice (in particular in the CA1 subregion) and allows for the study of how different •NO concentration dynamics affects O_2 consumption in the tissue. Figure 5 shows typical recording of •NO and O_2 in the CA1 subregion of the hippocampal slice stimulated with 10, 50, and 100 μM NMDA.
6. Pharmacological modulation can be achieved either by pretreating slices or by addition to perfusion media (*see* **Note 13**).

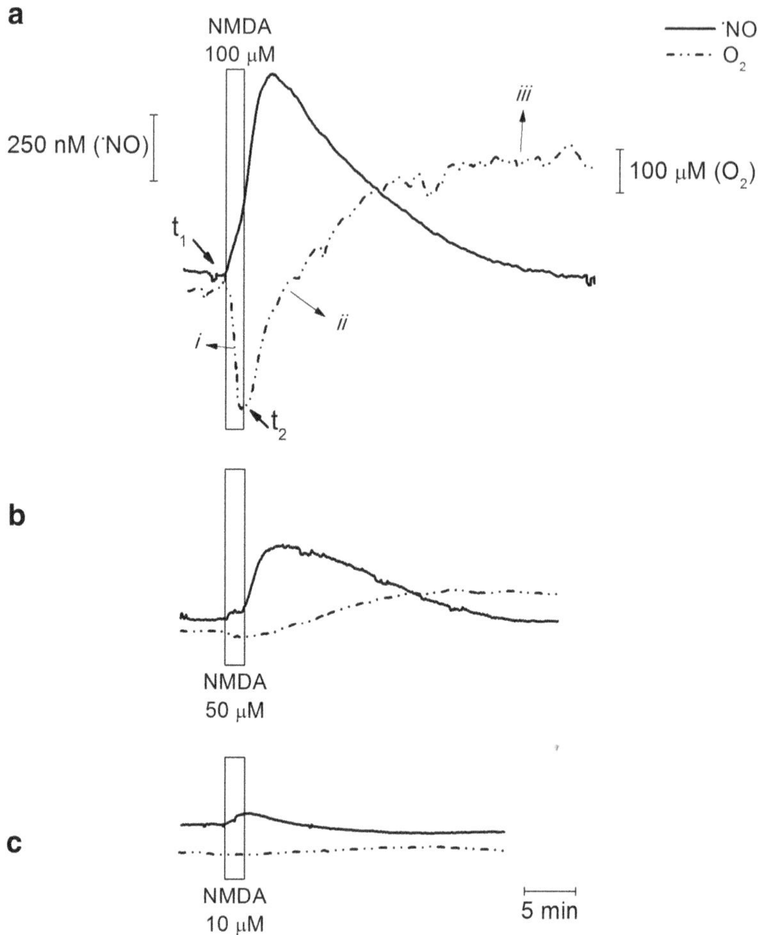

Fig. 5 Typical recording of •NO and O_2 in the CA1 subregion of the hippocampal slice challenged with NMDA 100 μM (**a**), 50 μM (**b**), and 10 μM (**c**) applied for a 2-min period in perfusion (box). The solid line is the •NO oxidation current and dashed line is the O_2 reduction current. Stimulation of the tissue with NMDA induces concentration-dependent increase in •NO. The evolution of the O_2 profile is more complex: activation of the neuronal tissue induces an initial increase in O_2 consumption (*i*) which is followed by a decrease in consumption (*ii*) until a new steady state is reached (*iii*). The analysis of these and other simultaneous recordings (for more information, *see* ref. 18) reveals that the maximal variation in [O_2] is directly related to the peak [•NO] and that a threshold concentration of •NO is required for inhibition of O_2 consumption to be observed. Taken together, the results indicate that NMDA-evoked •NO production inhibits O_2 consumption in the CA1 subregion of the hippocampal slice in a concentration-dependent fashion (Reproduced from ref. 18 with permission from Elsevier)

3.4 Concluding Remarks

The approach here described allows for the study of •NO regulation of cellular respiration in acute hippocampal slices, a tissue model closer to the in vivo setting, thus offering several advantages over simpler systems such as mitochondrial suspensions and cell cultures [18]. The simultaneous, direct, and real-time recording of •NO and O_2 concentration profiles in a biological preparation which retains high cytoarchitectural integrity is of great interest in gaining deeper insight of the mechanism by which endogenous •NO may regulate mitochondrial respiration in vivo. Among other advantages of using tissue preparations such as the hippocampal slice is the fact that under these conditions both compounds diffuse in the tissue in a fashion similar to in vivo and pO_2 in the tissue core is within physiological range [32]. This last aspect is critical, as the high pO_2 observed in conditions of cell culture and isolated mitochondria preparations can detour •NO bioactivity by facilitating production of reactive oxygen and nitrogen species.

The approach described here may be applied not only for the study of mitochondrial metabolism but also in the context of mitochondrial pharmacological studies. Understanding how different •NO concentration dynamics impact mitochondrial respiration via the reversible inhibition of CcO, the irreversible inhibition of other mitochondrial complexes by peroxynitrite or even mitochondrial production of reactive oxygen species with signaling properties (such as H_2O_2) is interesting not only from the standpoint of fundamental research but may also be useful in pharmacological studies and even in disease models where mitochondrial dysfunction is likely to occur.

4 Notes

1. For the isolation of rat hippocampal slices, ice chips can be allowed to form by placing aCFS at −20 °C for a short period. The low temperature is essential in order to guarantee higher viability of tissue during isolation process.
2. The acetone serves two purposes: firstly, it facilitates the insertion of the fiber into the capillary (this is essential when using the 10 μm fiber, otherwise it will not slide into the capillary); secondly, it cleans the carbon fiber of residues.
3. Pulling of the capillary/fiber array is a critical step in the fabrication of carbon fiber microelectrodes. The strength and heat of the pulling must be adjusted in order to obtain a tight seal between the capillary and the carbon fiber. Also, the pulled portion of the glass should not be excessively long in order to avoid bending the microelectrode tip.
4. The carbon fibers are cut at different lengths in accordance to the purpose of use. In the case of •NO microsensors, the fiber is cut longer to increase recording surface and thus sensitivity. For O_2 microelectrodes, the fiber is cut short to increase spatial resolution of the recording.
5. Lower temperature curing produces thicker films compared to higher temperature curing [42], so it is important to guarantee that Nafion® is dried at 170 °C. The application of Nafion®,

while increasing electrode selectivity, will also decrease sensitivity and response time. As such, a compromise must be reached between these properties. Others and we have established that the application of two layers of Nafion® results in good selectivity without significantly compromising sensitivity and response time [30, 31, 40].

6. After the Nafion® film has been hydrated, it must not be allowed to dry. Thus, after polymerization of *o*-PD, the electrode must always remain with the tip immersed in PBS lite.
7. Several chemical species found in the brain extracellular fluid are electroactive and can be oxidized at the electrode surface at the working potential used to record •NO (+0.9 V vs Ag/AgCl). Among others, ascorbate is a major interferent as extracellular concentrations in the brain tissue are around 250–500 μM [43–45]. Other putative interferents include catecholamine and indolamine neurotransmitters and their metabolites, urate, and hydrogen peroxide. Coating of the carbon fiber microelectrode with Nafion® and o-PD is meant to decrease sensitivity toward such interferents. The relative abundance of each chemical entity varies as a function of the brain region.
8. Animals that are older than 5–6 weeks and over 150 g may need to be anesthetized prior to cervical displacement, in accordance to the animal experimentation regulation observed in each Laboratory.
9. It is critical to reduce to a minimum the time between the death of the animal and having the brain submerged in ice-cold aCSF. Ideally, this procedure should last no longer than 1 min. During this period, the nervous tissue is not being oxygenated, and elevated temperature leads to high metabolic activity.
10. Using sections at 400 μm of thickness guarantees that the core of the tissue is healthy and sufficiently oxygenated during recordings at 32 °C. During the sectioning process, about 100 μm of tissue on each side of the slice is damaged [37, 38].
11. Holding the slices in the nylon mesh may be a difficult task. It is useful to gently scratch the nylon mesh with a needle. To fix the slice the fimbria can be crushed against the mesh.
12. It is important to perform recordings in the tissue core as the upper and lower 100 μm of tissue are damaged due to slice preparation. Also, at the tissue core, the basal [O_2] is at physiological levels [32, 38, 46].
13. Any compound applied to the perfusion media must be tested for interference at the microelectrodes.

Acknowledgments

This work was funded by FEDER via the COMPETE 2020 Program and National Funds via FCT (Fundação para a Ciência e Tecnologia, Portugal) through the strategical project POCI-01-0145-FEDER-007440.

References

1. Laranjinha J, Santos RM, Lourenço CF, Ledo A, Barbosa RM (2012) Nitric oxide signaling in the brain: translation of dynamics into respiration control and neurovascular coupling. Ann N Y Acad Sci 1259(1):10–18
2. Santos RM, Lourenco CF, Gerhardt GA, Cadenas E, Laranjinha J, Barbosa RM (2011) Evidence for a pathway that facilitates nitric oxide diffusion in the brain. Neurochem Int 59(1):90–96. https://doi.org/10.1016/j.neuint.2011.05.016
3. Moncada S, Higgs A (1993) The L-arginine-nitric oxide pathway. N Engl J Med 329:2002–2012
4. Poulos TL (2006) Soluble guanylate cyclase. Curr Opin Struct Biol 16(6):736–743. https://doi.org/10.1016/j.sbi.2006.09.006
5. Kleppisch T, Feil R (2009) cGMP signalling in the mammalian brain: role in synaptic plasticity and behaviour. Handb Exp Pharmacol 191:549–579. https://doi.org/10.1007/978-3-540-68964-5_24
6. Moncada S, Higgs EA (2006) Nitric oxide and the vascular endothelium. Handb Exp Pharmacol 176(Pt 1):213–254
7. Lourenco CF, Ledo A, Barbosa RM, Laranjinha J (2017) Neurovascular-neuroenergetic coupling axis in the brain: master regulation by nitric oxide and consequences in aging and neurodegeneration. Free Radic Biol Med 108:668–682. https://doi.org/10.1016/j.freeradbiomed.2017.04.026
8. Poderoso JJ, Carreras MC, Lisdero C, Riobo N, Schopfer F, Boveris A (1996) Nitric oxide inhibits electron transfer and increases superoxide radical production in rat heart mitochondria and submitochondrial particles. Arch Biochem Biophys 328(1):85–92
9. Sarti P, Arese M, Forte E, Giuffre A, Mastronicola D (2012) Mitochondria and nitric oxide: chemistry and pathophysiology. Adv Exp Med Biol 942:75–92. https://doi.org/10.1007/978-94-007-2869-1_4
10. Bolanos JP, Peuchen S, Heales SJ, Land JM, Clark JB (1994) Nitric oxide-mediated inhibition of the mitochondrial respiratory chain in cultured astrocytes. J Neurochem 63:910–916
11. Brown GC, Cooper CE (1994) Nanomolar concentrations of nitric oxide reversibly inhibit synaptosomal respiration by competing with oxygen at cytochrome oxidase. FEBS Lett 356:295–298
12. Cleeter MW, Cooper JM, Darley-Usmar VM, Moncada S, Schapira AH (1994) Reversible inhibition of cytochrome c oxidase, the terminal enzyme of the mitochondrial respiratory chain, by nitric oxide: implications for neurodegenerative diseases. FEBS Lett 345:50–54
13. Schweizer M, Richter C (1994) Nitric oxide potently and reversibly deenergizes mitochondria at low oxygen tension. Biochem Biophys Res Commun 204:169–175
14. Mason MG, Nicholls P, Wilson MT, Cooper CE (2006) Nitric oxide inhibition of respiration involves both competitive (heme) and noncompetitive (copper) binding to cytochrome c oxidase. Proc Natl Acad Sci U S A 103(3):708–713. https://doi.org/10.1073/pnas.0506562103
15. Antunes F, Boveris A, Cadenas E (2004) On the mechanism and biology of cytochrome oxidase inhibition by nitric oxide. Proc Natl Acad Sci U S A 101:16774–16779
16. Antunes F, Cadenas E (2007) The mechanism of cytochrome C oxidase inhibition by nitric oxide. Front Biosci 12:975–985
17. Cooper CE, Mason MG, Nicholls P (2008) A dynamic model of nitric oxide inhibition of mitochondrial cytochrome c oxidase. Biochim Biophys Acta 1777:867–876
18. Ledo A, Barbosa R, Cadenas E, Laranjinha J (2010) Dynamic and interacting profiles of •NO and O2 in rat hippocampal slices. Free Radic Biol Med 48(8):1044–1050. https://doi.org/10.1016/j.freeradbiomed.2010.01.024
19. Jarrard LE (1995) What does the hippocampus really do? Behav Brain Res 71(1–2):1–10

20. Garthwaite J, Garthwaite G, Palmer RM, Moncada S (1989) NMDA receptor activation induces nitric oxide synthesis from arginine in rat brain slices. Eur J Pharmacol 172 (4–5):413–416
21. Prast H, Philippu A (2001) Nitric oxide as modulator of neuronal function. Prog Neurobiol 64(1):51–68
22. Hopper RA, Garthwaite J (2006) Tonic and phasic nitric oxide signals in hippocampal long-term potentiation. J Neurosci 26 (45):11513–11521. https://doi.org/10.1523/JNEUROSCI.2259-06.2006
23. Laranjinha J, Ledo A (2007) Coordination of physiologic and toxic pathways in hippocampus by nitric oxide and mitochondria. Front Biosci 12:1094–1106
24. Zhuo M, Hawkins RD (1995) Long-term depression: a learning-related type of synaptic plasticity in the mammalian central nervous system. Rev Neurosci 6:259–277
25. Zorumski CF, Izumi Y (1998) Modulation of LTP induction by NMDA receptor activation and nitric oxide release. Prog Brain Res 118:173–182
26. Amaral DG, Witter MP (1989) The three-dimensional organization of the hippocampal formation: a review of anatomical data. Neuroscience 31(3):571–591
27. Anderson P, Bliss TV, Skrede KK (1971) Lamellar organization of hippocampal pathways. Exp Brain Res 13(2):222–238
28. Bliss TV, Collingridge GL (1993) A synaptic model of memory: long-term potentiation in the hippocampus. Nature 361(6407):31–39. https://doi.org/10.1038/361031a0
29. Gerhardt GA, Oke AF, Nagy G, Moghaddam B, Adams RN (1984) Nafion-coated electrodes with high selectivity for CNS electrochemistry. Brain Res 290 (2):390–395
30. Santos RM, Lourenco CF, Piedade AP, Andrews R, Pomerleau F, Huettl P, Gerhardt GA, Laranjinha J, Barbosa RM (2008) A comparative study of carbon fiber-based microelectrodes for the measurement of nitric oxide in brain tissue. Biosens Bioelectron 24 (4):704–709. https://doi.org/10.1016/j.bios.2008.06.034
31. Ferreira NR, Ledo A, Frade JG, Gerhardt GA, Laranjinha J, Barbosa RM (2005) Electrochemical measurement of endogenously produced nitric oxide in brain slices using Nafion/o-phenylenediamine modified carbon fiber microelectrodes. Anal Chim Acta 535 (1–2):1–7
32. Ledo A, Barbosa RM, Gerhardt GA, Cadenas E, Laranjinha J (2005) Concentration dynamics of nitric oxide in rat hippocampal subregions evoked by stimulation of the NMDA glutamate receptor. Proc Natl Acad Sci U S A 102(48):17483–17488. https://doi.org/10.1073/pnas.0503624102
33. Dias C, Lourenço CF, Ferreiro E, Barbosa RM, Laranjinha J, Ledo A (2016) Age-dependent changes in the glutamate-nitric oxide pathway in the hippocampus of the triple transgenic model of Alzheimer's disease: implications for neurometabolic regulation. Neurobiol Aging 46:84–95. https://doi.org/10.1016/j.neurobiolaging.2016.06.012
34. Ledo A, Lourenco CF, Caetano M, Barbosa RM, Laranjinha J (2015) Age-associated changes of nitric oxide concentration dynamics in the central nervous system of fisher 344 rats. Cell Mol Neurobiol 35(1):33–44. https://doi.org/10.1007/s10571-014-0115-0
35. Lourenço CF, Santos R, Barbosa RM, Gerhardt G, Cadenas E, Laranjinha J (2011) In vivo modulation of nitric oxide concentration dynamics upon glutamatergic neuronal activation in the hippocampus. Hippocampus 21(6):622–630
36. Ledo A, Lourenco CF, Laranjinha J, Brett CM, Gerhardt GA, Barbosa RM (2017) Ceramic-based multisite platinum microelectrode arrays: morphological characteristics and electrochemical performance for extracellular oxygen measurements in brain tissue. Anal Chem 89 (3):1674–1683. https://doi.org/10.1021/acs.analchem.6b03772
37. Jiang C, Agulian S, Haddad GG (1991) O2 tension in adult and neonatal brain slices under several experimental conditions. Brain Res 568(1–2):159–164
38. Mulkey DK, Henderson RA 3rd, Olson JE, Putnam RW, Dean JB (2001) Oxygen measurements in brain stem slices exposed to normobaric hyperoxia and hyperbaric oxygen. J Appl Physiol 90(5):1887–1899
39. Barbosa RM, Lopes Jesus AJ, Santos RM, Pereira CL, Marques CF, Rocha BS, Ferreira NR, Ledo A, Laranjinha J (2011) Preparation, standardization and measurement of nitric oxide solutions. Global J Anal Chem 2(6):272–284
40. Friedemann MN, Robinson SW, Gerhardt GA (1996) o-Phenylenediamine-modified carbon fiber electrodes for the detection of nitric oxide. Anal Chem 68(15):2621–2628
41. Sander R (2015) Compilation of Henry's law constants (version 4.0) for water as solvent. Atmos Chem Phys 15(8):4399–4981. https://doi.org/10.5194/acp-15-4399-2015

42. Gerhardt GA, Hoffman AF (2001) Effects of recording media composition on the responses of Nafion-coated carbon fiber microelectrodes measured using high-speed chronoamperometry. J Neurosci Methods 109(1):13–21
43. Rice ME (2000) Ascorbate regulation and its neuroprotective role in the brain. Trends Neurosci 23(5):209–216. https://doi.org/10.1016/S0166-2236(99)01543-X
44. Ferreira NR, Santos RM, Laranjinha J, Barbosa RM (2013) Real time in vivo measurement of ascorbate in the brain using carbon nanotube-modified microelectrodes. Electroanalysis 25 (7):1757–1763. https://doi.org/10.1002/elan.201300053
45. Ferreira NR, Lourenço CF, Barbosa RM, Laranjinha J (2015) Coupling of ascorbate and nitric oxide dynamics in vivo in the rat hippocampus upon glutamatergic neuronal stimulation: a novel functional interplay. Brain Res Bull 114:13–19. https://doi.org/10.1016/j.brainresbull.2015.03.002
46. Erecińska M, Silver IA (2001) Tissue oxygen tension and brain sensitivity to hypoxia. Respir Physiol 128(3):263–276

Chapter 6

Mitochondrial Membrane Potential ($\Delta\Psi$) Fluctuations Associated with the Metabolic States of Mitochondria

João Soeiro Teodoro, Carlos Marques Palmeira, and Anabela Pinto Rolo

Abstract

The proton electrochemical gradient generated by the respiratory chain activity accounts for over 90% of the available respiratory energy, and, as such, its evaluation and accurate measurement regarding total values and fluctuations are an invaluable component of the understanding of mitochondrial function. Consequently, alterations in electric potential across the inner mitochondrial membrane generated by differential protonic accumulation and transport is known as the mitochondrial membrane potential, or $\Delta\Psi$, and is reflective of the functional metabolic status of mitochondria. There are several experimental approaches to measure $\Delta\Psi$, ranging from fluorometric evaluations to electrochemical probes. Here, we will expose a particular method for $\Delta\Psi$ evaluation, which is dependent on the movement of a particular ion, tetraphenylphosphonium (TPP^+) with a selective electrode. The evaluation of the accumulation and movements of TPP^+ across the inner mitochondrial membrane is a sensitive, immediate, accurate, and simple method of evaluation of $\Delta\Psi$ in isolated, respiring mitochondria.

Key words TPP^+-selective electrode, Membrane potential, Mitochondria, Metabolic states

1 Introduction

The inner mitochondrial membrane transduces energy through oxidative phosphorylation. This is the main process responsible for the production of ATP in eukaryotic cells [1]. The whole sequence of biochemical events and reactions leading to ATP synthesis has been known since Mitchell proposed the chemiosmotic theory [2]. It elegantly described how mitochondrial respiration creates an electrochemical gradient of protons across the inner mitochondrial membrane, which in turn is the driving force behind ATP synthesis through the activity of the mitochondrial F1-F0 ATP synthase. However, the way how oxidative phosphorylation is regulated is yet to be fully understood, for there are virtually countless effectors and modulators that act directly or indirectly on this system, altering its activity in response to the variety of alterations in cellular homeostasis.

Carlos M. Palmeira and António J. Moreno (eds.), *Mitochondrial Bioenergetics: Methods and Protocols*, Methods in Molecular Biology, vol. 1782, https://doi.org/10.1007/978-1-4939-7831-1_6,

Besides ATP generation, other cellular activities are dependent on mitochondria and on the ATP they generate. Examples relating directly to mitochondria include electrophoretic or protonophoric transport of ions and metabolic substrates or protein activity and transport, which are directly supported by this primary form of energy, the electrochemical proton gradient. Interference with the generation of this gradient or its induced dissipation directly alters mitochondrial function and its capacity to serve the key player in cellular bioenergetics. The initial event of energy conservation is charge separation at the mitochondrial inner membrane. The electrochemical proton gradient is generated by means of electrogenic pumping of protons, from the mitochondrial matrix to the intermembrane space. This pumping is catalyzed by the activity of the respiratory chain complexes (Fig. 1). Electrons deriving from the oxidation of substrates are funneled through the redox carriers of the respiratory chain [3], with protons being ejected toward the intermembrane space at the level of complexes I (NADH: ubiquinone reductase), III (ubiquinol: cytochrome *c* reductase), and IV (cytochrome *c* oxidase). Electrons transported through the respiratory chain are transferred to the final electron acceptor, molecular oxygen, which, upon receiving four electrons, binds protons to

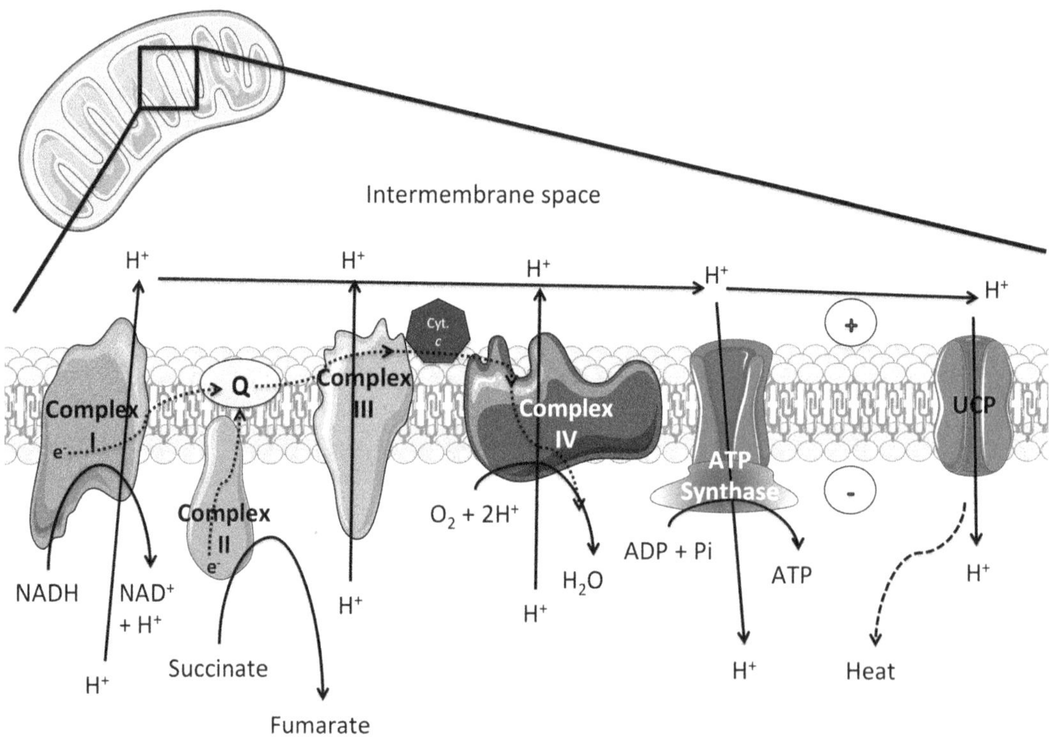

Fig. 1 Mitochondrial electron transport chain and dissipation of the proton gradient by uncoupling proteins, converting electrochemical energy into heat. *Q*, ubiquinone

form a molecule of water. The amplitude of the electrochemical proton gradient, which is known as the respiratory control, regulates the overall rate of electron transport through the respiratory chain. Complex II (succinate dehydrogenase) is a membrane-bound enzymatic complex that is part of both the citric acid cycle, or Krebs cycle, and of the respiratory chain. The dehydrogenation of succinate to fumarate generates the reduction of FAD^+ to $FADH_2$, which reduces ubiquinone to ubiquinol. Similarly, the transport of electrons from complexes I to III requires the reduction of ubiquinone, but this time the electrons are being provided by NADH, through the activity of complex I. Electronic transfer occurs from complex I (from NADH) or II (from succinate) to III and from III to IV and from here to O_2. Ubiquinone is a mobile electron carrier, dissolved in the lipid phase of the inner mitochondrial membrane and is responsible (through a cycle of reduction/oxidation) by shuttling electrons from complexes I and II toward III [4]. Distinctively, cytochrome *c* is a mobile protein attached to the cytosolic leaflet of the inner mitochondrial membrane and serves as the connection between complexes III and IV, transporting electrons between these.

Complex I is formed by more than 40 individual polypeptides, seven of which are encoded in the mitochondrial genome. It contains a prosthetic flavin mononucleotide and six Fe-S centers, as well as a binding site for ubiquinone, which receives NADH; electron transfer at this stage results in the formation of the transient semiquinone radical, which can be fully reduced to ubiquinol by further electron transfer. Rotenone is a well-known complex I inhibitor, a lipophilic pesticide that binds with tremendous affinity to complex I, hindering its catalytic activity by preventing the transfer of electrons toward ubiquinone [5]. Since complex I is the main entry point of electrons into the respiratory chain, its inhibition results in the blockade of most of the oxidative metabolic reactions within mitochondria.

Complex III is comprised of 11 polypeptides, from which only one is encoded in the mitochondrial genome. The binding site of ubiquinol where it becomes oxidized can be inhibited by myxothiazol, preventing electron transfer from ubiquinol toward a Rieske Fe-S protein. On the other end of the complex, the prevention of electron transfer toward cytochrome *c* can be prevented by antimycin A. Cytochrome *b*, the Rieske Fe-S protein, and the cytochrome *c1* are the main polypeptides that anchor the redox centers.

Complex IV, which contains three subunits encoded by the mitochondrial genome, can be inhibited by cyanide. Nitric oxide also inhibits complex IV by directly competing with O_2, but in a reversible fashion, since it is here that over 90% of the cell's oxygen consumption takes place. This final reduction of O_2 occurs in two steps: first, there is the formation, in the enzyme's active site, of transient oxide anions (O_2^-); then, these anions react with matrix protons, leading to the formation of H_2O. This way, the formation of the superoxide radical anion ($O_2^{\cdot -}$) is greatly avoided, since it prevents the release of partially reduced oxygen species, due to the high affinity of the complex IV for these species. However, despite all this, 1–5% of all consumed oxygen still ends up in superoxide; this formation can be greatly increased by the inhibition of complexes III or IV. The reduction of O_2 to H_2O also induces matrix alkalization, promoting transmembrane electrochemical proton gradient stability. This gradient is present across the inner mitochondrial membrane, forms the aptly named protonmotive force (Δp), and is comprised of the electrical membrane potential ($\Delta \Psi$) and the pH gradient (ΔpH). This gradient has a magnitude of roughly −220 mV, and under normal physiological conditions, most of it is in the form of the $\Delta \Psi$ [6]. Since the matrix side of the inner mitochondrial membrane is negatively charged and slightly alkaline, mitochondria can accumulate huge quantities of positively charged lipophilic compounds and some acids.

Complex V or ATP synthase uses the electrochemical proton gradient as the motor to phosphorylate ADP to ATP with the use of phosphate ions [2, 7]. It consists of two main assemblies, both with a variety of polypeptide subunits (from which only two are encoded in the mitochondrial genome). The extrinsic, matrix-side F_1 subunit is the catalytic part of the complex, while the F_o transmembrane portion is a simple proton channel, directly inhibited by oligomycin. The movement of protons across the F_o portion in favor of the gradient drives the activity of the F_1 portion, activating it and causing ADP phosphorylation. As expected, both the ATP/ADP ratio (phosphorylation potential) and the NADH/NAD^+ ratio (redox potential) regulate the rate of mitochondrial ATP synthesis [8]. Given the right conditions (e.g., severe decrease of $\Delta \Psi$), complex V can also operate in reverse, as proton-translocating ATPase.

1.1 Estimation of $\Delta \Psi$

Since TPP^+ freely crosses the inner mitochondrial membrane and accumulates accordingly to the membrane potential (due to charge displacement), the $\Delta \Psi$ can be inferred from the Nernst equation, at 25 °C:

$$\Delta \Psi \ (\text{mV}) = -59 \log \frac{[\text{TPP}^+]_{\text{in}}}{[\text{TPP}^+]_{\text{out}}},$$

where $[\text{TPP}^+]_{\text{in}}$ and $[\text{TPP}^+]_{\text{out}}$, respectively, represent the concentrations of this ion in the matrix and the medium (Fig. 2). Possible differences in activity coefficients are negligible. It is typically assumed that the mitochondrial matrix has a volume of 1.1 μL/mg protein.

This equation can be further decomposed (once again, at 25 °C) into:

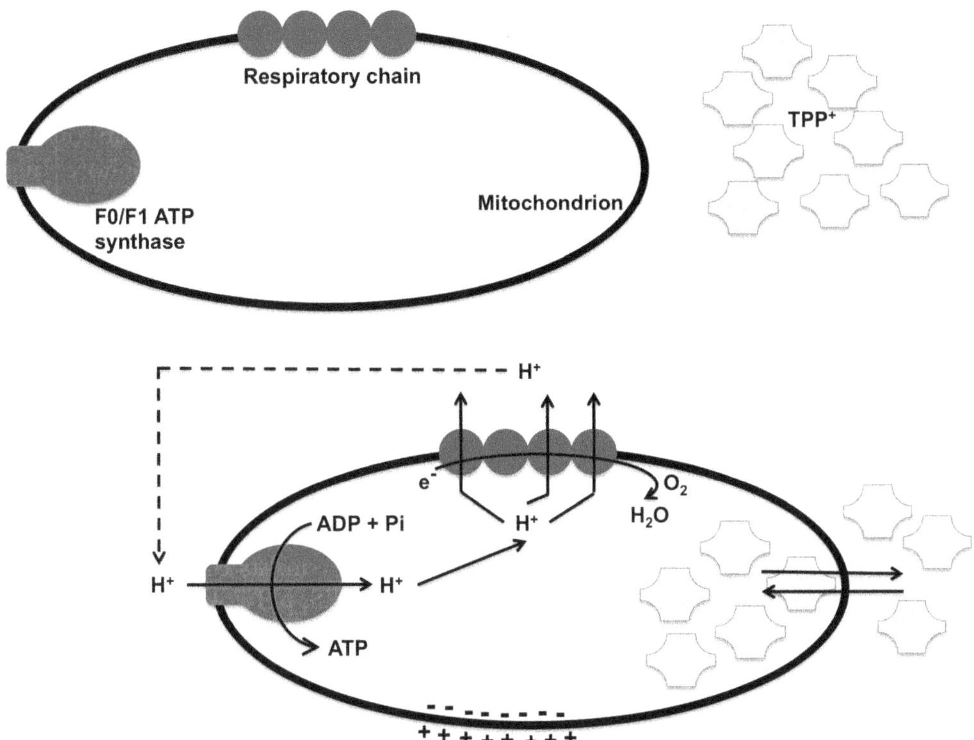

Fig. 2 Schematic representation of TPP^+ distribution across the mitochondrial membrane, according to the Nernst equation

$$\Delta\Psi\ (\text{mV}) = 59\log(v/V) - 59\log\left(10^{\Delta E/59} - 1\right),$$

where v, V, and ΔE are mitochondrial volume, reaction volume, and deflection of the electrode potential away from the baseline, respectively. This way, passive binding of TPP^+ to the membrane is disregarded, which might lead to an overestimation of the $\Delta\Psi$. However, this is a very small effect and could only be a problem if the total $\Delta\Psi$ was around −90 mV; since typical respiring, coupled mitochondria have a $\Delta\Psi$ of roughly −200 mV/mg protein; this is a negligible effect.

Alterations in the membrane potential when mitochondria go from "state a" to "state b" are given by:

$$\Delta\Psi/59 = \log\left(10^{\Delta Ea/59} - 1\right) - \log\left(10^{\Delta Eb/59} - 1\right),$$

which demonstrates that it is mandatory to measure ΔE"a" and ΔE"b" and not just the difference between E"a" and E"b."

TPP^+ should be added to the reaction medium at a concentration of 3 μM to achieve maximum sensitivity and, at the same time, prevent any possible toxic effects of this ion to mitochondria [9].

2 Materials

2.1 Reagents and Buffers

Tetraphenylphosphonium, tetraphenylboron, diisooctyl phthalate, substrates (glutamate/malate and succinate), and inhibitors of the respiratory chain (rotenone, potassium cyanide) ionophores (nigericin, FCCP), inhibitors of the TCA cycle (salicylate, Br-succinimide), inhibitor of the adenine nucleotide translocator (carboxyatractyloside), inhibitor of the calcium uniporter (ruthenium red).

Reaction medium (Medium C) consists of 130 mM sucrose, 50 mM KCl, 5 mM $MgCl_2$, 5 mM KH_2PO_4, and 10 mM HEPES, pH 7.4.

2.2 Isolation of Mitochondrial Fraction Solutions

Medium A (homogenization medium) consists of 225 mM mannitol, 75 mM sucrose, 0.5 mM EGTA, 0.5 mM EDTA, 0.1% fatty acid-free BSA, and 10 mM HEPES, pH 7.4.

Medium B (washing medium) consists of 225 mM mannitol, 75 mM sucrose, and 10 mM HEPES, pH 7.4.

2.3 Preparation of the TPP^+ Electrodes

$\Delta\Psi$ is evaluated with a TPP^+-sensitive electrode constructed by the use of a polyvinylchloride-based membrane containing tetraphenylboron as an ion exchanger, prepared according to [10] and using a calomel electrode as the reference electrode.

The polyvinylchloride membrane is typically prepared by allowing a solution of 0.34 mg of tetraphenylboron sodium salt, 16 mg of high-molecular-weight polyvinylchloride, 57 μL of diisooctyl phthalate, and tetrahydrofuran (to a final volume of 500 μL) to evaporate on a glass plate constrained by a glass ring of 1.9 cm of diameter. This ring must be covered with a glass beaker as tetrahydrofuran evaporates overnight at room temperature. This way, a clear 0.2 mm-thick membrane is obtained.

A piece of membrane is glued with tetrahydrofuran to a polyvinylchloride tube with an inner diameter of 2 mm. Extra care should be taken, as the tetrahydrofuran could lead to the dissolution of the solid membrane through which the TPP^+ concentration is going to be sensed. Very light and careful sucking and blowing into the tube should be performed to rapidly evaporate the tetrahydrofuran. Any membrane material overlapping the tubing should be cut with a surgical blade. The complete electrode is then filled with 0.1–0.2 mL of degassed TPP^+ 10 mM, as a reference solution. Finally, a silver wire coated with AgCl is inserted to the inside of the tube, without touching the inner side of the membrane. This silver wire should be connected to a suitable electrometer (**Note A**). Prior to use, the electrode should be conditioned overnight in a 10 mM TPP^+ solution. The electromotive force is then measured between the TPP^+ electrode and a calomel electrode in the sample solution.

A good TPP^+ electrode must have a linear voltage response to log [TPP^+], with a slope of 59, at 25 °C, in accordance with the Nernst equation. This can be easily and rapidly confirmed by successive additions of 1 mM TPP^+ solution, doubling the final TPP^+ chamber concentration each time. As

$$\Delta E = 2.3\mathrm{RT}/nF\log[C_1/C_2],$$

if C_1/C_2 is 2, and $2.3\mathrm{RT}/nF$ is 59 mV, then

$$\Delta E = 17.8\,\mathrm{mV}.$$

If the recorder coupled to the system has a scale of 20 mV, each pulse of doubling TPP^+ concentration should produce a similar response, of roughly 89% of the scale (**Note B**).

3 Methods

3.1 Isolation of Rat Liver Mitochondria

1. Take one rat (around 250 g), starved overnight, and sacrifice it by cervical dislocation and decapitation. Bleed the animal.
2. Using scissors and tweezers, cut across the abdomen under the rib cage and quickly remove the liver. Place it in a beaker containing ice-cold Medium A. Remove as much as possible the adhering fat and fibrous tissue and chop the liver into thin, small pieces, exchanging the Medium A to remove virtually all the blood.
3. Add approximately 6 mL of ice-cold Medium A for each gram of chopped liver and transfer all to a previously ice-cold Potter-Elvehjem glass homogenizer with a PTFE pestle.
4. Homogenize the tissue using 3–4 up-down strokes of the pestle rotating at 300 rpm. Make sure the pestle reaches the bottom of the homogenizer at the first or second stroke.
5. Transfer the homogenate to two previously cooled centrifuge tubes; balance them and centrifuge at 4 °C, 800 × *g* for 10 min, to sediment nuclei, red cells, and other larger particles.
6. Carefully decant the supernatant (waste a small amount of supernatant to ensure that no pelleted material is transported) to new centrifuge tubes and spin them at 4 °C, 10,000 × *g* for 10 min.
7. Discard all of the supernatant, and submerge the pelleted material in Medium B. Carefully but quickly resuspend the pellet using a light paint brush or a micropipette. The mitochondria form a soft brown pellet; if a red spot is seen in its center, it should be discarded as it is contamination of red blood cells. Spin the tubes at 4 °C, 10,000 × *g* for 10 min.
8. Repeat **step** 7.

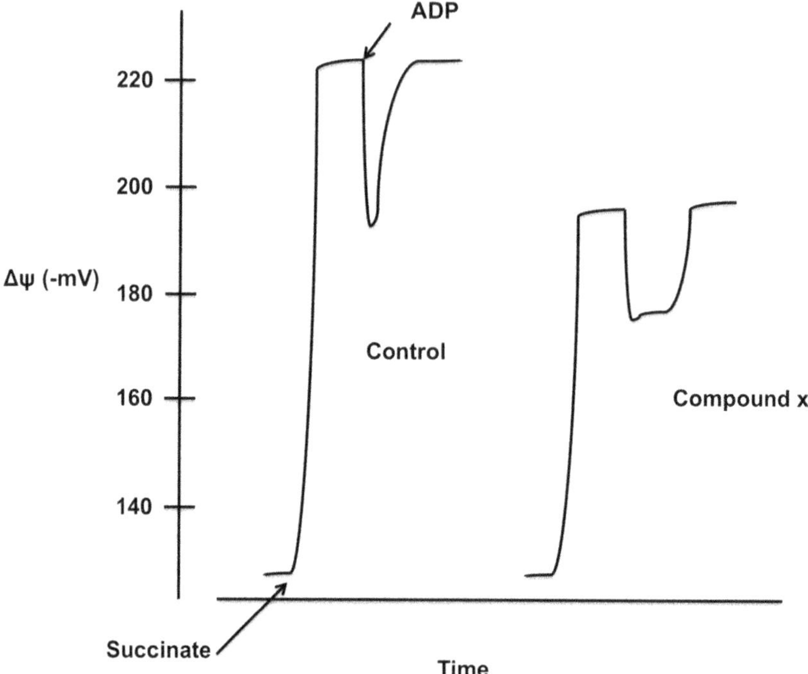

Fig. 3 Typical traces obtained with a TPP$^+$ electrode. After the addition of the compound *x*, a decrease in the initial mitochondrial membrane potential is noticeable, alongside an increase in the lag phase, i.e., the time it takes for the membrane potential stabilization to occur after ADP challenge

9. Resuspend the final pellet in roughly 1 mL of Medium B.
10. Quantify the mitochondrial protein content with a standard assay (typically, a biuret reagent method with BSA as a standard is ideal).

3.2 ΔΨ Fluctuations Associated to the Phosphorylation-Dephosphorylation Cycle

The membrane potential (ΔΨ) fluctuations (Fig. 3) are quantified in an open, thermostatized (25 °C) reaction chamber, under constant magnetic stirring (**Note C**).

3.2.1 Effect of FCCP or KCN

1. Add 1 mL of Medium C to the reaction chamber, supplemented with 5 mM succinate, 2 μM rotenone, and 3 μM TPP$^+$.
2. Set scale to 50 mV and wait for a stable signal. If using a paper recorder, set paper speed to 2 cm/min.
3. Add 1 mg of mitochondria and wait for a stable signal. Add 200 nmol ADP. The registered potential will rapidly decrease, but, after a short lag phase, it should recover to the same levels as before ADP was added. After a stable signal is once more registered, add 1 mM KCN (complex IV inhibitor) or FCCP (protonophoric uncoupler).

3.2.2 Effect of Phosphate

1. Repeat the experiment as in Subheading 3.2.1, **steps 1** and **2**, but using Medium C without KH_2PO_4. Add 1 mg mitochondria and wait for a stable signal, and then add 5 mM of KH_2PO_4.

3.2.3 Effect of Nigericin

1. Repeat the experiment as in Subheading 3.2.1, **steps 1** and **2**, but add nigericin 0.5 μg after the adding 1 mg mitochondria and waiting for a stable signal. Nigericin allows for the exchange of H^+ for K^+, and this means that the ΔpH is being converted to $\Delta\Psi$ (as noted by the small increase in recorded $\Delta\Psi$). From the degree of shift upon nigericin addition, the pH gradient can be determined. However, the ΔpH should be fairly low, at most 0.2–0.3 pH units.

3.2.4 Energization of Mitochondrial Membrane by ATP

1. Repeat the experiment as in Subheading 3.2.1, **steps 1** and **2**, but it is advisable to reduce recording speed to half and total scale voltage to 20 mV. Add 1 mg mitochondria, and then add 2 mM ATP after a stable signal is achieved. This causes TPP^+ to be further accumulated, reflecting an increase in $\Delta\Psi$. After new stabilization of the signal, add 2 μg oligomycin A and add a new dose of 2 mM ATP. In this particular experiment, rotenone must be used to prevent electronic transport reversal in complex I.

3.3 Effects of Inhibitors of the TCA Cycle on Membrane Potential

3.3.1 Effect of Salicylate or Br-Succinimide

1. Repeat the experiment as in Subheading 3.2.1, **steps 1** and **2**. Add 1 mg mitochondria, and then sequentially add 20 μM $CaCl_2$ and 1 mM acetoacetate.
2. Start the reaction by adding 1.5 μM palmitoyl-D,L-carnitine and record the fluctuation in $\Delta\Psi$. Once the signal has stabilized, add 1.5 mM salicylate or 1.5 mM Br-succinimide (inhibitors of the TCA cycle).

3.3.2 Effect of Malonyl-CoA

1. Repeat the experiment in Subheading 3.3.1, **step 1**. Add 2 mM of L-carnitine. After a stable signal is obtained, add 50 μM malonyl-CoA and record for roughly 3 more minutes.

3.3.3 Effect of CoASH

1. Repeat the experiment in Subheading 3.3.1, **step 1**. Add 1.5 μM palmitate + 25 μM CoASH. Wait for a stable signal and add 2 mM of L-carnitine. Finalize the assay with the addition of 1 mM ATP and record for roughly 3 more minutes.

3.4 Effects of Transportable Metabolites on $\Delta\Psi$

3.4.1 Effect of Carboxyatractyloside

1. Repeat the experiment in Subheading 3.2.1, **steps 1** and **2**. Add 1 mg mitochondria and wait for a stable signal.
2. Add 200 nmol ADP, and after the $\Delta\Psi$ has been restored, add 5 μM carboxyatractyloside (inhibitor of the mitochondrial adenine nucleotide translocator, ANT), followed by the addition of 200 nmol ADP. Record for an additional 3 min.

3.4.2 Effect of Ruthenium Red

1. Repeat the experiment in Subheading 3.4.1, **step 1**. Add 30 μM $CaCl_2$ and wait for signal stabilization.
2. Add 750 nM ruthenium red (inhibitor of the mitochondrial calcium uniporter), followed by 30 μM $CaCl_2$, and record for an additional 3 min.

3.4.3 Effect of Oligomycin A

1. Repeat the experiment in Subheading 3.4.1, **step 1**. Add 20 μM $CaCl_2$, followed by 1 mM acetoacetate, and finally 1.5 μM palmitate + 25 μM CoASH.
2. Add 2 mM L-carnitine, and, to finalize the assay, add 0.5 μg oligomycin A (ATP synthase inhibitor) and record for 3 more minutes.

3.4.4 Effect of Cotransport

1. Repeat the experiment in Subheading 3.4.1, **step 1**.
2. Start the reaction by adding 10 mM glutamate. Note the absence of $\Delta\Psi$ fluctuation. Add 10 mM malate and note the alteration in $\Delta\Psi$. Record for another 3 more minutes.

3.5 Effect of Oxidants on the Induction of the Mitochondrial Permeability Transition

3.5.1 Effect of NEM, DTT, or CyA

1. Repeat the experiment in Subheading 3.4.1, **step 1**, but without succinate supplementation nor adding mitochondria at this stage.
2. Sequentially add 5 mM succinate and 0.5 μg oligomycin A.
3. Start the reaction by adding 1 mg of mitochondria. Record the $\Delta\Psi$.
4. After the signal has stabilized, add 30 μM $CaCl_2$. After $\Delta\Psi$ has once more stabilized, add the oxidant (100 μM *tert*-butyl hydroperoxide or 50 μM menadione) and record for an additional 5 min.
5. When checking for protection with the reducing agents dithiothreitol (DTT, 1 mM) or *N*-ethylmaleimide (NEM, 20 μM) or with the permeability transition pore inhibitor cyclosporin A (CyA, 0.85 μM), add these compounds to the reaction medium at the beginning of the assay (at **step 2**).

4 Notes

Note A: Electrode lifetime: The parent membranes prepared in Subheading 2.3 can be stored dry for years and be used as stock. The average lifetime of the TPP^+ selective electrode is of roughly 2 months.

Note B: General rules for use of ion-specific electrodes—To obtain consistent stable electrode signals with low background noise, care should be taken in the following ways:

1. Shield all electrode connections with nonconducting material.
2. Protect the reaction chamber in a Faraday cage.

3. Maintain a constant temperature and magnetic stirring.
4. Use a medium with a reasonable ionic strength (e.g., 0.1 M KCl).
5. Maintain a sufficient pH buffer capacity.

Note C: Hydrophobic substances and reagents may cause problems because they might adhere to the PVC membrane. This can be overcomed by:

6. Mixing the hydrophobic substances with biological membrane material before bringing them into contact with the electrode.
7. Using these substances in the lowest possible amounts.
8. Washing the electrode after each experiment thoroughly with 70% ethanol or biological membrane material. Ethanol washing shortens the lifespan of the electrode by extracting some ionophore from the PVC membrane. However, this does not affect the potential or slope of the electrode, since the membrane contains about a thousand times more ionophore than needed for optimal function.

Acknowledgments

J.S.T. is a recipient of a postdoc scholarship from the Portuguese Fundação para a Ciência e a Tecnologia (SFRH/BPD/94036/2013).

References

1. Wallace DC (2009) Mitochondria, bioenergetics, and the epigenome in eukaryotic and human evolution. Cold Spring Harb Symp Quant Biol 74:383–393
2. Mitchell P (1966) Chemiosmotic coupling in oxidative and photosynthetic phosphorylation. Biol Rev 41:445–501
3. Jastroch M, Divakaruni AS, Mookerjee S, Treberg JR, Brand MD (2010) Mitochondrial proton and electron leaks. Essays Biochem 47:53–67
4. Crofts AR (2004) The Q-cycle – a personal perspective. Photosynth Res 80(1):223–243
5. Murai M, Miyoshi H (2016) Current topics on inhibitors of respiratory complex I. Biochim Biophys Acta 1857(7):884–891
6. Azzone GF, Petronilli V, Zoratti M (1984) 'Cross-talk' between redox- and ATP-driven H+ pumps. Biochem Soc Trans 12 (3):414–416
7. Gerle C (2016) On the structural possibility of pore-forming mitochondrial FoF1 ATP synthase. Biochim Biophys Acta 1857 (8):1191–1196
8. Brown GC (1992) Control of respiration and ATP synthesis in mammalian mitochondria and cells. Biochem J 284:1–13
9 Jensen BD, Gunther TE (1984) The use of tetraphenylphosphonium (TPPþ) to measure membrane potentials in mitochondria: membrane binding and respiratory effects. Biophys J 49:105–121
10. Kamo N, Muratsugu M, Hongoh R, Kobatake Y (1979) Membrane potential of mitochondria measured with an electrode sensitive to tetraphenyl phosphonium and relationship between proton electrochemical potential and phosphorylation potential in steady state. J Membr Biol 49(2):105–121

Chapter 7

Fluorescence Measurement of Mitochondrial Membrane Potential Changes in Cultured Cells

David G. Nicholls

Abstract

The mitochondrial membrane potential is the dominant component of the proton-motive force that is the potential term in the proton circuit linking electron transport to ATP synthesis and other energy-dependent mitochondrial processes. Cationic fluorescent probes have been used for many years to detect gross qualitative changes in mitochondrial membrane potentials in intact cell culture. In this chapter I describe how these fluorescence signals may be used to obtain a semiquantitative measure of changes in mitochondrial membrane potential.

Key words Mitochondria, Membrane potential, Neuron, Glutamate, Proton-motive force, Proton electrochemical potential, Tetramethylrhodamine methyl ester, Rhodamine 123, JC-1

1 Introduction

The chemiosmotic proton circuit links the mitochondrial electron transport chain to the ATP synthase and other pathways (proton leaks, nicotinamide nucleotide transhydrogenase, metabolite transporters, etc.) that utilize the proton current generated by proton-translocating complexes I, III, and IV [1]. The proton electrochemical potential of the proton gradient across the inner membrane is given (in millivolts) by the proton-motive force (Δp):

$$\Delta p = \Delta\psi - 61\Delta pH \text{ at } 37^{\circ}C$$

where Δψ is the membrane potential (i.e., the difference in electrical potential between the matrix and cytoplasm) in millivolts and ΔpH is the pH difference across the inner membrane. Under most physiological conditions, the contribution of the membrane potential to the total Δp is dominant (roughly 150 mV), and the

Electronic supplementary material: The online version of this chapter (https://doi.org/10.1007/978-1-4939-7831-1_7) contains supplementary material, which is available to authorized users.

Carlos M. Palmeira and António J. Moreno (eds.), *Mitochondrial Bioenergetics: Methods and Protocols*, Methods in Molecular Biology, vol. 1782, https://doi.org/10.1007/978-1-4939-7831-1_7,

ΔpH of typically −0.5 units contributes about 30 mV. Absolute determinations of the components of Δp in intact cells are exceedingly complex [2] and reliant on multiple assumptions, and virtually all studies not only focus on changes rather than absolute measurements but also do not consider the ΔpH component of Δp. Even with these simplifications in mind, the determination of relative changes in mitochondrial membrane potential is far from trivial.

An extremely wide range of cells and experiments are performed where mitochondrial membrane potential changes are monitored. For the purposes of this chapter, I shall focus on a single-cell type, neurons cultured from rat cerebellum, and detail how semiquantitative measurements may be made of changes in mitochondrial membrane potential ($\Delta\psi_m$) under a range of conditions using fluorescent membrane-permeant cations (so-called mitochondrial membrane potential indicators). An important proviso is that the fluorescence of these indicators is affected by changes in plasma membrane potential, and the consequences of this will be discussed.

Interpretation of the fluorescence traces is not trivial, particularly if there are indications that $\Delta\psi_p$ is changing during the experiment (following, e.g., activation of plasma membrane ion channels). For that reason we developed an Excel spreadsheet to deconvolute the traces in terms of dynamic changes in both $\Delta\psi_p$ and $\Delta\psi_m$ [3]. Access to this is described in the text. More recently the technique has been expanded to include an independent anionic indicator of plasma membrane potential in parallel with the cationic "mitochondrial" indicator [4]. The constraints of this chapter do not permit a detailed description of this additional technique, which can be accessed online in the paper and accompanying supplementary material. This semiquantitative technique has subsequently being refined to improve quantification [5]. The approach taken in this chapter is not unique; additional programs to interpret fluorescent membrane potential traces exist [6, 7].

2 Materials

Unless otherwise stated, all chemicals may be purchased from Sigma-Aldrich and fluorescent probes from Invitrogen Molecular Probes. Disposables are from VWR or similar suppliers.

2.1 Cell Culture

(It is assumed that readers are familiar with the conditions required to culture their cells of interest.)

1. *Culture chambers*—Labtech 8-well chambered coverslip, sterile (NUNC) VWR product # 43300-774. Wells are previously coated with 33 mg/ml polyethyleneimine and allowed to dry overnight.

2.2 Incubation Media

1. *Ambient CO_2*—For microscopes that do not have the ability to maintain a 5% CO_2 environment during the experiment. 3.5 mM KCl, 120 mM NaCl, 1.3 mM $CaCl_2$, 1.2 mM Mg Cl_2, 0.4 mM KH_2PO_4, 5 mM $NaHCO_3$, 1.2 mM Na_2SO_4, 15 mM D-glucose, 20 mM Na-TES, pH 7.4 at 37 °C (*see* **Notes 1** and **2**).

2.3 Fluorescent Indicators (See Note 3)

1. Tetramethylrhodamine methyl ester (TMRM perchlorate, Invitrogen, T-668) is dissolved in DMSO at a concentration 100–500 times greater than the final concentration to be used in the assay (which is typically 2–100 nM depending on mode, see below) and is protected from light. Cells in incubation medium are equilibrated with the probe for 45–60 min. In contrast to R123, the experiment is performed *without* washing away the probe.
2. Rhodamine 123 (R123, Invitrogen R-302) is dissolved in DMSO at a concentration of 2.6 mM. Cells in incubation medium are equilibrated with the probe for 15 min at 22 °C and are washed with fresh incubation medium not containing R123. In contrast to TMRM, the probe is *not* added to the experimental medium.

2.4 Fluorescence Imaging

1. For confocal imaging the minimal requirements are for an inverted microscope equipped with a 20× objective with argon and HeNe lasers capable of excitation at 488 nm (or 514 nm) and 543 nm. For single probe imaging, simple long-pass (LP) emission filters may be used. Other possibilities of course exist.
2. A computer-controlled stage is an advantage in that, if the microscope's software allows (as in the Zeiss equipped with the MultiTimer option), fields can be defined in up to eight separate wells of the Labtech chamber allowing up to eight long-term experiments to be performed in parallel.
3. Autofocus control is important in long-term experiments. With suitable attenuation the autofocus configuration can be set to detect the reflected light from the glass sample interface and then to move a predetermined distance into the sample to optimize the focal plane.
4. For short-term (<30 min) experiments, single-field time-course imaging (rather than the MultiTimer option of the Zeiss confocal) is to be preferred to allow greater time resolution.
5. Stringent confocality is not important when single-cell (rather than single mitochondrial) resolution is employed. Indeed with the 20 × 0.95 n.a. air objective employed for these studies, a sufficiently wide pinhole is used to allow a 5–10 nm depth of focus. Wide-field imaging with suitable excitation and emission filters is equally suitable for most studies.

2.5 Temperature Control

1. Accurate temperature regulation is critical for metabolic studies of membrane potentials. The best solution is to enclose the entire microscope above the focus controls within a temperature-controlled enclosure that may be obtained commercially for most microscopes. The author has constructed enclosures from acrylic (plexiglass) sheets that allow access through multiple doors to the microscope, with the focus control and the heat-generating lamp enclosure outside.
2. Temperature control is obtained with a Warner TC-344B Dual Automatic Temperature Controller with the temperature sensors located close to the stage. Since the controller has a maximum output of 18 W per channel, the output from one channel is amplified via an assembled Velleman (vellemanusa.com) DC Controlled Dimmer K8003 (now replaced by the equivalent K8064) to power a 150w vivarium ceramic heater. The second channel of the TC-344B is used simply to control three internal fans to ensure air circulation. This assembly maintains the microscope, stage, and objectives within 0.5 °C of the desired temperature.
3. If a full enclosure is not available, conventional heated stage inserts work, but particularly if oil immersion objectives are employed, it is essential to heat the objective to prevent it acting as a cold sink. This last should be done cautiously, as frequent warming and cooling of the objective can introduce internal strains in the optics.

3 Methods

In this section I describe a representative set of applications to monitor membrane potential changes in primary neuronal cultures. In order to interpret the resulting, frequently complex, fluorescence traces, it is essential to have some understanding of the basic principles that govern whole-cell fluorescence. These have been covered in publications [3, 4, 8] but will be restated here. These simple principles allowed us to formulate Excel spreadsheets that model with a surprising degree of accuracy the predicted responses of cells loaded with different concentrations of R123 or TMRM in response to simultaneous changes in $\Delta\psi_m$ and $\Delta\psi_p$ (*see* **Note 4**). Since many experimental protocols involve addition of agents that will alter $\Delta\psi_p$ (e.g., ionotropic receptor and channel activators), it is essential to be aware of the influence this will have on the fluorescence signal.

3.1 Principles

1. Lipophilic cations and anions seek to achieve a Nernstian equilibrium across both the plasma and mitochondrial membranes. The equilibrium concentration of $TMRM^+$ in the cytoplasm (*c*)

and mitochondrial matrix (m) relative to the external medium (e) is given at 37 °C by

$$[TMRM^+]_m = [TMRM^+]_c.\ 10^{\Delta\psi m/61}$$
$$[TMRM^+]_c = [TMRM^+]_e.\ 10^{-\Delta\psi p/61}$$
$$[TMRM^+]_m = [TMRM^+]_e.\ 10^{(\Delta\psi m\ -\ \Delta\psi p)/61}$$

It is therefore apparent that the accumulation of TMRM in the matrix is equally sensitive to changes in either potential.

2. While the probes are nonselectively permeable across lipid bilayer regions of both the plasma and mitochondrial membranes, the far greater surface-to-volume ratio of the mitochondrial matrix compared to the cell body means that the probes will redistribute across the former in response to a perturbation much faster than across the plasma membrane. For most practical rates of data acquisition, it can be assumed that redistribution across the inner mitochondrial membrane is instantaneous.
3. At a critical concentration in the matrix, the probes undergo aggregation. The aggregated probe is nonfluorescent, and the mitochondrial fluorescence therefore saturates. It is critically important to decide in advance whether this aggregation (or quenching) is to be exploited or avoided.

3.2 Preliminary Considerations

1. *Does your cell possess a multidrug transporter?*—Before performing any experiments, it is essential to determine from the literature or by experiment whether the cells possess an MDR that can disturb and invalidate the equilibration of the probe across the plasma membrane. R123 is a common substrate used to assay MDR activity. If activity exists it must be inhibited by addition of an MDR inhibitor that will not itself alter the metabolic processes that the experiment is designed to investigate.
2. *Quench or non-quench?*—Quench mode is a sensitive means to detect rapid changes in $\Delta\psi_m$ that occur *during* the experiment, for example, if ATP synthesis is inhibited by oligomycin or if a protonophore is added. Quench mode does *not* detect preexisting differences in $\Delta\psi_m$ between two populations (e.g., comparing two cell types) and incidentally must therefore never be used for flow cytometry. Changes in $\Delta\psi_m$ and $\Delta\psi_p$ can frequently be distinguished even with a single indicator based on the direction and rapidity of the fluorescent changes. In non-quench mode at single-cell resolution, redistribution of probe from matrix to cytoplasm does not produce a change in signal, and what is detected is the redistribution of probe across the plasma membrane to reestablish the $\Delta\psi_p$ Nernstian equilibrium. In this mode, but *not* for quench mode, it is therefore

advantageous to use the more permeant TMRM and also to include 1 μM tetraphenyl boron throughout the protocol [3]. The TPB anion accelerates the equilibration of TMRM across the plasma membrane without affecting the final equilibrium.

3. *TMRM or R123?*—The two probes are structurally related. R123 is more hydrophilic and equilibrates across the plasma membrane about 20 times more slowly than TMRM. For that reason R123 is loaded by brief exposure to a high concentration of probe, followed by washing (in contrast to TMRM which can achieve equilibrium within 60 min). This means that R123 loading is difficult to control, and in practice R123 is always used in quench mode. However because of its slower equilibration across the plasma membrane, R123 is less sensitive to $\Delta\psi_p$ changes than TMRM (*see* Fig. 2).

4. *TMRM alone or in combination with an anionic plasma membrane potential indicator (PMPI)?*—Because non-quench mode is dependent on equilibration across the plasma membrane, it follows that the single-cell fluorescence is sensitive to changes in both $\Delta\psi_p$ and $\Delta\psi_m$. If the experimenter can be confident that one of these is not changing during the experiment, this will not be a problem; otherwise it may be necessary to perform a dual-label experiment with both TMRM and the anionic probe PMPI [4, 5], the main proprietary component of the Molecular Devices "membrane potential assay kit, explorer format" (R-8042). If the microscope possesses suitable filters, this does not pose a technical problem, although it is important to adjust the concentrations of PMPI and TMRM (non-quench mode) to roughly equalize fluorescent intensities to facilitate subsequent deconvolution. This brief chapter is restricted to the use of single cationic indicators (TMRM and R123); full details of the use of PMPI and TMRM in concert are contained in [4], together with an Excel spreadsheet and a description of the underlying mathematics that may be accessed online at the *Journal of Biological Chemistry*. Reference [5] contains full details of the quantitative application of the dual probe technique.

5. *Other probe combinations*—Monitoring of $\Delta\psi_m$ can readily be combined with other functional indicators as long as sufficient spectral resolution permits. An extremely valuable combination in addition to $\Delta\psi_m + \Delta\psi_p$ is $\Delta\psi_m$ + cytoplasmic free Ca^{2+} (Fluo-4 AM, etc.) since this can be used to establish single-cell correlations between stochastic changes in both parameters [3].

3.3 Predictive Modeling of Responses

We have published two Excel spreadsheets with the dual goal of helping with both the design and interpretation of in situ mitochondrial membrane potential experiments. The first, simpler, spreadsheet [3] is relevant to the present chapter and may be obtained from the author (*see* **Note 4**). It facilitates interpretation of the fluorescence changes produced by TMRM or R123 in both quench and non-quench modes in response to changes in either membrane potential. In this model $\Delta\psi_p$ is not determined directly but is inferred from the curve-fitting traces. The second spreadsheet [4] is more complex because it is designed to interpret parallel changes in PMPI and TMRM (in non-quench mode) fluorescence. Full details of the use, calibration, and interpretation of these dual-probe experiments can be accessed in the online journal. Either spreadsheet can be used in two ways. In the *predictive* mode, expected changes in potential are fed into the program which generates a single-cell fluorescence trace approximating to the subsequent experimental trace. This is valuable for designing experiments, deciding on optimal concentrations of probe, and reinterpreting existing results in the literature. In the *analytical* mode, the experiment is first performed, and then the computer simulation is adjusted manually to provide a best fit of the experimental trace, the readout being the membrane potential time-course that produces the best fit. To reduce the degrees of freedom, a number of constants have to be determined or assumed (Subheading 3.4).

It is important to emphasize that the techniques described in this chapter do not *determine* the mitochondrial membrane potential but rather produce a semiquantitative estimate of *changes* from an initial defined membrane potential. The value for the starting potential can be taken from the very limited attempts to determine an absolute value. Alternatively this group has tended to start with an arbitrary 150 mV for $\Delta\psi_m$, since the absolute value of the potential does not affect the interpretation of the changes in potential seen during an experiment. Perhaps more importantly, a "null-point" technique can be incorporated into the experiment to determine whether the mitochondria are net generators or consumers of ATP (Subheading 3.5, **step 2**).

3.4 Spreadsheet: Single-Labeled TMRM or R123 Cells

The spreadsheet utilizes the three basic principles defined in the introduction to Subheading 3.1. A full description of the mathematical assumptions inherent in the calculations is found in the Appendix of ref. 3. The interface is shown in Fig. 1. Starting values are required for $\Delta\psi_p$ (from the appropriate electrophysiology literature) and $\Delta\psi_m$ (*see* above), the fractional volume of the cell occupied by the mitochondrial matrix, the first-order rate constant for the re-equilibration of the probe across the plasma membrane of the cell being investigated and the quench limit. Each of these may be determined experimentally as described below, or for more

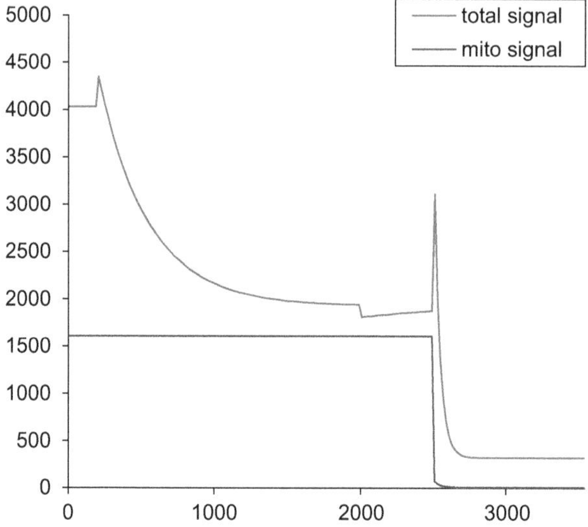

Fig. 1 Excel spreadsheet interface for the predictive and analytic modeling of the single-cell fluorescence response of TMRM or R123 to changes in $\Delta\psi_m$ or $\Delta\psi_p$. The setting of the parameters is described in the text. The red trace models the total cell fluorescence, and the black trace the mitochondrial contribution to the total signal. Values in blue may be modified to simulate a range of experiments and conditions. The spreadsheet may be obtained from the author

qualitative purposes, the values predetermined for rat cerebellar granule neurons [4] may be used as starting values.

1. *Determination of the quench limit (Q, the concentration at which probe aggregation is initiated in the mitochondrial matrix)*—Once starting values for $\Delta\psi_m$ and $\Delta\psi_p$ are decided, cells are equilibrated with a range of external TMRM concentrations and an imaging time-course is initiated for each. 2 μM FCCP is added, and if a transient "spike" in fluorescence is seen followed by a decay (see, e.g., Fig. 2), then the experiment is in quench mode. The experiment is repeated with decreasing TMRM concentrations until the *spike* is no longer seen (see ref. 4, Fig. 4). The matrix quench limit Q is then calculated from this external probe concentration $[\text{TMRM}]_e$

   $$Q = [\text{TMRM}]_e \times 10^{(\Delta\psi p\,+\,\Delta\psi m)/60}$$

2. *Determination of the plasma membrane rate constant V for TMRM*—To allow interpretation of dynamic, rather than steady-state, changes in potential, it is necessary to establish the rate constant for the equilibration of TMRM across the plasma membrane. Because of differing surface/volume ratios of cells, it is advisable to obtain this data for each cell type investigated, although semiquantitative results can be obtained using a value of V obtained in other cells allowing for altered cell body diameter. In practice, the same experiment that was

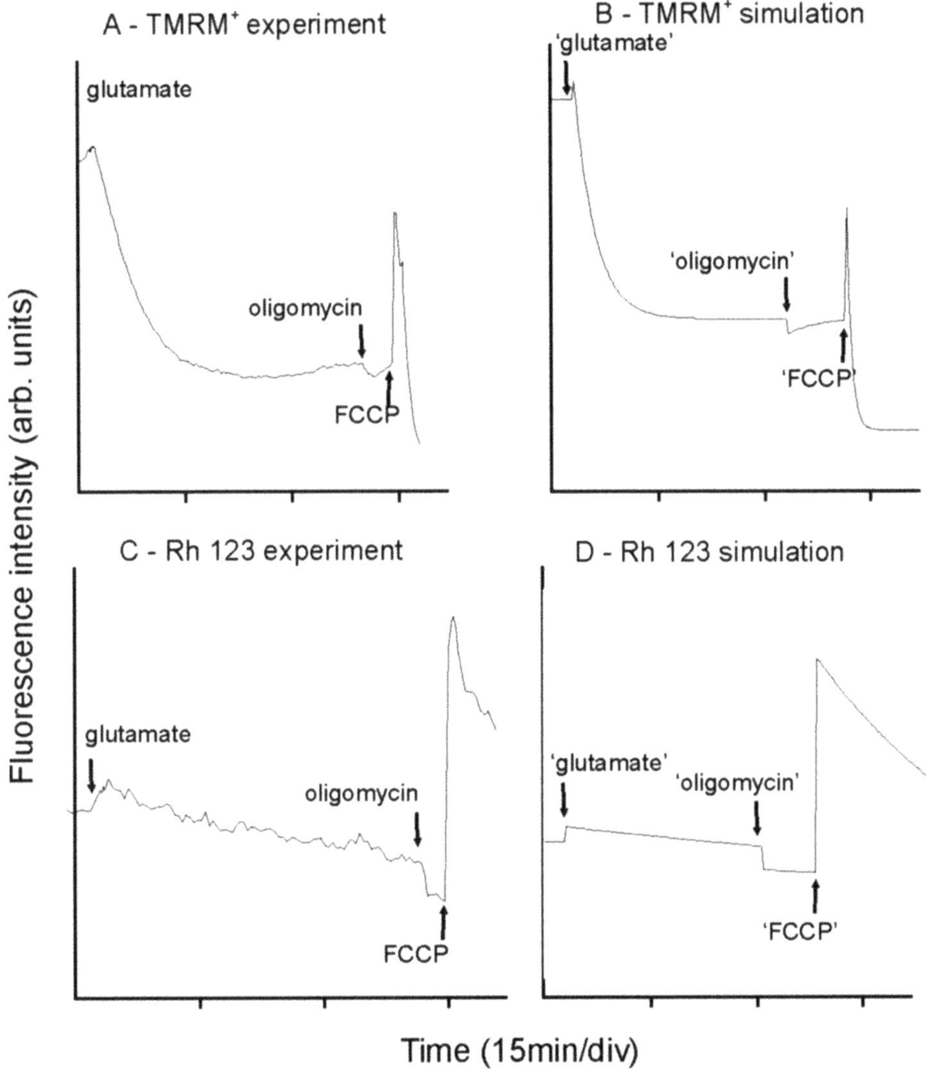

Fig. 2 Single-soma fluorescence of cerebellar granule neurons equilibrated with 50 nM TMRM (**a**) or loaded for 15 min at 22 °C with 2.6 μM R123 (**c**) were exposed to 100 μM glutamate plus 10 μM glycine. Where indicated, 2 μg/ml oligomycin and 1 μM FCCP were added. (**b**, **c**) Simulated traces fitted to the following potential changes: $\Delta\psi_p$, depolarization from −60 mV to −20 mV on addition of glutamate; $\Delta\psi_m$, depolarization from 150 mV to 145 mV on addition of glutamate, hyperpolarization to 155 mV on addition of oligomycin, and collapse of potential with FCCP. Data from [3]

used to determine Q above can be used to obtain V. Using the spreadsheet, adjust the value of the plasma membrane rate constant V until the decay rate in the simulated fluorescence matches that of the quenched experiments. Some adjustment will be necessary depending on the experimental conditions. From experience the rate constant for R123 is 10- to 20-fold slower than for TMRM [3].

3. *Fractional matrix volume*—The volume fraction x of the soma occupied by the mitochondrial matrix can be estimated by determining the residual cytoplasmic fluorescence after mitochondrial depolarization by calculating the ratio of the whole-cell $TMRM^+$ fluorescence (in non-quench mode) for a cell with depolarized mitochondria (e.g., in the presence of myxothiazol to inhibit the respiratory chain and oligomycin to block the ATP synthase) relative to the same cell with polarized mitochondria prior to the addition of inhibitors. If the fraction of the cell occupied by the mitochondrial matrices is x, then the ratio for the whole-cell fluorescence for depolarized vs. polarized mitochondria, $(\Sigma TMRM^+)_{depol}/(\Sigma TMRM^+)_{pol}$, will be given by

$$(\Sigma TMRM)_{depol}/(\Sigma TMRM)_{pol} = 1/(1 + x.10^{\Delta\psi m/60})$$

The ratio determined for cerebellar granule neurons [4] was 0.12; substituting this value into the above equation gives a value for the volume fraction x of 2.3% when $\Delta\psi_m$ is 150 mV. It must be emphasized that this assumes that there is no potential-independent binding of TMRM to components of the cell. Allowing for such binding would indicate an increased matrix volume fraction; however the only difference this would make to the simulation of the $\Delta\psi_m$ time-course would be to decrease further the small contribution of the cytoplasmic TMRM to the whole-cell signal.

3.5 Predictive Modeling of a Neuronal Experiment with the Spreadsheet

The spreadsheet is no longer available as a supplement to ref. 3 but may be obtained freely from the author (*see* **Note 4**). The sample experiment will investigate the effect of NMDA glutamate receptor activation on cultured cerebellar granule neurons and will test whether the mitochondria in the exposed cell is still generating ATP. The program can use both quench and non-quench modes and simulate TMRM and R123. The significance of this approach is that it allows changes in $\Delta\psi_m$ and $\Delta\psi_p$ to be distinguished on the basis both of the direction and the kinetics of the single-cell fluorescence response *but only in quench mode.*

1. In the sample experiment, granule neurons loaded with TMRM in quench mode are exposed to glutamate/glycine to activate their NMDA receptors. An initial slight mitochondrial depolarization results in a transient increase in fluorescence. The parallel plasma membrane depolarization produces a decrease in cell fluorescence with a much slower time-course due the rate constraints of equilibration across the plasma membrane.
2. Addition of oligomycin performs an important role: this "oligomycin null-point test" determines whether the mitochondria in the cell are synthesizing ATP at the moment the

inhibitor is added or whether, due to damage or partial uncoupling, they are hydrolyzing cytoplasmic ATP by reversal of the ATP synthase. In the former case, oligomycin will hyperpolarize the mitochondria (decreased cell fluorescence in quench mode); if they are damaged, they will further depolarize on addition of oligomycin (increased fluorescence in quench mode); see [9].

3. Finally FCCP is added to complete depolarization and confirm that the experiment was still in quench mode prior to its addition. (If there is no FCCP spike, it is best to repeat the simulation (and hence the actual experiment) at a higher TMRM concentration.)
4. Load the spreadsheet into Excel. Save a "master copy."
5. The only editable cells are in blue. The default values are for rat cerebellar granule neurons as follows: mitochondrial volume fraction 0.023 [4], external probe concentration 100 nM (for quench mode), 5 nM may be used for non-quench mode, plasma membrane rate constants 0.02 (TMRM) or 0.001 (R123), and quench limit 70 μM. The interval (to set the time range of the "experiment") is set to 20s (giving a total time-course of 1 h). The program allows for changes in $\Delta\psi_m$ and $\Delta\psi_p$ at four time points during the simulation. The default values for these neurons are set to −83 mV and −150 mV, respectively. Lines 31–219 are the iterative calculations and are not editable. The graph displays the total cell fluorescent signal and the mitochondrial fluorescence.
6. To model the protocol for determining the quench limit, set the final $\Delta\psi_m$ to zero to simulate protonophore addition and adjust the external probe concentration until the "FCCP" spike just disappears.
7. To model the effects of partial mitochondrial depolarization in quench and non-quench modes, set the final $\Delta\psi_m$ to −120 mV and compare a TMRM concentration of 100 nM and 5 nM. This shows up a very important limitation of quench mode, namely, that small changes in $\Delta\psi_m$ produce only a transient single-cell signal – the FCCP "spike" rapidly decays. Thus quench mode cannot be used to detect preexisting changes in $\Delta\psi_m$. For example, it is inappropriate for flow cytometry where $\Delta\psi_m$ in two conditions or cell types are to be compared. This is frequently not appreciated.
8. Repeat **step 7** with both 5 nM and 100 nM R123 simulated (set Q to 0.001).
9. To demonstrate the ambiguity of non-quench mode to changes in the two potentials, compare the response to $\Delta\psi_m$ depolarization to −120 mV with $\Delta\psi_p$ depolarization to −40 mV at 5 nM and 100 nM TMRM.

10. In the sample experimental protocols below for TMRM and R123, it would be predicted that NMDA receptor activation would depolarize the plasma membrane (say to −30 mV). The mitochondria would partially depolarize in response to the increased Ca^{2+} uptake and ATP turnover (say to −140 mV). The third time point simulates the mitochondrial hyperpolarization in response to oligomycin ($\Delta\psi_p$ −30 mV, $\Delta\psi_m$ −160 mV), and in the final time point, FCCP is used to collapse $\Delta\psi_m$.
11. Keeping this membrane potential protocol, compare quench and non-quench concentrations of TMRM and R123. Figure 2 shows an actual experiment, which is detailed below) and the accompanying simulation [3].

3.6 Sample Experimental Protocol: Calibration and NMDA Receptor Activation of Cerebellar Granule Neurons with TMRM in Quench Mode

This is the actual experiment that was used in the above predictive simulation.

1. Culture neurons at about 100,000–350,000 cells per well into the appropriate number of wells of the 8-well Labtech chambered coverslip.
2. Take Labtech chamber from incubator. Remove culture medium and immediately replace with 400ul of incubation medium) without $MgCl_2$ (for NMDA receptor studies) but containing 50 nM TMRM (*see* **Note 5**). Incubate the cells in an air incubator (or the temperature-controlled enclosure of the microscope) for 60 min to allow equilibration of the probe.
3. Insert the Labtech chamber onto the microscope stage (*see* **Note 6**). With the binoculars and visible transmission, locate a suitable field and focus.
4. Set up a confocal configuration with 543 nm excitation (although 514 or even 488 nm will also work) and a 560 long-pass (or similar) emission filter. 512×512 resolution is adequate. Ensure that laser power is low (1–2%), and use an amplifier gain that is high but does not produce noise in the image.
5. To make additions during an experiment, pipette the required volume of reagent into a 0.5 ml Eppendorf Tube. Equip a 100 μl micropipette with a long gel-loading pipette tip (Microflex 0.1–200 μl VWR # 53503-189) whose last 1 cm is bent downward at 45°. Carefully withdraw 50 μl of incubation medium from the well that is being imaged, mix thoroughly with the reagent, and smoothly re-add to the well, ensuring full mixing by 3–4 slow up and down strokes of the micropipette. With practice it is easy to do this without disturbing the cells or the focus.
6. To investigate $\Delta\psi_m$ changes in response to NMDA receptor activation, the following additions may be made during a 60 min experiment:

(a) 100 μM glutamate plus 10 μM glycine at 5 min (*see* **Note 7**)

(b) 2 μg/ml oligomycin at 40 min (*see* **Note 8**)

(c) 1 μM FCCP at 50 min (*see* **Note 9**)

7. Save the time-course and define regions of interest (ROI) around individual cells. Export the resulting time-courses to Excel.
8. Interpretation – While the responses of individual cells vary stochastically, an actual trace from a single cell that survived the exposure to glutamate/glycine without undergoing delayed Ca^{2+} deregulation is shown in Fig. 2a. In Fig. 2b the simulated trace is shown that was generated using the spreadsheet on the assumptions that glutamate caused an immediate $\Delta\psi_p$ depolarization from −60 to −20 mV. Simultaneously the mitochondria depolarized from 150 mV to 145 mV (due to enhanced ATP demand and Ca^{2+} uptake) [3]. The small decrease in signal with oligomycin is consistent with a 10 mV mitochondrial hyperpolarization (showing that they were generating ATP). Finally the FCCP spike confirms that the cell remain in quench mode.
9. Alternative: selected imaging of mitochondrial poor (nuclear) regions – If resolution permits the regions of interest can be limited to the nuclear regions of individual cells. TMRM and R123 appear to equilibrate freely between the cytoplasm and nucleus. In quench mode this improves the sensitivity of the technique by removing the (constant) mitochondrial signal. In the simulation this cytoplasmic signal is given by the difference between the whole cell and matrix traces.

3.7 Sample Experimental Protocol: NMDA Receptor Activation of Cerebellar Granule Neurons with R123 in Quench Mode

Rhodamine 123 is structurally related to TMRM but is more hydrophilic and therefore equilibrates more slowly across the plasma membrane (although redistribution across the mitochondrial membrane is still very rapid). R123 fluorescence is therefore more slowly affected by redistribution across the plasma membrane, and this can be useful in simplifying the interpretation. R123 is always used in quench mode and is loaded in a different manner (see below). The R123 experiment is identical apart from the loading and fluorescence wavelengths.

1. *R123 loading*—Cells are equilibrated with 2.6 μM R123 for 15 min at 22 °C after which they are washed in incubation medium in the absence of R123. Trial and error is advised to achieve the optimal loading. Be very aware of the risks of phototoxicity; use minimal laser power, and avoid exposing the cells to unnecessary excitation. Remember that no R123 is added after washing or during the experiment.

2. Set up a confocal configuration with 488 nm or 514 nm excitation and a 530 nm long-pass (or similar) emission filter. 512×512 resolution is adequate. Ensure that laser power is low (1–2%) and use an amplifier gain that is high but does not produce noise in the image.
3. The identical addition protocol to the TMRM experiment is performed.
4. Interpretation—the response of a representative cell that survived glutamate exposure is shown in Fig. 2c. Note that the slopes of the signal following glutamate and FCCP are much shallower than for TMRM, due to the more hydrophilic nature of R123 and its consequent slower redistribution across the plasma membrane. The simulated trace (Fig. 2d) was obtained using the spreadsheet with all parameters identical to the TMRM experiment, except that the rate constant V was lowered from $0.02\ s^{-1}$ to $0.001\ s^{-1}$ [3].
5. Employ spreadsheet 1 as detailed in Subheading 3.4 to generate a curve fit to the experiment.

4 Notes

1. For experiments investigating NMDA glutamate receptors, $MgCl_2$ is omitted to prevent the voltage block of the receptor under polarized conditions.
2. To work in an ambient CO_2 atmosphere, HCO_3 concentration is lowered to 5 mM, and a high buffering capacity is introduced (20 mM Na-Tes, pH 7.4) to minimize alkalinization of the medium as CO_2 is slowly lost to the atmosphere.
3. TMRM (or the closely related ethyl ester TMRE) and R123 are the accepted "least bad" indicators, in that they give generally reliable results with little interference with normal mitochondrial function. The seductive JC-1, where the aggregate instead of being nonfluorescent fluoresces red, is inappropriate firstly because it is not valid to ratio the red fluorescence of the aggregate in the matrix and the green monomer fluorescence in the cytoplasm as is frequently done. Secondly the aggregates can fail to redissolve when $\Delta\psi_m$ is decreased. Finally the very slow permeation of the rather hydrophilic probe across the plasma membrane means that it is possible to be misled that, for example, mitochondria in thin processes possess a higher $\Delta\psi_m$ than mitochondrial in the cell bodies, when this is simply because insufficient time has been allowed for equilibration into the large cell body. Cyanine dyes such as $DiOC_6$ [3] need to be avoided since many are potent inhibitors of mitochondrial electron transport.

4. The spreadsheet may be obtained free of charge from the author by emailing dnicholls@buckinstitute.org
5. The concentration of TMRM required to remain in quench mode until the final FCCP depolarization may need to be determined in pilot experiments by experimenting with different concentrations and confirming that addition of FCCP gives a transient increase in fluorescence (dequenching).
6. To minimize evaporation, the Labtech chambers have a loose-fitting lid. Taking on and off this lid can disturb the focus slightly. To allow four experiments to be performed sequentially without evaporation, drill a 8 mm hole in spare lids such that additions can be made to a single well while the remainder are covered.
7. Glycine is a co-activator of the NMDA receptor. To obtain maximal activation of the receptor, glycine is added, and Mg^{2+} is omitted from the medium.
8. Oligomycin inhibits the mitochondrial ATP synthase. If the mitochondria in a cell are still generating ATP at this stage, oligomycin will cause a *hyperpolarization* in $\Delta\psi_m$ as the proton reentry will be slowed. Conversely, if the mitochondria are proton leaky or with a compromised electron transport chain, the ATP synthase can reverse and hydrolyze cytoplasmic glycolytic ATP. In this case oligomycin will *depolarize* the mitochondria.
9. It is useful to conclude the experiment by adding FCCP to collapse $\Delta\psi_m$ and to confirm that the experiment was performed in quench mode.

References

1. Nicholls DG, Ferguson SJ (2002) Bioenergetics 3. Academic Press, London
2. Hoek JB, Nicholls DG, Williamson JR (1980) Determination of the mitochondrial protonmotive force in isolated hepatocytes. J Biol Chem 255:1458–1464
3. Ward MW, Rego AC, Frenguelli BG, Nicholls DG (2000) Mitochondrial membrane potential and glutamate excitotoxicity in cultured cerebellar granule cells. J Neurosci 20:7208–7219
4. Nicholls DG (2006) Simultaneous monitoring of ionophore- and inhibitor-mediated plasma and mitochondrial membrane potential changes in cultured neurons. J Biol Chem 281:14864–14874
5. Gerencser AA, Chinopoulos C, Birket MJ, Jastroch M, Vitelli C, Nicholls DG, Brand MD, (2012) Quantitative measurement of mitochondrial membrane potential in cultured cells: calcium-induced de- and hyperpolarization of neuronal mitochondria. Am J Physiol 590:2845–2871
6. Ward MW, Huber HJ, Weisova P, Duessmann H, Nicholls DG, Prehn JHM (2007) Mitochondrial and plasma membrane potential of cultured cerebellar neurons during glutamate induced necrosis, apoptosis and tolerance. J Neurosci 27:8238–8249
7. Lemasters JJ, Ramshesh VK (2007) Imaging of mitochondrial polarization and depolarization with cationic fluorophores. Methods Cell Biol 80:283–295
8. Nicholls DG, Ward MW (2000) Mitochondrial membrane potential and cell death: mortality and millivolts. Trends Neurosci 23:166–174
9. Rego AC, Vesce S, Nicholls DG (2001) The mechanism of mitochondrial membrane potential retention following release of cytochrome *c* in apoptotic GT1-7 neural cells. Cell Death Differ 8:995–1003

Chapter 8

Comparison of Mitochondrial Incubation Media for Measurement of Respiration and Hydrogen Peroxide Production

Timea Komlódi, Ondrej Sobotka, Gerhard Krumschnabel, Nicole Bezuidenhout, Elisabeth Hiller, Carolina Doerrier, and Erich Gnaiger

Abstract

High-Resolution FluoRespirometry is a well-established and versatile approach to study mitochondrial oxygen uptake amperometrically in combination with measurement of fluorescence signals. One of the most frequently applied fluorescent dyes is Amplex UltraRed for monitoring rates of hydrogen peroxide production. Selection of an appropriate mitochondrial respiration medium is of crucial importance, the primary role of which is to support and preserve optimum mitochondrial function. For harmonization of results in a common database, we compared respiration and H_2O_2 production of permeabilized HEK 293T cells measured in MiR05 (sucrose and K-lactobionate), Buffer Z (K-MES and KCl), MiR07 (combination of MiR05 and Buffer Z), and MiRK03 (KCl). Respiration in a simple substrate-uncoupler-inhibitor titration protocol was identical in MiR05, Buffer Z, and MiR07, whereas oxygen fluxes detected with MiRK03 were consistently lower in all coupling and electron transfer-pathway states. H_2O_2 production rates were comparable in all four media, while assay sensitivity was comparatively low with MiR05 and MiR07 and higher but declining over time in the other two media. Stability of assay sensitivity over experimental time was highest in MiR05 but slightly less in MiR07. Taken together, MiR05 and Buffer Z yield comparable results on respiration and H_2O_2 production. Despite the lower sensitivity, MiR05 was selected as the medium of choice for FluoRespirometry due to the highest stability of the sensitivity or calibration constant observed in experiments over periods of up to 2 h.

Key words Amplex UltraRed, High-Resolution FluoRespirometry, Oxygraph-2k, Respiration media, Substrate-uncoupler-inhibitor titration, HEK 293T cells, Permeabilized muscle fibers, DTPA

1 Introduction

Over the past 25 years, high-resolution respirometry (HRR) has been firmly established as a prime way to study mitochondrial function [1]. The extension of respirometric analysis by High-Resolution FluoRespirometry (HRFR) allows the simultaneous

Carlos M. Palmeira and António J. Moreno (eds.), *Mitochondrial Bioenergetics: Methods and Protocols*, Methods in Molecular Biology, vol. 1782, https://doi.org/10.1007/978-1-4939-7831-1_8,

measurement of respiration with an additional parameter such as H_2O_2 production or mitochondrial membrane potential. Mitochondrial preparations are isolated mitochondria or tissue and cellular preparations in which the barrier function of the plasma membrane is disrupted [2]. In all applications with mitochondrial preparations, mitochondrial respiration media are used to support and preserve optimum mitochondrial function. The most important variables to be considered are osmotic pressure, ionic strength, ion composition, binding of Ca^{2+} and free fatty acids, and antioxidant capacity. The media must not contain components which are unstable in the presence of oxygen, causing chemical background O_2 consumption [3]. A new factor for evaluation of mitochondrial respiration media emerged in HRFR, when comparing the sensitivity of the fluorescence signal in response to H_2O_2 calibrations and the magnitude of the chemical background, i.e., the change of the signal over time without biological sample, in media of different composition [4]. High sensitivity increases the signal-to-noise ratio of the assay, and low chemical background effects tend to minimize the effect of chemical background correction of H_2O_2 flux expressed per volume of the instrumental chamber [pmol $H_2O_2{\cdot}s^{-1}{\cdot}mL^{-1}$]. Stability of sensitivity over experimental periods of up to 2 h and of chemical background slope of the fluorescence signal are of even higher importance for resolution of H_2O_2 flux in the sample and for practical reasons, minimizing the number of sequential calibrations required during the experiment. The present comparison of mitochondrial respiration media widely used for measurement of H_2O_2 flux allows for harmonization of results reported by research groups using different media. Importantly, our study also provides a rationale for selecting the most appropriate mitochondrial respiration medium for combined measurement of respiration and H_2O_2 production.

2 Materials

2.1 System for High-Resolution FluoRespirometry

The Oxygraph-2k (O2k, Oroboros Instruments, Austria; http://www.oroboros.at) is a closed-chamber respirometer for high-resolution respirometry (HRR) and has been described in detail [1]. It is a modular system for extension to the O2k-FluoRespirometer including optical sensors with a light-emitting diode, a photodiode, and specific optical filters. In combination with fluorescence probes, it is possible to determine mitochondrial membrane potential (with Safranin or TMRM, [5, 6]), ATP production (with Magnesium Green™; [7]), mitochondrial calcium release (with Calcium Green; [8, 9]), or hydrogen peroxide production [4, 10]. For detection of H_2O_2 production with Amplex UltraRed (AmR), we use the Fluorescence-Sensor Green with maximum

Table 1
Composition of mitochondrial respiration media used in FluoRespirometry

	MiR05	MiRK03	Buffer Z	MiR07-T1	KCl-med	DPBS
mM						
Sucrose	110	–	–	110	25	–
K-lactobionate	60	–	–	–	–	–
K-MES	–	–	105	105	–	–
HEPES free acid	–	20	–	–	–	–
K-HEPES	20	–	–	–	–	–
Taurine	20	–	–	–	–	–
KCl	–	130	30	–	125	2.68
KH_2PO_4	10	10	10	10	5	1.42
$MgCl_2$	3	3	3	3	5	–
NaCl					–	136.89
$Na_2HPO_4.7H_2O$					–	8.06
EGTA	0.5	0.5	1	0.5	–	–
BSA	1 mg/mL	1 mg/mL	5 mg/mL	1 mg/mL	0.5 mg/mL	–
pH	7.1	7	7.1	7.1	7.4	7.0–7.3

excitation at approximately 525 nm and emission detection with a 600 nm long-pass filter [4].

2.2 Media and Reagents

2.2.1 Mitochondrial Respiration Media for Combined FluoRespirometry

Two experimental series were conducted examining respiration and simultaneous H_2O_2 production in HEK 293T cells. The first experiments compared measurements employing three respiration media widely differing in chemical composition (Table 1). MiR05 has been specifically designed for use with the O2k to optimally support mitochondrial respiration during prolonged substrate-uncoupler-inhibitor titration (SUIT) protocols [3]. MiR05-Kit is a ready-made powder of solid chemicals for the preparation of MiR05 as described elsewhere [11]. MiRK03 has been modified according to a medium described by Komary et al. [12], for use in H_2O_2 production measurements with AmR. Buffer Z is a medium introduced by Neufer and colleagues [13], which is primarily used with permeabilized muscle fibers. Two other media which were only tested for their use in fluorometry are Dulbecco's Phosphate-Buffered Saline (DPBS), a commonly used buffer solution applied in biological research including cell culture, and a

KCl-based medium applied by Hoffman and Brookes [14] to study the oxygen dependence of respiration and H_2O_2 production.

2.2.2 Cells and Media for Cell Culture and Cryopreservation

HEK 293T cells were cultured in Dulbecco's Modified Eagle Medium (DMEM) supplemented with 10% fetal bovine serum and 1% penicillin (50 units/mL) and streptomycin (50 μg/mL). For cryopreservation cells were suspended in fetal bovine serum containing 10% dimethyl sulfoxide (DMSO).

2.2.3 Substrates, Uncouplers, and Inhibitors for Titrations (SUIT Chemicals)

Stock solution stored at −80 °C, prepared in distilled water: ADP (D) 0.5 M, with 300 mM $MgCl_2$.

Stock solutions stored at −20 °C, prepared in distilled water: octanoylcarnitine (Oct) 100 mM; malate (M) 400 mM; glutamate (G) 2 M; cytochrome c (c) 4 mM; succinate (S) 1 M; ascorbate (As) 800 mM, pH adjusted to approx. 6 with ascorbic acid; *N,N,N′,N′*-Tetramethyl-*p*-phenylenediamine dihydrochloride (TMPD, Tm) 200 mM; azide (Azd) 4 M.

Stock solutions stored at −20 °C, prepared in ethanol: CCCP (U) 1 mM; rotenone (Rot) 1 mM; antimycin A (Ama) 5 mM; auranofin (AF) 10 mM; dinitrochlorobenzene (DNCB) 2 mM.

Stock solutions prepared fresh daily in distilled water: pyruvate (P) 2 M.

2.2.4 Solutions for Hydrogen Peroxide Assay

Amplex UltraRed: a 2 mM stock solution is prepared in DMSO, and aliquots of 20 μL are stored at −20 °C in the dark and protected from moisture. The stock solution can be used for at least 6 months.

Horseradish peroxidase (HRP): a stock solution with 500 U HRP/mL in MiR05 is prepared and aliquots are stored at −20 °C.

Superoxide dismutase (SOD): a commercial stock, typically containing between 2000 and 6000 U/mg protein, is used as provided and injected to have a final concentration of 5 U/mL in the respiration chamber.

Calibration standard of H_2O_2: a commercial H_2O_2 stock solution is diluted with 10 μM HCl to obtain an 80 μM H_2O_2 solution ready for injection. The calibration solution must be kept on ice or refrigerated during the experiments.

DTPA (diethylentriaminepentaacetic acid): the commercial powder is dissolved in warm (33 °C) distilled water, and 5 M KOH is added to facilitate the dissolution in a glass tube to obtain a stock concentration of 5 mM. Aliquots are stored at −20 °C.

2.3 Animals

Experimental animals were C57BL/6 mice which were sacrificed at approximately 4 months of age.

3 Methods

3.1 Cell Culture, Freezing, and Thawing

Cells were cultured in DMEM at 37 °C, in 5% CO_2 and 98% humidity, and were passaged according to standard procedures. For freezing, cells were collected by centrifugation, the pellet resuspended at $60 \cdot 10^6$ cells/mL in 250 µL freeze medium in standard CryoTubes and immediately transferred to a freezer at −80 °C. For quick thawing the cells were removed from the freezer, 500 µL pre-warmed MiR05 added, and the cells were brought into suspension by gentle pipetting. Cells were thawed within 5 min before injection into the respirometer chamber.

3.2 SUIT protocol for Permeabilized Cells

1. Experimental temperature is set to 37 °C; the O2k-glass chambers are washed, filled with 2.3 mL respiration medium (Table 1), and calibrated according to standard operating procedures. The final chamber volume after closing the stoppers is 2.0 mL.
2. The internal chamber illumination is switched off, the voltage of the amperometric channel is set to 500 mV at and gain 1000 (signal amplification), and the fluorescence sensors are inserted to the front windows of the O2k-Chambers.
3. 2 µL of the AmR stock solution are injected into each respiration chamber, yielding a final concentration (f.c., unless otherwise stated) of 10 µM, followed by 4 µL of HRP stock (1 U/mL) and 2 U/mL SOD. The fluorescence change in the absence of cells is observed to measure the chemical background flux of the AmR assay.
4. Two times 5 µL H_2O_2 calibration solution: step changes in the fluorescence signal are caused by the rapid reaction of the added H_2O_2 with AmR to the fluorescent product UltroxRed/resorufin (*see* **Note 1**).
5. 150 µL of freshly thawed cells: O_2 and H_2O_2 flux are measured in the ROUTINE state.
6. Digitonin (Dig; 10 µg/mL): permeabilization of the plasma membranes.
7. 5 µL H_2O_2 calibration solution: determine the change of AmR assay sensitivity after addition of sample.
8. Pyruvate (5 mM) and malate (2 mM): NADH-linked LEAK respiration and H_2O_2 production; $N(PM)_L$.
9. ADP (2.5 mM): induce NADH-linked oxidative phosphorylation; N-OXPHOS state; $N(PM)_P$.
10. 5 µL H_2O_2 calibration solution: calibrate AmR assay sensitivity in the N-OXPHOS state.

11. Succinate (10 mM): convergent NADH- and succinate-linked OXPHOS; NS-OXPHOS; NS(PM)$_P$.
12. Auranofin (AF) is injected in two consecutive steps (10 and 20 μL) to inhibit thioredoxin reductase and observe a potential increase in net H_2O_2 flux.
13. Dinitrochlorobenzene (DNCB) is injected in two steps (2 and 4 μL) to inhibit glutathione peroxidase and see a predicted increase in net H_2O_2 flux.
14. The uncoupler CCCP is titrated in 1 μM steps to obtain maximum noncoupled respiration representing NADH- and succinate-linked electron transfer capacity; NS-ET capacity; NS (PM)$_E$.
15. 5 μL H_2O_2 calibration solution: calibrate AmR assay sensitivity in the NS-ET state.
16. Rotenone (0.5 μM) is injected to inhibit Complex I and measure succinate-linked ET capacity; S$_E$.
17. Antimycin A (2.5 μM): inhibit CIII and measure residual oxygen consumption (ROX state).
18. 5 μL H_2O_2 calibration solution: calibrate AmR assay sensitivity in the ROX state.

A second series of experiments was conducted as described, for comparison of MiR05 and MiR07, omitting the use of AF and DNCB.

3.3 Fluorometric Background and Sensitivity

1. Background fluorescence changes in the AmR assay: Fluorescence measurements in different media showed similarities in the general patterns, but also some interesting differences. In the absence of cells, background fluorescence changes detected upon addition of the AmR assay components were minor in MiRK03, DPBS, KCl-based medium, and Buffer Z, whereas fluorescence showed a considerable steady increase in MiR05 (Fig. 1 and Fig. 2; data for DPBS and KCl-based medium not shown). This higher background fluorescence change persisted in the presence of cells resulting in a much larger overall fluorescence increase during an experiment in MiR05 as compared to the other media. Since this high background slope could be the result of iron-mediated side reactions, we applied the iron chelator diethylenetriaminepentaacetic acid (DTPA; [15]) (*see* **Note 2**). DTPA reduced the background slope of the AmR assay significantly in MiR05 (Fig. 3) and nonsignificantly in Buffer Z and KCl-based medium. The effect of DTPA was dose-dependent, saturating at a concentration of 15 μM. Thus, we recommend addition of 15 μM DTPA to MiR05 when examining H_2O_2 production.

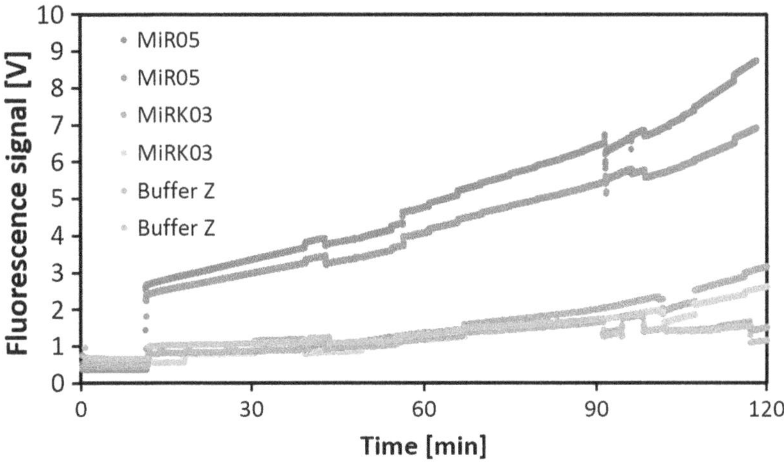

Fig. 1 Changes of the fluorescence signal measured with the Amplex UltraRed assay and detected using the O2k-FluoRespirometer in experiments with MiR05, MiRK03, and Buffer Z. For each medium, representative traces are shown of two O2k-Chambers run in parallel

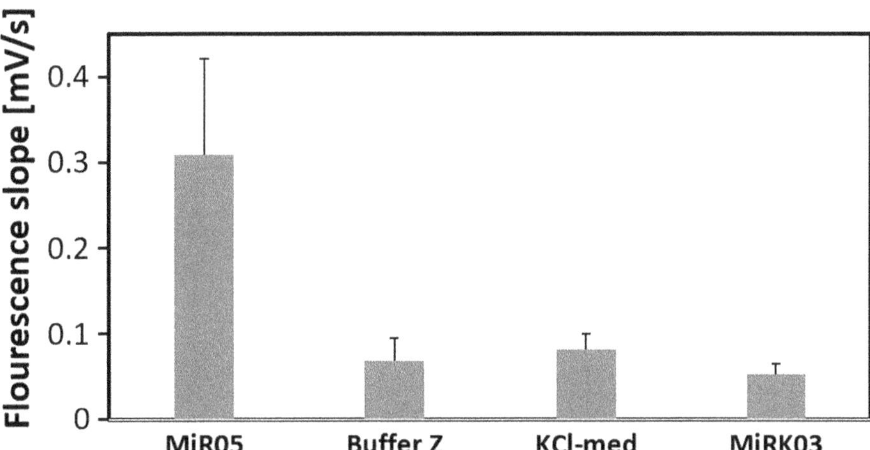

Fig. 2 Slope of chemical background fluorescence in different media, expressed as apparent UltroxRed production, in the absence of biological sample. UltroxRed fluxes were detected in the O2k-FluoRespirometer with the AmR assay. Bars are means + SD of four chambers used in four to five experiments

2. Sensitivity changes of the AmR assay: We have previously shown that the AmR assay sensitivity, *i.e.*, the fluorescence change upon addition of H_2O_2 calibration solution, changes considerably during an experiment [4]. This change may be a function of experimental time, changes in optical properties, radical scavenging power added with the sample, and the concentration of accumulated UltroxRed. To determine the extent of these changes in different respiratory media, multiple sequential additions of a H_2O_2 calibration solution were made in the presence of AmR, HRP, and SOD, in the absence of biological sample (Figs. 4 and 5). Sensitivity, expressed as

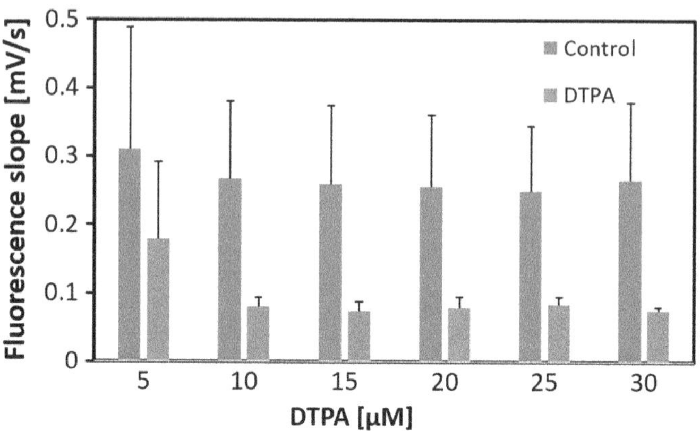

Fig. 3 The effect of DTPA on the chemical background of the AmR assay with MiR05 in the absence of biological sample. UltroxRed fluxes were measured in the O2k-FluoRespirometer with the AmR assay and are shown as fluorescence slope [mV/s]. DTPA was added every 10 min in 5 μM steps and compared to a control curve, where the solvent for DTPA, distilled water, was injected. Note that the x-axis does not only represent different DTPA concentrations but also experimental time, so that the changes in chemical background reflect both the impact of DTPA and of time. Bars are means + SD of four to five O2k-Chambers in four to five experiments. In each experiment the same Smart FluoSensor was applied in the same O2k-Chamber

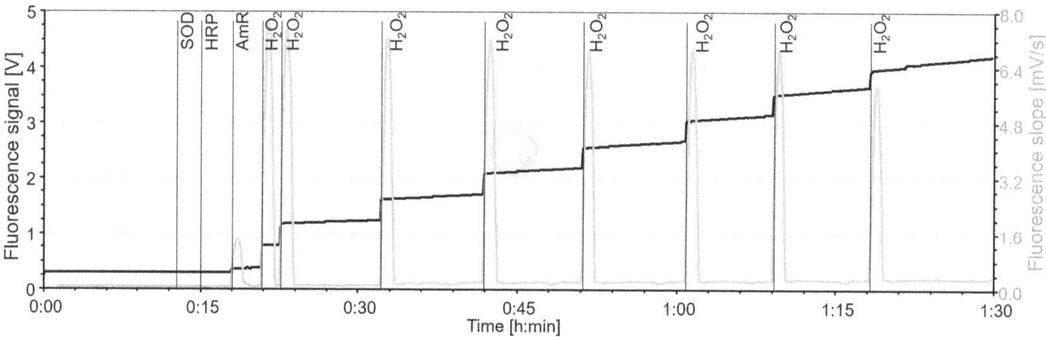

Fig. 4 Representative traces of chemical background and sensitivity estimation of the AmR assay over time in KCl-based medium in the absence of biological sample. UltroxRed fluxes were measured in the O2k-FluoRespirometer with the AmR assay. Sensitivity was determined from the step change in the fluorescence raw signal in volts after injection of 0.1 μM H_2O_2 calibration solution. The fluorescence signal emitted from the AmR assay is shown in black, the raw UltroxRed fluorescence slope calculated from the black line is shown in green and expressed in [mV/s]

V/μM, decreased over experimental time in all media (Fig. 5b). Although absolute sensitivities of KCl-based medium and DPBS were approximately threefold higher compared to the other media (Fig. 5b), the changes in sensitivity relative to the initial value, corrected for changes in background fluorescence, were similar in all media (Fig. 5c).

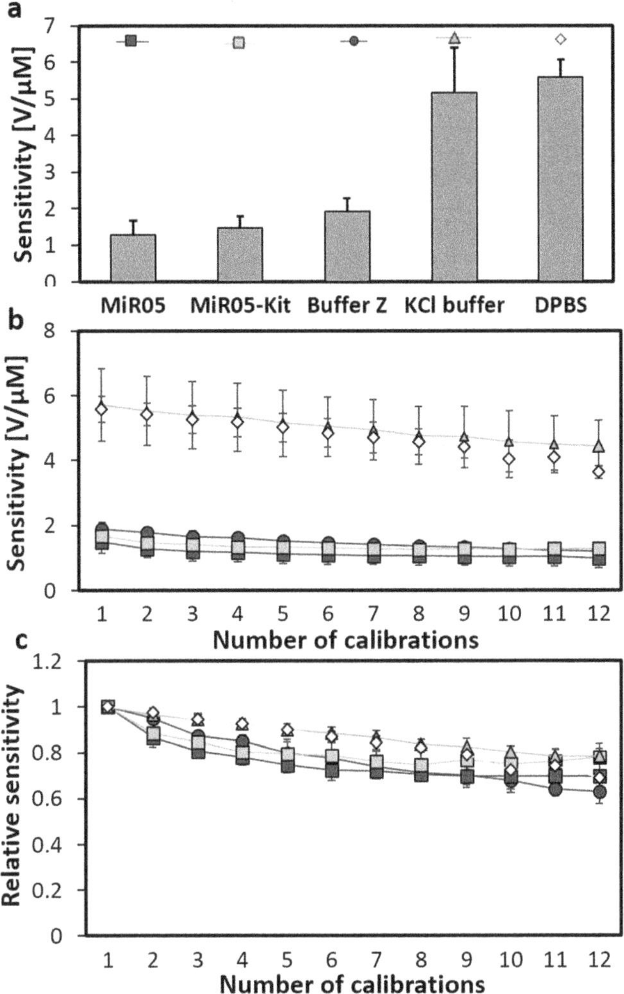

Fig. 5 Sensitivity of the AmR assay in different media in the absence of biological sample. Sensitivity was determined from the step change in the fluorescence raw signal in volts after injection of 0.1 µM H_2O_2 calibration solution. (**a**) Sensitivity based on the first calibration in MiR05, MiR05-Kit, Buffer Z, KCl-medium, and DPBS. (**b**) Absolute sensitivity in consecutive H_2O_2 calibration steps of 0.1 µM H_2O_2. (**c**) Relative sensitivity in consecutive H_2O_2 calibration steps, normalized for the first calibration step in each medium. Means ± SD calculated from the medians ($n = 4$) of four independent experiments ($n = 4$)

With the addition of cells, a particularly pronounced decrease of assay sensitivity is observed in MiR05, MiRK03, Buffer Z, and KCl-based medium, presumably due to the introduction of additional radical scavenging capacity into the respirometric chamber (*see* **Note 3**). In addition, sensitivity may also change over experimental time, although this further change is usually less pronounced. Thus, to be able to account for these changes, multiple

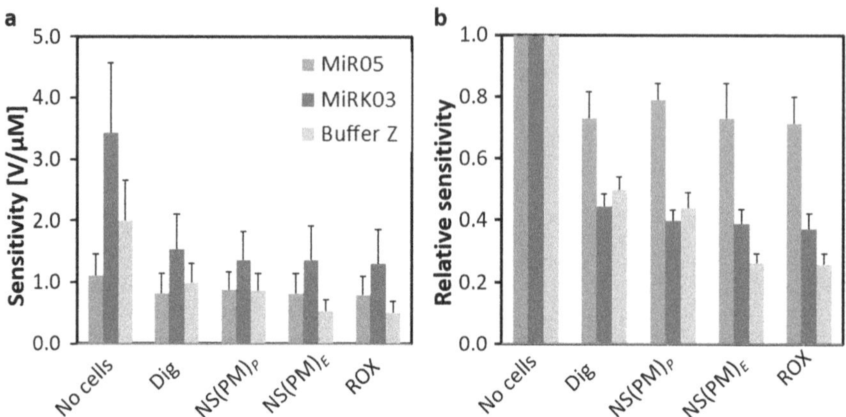

Fig. 6 AmR assay sensitivity measured in progressive states of a SUIT protocol in MiR05, MiRK03, and Buffer Z. Bars are means + SD of eight O2k-Chambers used in four experiments. (**a**) Sensitivity determined from the step change in the fluorescence raw signal in volts after titrating 0.1 μM H_2O_2 calibration solution. (**b**) Relative assay sensitivity, calculated by normalizing sensitivities to the value obtained in the first titration of H_2O_2 calibration solution

injections of H_2O_2 solution for repeated calibration were made during the experiments, including time points before and after addition of the sample and at different respiratory states. A comparison of sensitivity changes in Mir05, MirK03, and Buffer Z is summarized in Fig. 6a. Normalized sensitivity values are shown in Fig. 6b. Sensitivity is initially higher in MiRK03 and Buffer Z as compared to MiR05, but the addition of the sample resulted in a massive decrease in sensitivity in MiRK03 and Buffer Z by more than 50%, followed by a further decline to between 25% and 40% over time (Fig. 6b). In contrast, sensitivity decreased by approximately 25% after addition of cells in MiR05 and remained relatively stable throughout the experiment thereafter, in agreement with previous findings [4].

3.4 FluoRespirometry with Permeabilized Cells in MiR05, MiRK03, and Buffer Z

1. Representative experimental traces of combined FluoRespirometric measurements are shown in Figs. 7 and 8. The respiratory patterns are very similar in all three media, with a strong response to substrates supporting the NADH pathway, a further increase of O_2 flux upon addition of succinate, and another stimulation by uncoupler titration (Fig. 8). The uncoupler concentration required to obtain maximum respiration was significantly lower in MiRK03 compared to the other media. The mean concentration of CCCP applied in eight respiration chambers (four independent experiments with duplicate measurements) amounted to 3.7 ± 0.71 μM (mean ± SD), 2.25 ± 0.70 μM, and 5.0 ± 0.76 μM in MiR05, MiRK03, and Buffer Z, respectively. The high uncoupler concentration

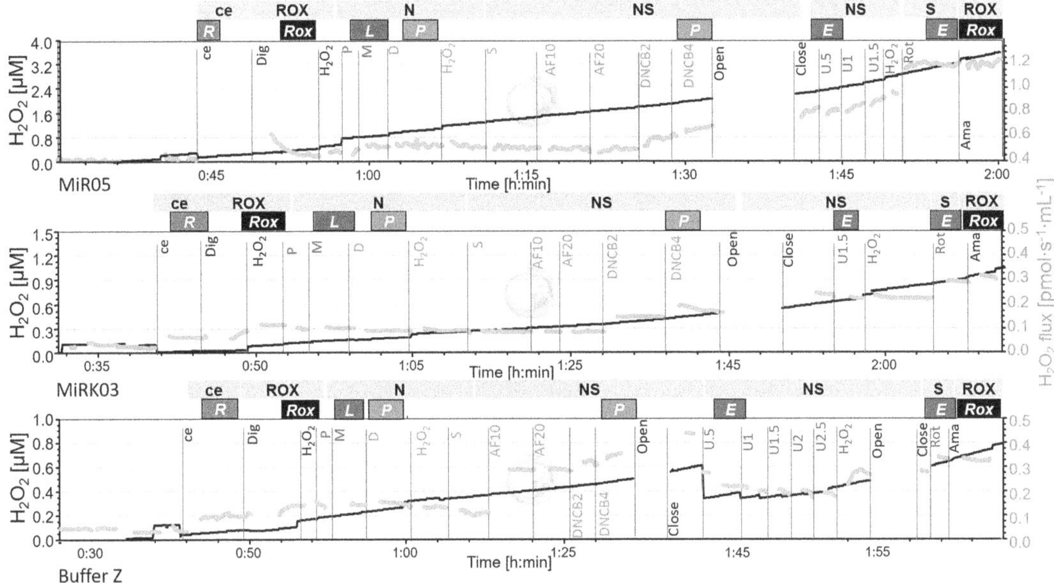

Fig. 7 H_2O_2 fluxes measured in the O2k-FluoRespirometer. Panels show measurements in the same experiments as shown in Fig. 8. In each panel, the fluorescence signal emitted from the AmR assay is shown in black, calibrated with the first addition of H_2O_2 calibration solution. The volume-specific H_2O_2 flux is shown in green (right *Y*-axis [pmol·s^{-1}·mL^{-1}]) calculated as the positive time derivative of the fluorescence signal. Additions are cells (ce), digitonin (Dig), pyruvate (P), malate (M), ADP (D), hydrogen peroxide (H_2O_2), succinate (S), auranofin (AF), dinitrochlorobenzene (DNCB), CCCP (U), rotenone (Rot), and antimycin A (Ama) at concentrations given in the text or indicated along with the abbreviation (i.e., U1 is CCCP at 1 μM)

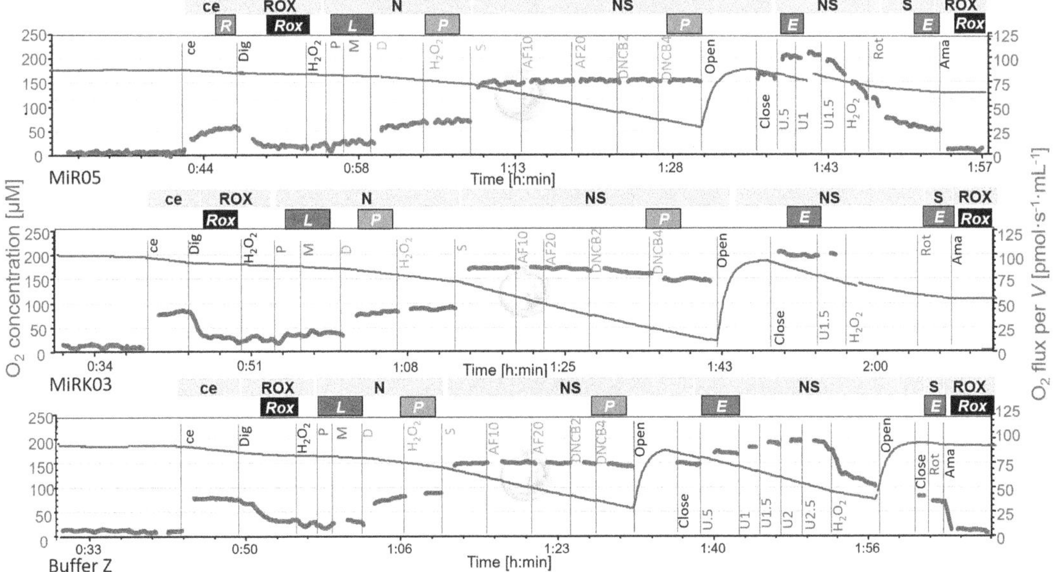

Fig. 8 Respiration in permeabilized HEK 293T cells in different mitochondrial media. Representative traces in MiR05 (upper panel), MiRK03 (middle panel), and Buffer Z (lower panel), showing oxygen concentration (blue plots; left *Y*-axis [μM]) and volume-specific oxygen flux (red plots; right *Y*-axis [pmol·s^{-1}·mL^{-1}]). For details on additions *see* Fig. 7

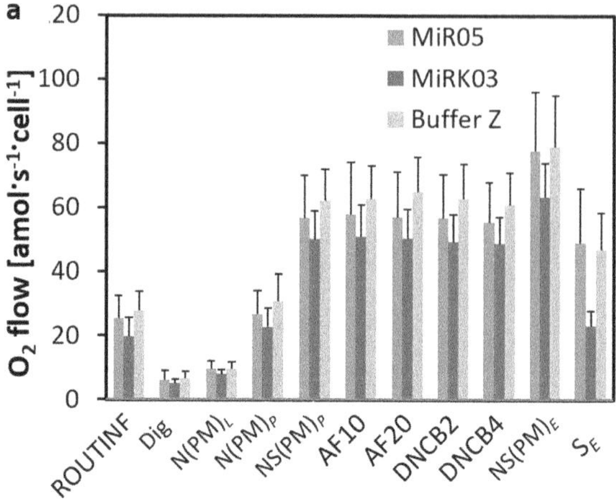

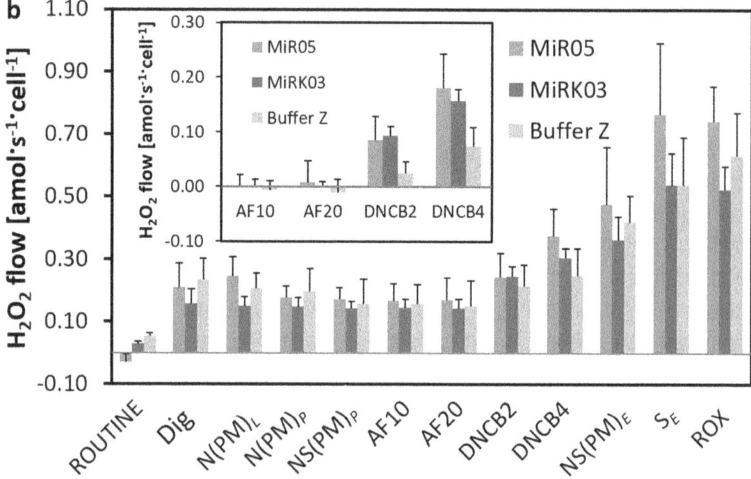

Fig. 9 Respiratory O_2 flow (**a**) and H_2O_2 flow (**b**) of HEK 293T cells in different mitochondrial media. Bars are means + SD of four independent experiments, each measured in duplicate. Respiratory fluxes are corrected for *Rox*; H_2O_2 flows are corrected for chemical background changes and for alterations of assay sensitivity over time. The inset in b shows the increase of H_2O_2 flow upon injection of 10 and 20 nM AF and 2 and 4 μM DNCB

required in Buffer Z is probably related to the high BSA concentration (Table 1).

2. The results on respiration and H_2O_2 production in three respiration media are summarized in Fig. 9. O_2 flows were identical in MiR05 and Buffer Z throughout the SUIT protocol, whereas respiration with MiRK03 was consistently lower (Fig. 9a) (*see* **Note 4**). In contrast, H_2O_2 flows were higher in MiR05 throughout the entire experiment as compared to MiRK03 and Buffer Z (Fig. 9b). The background fluorescence slopes determined in the absence of sample were subtracted from total slopes to calculate H_2O_2 flows, assuming a constant

background throughout the experiment. For baseline correction, the apparent H_2O_2 flow observed in the ROUTINE state (before permeabilization by digitonin) was subtracted [10]. The addition of the thioredoxin reductase inhibitor AF did not produce an increase of H_2O_2 production in any medium, whereas the glutathione peroxidase inhibitor DNCB exerted a similar stimulatory effect in MiR05 and MiRK03, which was less pronounced in Buffer Z (Fig. 9b, inset) (*see* **Note 5**).

3.5 FluoRespirometry with Permeabilized Cells in MiR05Cr and MiR07Cr

Compounds of MiR05 and Buffer Z were combined in MiR07 (Table 1) in an attempt to further optimize a respiration medium for the AmR assay. Since MiR05 and Buffer Z supported identical respiration rates and the AmR assay sensitivity was rather stable in MiR05 but initially much higher in Buffer Z, we tested MiR07 as a combination of both media to potentially combine their advantages with a slightly simpler chemical composition compared to MiR05. 20 mM creatine was added to MiR05 and MiR07 directly prior to experiments (MiR05Cr and MiR07Cr; *see* **Note 6**). A slightly simplified SUIT protocol was applied (Fig. 10).

1. Respiration was identical in both media (Fig. 10a). H_2O_2 production was slightly but not significantly lower in MiR05Cr compared to MiR07Cr in all respiratory states (Fig. 10b). Compared to series 1, addition of rotenone and antimycin A induced only minor elevations of H_2O_2 production in the absence of AF and DNCB.
2. The changes in AmR assay sensitivity were reproducible in comparison with series 1 for MiR05Cr, with a drop of about 25% after sample injection and subsequent stability (Fig. 11). In MiR07Cr, the initial sensitivity was more than twice that obtained in MiR05Cr but showed a more than 60% decline in response to sample addition and a further slight progressive reduction over time.

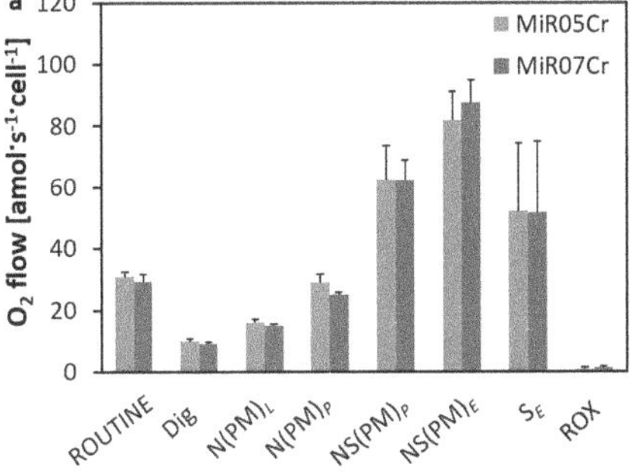

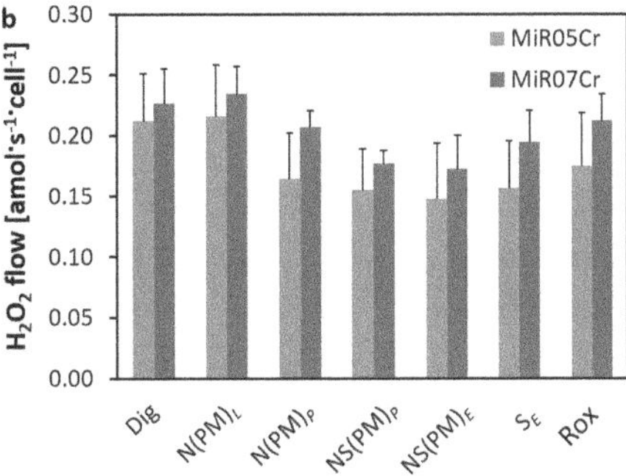

Fig. 10 O_2 flow (**a**) and H_2O_2 flow (**b**) of HEK 293T cells in MiR05Cr and MiR07Cr. Bars are means + SD of four independent experiments, three of which were measured in duplicate. Respiratory flows are corrected for *Rox*; H_2O_2 flows are corrected for chemical background changes and for alterations of AmR assay sensitivity over time

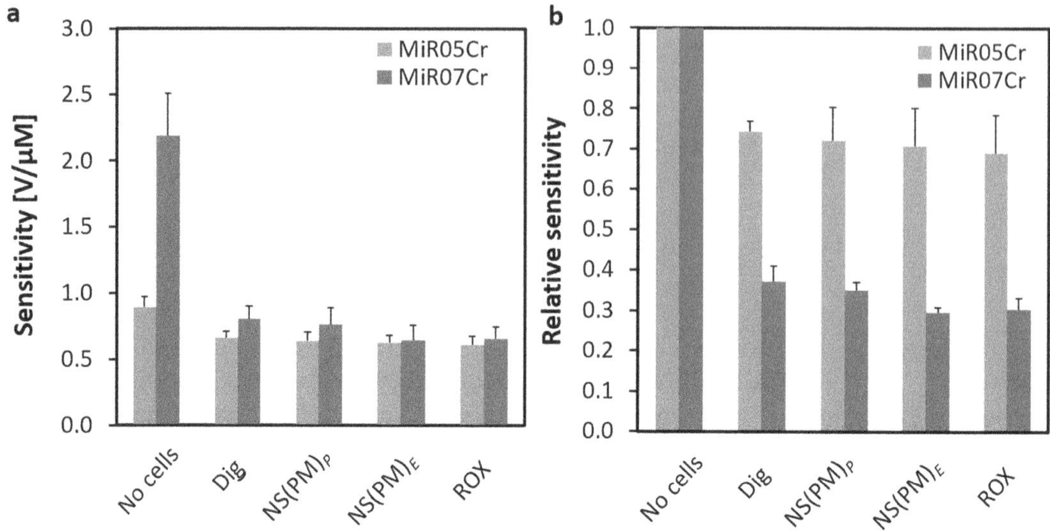

Fig. 11 Changes in AmR assay sensitivity over time assessed in MiR05Cr and MiR07Cr. Bars are means + SD of seven chambers used in four experiments. (**a**) Sensitivity determined from the step change in the fluorescence raw signal in volts upon injecting 0.1 μM H_2O_2 calibration solution. (**b**) Relative assay sensitivity, calculated by normalizing sensitivities for the value obtained from the first injection of calibration solution

Summarizing our observations in permeabilized HEK 293T cells, MiR05, MiR07, and Buffer Z support identical respiratory capacities in all respiratory states, whereas respiration was lower in KCl-based MiRK03. For assessment of H_2O_2 production with the AmR assay, MiR05 seems superior compared to the other media, owing to a better stability of the calibration factor (sensitivity), which provides the decisive methodological criterium in the present comparison. Although background fluorescence changes were much lower and assay sensitivity in the absence of sample much higher in Buffer Z, KCl-based medium, and DPBS as compared to MiR05, the decline and subsequent instability of assay sensitivity in Buffer Z upon sample addition appear to favor MiR05. Thus, although we recommend to account for changes in assay sensitivity over time even in MiR05, the impact of the required corrections, and correspondingly the potential errors of calibration, are lower in MiR05 (*see* **Note 7**).

3.6 Preparation of Permeabilized Mouse Skeletal Muscle Fibers

MiR06Cr and Buffer ZCr are widely used in studies of permeabilized muscle fibers [1, 3, 12, 16–18], but only few comparisons are available [13]. Therefore, we compared respiration of permeabilized mouse skeletal muscle fibers in MiR06Cr and Buffer ZCr. Fiber preparation is roughly divided into (1) tissue sampling and storage, (2) tissue preparation including mechanical separation and chemical permeabilization, and (3) tissue wet weight determination before adding the fibers into the O2k-Chambers [1] or dry weight determinations after retrieving the fibers quantitatively from the O2k-Chamber. Skeletal muscle was isolated from male mice C57BL6/N, aged 3–4 months. Permeabilized fibers with 1.5–2.5 mg wet weight were transferred into the O2k-Chamber.

3.7 SUIT Protocol for Fiber Respiration

1. Permeabilized muscle fibers were transferred into the O2k-Chamber containing either MiR06Cr or Buffer ZCr containing 20 mM creatine and 280 U catalase/mL equilibrated with air. Buffer ZCr contained 25 μM blebbistatin [13].
2. The O2k-Chambers are closed by inserting the stoppers, and an initial ROX state of respiration is recorded, without external fuel substrates.
3. Octanoylcarnitine (Oct, 0.4 mM) and malate (M, 2 mM): fatty acid oxidation, initiating F-LEAK respiration.
4. ADP (D, 5 mM): F-OXPHOS capacity.
5. Pyruvate (P, 5 mM): combined FN-OXPHOS capacity.
6. Cytochrome *c* (c, 10 μM): test of the integrity of the outer mitochondrial membrane. If this membrane is damaged during preparation of the sample, cytochrome *c* is released from the intermembrane space, and respiration is limited by electron transfer from CIII to CIV, indicated by stimulation of respiration by addition of cytochrome *c*.
7. Oxygen dependence of FN-OXPHOS capacity: respiration proceeds until a lower limit of 80 μM O_2 is reached, at which level the chamber stoppers are gently removed into a position guided by the stopper spacer. Pure O_2 gas is injected through the capillaries of the stoppers into the gas phase on top of the stirred respiration medium. The O_2 concentration in the medium increases quickly to 250 μM, when the stoppers are inserted (closed chamber), and respiration is recorded in the range of 250–200 μM O_2. This is followed by another re-oxygenation up to 400 μM O_2 and recording respiration in the high-oxygen range of 400–380 μM O_2 (Fig. 12).
8. Glutamate (G, 10 mM): additional substrate supporting N-OXPHOS capacity, maintaining O_2 concentration between 400 and 300 μM.
9. Succinate (S, 10 mM): FNS-OXPHOS.
10. Rotenone (Rot, 0.5 μM): inhibition of CI, S-OXPHOS.
11. Succinate (S50, 50 mM): evaluation of kinetically saturating concentration of S.
12. Antimycin A (Ama, 2.5 μM): inhibition of CIII, ROX state.
13. Ascorbate (As, 2 mM) and TMPD (Tm, 0.5 mM): Complex IV assay.

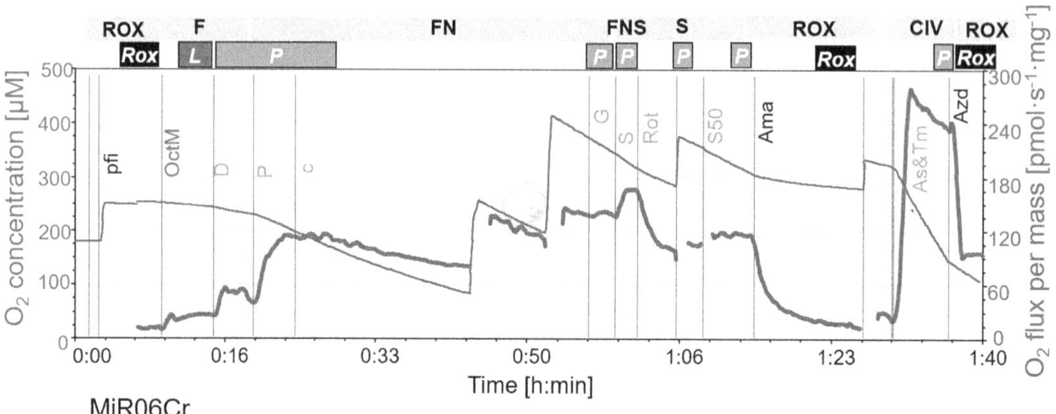

Fig. 12 Respiration in permeabilized mouse skeletal muscle fibers in MiR06Cr with blebbistatin. Representative trace obtained with O_2 concentration shown as the blue plot (left *Y*-axis [μM]) and mass-specific oxygen flux as the red plot (right *Y*-axis [$pmol{\cdot}s^{-1}{\cdot}mg^{-1}$]). Additions are permeabilized fibers (pfi), octanoylcarnitine and malate (OctM), ADP (D), pyruvate (P), cytochrome *c* (c), glutamate (G), succinate at 10 mM (S), rotenone (Rot), succinate at 50 mM (S50), antimycin A (Ama), TMPD and ascorbate (Tm, As), and azide (Azd) at concentrations given in the previous section

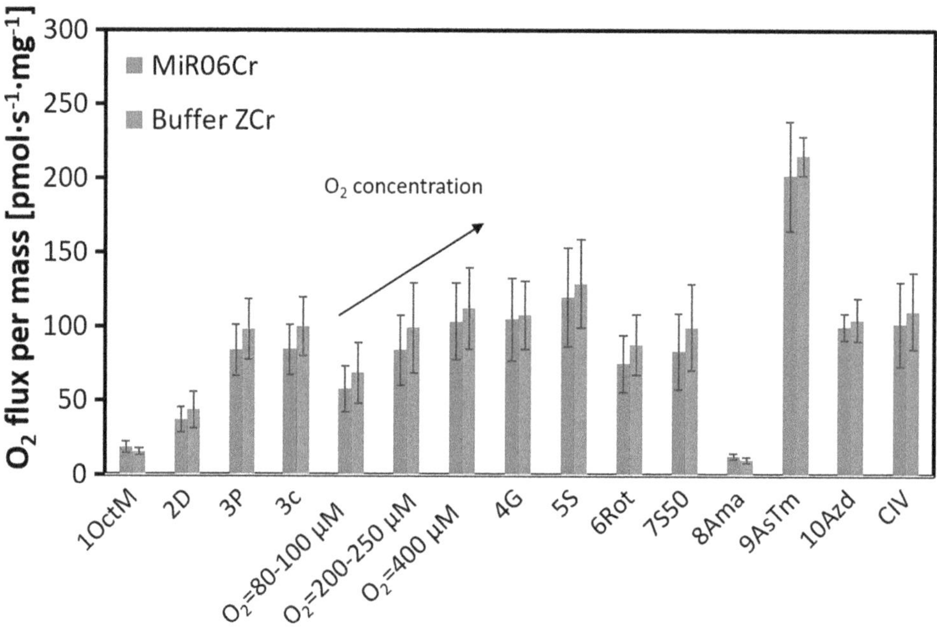

Fig. 13 Respiratory fluxes of permeabilized mouse skeletal muscle fibers in MiR06Cr and Buffer ZCr with blebbistatin. Bars are means + SD of three independent experiments on fibers from three different mice

14. Azide (Azd, 200 mM): inhibition of CIV, residual and chemical O_2 consumption due to autoxidation of ascorbate, TMPD, and cytochrome *c*.

3.8 Fiber Respiration MiR06Cr and Buffer ZCr

A summary of respiratory experiments on fibers from three mice conducted in parallel in MiR06Cr and Buffer ZCr is shown in Fig. 13. In all coupling/ET-pathway control states O_2 flux was nearly identical in the two media, without significant differences. Independent of the medium used, respiration of the muscle fibers was clearly oxygen-dependent in the range of high O_2 concentrations (100–400 μM), independent of blebbistatin added to Buffer ZCr. Diffusion limitation of respiration can thus be avoided by a hyperoxic experimental regime with permeabilized fibers.

4 Notes

1. The product of the reaction of Amplex Red with H_2O_2 is resorufin; that of Amplex UltraRed is referred to as UltroxRed by the producer of the chemical.
2. During the organic synthesis of the medium component or other chemicals such as Amplex UltraRed or horseradish peroxidase, iron contamination is possible. This contamination can interfere with several compounds in the respiratory medium or the biological sample and may lead to artificially elevated ROS formation [15]. In case of the Amplex UltraRed assay employed here, this artifact can be also expressed as higher chemical background. Therefore, iron chelators, e.g., DTPA (diethylentriaminepentaacetic acid), are used to reduce this side effect. Addition of DTPA did not only reduce the chemical background but also the sensitivity of the AmR assay. However, the decrease in sensitivity (17 ± 9%) was less pronounced than that of the chemical background (57 ± 12%), suggesting to include DTPA in the respiration medium.
3. We observed a decrease of AmR assay sensitivity over time, using MiR05 in HEK 293T cells, HeLa cells, and Huh7 liver cells. Unexpectedly, however, an increase of AmR assay sensitivity over time was observed in experiments with isolated mouse cardiac mitochondria and permeabilized mouse skeletal muscle fibers.
4. The particularly pronounced difference between media in state S-ET, i.e., after the addition of rotenone, may in part be attributed to the fact that in MiRK03 respiration was inhibited faster than in the other media. In these, correct estimation of S-ET capacity requires much longer periods of inhibition. In general, the time required to reach a new steady level of respiration after addition of rotenone varies widely between diverse types of cells and tissues and may range from a few minutes up to 45 min. In the latter case, however, it becomes difficult to distinguish the inhibitory effect of rotenone from a general deterioration of the sample over time.

5. The concentration of AF and DNCB were selected according to published values eliciting elevated H_2O_2 production in cardiac mitochondria from various model species [19]. Our own experiments with cardiac mitochondria from mouse confirmed the efficacy of the drugs, suggesting that either different concentrations may be required for permeabilized HEK 293T cells or that thioredoxin reductase is less important for preserving redox balance in these cells.
6. Creatine is typically included in media used for measuring respiration of muscle fibers, as it may help reducing diffusion limitation for ADP/ATP. In HEK 293T cells, its affect was not systematically assessed, but preliminary experiments do not indicate any effect. Creatine was nonetheless included to ascertain the general applicability and rule out any interference of the chemical with the AmR assay system.
7. We conducted a comparative analysis of H_2O_2 production rates measured in MiR05 using permeabilized HeLa cells, where assay sensitivity changes were the same as those seen with HEK 293T cells. We found that calibrations based on the first H_2O_2 injection after addition of cells produced nearly identical results as those obtained when correcting for changes in assay sensitivity at each respiratory state or simply taking an average sensitivity from these repeated H_2O_2 injections. However, given that individual experiments may show fluctuations and that the impact of chemicals has not yet been tested for all substrates and inhibitors, we recommend to apply serial calibrations.

Acknowledgments

We thank Roland Stocker for guiding us to the use of DTPA. This work was supported by K-Regio project MitoFit and is a contribution to COST Action CA15203 MitoEAGLE.

Competing Financial Interests

E.G. is founder and CEO of Oroboros Instruments, Innsbruck, Austria.

References

1. Doerrier C, Garcia-Souza LF, Krumschnabel G, Wohlfarter Y, Mészáros AT, Gnaiger E (2017) High-resolution FluoRespirometry and OXPHOS protocols for human cells, permeabilized fibres from small biopsies of muscle and isolated mitochondria. Methods Mol Biol 810:25–58
2. MitoEAGLE_preprint_2018-02-08 (37). Mitochondrial respiratory states and rates: Building blocks of mitochondrial physiology Part 1. http://www.mitoeagle.org/index.php/MitoEAGLE_preprint_2018-02-08
3. Gnaiger E, Kuznetsov AV, Schneeberger S, Seiler R, Brandacher G, Steurer W, Margreiter R (2000) Mitochondria in the cold. In:

Heldmaier G, Klingenspor M (eds) Life in the cold. Springer, Heiderlberg, Berlin, New York, pp 431–442

4. Krumschnabel G, Fontana-Ayoub M, Sumbalova Z, Heidler J, Gauper K, Fasching M, Gnaiger E (2015) Simultaneous high-resolution measurement of mitochondrial respiration and hydrogen peroxide production. Methods Mol Biol 1264:245–261
5. Krumschnabel G, Eigentler A, Fasching M, Gnaiger E (2014) Use of safranin for the assessment of mitochondrial membrane potential by high-resolution respirometry and fluorometry. Methods Enzymol 542:163–181
6. Roy Chowdhury SK, Djordjevic J, Albensi B, Fernyhough P (2015) Simultaneous evaluation of substrate-dependent oxygen consumption rates and mitochondrial membrane potential by TMRM and safranin in cortical mitochondria. Biosci Rep 36(1):e00286
7. Pham T, Loiselle D, Power A, Hickey AJ (2014) Mitochondrial inefficiencies and anoxic ATP hydrolysis capacities in diabetic rat heart. Am J Phys 307:C499–C507
8. Elustondo PA, Negoda A, Kane CL, Kane DA, Pavlov EV (2014) Spermine selectively inhibits high-conductance, but not low-conductance calcium-induced permeability transition pore. Biochim Biophys Acta 1847:231–240
9. Liepinsh E, Makrecka-Kuka M, Volska K, Kuka J, Makarova E, Antone U, Sevostjanovs E, Vilskersts R, Strods A, Tars K, Dambrova M (2016) Long-chain acylcarnitines determine ischaemia/reperfusion-induced damage in heart mitochondria. Biochem J 473:1191–1202
10. Makrecka-Kuka M, Krumschnabel G, Gnaiger E (2015) High-resolution respirometry for simultaneous measurement of oxygen and hydrogen peroxide fluxes in permeabilized cells, tissue homogenate and isolated mitochondria. Biomol Ther 5:1319–1338
11. Oroboros Instruments, Innsbruck, Austria. Mitochondrial respiration meidum-MiR05-Kit. http://bioblast.at/index.php/MiR05-Kit
12. Komary Z, Tretter L, Adam-Vizi V (2010) Membrane potential-related effect of calcium on reactive oxygen species generation in isolated brain mitochondria. Biochim Biophys Acta 1797:922–928
13. Perry CG, Kane DA, Lin CT, Kozy R, Cathey BL, Lark DS, Kane CL, Brophy PM, Gavin TP, Anderson EJ, Neufer PD (2011) Inhibiting myosin-ATPase reveals a dynamic range of mitochondrial respiratory control in skeletal muscle. Biochem J 437:215–222
14. Hoffman DL, Brookes PS (2009) Oxygen sensitivity of mitochondrial reactive oxygen species generation depends on metabolic conditions. J Biol Chem 284:16236–16245
15. Bosworth CA, Toledo JC Jr, Zmijewski JW, Li Q, Lancaster JR (2009) Dinitrosyliron complexes and the mechanism(s) of cellular protein nitrosothiol formation from nitric oxide. Proc Natl Acad Sci U S A 106:4671–4676
16. Pesta D, Gnaiger E (2012) High-resolution respirometry. OXPHOS protocols for human cells and permeabilized fibres from small biopsies of human muscle. Methods Mol Biol 810:25–58
17. Whitfield J, Ludzki A, Heigenhauser GJ, Senden JM, Verdijk LB, van Loon LJ, Spriet LL, Holloway GP (2015) Beetroot juice supplementation reduces whole body oxygen consumption but does not improve indices of mitochondrial efficiency in human skeletal muscle. J Physiol 594:421–435
18. Ludzki A, Paglialunga S, Smith BK, Herbst EA, Allison MK, Heigenhauser GJ, Neufer PD, Holloway GP (2015) Rapid repression of ADP transport by palmitoyl-CoA is attenuated by exercise training in humans; a potential mechanism to decrease oxidative stress and improve skeletal muscle insulin signaling. Diabetes 64:2769–2779
19. Aon MA, Stanley AS, Sivakumaran V, Kembro JM, O'Rourke B, Paolocci N, Cortassa S (2012) Glutathione/thioredoxin sytems modulate mitochondrial H_2O_2 emission: an experimental-computational study. J Gen Physiol 139(6):479–491

Chapter 9

Measurement of Proton Leak in Isolated Mitochondria

Charles Affourtit, Hoi-Shan Wong, and Martin D. Brand

Abstract

Oxidative phosphorylation is an important energy-conserving mechanism coupling mitochondrial electron transfer to ATP synthesis. Coupling between respiration and phosphorylation is not fully efficient due to proton leaks. In this chapter, we present a method to measure proton leak activity in isolated mitochondria. The relative strength of a modular kinetic approach to probe oxidative phosphorylation is emphasized.

Key words Mitochondria, Oxygen consumption, Membrane potential, Oxidative phosphorylation, Modular kinetic analysis

1 Introduction

Mitochondria conserve energy as ATP by oxidative phosphorylation. Reducing equivalents derived from the cellular breakdown of carbon-based substrates are donated to the mitochondrial electron transport chain and eventually fully reduce molecular oxygen to water. The energy liberated during this mitochondrial electron transport is used to establish an electrochemical proton gradient across the mitochondrial inner membrane that in turn is used to drive ATP synthesis [1]. The coupling between electron transport and ATP synthesis is not absolute: protons can flow back into the mitochondrial matrix by mechanisms that bypass the ATP synthase [2]. Although proton leak lowers the coupling efficiency of oxidative phosphorylation and thereby decreases the ATP that is generated using carbon fuels, it is not necessarily a wasteful process. Modulation of coupling efficiency by proton leak activity is likely an important mechanism to control cell physiology.

It is clear that mitochondrial proton leak is an important process that, therefore, requires accurate and reliable measurement. In this chapter, we present a protocol to determine proton leak activity in isolated mitochondria. Moreover, we describe a procedure to obtain a system-level, modular kinetic description of oxidative

Carlos M. Palmeira and António J. Moreno (eds.), *Mitochondrial Bioenergetics: Methods and Protocols*, Methods in Molecular Biology, vol. 1782, https://doi.org/10.1007/978-1-4939-7831-1_9,

phosphorylation. The methods described may be readily adapted for use with mitochondria from many different sources, but some caution is warranted when applying them to mitochondria from a new source for the first time. The electron leak activity of mitochondria is described elsewhere in this volume under "Plate-based measurement of superoxide and hydrogen peroxide production by isolated mitochondria."

2 Materials

Unless stated otherwise, all chemicals may be purchased from Sigma-Aldrich (St. Louis, MO).

1. Triphenylmethylphosphonium- (TPMP)-selective electrode sleeve:
   - (a) *Solutions*—Tetraphenylboron (10 mM) in 6 mL tetrahydrofuran (THF); 1 g high-molecular-weight polyvinyl chloride (PVC) in 20 mL THF; dioctyl phthalate (a plasticizer); TPMP (10 mM in water stored at room temperature).
   - (b) *Equipment*—Five glass Petri dishes (100 mm diameter); PVC tubing (4 mm outside diameter); sharp razor blades and scissors.
2. Mitochondrial oxygen consumption and membrane potential:
   - (a) *Oxygen consumption*—A water-jacketed Clark oxygen electrode; oxygen-permeable Teflon membrane (*see* **Note 1**); an electrode controller unit and an electronic stirrer, which may all be obtained from Rank Brothers Ltd. (Cambridge, UK). A circulating water bath.
   - (b) *Membrane potential ($\Delta\Psi$)*—A TPMP-selective electrode sleeve; a 2–3 cm piece of platinum wire soldered to a screened cable; a solid-state Ag/AgCl reference electrode (World Precision Instruments Inc., USA); an adapted oxygen electrode plunger with two additional holes that allow insertion of the $TPMP^+$ and Ag/AgCl reference electrodes, and a pH meter (we use a pH-Amp front end from ADInstruments, UK).
   - (c) *Data acquisition*—A digital recording system comprising a PowerLab analog-to-digital signal converter (ADInstruments, UK) linked to any personal computer running LabChart software (ADInstruments, UK).
   - (d) *Assay medium*—KHEP (115 mM KCl, 10 mM KH_2PO_4, 3 mM HEPES, 1 mM EGTA, 2 mM $MgCl_2$, 0.3% w/v bovine serum albumin (BSA), pH 7.2 at 37 °C) stored at 4 °C (*see* **Note 2**).

(e) *Calibration*—Sodium dithionite (powder) and TPMP (1 mM in H_2O), both kept at room temperature, to calibrate the oxygen and TPMP-selective electrodes, respectively.

(f) *Respiratory substrates and effectors*—Succinate, malate, and malonate (stocks at 1 M), ADP (100 mM), all prepared in KHEP with pH adjusted to 7.2 at 37 °C. All stocks are stored at −20 °C. Pyruvate stock at 1 M in KHEP with pH adjusted to 7.2 at 37 °C is best made up fresh each day.

(g) *Effectors*—1 mg/mL nigericin, 4 mM rotenone, 1 mg/mL oligomycin, and 1 mM carbonyl cyanide p-trifluoromethoxyphenylhydrazone (FCCP), all prepared in ethanol (96% v/v). All stocks are stored at −20 °C.

(h) *Assays*—Automatic pipettes to add KHEP and mitochondria and microsyringes (5–25 μL) to add respiratory substrates and effectors.

3 Methods

In this chapter, we provide a method to measure proton leak during oxidative phosphorylation in isolated mitochondria. The standard operating procedure in our laboratory is to determine the overall rate of proton leak at a range of values of $\Delta\Psi$ to yield the kinetic response of proton leak rate to its driving force, $\Delta\Psi$. Such kinetics are an integral part of a more complete, system-level kinetic description of oxidative phosphorylation that is described below.

1. Mitochondrial isolation from rat skeletal muscle is described elsewhere in this volume under "Plate-based measurement of respiration by isolated mitochondria."
2. Preparation of TPMP-selective electrode sleeves:
   (a) Dissolve 1 g PVC in 20 mL THF: stir in a covered 50 mL conical flask until fully dissolved, which takes approximately 1 h.

   (b) Combine 6 mL of 10 mM tetraphenylboron and 20 mL PVC (both in THF) in a 100 mL flask; mix vigorously.

   (c) Add 3 mL dioctyl phthalate and continue mixing.

   (d) Divide this mixture equally into five Petri dishes (100 mm diameter) and allow it to dry on an absolutely level surface in a switched-off fume hood (*see* **Note 3**).

   (e) Complete evaporation of the solvent, which will take 24–48 h, yields a colorless and fairly robust membrane that needs to be attached to an electrode sleeve that should be prepared from a piece of PVC tubing with a 4 mm outside diameter and approximately 4 cm length; we use domestic earth (ground) sleeve from an electrical supplier for this purpose (*see* **Note 4**).

(f) Put a drop of THF on the membrane and place the PVC sleeve squarely on the drop (alternatively, dip one end of the PVC sleeve in THF and then place on the membrane); support the sleeve until the THF has evaporated and the sleeve is stuck firmly to the membrane. Multiple sleeves can be stuck in each Petri dish; they should be left for 24–48 h to cure.

(g) Cut around the sleeve with a very sharp razor blade and remove it, together with an attached membrane patch, from the Petri dish; trim away any excess membrane with sharp scissors.

(h) Fill the electrode sleeves with TPMP (10 mM) and check for any leaks; these sleeves should be immersed in TPMP (10 mM) for at least 48 h before use and may be stored dry or in this way for many years.

3. Mitochondrial oxygen consumption:

   (a) *Oxygen electrode assembly*—Set up the oxygen electrode following the manufacturer's instructions (*see* **Note 5**). To record dissolved oxygen tensions continuously, connect the electrode controller unit (which applies a polarizing voltage across the oxygen electrode's platinum cathode and Ag/AgCl anode and converts oxygen-induced currents into voltages) to the PowerLab analog-to-digital signal converter that, in turn, should be linked to a personal computer running LabChart software.

   (b) *Oxygen electrode calibration*—When the setup's platinum electrode is polarized at −0.6 V with respect to the silver electrode, the oxygen-induced current is linearly dependent on the dissolved oxygen tension, and a two-point calibration is sufficient to convert the electrical signal into a biologically meaningful unit. To calibrate, add 3.5 mL air-saturated and pre-warmed KHEP to a temperature-controlled (37 °C) and well-stirred oxygen electrode chamber, apply the plunger, and record a voltage that reflects the maximum dissolved oxygen tension; this tension has a value of 406 nmol atomic oxygen per mL at 37 °C [3]. To obtain a zero-oxygen signal, remove the plunger and add a few sodium dithionite crystals, which will remove any dissolved oxygen chemically (*see* **Note 6**). Between experimental traces, rinse the oxygen electrode vessel thoroughly with distilled water and ethanol (*see* **Note 7**).

(c) *Calculation of oxygen consumption rates*—When isolated mitochondria are incubated with a suitable electron donor, they will exhibit a respiratory activity that is directly proportional to the slope of the time-resolved oxygen trace. LabChart software provides a straightforward option to calculate and visualize the temporal derivative of the oxygen progress curves. The software thus facilitates rapid and objective determination of oxygen consumption rates and, importantly, allows accurate judgements as to whether *steady* respiratory states have been attained. Mitochondrial respiratory rates should be calculated and presented as specific activities normalized to the amount of mitochondrial protein in the assay to allow direct comparison with published activities.

(d) *Coupling of oxidative phosphorylation*—It is generally important to ascertain if and to what extent isolated mitochondria have retained coupling between oxygen consumption and ATP synthesis; mitochondrial samples that have lost this coupling are considered poor physiological models and should not be used for proton leak measurement. Oxygen uptake measurements during the various mitochondrial states defined by Chance and Williams [4, 5] provide insight into the degree of coupling, although a more complete understanding can be obtained from modular kinetic experiments that also take $\Delta\Psi$ values into account.

(e) *Data collection*—Incubate mitochondria (0.35 mg/mL if using rat skeletal muscle mitochondria) at 37 °C in 3.5 mL KHEP until a stable oxygen electrode signal is achieved (state 1). Then add, sequentially, rotenone (4 μM), succinate (5 mM), and ADP (50–100 μM) to, respectively, inhibit respiratory complex I and to provide an electron donor and a phosphorylation substrate. Well-coupled mitochondria will exhibit a substantial oxygen consumption rate under these conditions (state 3), which will decrease significantly after a short while when all ADP is depleted (state 4). To provoke further state 3/state 4 transitions, more ADP aliquots (50–100 μM) may be added. The experiment comes to an obvious end when all oxygen has been exhausted and an anaerobic state has been reached (state 5). A respiratory control ratio (RCR) can be derived from the recorded traces, which is a parameter defined as the quotient of oxygen uptake rates under state 3 and state 4 conditions. Good coupling of oxidative phosphorylation will be reflected by a high RCR, and we typically observe a value of 4 with rat skeletal muscle mitochondria oxidizing succinate. Alternatively, state 3 respiratory activity can be inhibited by the ATP synthase inhibitor oligomycin (0.7 μg/mL) and then be restimulated by the protonophore FCCP (1.5 μM). The ratio of FCCP-stimulated and oligomycin-inhibited oxygen uptake rates reflects the extent to which mitochondrial electron transfer is coupled to proton translocation. The FCCP-oligomycin rate ratio resembles an RCR as it indicates the intactness of the mitochondrial inner membrane, but it differs in that this ratio is not controlled to any extent by the ATP synthase.

The above experiments may also be performed with a combination of pyruvate and malate to engage respiratory complex I in oxidative phosphorylation. These substrates should both be added at 5 mM, and rotenone should be omitted from the assay medium.

4. Mitochondrial membrane potential measurements:
   (a) *TPMP electrode assembly* (Fig. 1)—Using an automatic pipette or a microsyringe, add TPMP (10 mM) to a TPMP sleeve, filling it to about 1 cm from the top; avoid trapping any air that would prevent proper electrical contact. Attach the filled sleeve to a screened cable such that the connected platinum wire sticks into the TPMP solution, but the soldered connection remains dry. A convenient way to achieve a tight fit is to use a yellow pipette tip, with the very end cut off, as a linker; puncture this pipette tip to prevent pressure changes that may damage the membrane during assembly. Insert both the TPMP sleeve and the solid-state reference electrode into the oxygen electrode chamber through an adapted plunger, and connect them to a voltmeter (a pH meter is ideal) that is linked to the same PowerLab digital data acquisition system that is used for the oxygen uptake measurements.
   (b) *TPMP electrode conditioning*—Add 3.5 mL pre-warmed KHEP to a temperature-controlled (37 °C) and well-stirred oxygen electrode chamber, insert the TPMP and reference electrodes, and wait until the signal is stable. Using the 1 mM stock solution, add TPMP to a concentration of 1 μM, which should cause a substantial electrode response; increasing the TPMP level in 1 μM increments should result in successively smaller responses. The signal should become stable and virtually noise-free at 5 μM TPMP. It may take several of these conditioning events and even an overnight incubation in 2–3 μM TPMP, until a sleeve exhibits drift- and noise-free behavior and responds to the TPMP dose in a truly logarithmic

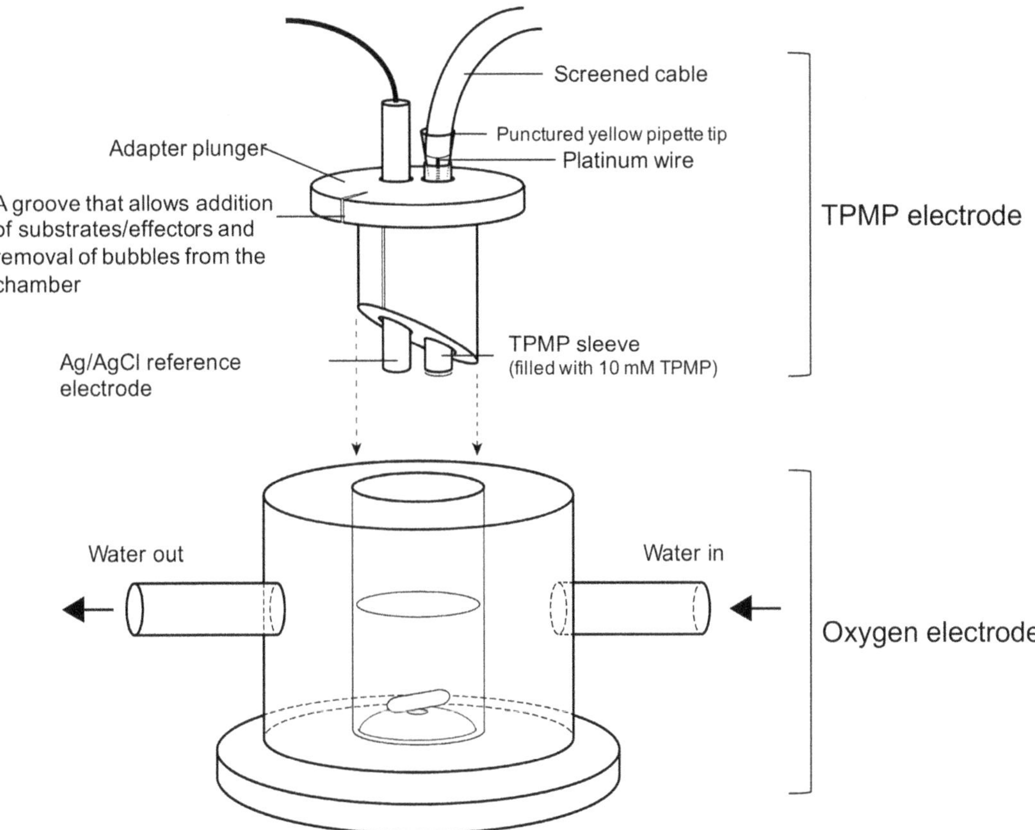

Fig. 1 TPMP electrode and oxygen electrode assembly. The TPMP electrode and oxygen electrode are set up as described in Subheading 3

fashion. Moreover, electrode behavior tends to improve with use. If such improvement is not observed rapidly, the easiest solution is to try another sleeve. Further troubleshooting suggestions are given in [6]. Following conditioning and between experiments, TPMP sleeves may be stored in medium containing 2–3 μM TPMP for many months.

(c) *TPMP electrode calibration and correction for small baseline drift*—Although the TPMP electrode exhibits fairly reproducible behavior within daily sets of experimental traces, we nevertheless recommend calibrating its response and assessing possible drift for each trace individually. For a typical experiment, incubate mitochondria (0.35 mg/mL) at 37 °C in 3.5 mL KHEP. Add rotenone (4 μM) and nigericin (350 ng/mL) to inhibit respiratory complex I and collapse the pH gradient across the mitochondrial inner membrane, respectively. Insert the TPMP and reference electrodes and wait until a stable signal is observed. Using the 1 mM solution, add TPMP in 0.5 μM increments to a concentration of 2.5 μM; await a stable signal before each successive addition (*see* **Note 8**). Energize the mitochondrial inner membrane by adding succinate (4 mM). If the mitochondria are well coupled, energization will result in a significant TPMP accumulation into the mitochondrial matrix and, therefore, a substantial decrease of the electrode signal. Effectors specific to a particular experiment can then be added; in general, wait until a steady signal is obtained before making additions. The experiment is completed by adding FCCP (1.5 μM), which will dissipate the mitochondrial protonmotive force (i.e., $\Delta\Psi$ when nigericin is present), will thus cause TPMP release from the mitochondria, and will bring the external TPMP concentration back to 2.5 μM within a minute or so. Comparison of the final electrode signal with that observed before succinate addition allows determination of a possible electrode drift rate. If deemed acceptably small, then this rate should be taken into account in the calibration of external TPMP levels. If electrode drift is substantial and/or persistent, its cause(s) should be identified and rectified.

(d) *Calculation of membrane potentials*—Lipophilic cations such as $TPMP^+$ equilibrate across the mitochondrial inner membrane according to the Nernst equation, $\Delta\Psi = 61.5 \times \log([TPMP^+]_{in}/[TPMP^+]_{out})$, at 37 °C. The membrane potential can thus be calculated from the external (i.e., extramitochondrial) and matrix TPMP concentrations. To enable this calculation, measure the signal deflections from the 2.5 μM TPMP baseline observed at the various applied TPMP levels, and plot this deflection as a function of log[TPMP]. This should yield a linear calibration relation, which can be used to calculate external TPMP concentrations throughout the experiment. Subtracting the external concentration from the 2.5 μM TPMP applied provides the TPMP level taken up by mitochondria, which is proportional to its concentration in the matrix. To calculate a reliable concentration, however, it is necessary to know how much of the hydrophobic TPMP is actually freely present within the matrix and how much has in fact bound to mitochondrial membranes. Excellent methods to assess probe binding to membranes and other mitochondrial and/or cellular components have been described [6]. For rat skeletal muscle mitochondria, TPMP binding has been determined as a function of mitochondrial matrix volume, and the obtained correction factor is 0.35 mg mitochondrial protein per μL matrix volume [7]. Multiplication of this correction factor by the TPMP taken up by mitochondria, normalized to the mitochondrial protein present in the assay (mg/mL), yields the free matrix TPMP concentration. $\Delta\Psi$ values in rat skeletal muscle mitochondria, incubated at 37 °C and 0.00035 mg/μL, can, therefore, be calculated from applied and external TPMP concentrations as:

$$61.5 \times \log\left(\frac{\left(\left[\mathrm{TPMP}^{+}\right]_{\mathrm{applied}} - \left[\mathrm{TPMP}^{+}\right]_{\mathrm{external}}\right) \times 0.35}{0.00035 \times \left[TPMP^{+}\right]_{\mathrm{external}}}\right)$$

5. Proton leak: A modular kinetic description of oxidative phosphorylation.

   Oxidative phosphorylation is a process that can be divided conceptually into events that either generate or dissipate the mitochondrial membrane potential [8]. Such a modular view of mitochondrial energy transduction (Fig. 2a) emphasizes that steady-state respiratory activities and $\Delta\Psi$ levels result from the kinetic interplay between activity of the mitochondrial electron transfer chain ($\Delta\Psi$-producer) and the combined activities of the ADP-phosphorylation machinery and the proton leak across the mitochondrial inner membrane ($\Delta\Psi$-consumers). The kinetic dependency of $\Delta\Psi$-establishing and $\Delta\Psi$-dissipating modules on $\Delta\Psi$ is modeled in Fig. 2b based on the assumption that the two $\Delta\Psi$-dissipating modules respond hyperbolically (this is not strictly correct, but serves to illustrate the method here). In mitochondria incubated under phosphorylating conditions, a respiratory steady state is reached when the $\Delta\Psi$-establishing rate is equal to the sum of the $\Delta\Psi$-dissipation rates (Fig. 2b, state 3). Steady states achieved upon either inhibition of phosphorylation with oligomycin or dissipation of the mitochondrial protonmotive force with excess FCCP are labeled "state 4" and "state F," respectively (Fig. 2b). Note that the oligomycin-inhibited state is generally comparable to the steady state reached in the absence of ADP (state 4). Below, we describe methods to determine modular kinetic relations experimentally.

   (a) *Theoretical principle*—The kinetic behavior of a "$\Delta\Psi$-producer" can be established by specific modulation of a "$\Delta\Psi$-consumer" (Fig. 2c), and, reciprocally, the behavior of a consumer is revealed upon modulation of a producer (Fig. 2d; [8]). In other words, if mitochondrial respiration is titrated under state 3 conditions with subsaturating amounts of an electron transfer chain inhibitor, the

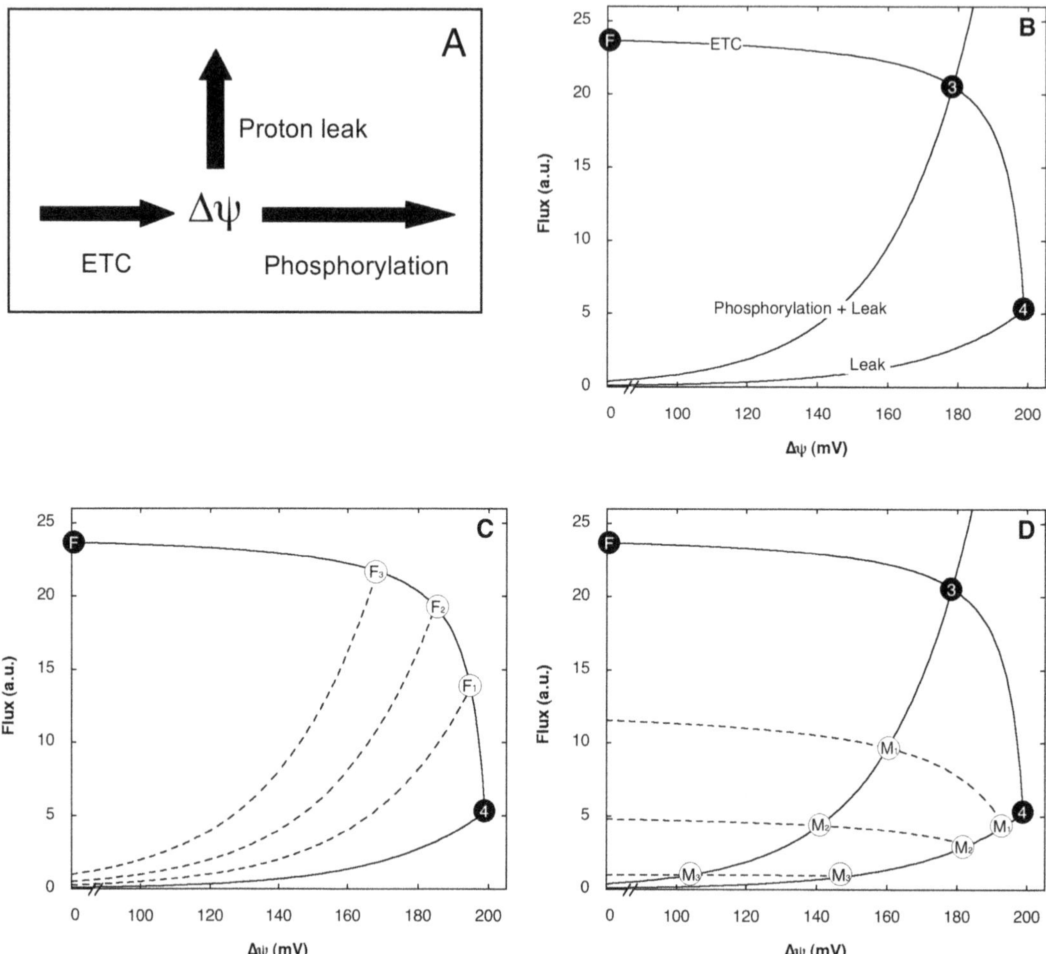

Fig. 2 Modeled modular kinetics during oxidative phosphorylation in isolated mitochondria. (**a**) Oxidative phosphorylation considered from a top-down perspective as an interaction between processes that establish $\Delta\Psi$ (electrode transport chain, ETC) and those that dissipate it (proton leak and phosphorylation). (**b**) The kinetic dependency of both the ETC and the sum of proton leak and phosphorylation rates (fluxes expressed in arbitrary units) on $\Delta\Psi$ was modeled using exponential (proton leak, phosphorylation) and hyperbolic (ETC) equations. The model assumes zero ΔpH (presence of nigericin) so that the protonmotive force equals $\Delta\Psi$. Steady states are achieved when $\Delta\Psi$-establishing flux equals total $\Delta\Psi$-dissipating flux, which is reflected by intersections of the curves describing the behavior of the respective rates. Modeled steady states include those reached in the presence (state 3) and the absence of ADP (state 4) as well as that achieved in the presence of FCCP (state F). (**c**) Titration of oligomycin-insensitive respiration rate with subsaturating amounts of FCCP increases proton leak activity specifically. New steady states will thus be achieved that reflect the altered proton leak kinetics and the *unaffected* ETC kinetics (state F1–F3). (**d**) Titration of succinate oxidation with subsaturating amounts of malonate inhibits ETC activity specifically. In this case, new steady states will be reached that reflect the altered ETC kinetics and the *unaffected* $\Delta\Psi$-dissipating kinetics (states M1–M3). Malonate titrations under phosphorylating conditions thus reveal the kinetic behavior of the sum of $\Delta\Psi$-dissipating behavior, and titrations in the presence of oligomycin reveal the kinetic behavior of proton leak alone

kinetics with respect to $\Delta\Psi$ of the total $\Delta\Psi$-dissipating activity (i.e., leak + phosphorylation) are revealed. If this titration is performed in the presence of oligomycin, the kinetics of proton leak alone are obtained, which may be subtracted from the overall $\Delta\Psi$-dissipating kinetics to reveal the behavior of the ADP-phosphorylation module. If respiration is titrated with FCCP, either in the presence or the absence of oligomycin, the kinetic dependency of the respiratory chain on $\Delta\Psi$ is established. During these respiration titrations, successive steady states will be attained in which both the oxygen uptake rate (flux) and $\Delta\Psi$ (intermediate concentration) should be measured simultaneously. Plotting oxygen consumption rates as a function of the concomitant $\Delta\Psi$ values allows empirical construction of modular kinetic plots.

(b) *Proton leak kinetics*—Add mitochondria (0.35 mg/mL) to 3.5 mL KHEP in an oxygen electrode vessel incubated at 37 °C in the presence of rotenone (4 μM), nigericin (350 ng/mL), and oligomycin (0.7 μg/mL). Apply the plunger, insert the TPMP and reference electrodes, and record both oxygen and TPMP signals. Make five sequential TPMP additions of 0.5 μM each. Add succinate (4 mM) to initiate oligomycin-resistant respiration and to energize the mitochondria. When the oxygen uptake rate and TPMP signal have stabilized, make at least five successive malonate additions increasing its concentration gradually to approximately 0.5 mM; finish the trace by adding FCCP (1.5 μM). Calculate specific oxygen consumption rates (*see* **Note 9**) and plot as a function of the concomitant $\Delta\Psi$ values. The leak rate dependency on $\Delta\Psi$ may be modeled approximately by an exponential expression.

(c) *Phosphorylation kinetics*—Perform an experiment identical to that described for the proton leak kinetics, but omit oligomycin from the incubation mixture. Instead, include sufficient ADP or an ADP-regenerating system to enable phosphorylating respiration throughout the trace. The data obtained in this experiment represent the kinetic dependency of the *sum* of phosphorylation and leak activities on $\Delta\Psi$. Subtraction of proton leak kinetics yields the $\Delta\Psi$-dependency of the phosphorylation module alone, which may also be approximated by an exponential expression.

(d) *Electron transfer chain kinetics*—Add mitochondria (0.35 mg/mL) to 3.5 mL KHEP in an oxygen electrode vessel incubated at 37 °C in the presence of rotenone (4 μM), nigericin (350 ng/mL), and oligomycin

(0.7 μg/mL). Add TPMP to 2.5 μM in five equimolar increments and add succinate (4 mM) to initiate respiration. When the oxygen uptake rate and TPMP signal have stabilized, make at least five successive FCCP additions, increasing its concentration gradually to approximately 0.8 μM; finish the trace by adding surplus FCCP (1.5 μM). Calculate specific oxygen consumption rates and plot as a function of the concomitant $\Delta\Psi$ values. The electron transfer chain rate dependency on $\Delta\Psi$ may be modeled approximately by a hyperbolic expression.

Simultaneous measurement of oxygen consumption and $\Delta\Psi$ provides considerably more insight into oxidative phosphorylation than individual determination of these parameters [1]. Mitochondrial coupling efficiency is better understood from modular kinetic information than from respiration alone since the kinetic behavior of the respective modules explains the exact nature of a particular RCR and, moreover, identifies the underlying cause(s) of possible differences in RCRs between experimental conditions and systems [2]. Mitochondrial proton leak activity can be compared in a meaningful manner between experimental systems and conditions, because the oligomycin-insensitive respiratory rates can be corrected for possible differences in the leak's driving force, $\Delta\Psi$. Similarly, activity of the other kinetic modules can be probed at identical $\Delta\Psi$ values [9]. Site(s) at which suspected effectors of oxidative phosphorylation act can be identified unambiguously [10]. Conclusive evidence can be obtained as to the role played by poorly characterized proteins in oxidative phosphorylation [11]. Elasticities of all kinetics modules to $\Delta\Psi$ can be calculated, which allows control and regulation of oxidative phosphorylation to be quantified [8].

4 Notes

1. The standard oxygen-permeable Teflon membrane may be substituted with high-sensitivity membrane (YSI Life Sciences, OH, USA) to decrease the response time of the oxygen electrode.
2. Bovine serum albumin should be added to the KHEP assay medium on the day of use; sprinkle the required amount (0.3%, w/v) on top of the solution and let it dissolve without stirring to prevent clotting.
3. Modern fume hoods may sometimes not be switched off; to prevent hasty evaporation and surface rippling of the membrane, cover the Petri dishes in this case with lids but leave small openings to enable drying.

4. When cutting PVC tubing, it is pivotal to get a very clean and straight cut as the sleeve will not stick to the membrane otherwise. The likelihood of obtaining such cuts is maximized when performed by two persons: one pulls the tubing as tightly as possible, and another cuts it very quickly and cleanly with a sharp razor blade.
5. Whether or not the assembled setup detects oxygen and responds rapidly to changes in oxygen concentration can be checked conveniently by switching off the electronic stirrer temporarily, which should cause a substantial and fast signal deflection. Lack of stirring means that oxygen level near the platinum cathode and the bulk assay medium are no longer kept in rapid equilibrium and, therefore, that the oxygen uptake at the cathode will lead quickly to localized oxygen deprivation.
6. Sodium dithionite cannot always be used to deplete oxygen from the assay medium as it may interfere with other measurements such as the detection of TPMP. An alternative way to obtain the zero-oxygen signal is to rely on mitochondrial respiration to consume all oxygen and use the anaerobic signal achieved at the end of an experimental trace. If this "biological zero" is known from previous experiments to be similar to the signal obtained from an unpolarized electrode, this "electrical zero" could be used as a quick-and-dirty approximation of a zero-oxygen signal.
7. Some respiratory effectors (e.g., rotenone and FCCP) are hard to wash away between experimental traces as they tend to stick to the plastic oxygen electrode vessel. Even when a glass vessel is used to reduce this sticking issue, it may sometimes be wise to rinse the electrode chamber between traces with a concentrated (3%, w/v) bovine serum albumin solution or with a suspension of frozen-thawed mitochondria that were left over from earlier experiments. Bovine serum albumin and mitochondria will bind many possible nasties, and using them as a mop will greatly reduce the likelihood of effector carry-over.
8. Once in a while, the TPMP electrode will exhibit small drift. It is important, however, to distinguish such drift from a lack of biological steady state as it is possible that $\Delta\Psi$ is changing slowly because of alteration in electron transfer chain, proton leak, and/or phosphorylation activities. Changes in these processes, however, will also likely be reflected in altered oxygen consumption rates. An important indication, therefore, that a slowly changing TPMP signal reflects biological reality and is not merely due to electrode drift, is a concomitantly changing oxygen electrode signal.

9. Oligomycin-insensitive oxygen uptake activities are converted to proton flux rates by multiplying them by appropriate H^+/O ratios, which are 6 and 10 during the oxidation of succinate and pyruvate/malate, respectively. ADP-phosphorylation rates are obtained by multiplying the leak-corrected state 3 respiratory rates by the appropriate P/O ratios, which are 1.64 during oxidation of succinate and 2.73 during oxidation of pyruvate plus malate [12].

Acknowledgments

C.A. thanks Julie Buckingham for the useful advice on the preparation of TPMP sleeves.

References

1. Nicholls DG, Ferguson SJ (2013) Bioenergetics, 4th edn. Academic Press, Elsevier, Amsterdam
2. Brand MD (2005) The efficiency and plasticity of mitochondrial energy transduction. Biochem Soc Trans 33:897–904. https://doi.org/10.1042/BST20050897
3. Reynafarje B, Costa LE, Lehninger AL (1985) O_2 solubility in aqueous media determined by a kinetic method. Anal Biochem 145:406–418
4. Chance B, Williams GR (1955) Respiratory enzymes in oxidative phosphorylation. III. The steady state. J Biol Chem 217:409–427
5. Brand MD, Nicholls DG (2011) Assessing mitochondrial dysfunction in cells. Biochem J 435:297–312. https://doi.org/10.1042/BJ20110162
6. Brand MD (1995) Measurement of mitochondrial protonmotive force. In: Brown GC, Cooper CE (eds) Bioenergetics: a practical approach. IRL, Oxford, pp 39–62
7. Rolfe DFS, Hulbert AJ, Brand MD (1994) Characteristics of mitochondrial proton leak and control of oxidative phosphorylation in the major oxygen-consuming tissues of the rat. Biochim Biophys Acta 1188:405–416. https://doi.org/10.1016/0005-2728(94)90062-0
8. Brand MD (1998) Top-down elasticity analysis and its application to energy metabolism in isolated mitochondria and intact cells. Mol Cell Biochem 184:13–20
9. Brand MD, Affourtit C, Esteves TC, Green K, Lambert AJ, Miwa S, Pakay JL, Parker N (2004) Mitochondrial superoxide: production, biological effects, and activation of uncoupling proteins. Free Radic Biol Med 37:755–767. https://doi.org/10.1016/j.freeradbiomed.2004.05.034
10. Affourtit C, Brand MD (2008) Uncoupling protein-2 contributes significantly to high mitochondrial proton leak in INS-1E insulinoma cells and attenuates glucose-stimulated insulin secretion. Biochem J 409:199–204. https://doi.org/10.1042/BJ20070954
11. Dröge W (2002) Free radicals in the physiological control of cell function. Physiol Rev 82:47–95. https://doi.org/10.1152/physrev.00018.2001
12. Mookerjee SA, Gerencser AA, Nicholls DG, Brand MD (2017) Quantifying intracellular rates of glycolytic and oxidative ATP production and consumption using extracellular flux measurements. J Biol Chem 292:7189–7207. https://doi.org/10.1074/jbc.M116.774471

Chapter 10

Imaging Mitochondrial Calcium Fluxes with Fluorescent Probes and Single- or Two-Photon Confocal Microscopy

Sean M. Davidson and Michael R. Duchen

Abstract

The concentration of calcium ions in the mitochondria has been shown to affect its function, modulating respiratory activity at low levels and causing lethal damage at high concentrations. The rhodamine series of dyes can be used to measure mitochondrial calcium concentration, but the reliability of measurements depends upon correct partitioning of the dye within to the mitochondria. Methods are described to aid verification and quantification of the mitochondrial calcium concentration using single- or two-photon confocal microscopy. The method of linear unmixing to separate fluorescent signals based on either differing excitation or emission spectra is outlined and for the purposes of illustration is applied to the separation of rhod-2 signals originating from the dye within the mitochondria and nucleoli.

Key words Calcium, Mitochondria, Rhod-2, Confocal microscopy, Multiphoton microscopy

1 Introduction

Ca^{2+} is not distributed evenly throughout the cell, but concentrated in microdomains, some of which are membrane-bound, such as the mitochondria [1]. Ca^{2+} enters mitochondria via the Ca^{2+} uniporter in accordance with its electrochemical gradient and accumulates in the mitochondrial matrix [2]. The concentration of Ca^{2+} in the mitochondria can affect mitochondrial respiration, since the rate-limiting enzymes of the citric acid cycle of mitochondria are activated by Ca^{2+} [2]. Physiological stimulation can increase $[Ca^{2+}]_{mito}$ in numerous cell types. For example, $[Ca^{2+}]_{mito}$ changes with each contraction in cardiomyocytes. Mitochondria may act as a spatial buffer in normal cellular calcium signaling [1, 3, 4].

Mitochondrial calcium uptake is an important determinant of cell death particularly in the heart and the brain [4]. In particular, mitochondrial calcium overload causes opening of a large nonspecific pore in the mitochondria—the mitochondrial permeability

Carlos M. Palmeira and António J. Moreno (eds.), *Mitochondrial Bioenergetics: Methods and Protocols*, Methods in Molecular Biology, vol. 1782, https://doi.org/10.1007/978-1-4939-7831-1_10,

transition pore (mPTP) [4]. This results in rapid depolarization of the mitochondrial membrane and cessation of mitochondrial respiration. Prevention of mPTP opening can prevent cells from dying during ischemia and reperfusion injury [5]. Given its pathological importance and the importance of understanding how it is regulated, it is therefore necessary to be able to measure $[Ca^{2+}]_{mito}$ accurately.

Direct measurement of intramitochondrial Ca^{2+} using electron probe X-ray microanalysis showed that, under physiological conditions, the concentration of free matrix Ca^{2+} in liver mitochondria in vivo is fairly low, at ~0.3 mM [6], though it should be kept in mind that up to 2500 times more calcium may be present in a non-ionized or bound form [7]. The development of dyes which increase fluorescence when they bind Ca^{2+} has greatly simplified the measurement of intracellular Ca^{2+}. Detection of mitochondrial Ca^{2+} was enabled in 1989 with the development of indicators based on rhodamine [8]. A range of these dyes are available with different Ca^{2+}-binding affinities (Table 1). With a K_D of 570 nM, rhod-2 is suitable for most studies. Intracellular accumulation is enhanced with the AM ester form of the dye (available from Invitrogen/Molecular Probes). In rhod-2/AM partitions in the mitochondria, but depending on the loading conditions used, there will normally be some dye present in the cytosol. With careful analysis, however, it can still be used to measure mitochondrial calcium.

There are other potential limitations with the use of rhod-2 that one should be aware of. Firstly, it is not ratiometric, i.e., since the emission spectrum does not change on binding to Ca^{2+}, there is no way to control for efficiency of loading of the dye by comparing Ca^{2+}-responsive and Ca^{2+}-nonresponsive emitted fluorescence. Secondly, the reaction products of AM ester hydrolysis are acetate and formaldehyde, which may affect cellular functioning.

Table 1
Kd (dissociation constant) values of a number of different mitochondrial Ca^{2+}-sensitive dyes (*see* Note 21)

Dye	Kd
Rhod-2	570 nM
X-rhod-1	700 nM
Rhod-FF	19 μM
Rhod-5 N	320 μM
X-rhod-FF	17 μM
X-rhod-5F	1.6 μM
X-rhod-5 N	350 μM

Thirdly, the presence of rhod-2 may affect Ca^{2+} buffering itself, since it binds to calcium, so the concentration of rhod-2/AM used should not be increased above that recommended here. Lastly, the dye may be sensitive to pH and may also accumulate in other cellular structure such as lysosomes [9]. In many cells, it also accumulates within nucleolar regions, appearing as bright regions within the nucleus, though this appears to be artifactual and there is no evidence it corresponds to Ca^{2+}. It may be possible to remove some of this contaminating signal by computational methods such as linear unmixing as described in the method below, using for the purposes of illustration the example of separating the fluorescent signal originating from the mitochondria from that originating from the nucleoli.

By being aware of these precautions and by following the steps outlined in the protocols below, it is possible to take accurate measurements of free mitochondrial calcium.

2 Materials

1. Rhod-2/AM (Molecular Probes/Invitrogen) is supplied in aliquots. Add 50 μl DMSO to a tube before use, and store at −20 °C for up to 1 month.
2. Imaging dish to contain coverslip and buffer while imaging cells.
3. Confocal microscope equipped with HeNe (543 nm wavelength) laser and inverted objective (*see* **Note 1**).
4. Imaging buffer: 156 mM NaCl, 3 mM KCl, 2 mM MgSO4.7H_2O, 1.25 mM K_2HPO_4, 2 mM $CaCl_2$, 10 mM HEPES, 10 mM D-glucose, adjust to pH 7.4 with NaOH.
5. Ca^{2+}-free buffer: imaging buffer + 5 μM A23187 (*see* **Note 2**) + 5 mM EGTA to obtain the minimum (Rmin) fluorescence value.
6. High Ca^{2+} buffer: imaging buffer +5 μM A23187 (*see* **Note 3**) in the presence of 3 mM Ca^{2+} to obtain the maximum (Rmax) fluorescence value.

Optional

1. Intracellular buffer: 130 mM KCl, 80 mM aspartate, 10 mM HEPES, 3 mM $MgCl_2$.6H_2O, 4 mM Na. pyruvate, 0.5 mM EGTA, 3 mM Na_2. ATP, 0.12 mM $CaCl_2$. Adjust pH to 7.3 using KOH.
2. Permeabilization buffer: intracellular buffer containing either 50 μg/ml saponin or 2.5 μM digitonin.

3 Methods

Rhod-2/AM is taken up into cells, and the AM ester is cleaved by cellular esterases. Various loading protocols can be used in order to increase the extent of the dye loading into the mitochondria. A common technique is to load the dye at low temperature (RT, or even 4 °C), which inhibits esterase activity and allows the dye to localize into mitochondria, and then move the cells to 37 °C where the dye is de-esterified and activated [9, 10]. The optimum protocol depends on the cell type and should be determined empirically. If loading of the dye can be achieved such that it is predominantly mitochondrial, then it may be possible to image fluorescence using a fluorescent microscope. If not, in order to reliably distinguish the mitochondrial signal from the cytosolic signal, it will be necessary to use a confocal microscope to obtain the images. Note that an "inverted microscope" with the objective approaching the coverslip from below is required so that the live cells can be imaged while bathed in the imaging buffer. However, even after optimization, there may still be some mislocalization of rhod-2 to intracellular structures such as lysosomes [9], and nucleoli, in which it may be possible to distinguish by using dye separation techniques such as linear unmixing. This involves taking a series of images of the same field over a range of excitation or emission wavelengths and then digitally separating the fluorescence into the two (or more) components and is described below.

Various approaches may be used to verify the extent of mitochondrial loading, including ensuring the majority of the dye remains in the mitochondria after plasma membrane permeabilization (Fig. 1d), loading cells with a cytosolic dye such as fluo-4/AM to confirm that the mitochondrial signal is distinct from that of the cytosol (Fig. 2), and addition of 1 μM FCCP to depolarize mitochondria causing a decrease in $[Ca^{2+}]_{mito}$.

It is very important to bear in mind that after loading cells with rhod-2, mitochondrial localization may not be immediately obvious, as the signal will be weak at resting levels of $(Ca^{2+})_m$. It may be necessary to stimulate the cells with an agonist to raise $(Ca^{2+})_m$ before the mitochondrial localization becomes evident (Fig. 2a (i and ii), b (i and ii)). Note also that calcium-independent variations in signal intensity through the cell—staining of lysosomes or the nucleoli, for example—disappear if the images are ratioed against a first "resting" image. This means extracting a baseline image from an image series and dividing the whole of the rest of the series by that first image (Fig. 2b). Under these conditions, any nonuniformities of labeling that do not change when calcium signals change disappear. To prevent the background between cells becoming very noisy (as here you are dividing very small numbers by other very small numbers, and so there will be a lot of noise),

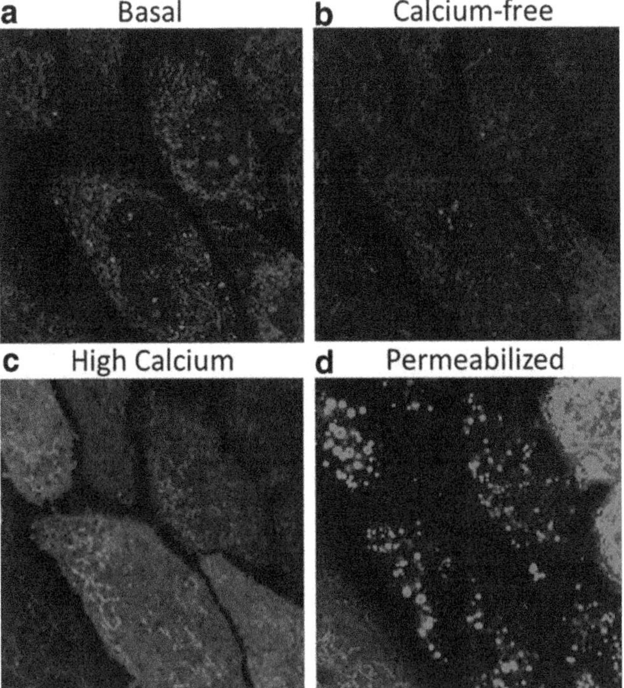

Fig. 1 Cells loaded with rhod-2/AM according to the standard protocol. Some fluorescence is visible originating from the mitochondria, though there is also some dye present throughout the cytosol and in the nucleoli (small spots within the nucleus) (**a**). By measuring the fluorescence on changing to the $Ca2^{+}$-free buffer (**b**) and then to high $Ca2^{+}$ buffer (**c**), it is possible to calculate the actual Ca^{2+} concentration; in this case, using Kd of 570 nM, we obtain a $[Ca^{2+}]_{mito}$ of 196 nM. From panel C, the relative concentration of rhod-2 in the mitochondria is determined to be 2.7-fold that in the cytosol. Finally, plasma membrane permeabilization (**d**) eliminates all but mitochondrial staining, allowing estimation of the proportion of the dye localized to mitochondria (here, ~30%)

one can multiply the image series by a binary mask. To achieve the binary mask is trivial in most imaging software—it requires selecting a threshold and setting all signals above the threshold to a value of unity and all pixels below the threshold to zero. Multiplying the image series by the mask makes all the signals below the threshold disappear as set to zero, while the image data of interest above threshold remains unchanged.

In the example in Fig. 2, HL-1 cells—a cardiac-derived cell line that shows spectacular spontaneous calcium signals—were dual loaded with rhod-2/AM and fluo-4/AM as described in the protocol above. The sequence of Fig. 2a shows the raw rhod-2 images, Fig. 2b shows the ratioed rhod-2 images, while Fig. 2c shows the sequence of the fluo-4 signals (ratioed against the first image). Figure 2d shows plots of the intensity with time for each fluorophore from two regions of interest as marked (arrowhead and asterisk). Spontaneous cytosolic signals can be seen (1), and only when the cytosolic signals reached a high enough level did

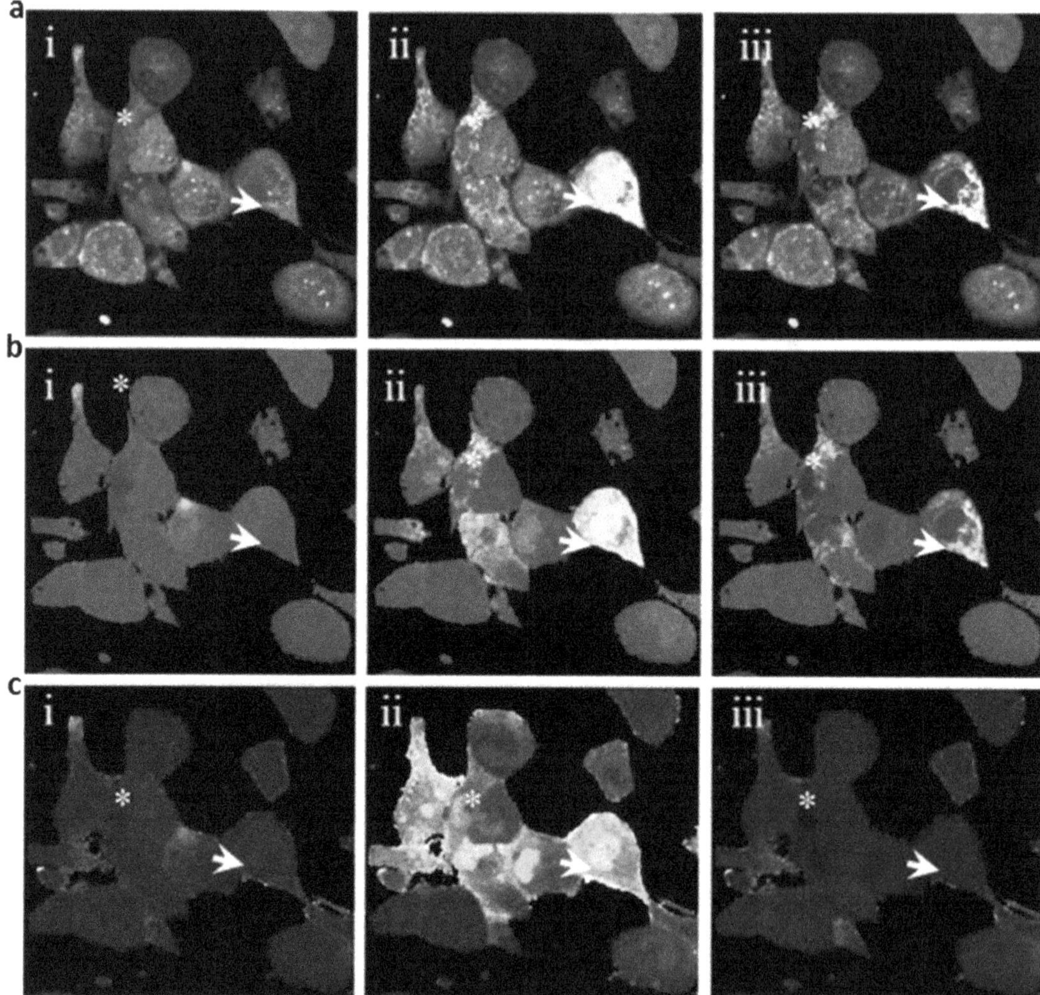

Fig. 2 HL-1 cells were dual loaded with fluo-4/AM and rhod-2/AM (5 μM each for 20 min at room temperature) followed by washing. They were mounted on the stage of a Zeiss 700 confocal microscope and images acquired sequentially exciting at 488 and 555 nm, measuring emitted fluorescence at 505–550 nm (fluo-4) and at >570 nm (rhod-2). Both localized and global spontaneous calcium signals occur in these cells. Images extracted from the time sequence show a "resting" signal showing small local calcium signals in two cells (1), the peak global cytosolic signal (2), and the residual mitochondrial signal after the recovery of the cytosolic signal (3). The sequence (**a**) shows the raw rhod-2 data, in (**b**) the rhod-2 images ratioed against the resting signal and in (**c**) the corresponding fluo-4 images ratioed against the resting images. In (**d**), two plots are shown with the points indicated by arrowhead and asterisk showing the divergence of the time course of the cytosolic calcium signal (green) and the mitochondrially localized calcium signal (rhod-2, red)

mitochondrial calcium rise. There are several important points illustrated here: (1) at the start of the imaging, no mitochondrial signal is detectable, but mitochondria appear when calcium rises; (2) significant cytosolic rhod-2 signal remains but is distinguishable from the mitochondrial signal as the latter shows a distinct time course which is quite different to the cytosolic signal—which is in

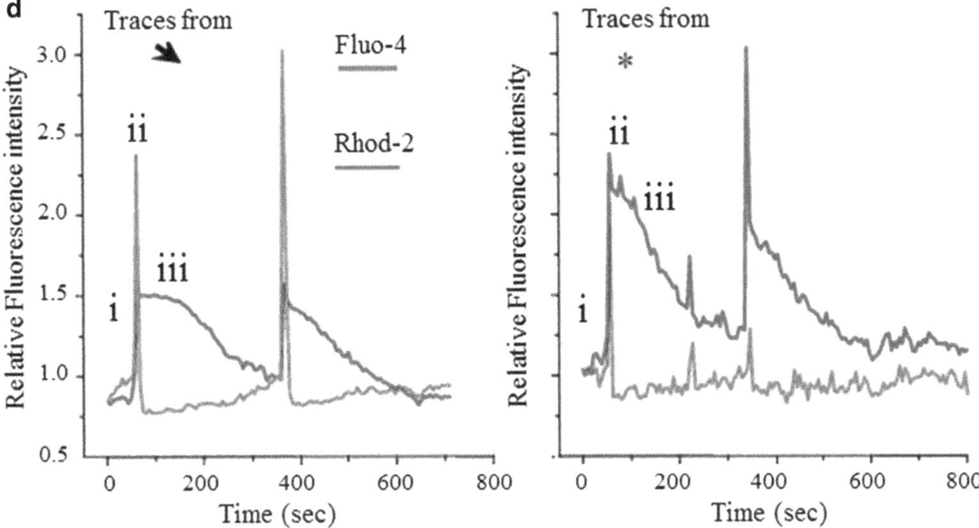

Fig. 2 (continued)

turn confirmed by the fluo-4 signal; (3) in the ratioed image, local variations in signal disappear, highlighting only the areas in which the calcium-specific signal is changing.

It is possible to convert the fluorescent values obtained from the microscope into actual Ca^{2+} concentration. This requires calibration of the fluorescence values obtained when the dye is saturated (i.e., in the presence of high [Ca^{2+}]) (Fig. 1c) and background (i.e., in the presence of a chelator of Ca^{2+} such as EGTA) (Fig. 1b). These measurements are performed in the presence of a Ca^{2+} ionophore which equalizes [Ca^{2+}] inside and outside the cell.

The following procedure describes the use of a Leica SP5 confocal microscope, but the procedure is similar on any similar confocal microscope.

3.1 Preparing the Cells

1. Place glass coverslips in the bottom of 6-well tissue culture dishes or plates (*see* **Note 4**).
2. Trypsinize the cells and plate them in the wells containing coverslips at a density sufficient to achieve ~70% confluence when they are imaged.
3. Return cells to tissue culture incubator for at least 24 h to allow them to attach and recover (*see* **Note 5**).

3.2 Loading Rhod-2/AM into the Cells

1. In a 1.5 ml Eppendorf tube, add 5 μM rhod-2/AM + 5 μl 20% pluronic, and then add 1 ml imaging buffer.
2. Replace the buffer on the cells with imaging buffer containing rhod-2/AM and pluronic.
3. Leave cells 30 min RT to take up the dye.

4. Place cells in tissue culture incubator for 20 min for de-esterification of dye.
5. Leave the cells 20 min RT to complete dye de-esterification (*see* **Note 6**).

3.3 Confocal Imaging of Rhod-2/AM Fluorescence

1. Transfer the coverslip containing the cells to the imaging chamber.
2. Wipe any liquid from underneath the coverslip using a tissue, and place the chamber on the objective of a microscope, adding a drop of oil to the objective if necessary.
3. Using phase contrast or transmitted light, adjust the focus until the cells are clearly visible (*see* **Note 7**).
4. In the confocal microscope software, choose an appropriate default imaging option for imaging of a red fluorescent dye, (*see* **Note 8**) i.e., excitation using the 543 nm line of the laser and emitted light collected between 560 and 630 nm (or if a band-pass filter is not available, use a longpass filter of 560 nm).
5. Decrease the laser power to ~5% to avoid damaging the cells, and increase the gain setting to maximum.
6. Start the continuous scan, and adjust the focus until the mitochondria in the cell are clearly visible (*see* **Note 9**).
7. Open the pinhole setting in the software to approximately 3 Airey Units (AU), (*see* **Note 10**) and decrease the gain until a signal of ~50% saturated intensity is obtained (*see* **Note 11**).
8. Stop continuous scanning.
9. Increase the number of images averaged to 4, and/or decrease the scan speed to obtain a higher-quality image.
10. Start the scan to obtain the image of rhod-2 fluorescence in the cells.

3.4 Calibrating the Fluorescence to Calcium Concentration

Incubating the cells with a calcium ionophore will cause [Ca^{2+}] in all cellular compartments to equalize to the concentration in the buffer. In a buffer containing no Ca^{2+} and EGTA to chelate all Ca^{2+}, the minimum rhod-2 fluorescence value can be obtained (Fig. 1b). In a buffer containing a saturating concentration of Ca^{2+}, the maximum rhod-2 fluorescence value can be obtained (Fig. 1c). Using these images and the Kd for the dye being used, it is possible to estimate the organellar concentration of calcium (Fig. 1a). However, this will necessarily be an estimate, since intracellular components, particularly pH and the presence of heavy metals, can influence the dye response.

1. Replace the buffer on the cells with calcium-free buffer and wait 2 min (*see* **Notes 12** and **13**).
2. Image the cells using the same parameters as above (*see* **Note 14**).

3. Replace the buffer on the cells with high-calcium buffer and wait 2 min.
4. Image the cells using the same parameters as above.
5. Using the analysis part of the software, draw a region of interest (ROI) containing a single mitochondrion or close group of mitochondria. Measure the average fluorescence intensity in the ROI to obtain F_{min} (calcium-free image) and F_{max} (high-calcium image).
6. To calculate free $[Ca^{2+}]_{mito}$, use F_{min}, F_{max}, and the Kd for the dye being used (*see* Table 1), in the following formula from Grynkiewicz et al. [11].

$$[Ca^{2+}]_{free} = Kd\ [F - F_{min}]/[F_{max} - F]$$

3.5 Assessing the Extent of Dye Compartmentalization by Plasma Membrane Permeabilization

One approach to estimate the extent of dye that is compartmentalized in the mitochondria is to permeabilized the plasma membrane, resulting in the loss of cytosolic dye into the buffer (Fig. 1d), and calculate the proportion remaining. This does not account for mitochondria volume however. An alternative approach is to determine the relative effective concentration of rhod-2 in the mitochondria compared to in the cytosol by determining the ratio of rhod-2 fluorescence in the mitochondria compared to the cytosol in the presence of saturating (high) calcium.

1. Use the image of the cells in high calcium from **step 4** of Subheading 3.4.
2. The relative effective concentration of rhod-2 in the mitochondria compared to in the cytosol can be determined by drawing one ROI in the cytosol and one ROI around the perimeter of a mitochondria and a third ROI in a region with no cell ("background"). Calculate the ratio according to mitochondria-background/cytosol-background.
3. Continuing from **step 4** of Subheading 3.4, replace the buffer on the cells with permeabilization buffer, which permeabilizes plasma membrane while leaving organelle membranes intact.
4. Start continuous scanning.
5. After several minutes, (*see* **Note 15**) the cytosolic signal will rapidly disappear over 10–20 s. This will happen at approximately the same time in cells. Wait until all cells in the field are permeabilized and signal intensity is constant.
6. Image the cells using the same parameters as **step 1** ("organelle fluorescence").
7. Add 1 mM $MnCl_2$ + 5 μM A23187 to the buffer to quench all remaining dye fluorescence from the mitochondria and other organelles.

8. Image the cells using the same parameters as **step 1** ("background").
9. Using a region of interest surrounding an entire cell, calculate the average fluorescence intensity of each image.
10. Express ("organelle fluorescence"-"background")/(total fluorescence"-"background") as a percentage. This represents the extent of dye compartmentalization (i.e., the percentage of rhod-2 in the cell localized to mitochondria).

3.6 Using Linear Unmixing of Emitted Wavelengths to Remove Unwanted Fluorescence Artifacts

There can be occasions when it is useful to improve the image by removing unwanted fluorescence originating from cellular autofluorescence or other artifacts. It is also possible to remove or separate the fluorescent signal of other added dyes even if they have very similar or largely overlapping fluorescent emission spectra (*see* [12] for an example of separating rhod-2 from TMRM signal) (Fig. 3b). To achieve this, it is necessary to have a confocal microscope equipped with a finely graded emission filter. For example, the Leica SP5 includes as standard the ability to perform a "lambda" scan—collecting fluorescence over a range of wavelengths. The Zeiss META addition confers a similar capability. The software then allows "linear unmixing" in which the signal is separated into that originating from the different dyes. The basic principle that follows is for a Zeiss 510 META confocal and demonstrates how the nonspecific rhod-2 fluorescence from the nucleoli can be removed from the image (Fig. 4a).

1. Perform the steps outlined in Subheadings 3.1 and 3.2 to obtain a standard image of rhod-2 fluorescence.
2. In lambda mode (selected under "configuration control" button), using the 543 nm laser, set the start and end of the

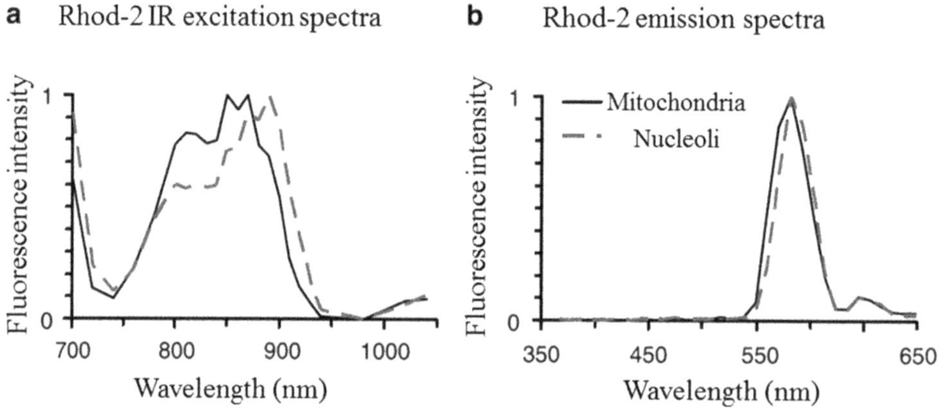

Fig. 3 The infrared excitation spectra (**a**) and emission spectra (**b**) for rhod-2 localized to the mitochondria (solid line) and nucleoli (dashed line). Despite the spectra for rhod-2 being very similar in both compartments, they are sufficiently different to allow separation of the signals when using multiphoton microscopy

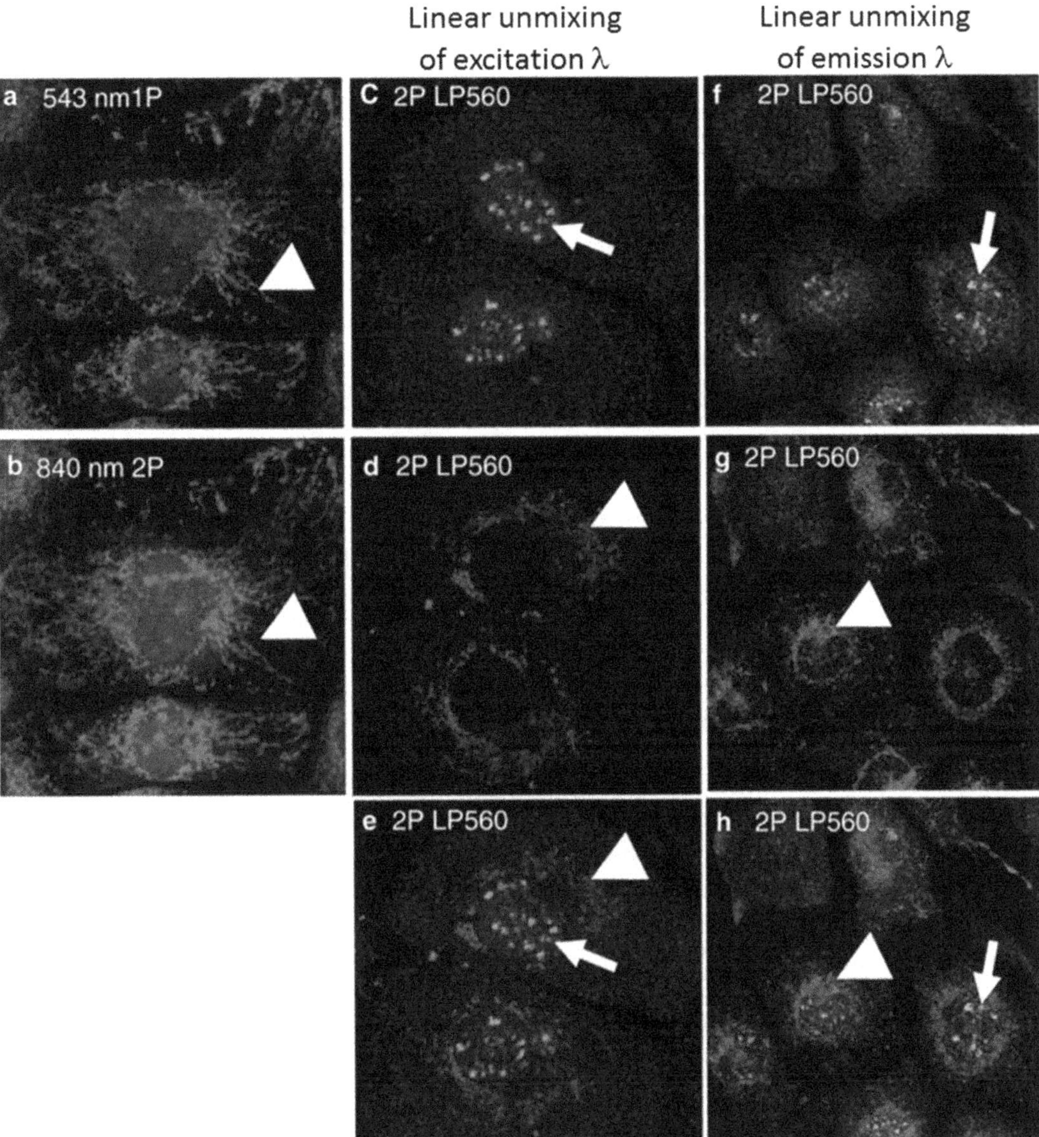

Fig. 4 Either single-photon confocal microscopy (**a**) or two-photon confocal microscopy (**b**) can be used to visualize rhod-2/AM fluorescence. The mitochondrial signal (arrowheads) can be separated from other contributing fluorescence sources such as nucleoli (arrows), by using linear unmixing of a series of images taken using 800–950 nm excitation wavelengths (**c**–**e**) or by linear unmixing of a series of images taken with 840 nm excitation wavelength and a lambda series of emission wavelengths ranging from 506 to 635 nm (**f**–**h**)

lambda stack at 550 nm and 650 nm, respectively, and the interval to 10.7 nm. Set the color palette to "Range indicator," and using continuous scan, adjust the gain so that the signal does not saturate at any pixels.

3. View the image by clicking the "mean of ROIs" display method.

4. Using the region of interest (ROI) tools, draw a ROI exactly surrounding the rhod-2 fluorescence in a nucleolus, a second ROI around a region where the mitochondria are bright, and a third large ROI in a region outside the cell where there is background signal.
5. Click on "Linear unmixing," and the software will separate the image into a new image with four panels. Panel 1 contains all the nonspecific and nucleolar fluorescence, panel 2 contains the mitochondrial rhod-2 signal, panel 3 contains the background (which should be dark), and panel 4 contains an overlay of the other three panels. Adjust the colors as desired (e.g., red for mitochondria, green for nonspecific) (Fig. 4c–e).

3.7 Multiphoton Confocal Imaging of Rhod-2 Fluorescence

Multiphoton confocal imaging is similar to confocal imaging except it is designed to use a tuneable laser in the infrared range of wavelengths, and excitation of the fluorophore (i.e., rhod-2 in this case) occurs only when two independent photons arrive simultaneously at the focal point. This design confers a number of advantages including lower toxicity, less photobleaching, and greater depth penetration in the case where imaging is performed in whole tissues. The microscope must be designed or adapted specifically for multiphoton imaging, usually including a non-descanned detector for greater light sensitivity. When using the non-descanned detector, it is necessary to exclude all extraneous lights by covering the microscope in a lightproof sheet or cover and normally by turning out the room lights as well. To image rhod-2 fluorescence in cells, load the dye as per the procedure above, but instead of the instructions in Subheading 3.3, perform the following steps (described for a Zeiss LSM NLO microscope). Rhod-2 can be excited by any of a range of infrared excitation wavelengths (Fig. 3a), though usually 840 nm is appropriate.

1. Transfer the coverslip containing the cells to the imaging chamber.
2. Place the chamber on the objective of a microscope.
3. Using phase contrast or transmitted light, adjust the focus until the cells are clearly visible (*see* **Note 16**).
4. In the confocal microscope software, adjust the excitation wavelength to an appropriate value (e.g., 840 nm).
5. Collect emitted light using a band-pass filter such as 575–640 nm (or if a band-pass filter is not available, use a longpass filter of 560 nm).
6. Using the least power possible, adjust the gain and power to obtain a satisfactory image.
7. Start the continuous scan, and adjust the focus until the mitochondria in the cell are clearly visible (*see* **Note 17**).

8. Decrease the gain until a signal of ~50% saturated intensity is obtained (*see* **Note 18**).
9. Stop continuous scanning.
10. Increase the number of images averaged to 4, and/or decrease the scan speed to obtain a higher-quality image.
11. Start the scan to obtain the image of rhod-2 fluorescence in the cells.

3.8 Using Linear Unmixing of Excitation Wavelengths to Remove Unwanted Fluorescence Artifacts

As describe in Subheading 3.5, it is possible to separate out the fluorescence from different dyes, not only by their emitted spectra but by their excitation spectra. Since it is possible to tune the excitation wavelength of the infrared laser, using a multiphoton microscope, one can also perform linear unmixing of the images as described below for the Zeiss 510 NLO.

1. Begin by obtaining a standard image at 840 nm as described in Subheading 3.7.
2. Configure the software so that it is imaging relatively fast over the entire area of the cell (<4 s per scan) and so that there are no saturated pixels.
3. Select and run the macro "XPrint" (*see* **Note 19**). Select the "Excitation lambda stack" tab."
4. Load the data table for the appropriate objective (*see* **Note 20**).
5. Set the start and end wavelengths to 550 nm to 650 nm, respectively, at 10 nm intervals.
6. Click "Start."
7. Using the region of interest (ROI) tools, draw a ROI exactly surrounding the rhod-2 fluorescence in a nucleolus, a second ROI around a region where the mitochondria are bright, and a third large ROI in a region outside the cell where there is background signal.
8. Click on "Linear unmixing," and the software will separate the image into four panels. Panel 1 contains all the nonspecific and nucleolar fluorescence, panel 2 contains the mitochondrial rhod-2 signal, panel 3 contains the background (which should be dark), and panel 4 contains an overlay of the other three panels. Adjust the colors as desired (e.g., red for mitochondria, green for nonspecific) (Fig. 4f–h).

4 Notes

1. An inverted microscope is more convenient for imaging live cells in buffer since the objective approaches from below the coverslip; however, it may be possible to use a microscope in a standard orientation by using an appropriate imaging chamber

such as a perfusion chamber in which the cells on the coverslip are placed at the top (facing in toward the buffer in the chamber).

2. Ionomycin may be used instead of A23187.
3. Ionomycin may be used instead of A23187.
4. In order to obtain images on sufficient resolution to distinguish mitochondria, it is essential to image the cells grown on a glass coverslip rather than imaging directly in the plastic tissue culture dish. An alternative is to grow cells on glass bottom tissue culture dishes (MatTek Corporation, MA, USA).
5. If the cells are not well flattened, the mitochondria will tend to remain around the nucleus making them difficult to image. It may be necessary to allow a longer time in order for some cell types to spread out on the coverslip or to pre-coat the coverslip with 0.1% gelatin or fibronectin.
6. Over longer periods, the dye will be gradually extruded from the cytosol but remains in the mitochondria. It can be advantageous at this stage to return the cells to normal medium overnight, replacing the imaging buffer the following day, resulting in the fluorescence being close to 100% mitochondrial in origin.
7. Do not use a fluorescent lamp to focus the cells as this tends to cause oxidative damage and a progressive increase in dye fluorescence.
8. On the Leica confocal, "texas red" is suitable. Other equivalent dyes are TMRE or MitoTracker Red.
9. It is usually easiest to focus on the mitochondria spread out around the middle of the cell, rather than those clustered around the nucleus.
10. This increases the "thickness" of the optical slice being imaged, thus decreasing the likelihood that mitochondria will move up or down out of the imaging plane.
11. The presence of saturated (i.e., maximum intensity) pixels is determined by changing the color scale to 1 indicating saturated pixels as blue. ~50% intensity is selected so that there is "overhead" room for the signal to increase without saturating the measurement.
12. To replace the buffer without moving the imaging field, use a 2 ml syringe with a 20 cm piece of flexible narrow tubing slipped tightly over the needle to remove the buffer, and then, before cells dry out, rapidly replace the buffer with a 1 ml Gilson pipette.
13. Some cells such as muscle cells may contract when calcium is increased, making analysis difficult. To prevent morphological

alterations during calibration, cells can be depleted of ATP by 10 minute pretreatment with 1 mM cyanide.

14. In order to make valid comparisons and calculations between the different treatments, it is essential that the imaging parameters (laser power, gain, objective, zoom, etc.) remain the same.
15. The exact time will depend upon the cell type.
16. Do not use a fluorescent lamp to focus the cells as this tends to cause oxidative damage and a progressive increase in dye fluorescence.
17. It is usually easiest to focus on the mitochondria spread out around the middle of the cell, rather than those clustered around the nucleus.
18. The presence of saturated (i.e., maximum intensity) pixels is determined by changing the color scale to 1 indicating saturated pixels as blue. ~50% intensity is selected so that there is "overhead" room for the signal to increase without saturating the measurement.
19. If not printed on one of the macro buttons, it may be necessary to install the macro first (refer to the Zeiss LSM software manual).
20. This data table contains a list of laser attenuations necessary to ensure that the power is the same at each laser wavelength (since power varies with wavelength) and must be calibrated to each objective normally used.
21. The Kd of dyes in vivo may be different from these values determined in vitro, largely due to interfering interactions with cellular components (e.g., [13]). For example, the *K*d for rhod-2 is 720 nM in intact cardiomyocytes [14].

References

1. Davidson SM, Duchen MR (2006) Calcium microdomains and oxidative stress. Cell Calcium 40:561–574
2. Jacobson J, Duchen MR (2004) Interplay between mitochondria and cellular calcium signalling. Mol Cell Biochem 256-257:209–218
3. Davidson SM, Duchen MR (2007) Endothelial mitochondria: contributing to vascular function and disease. Circ Res 100:1128–1141
4. Duchen MR (2004) Roles of mitochondria in health and disease. Diabetes 53(Suppl 1): S96–102
5. Hausenloy DJ, Yellon DM (2003) The mitochondrial permeability transition pore: its fundamental role in mediating cell death during ischaemia and reperfusion. J Mol Cell Cardiol 35:339–341
6. Somlyo AP, Bond M, Somlyo AV (1985) Calcium content of mitochondria and endoplasmic reticulum in liver frozen rapidly in vivo. Nature 314:622–625
7. Coll KE, Joseph SK, Corkey BE, Williamson JR (1982) Determination of the matrix free Ca2+ concentration and kinetics of Ca2+ efflux in liver and heart mitochondria. J Biol Chem 257:8696–8704
8. Minta A, Kao JP, Tsien RY (1989) Fluorescent indicators for cytosolic calcium based on rhodamine and fluorescein chromophores. J Biol Chem 264:8171–8178

9. Trollinger DR, Cascio WE, Lemasters JJ (2000) Mitochondrial calcium transients in adult rabbit cardiac myocytes: inhibition by ruthenium red and artifacts caused by lysosomal loading of Ca(2+)-indicating fluorophores. Biophys J 79:39–50
10. Trollinger DR, Cascio WE, Lemasters JJ (1997) Selective loading of Rhod 2 into mitochondria shows mitochondrial Ca2+ transients during the contractile cycle in adult rabbit cardiac myocytes. Biochem Biophys Res Commun 236:738–742
11. Grynkiewicz G, Poenie M, Tsien RYA (1985) New generation of Ca2+ indicators with greatly improved fluorescence properties. J Biol Chem 260:3440–3450
12. Davidson SM, Yellon D, Duchen MR (2007) Assessing mitochondrial potential, calcium, and redox state in isolated mammalian cells using confocal microscopy. Methods Mol Biol 372:421–430
13. Harkins AB, Kurebayashi N, Baylor SM (1993) Resting myoplasmic free calcium in frog skeletal muscle fibers estimated with fluo-3. Biophys J 65:865–881
14. Du C, MacGowan GA, Farkas DL, Koretsky AP (2001) Calibration of the calcium dissociation constant of Rhod(2)in the perfused mouse heart using manganese quenching. Cell Calcium 29:217–227

Chapter 11

Mitochondrial Permeability Transition Pore and Calcium Handling

Randi J. Parks, Elizabeth Murphy, and Julia C. Liu

Abstract

The opening of a large conductance channel in the inner mitochondrial membrane, known as the mitochondrial permeability transition pore (PTP), has been shown to be a primary mediator of cell death in the heart subjected to ischemia-reperfusion injury. Inhibitors of the PTP have been shown to reduce cardiac ischemia-reperfusion injury in many animal models. Furthermore, most cardioprotective strategies appear to reduce ischemic cell death either by reducing the triggers for the opening of the PTP, such as reducing calcium overload or reactive oxygen species, or by inhibiting PTP modulators. This chapter will focus on key issues in the study of the PTP and provide some methods for measuring PTP opening in isolated mitochondria.

Key words Calcium, Mitochondria, Pore opening, Swelling, Uptake

1 Introduction

The ability of mitochondria to take up and release calcium (Ca^{2+}) is well established. The response of mitochondria to the addition of Ca^{2+} depends on the amount of Ca^{2+} added and the level of matrix Ca^{2+}. Nicholls first reported that mitochondria buffer extramitochondrial Ca^{2+} at pCa^{2+} (the negative logarithm of free extramitochondrial Ca^{2+} concentration) of ~6.1 until matrix Ca^{2+} reaches a level of ~60 nmol/mg protein [1]. He showed that when Ca^{2+} is added to isolated mitochondria, they accumulate extramitochondrial Ca^{2+} until they reach this set point and when EGTA is added to lower Ca^{2+} below that set point, mitochondria release Ca^{2+} (via a Ca^{2+} efflux pathway such as the Na^{+}-Ca^{2+} or H^{+}-Ca^{2+} exchanger) and return to the same extramitochondrial Ca^{2+} set point. At matrix Ca^{2+} levels above ~4 nmol/mg mitochondrial protein, extracellular Ca^{2+} buffering occurs because the Ca^{2+} efflux pathway is saturated and runs at a constant rate (e.g., 5 nmol/min/mg protein in rat liver mitochondria) that is independent of matrix or

Carlos M. Palmeira and António J. Moreno (eds.), *Mitochondrial Bioenergetics: Methods and Protocols*, Methods in Molecular Biology, vol. 1782, https://doi.org/10.1007/978-1-4939-7831-1_11,

extramitochondrial Ca^{2+} [1]. When Ca^{2+} is added outside the mitochondria, the mitochondria uptake Ca^{2+} through the mitochondrial Ca^{2+} uniporter until the extramitochondrial Ca^{2+} falls to a level at which the rate of uptake by the uniporter equals the constant rate of efflux. At this point (i.e., pCa^{2+} ~6.1 in rat liver mitochondria in the absence of Mg^{2+} and Na^{+}), influx equals efflux, and the extramitochondrial Ca^{2+} is maintained at a constant level. However, this set point only applies when the level of total mitochondrial Ca^{2+} is above ~4 nmol/mg mitochondrial protein [2]. At levels below this, the efflux pathway is not constant (i.e., the Ca^{2+} efflux pathway is not at V_{max}), and a different steady-state extramitochondrial Ca^{2+} concentration will be achieved. McCormack and others [3, 4] have pointed out that during isolation, mitochondria accumulate Ca^{2+} and that unless care is taken, the levels of Ca^{2+} found in isolated mitochondria exceed those that occur in vivo. They further note that at matrix Ca^{2+} levels above 4 nmol/mg protein, the mitochondrial dehydrogenases are saturated. Therefore, if the mitochondrial dehydrogenases are regulated by changes in cellular and mitochondrial Ca^{2+}, then the basal, unstimulated matrix Ca^{2+} level is expected to be below a level that activates the dehydrogenases. Possible ways to address the concern of high matrix Ca^{2+} in isolated mitochondria include adding a Ca^{2+} uptake inhibitor (i.e., Ru360) or a Ca^{2+} chelator (EGTA) to the isolation buffer or performing a Ca^{2+} depletion step [5] following mitochondria isolation.

If sufficient Ca^{2+} is accumulated by mitochondria, this leads to a large increase in permeability of the inner mitochondrial membrane and the loss of solutes with a molecular weight less than 1.5 kDa. This phenomenon, termed mitochondrial permeability transition pore (PTP) opening, was described by Haworth and Hunter in 1979 [6–8]. In an isolated mitochondria preparation, opening of the PTP leads to swelling of the mitochondria, and thus swelling is commonly used as a measure of PTP opening. However, although mitochondrial swelling occurs with isolated mitochondria undergoing PTP opening, mitochondrial swelling does not appear to occur in situ because of the high cytosolic level of proteins and solutes. The PTP was largely a curiosity only of interest to mitochondrial experts until it was implicated as playing a role in apoptotic and necrotic cell death. Studies in the mid-1990s suggested that inhibition of the PTP with cyclosporin A reduces cell dysfunction and death following cardiac ischemia-reperfusion [9]. Indeed, over the ensuing decades, numerous studies in animal models have confirmed that inhibition of the PTP is cardioprotective, and the cardioprotective effects of PTP inhibition have been demonstrated in humans in a small proof of concept clinical trial [10]. However, in a recent clinical trial, patients with anterior ST-elevation myocardial infarction who had been referred for primary percutaneous coronary intervention, intravenous cyclosporine did not result in better clinical outcomes than those with placebo and did not prevent adverse left ventricular remodeling at 1 year [11]. Since cyclosporin A is merely a regulator and desensitizer of the PTP, it has been proposed that a direct inhibitor of the PTP might prove more beneficial. Thus, identification of the PTP is crucial to designing cardioprotective drugs.

Although there is considerable interest in the PTP, the molecular identity of the PTP remains controversial. Inhibitors of the adenine nucleotide translocator (ANT) can both inhibit and activate the PTP. Bonkregic acid, which locks ANT in a conformation that faces the matrix, inhibits the PTP, whereas atractyloside, which inhibits ANT by locking it in a conformation that faces the cytosol, activates the PTP. Based on these studies and others, ANT was suggested as a possible component of the PTP. It was speculated that under high matrix Ca^{2+} levels, ANT and the voltage-dependent anion channel (VDAC) come together at contact sites to adopt an alternative conformation and form a large permeability channel (i.e., the PTP). However, PTP opening still occurs in liver mitochondria from mice in which ANT is genetically ablated [12]. Similarly, studies of VDAC ablation showed that PTP remains in these genetically altered mice [13]. In contrast, similar studies showed that loss of cyclophilin D results in a decreased sensitivity of PTP activation to calcium or during ischemia-reperfusion [14]. These studies strongly support the concept that cyclophilin D (the binding partner for the PTP inhibitor cyclosporin) is an important regulator of the PTP. It has more recently been suggested that the phosphate carrier might be a component of the PTP [15, 16]. However, PTP opening was not lost when the phosphate carrier was genetically deleted [17, 18]. Recent studies by Bernardi's group have suggested a role for the F1-Fo-ATPase (complex V) in forming the PTP [19].

In addition to a lack of understanding of the composition of the PTP, we have little understanding of whether it has any role in cell physiology or whether it exists only to mediate cell death. The PTP can function in subconductance states, and it has been suggested that these transient and/or subconductance openings might be important in regulating mitochondrial Ca^{2+} [20]. The purpose of transient pore opening has not been established, but it has been speculated that it might serve to release matrix Ca^{2+} [21–23]. It is generally thought that Ca^{2+} uptake into the matrix via the Ca^{2+} uniporter is more rapid than Ca^{2+} release via the mitochondria Na^{+}-Ca^{2+} exchanger. Currently, there is controversy as to whether mitochondrial matrix Ca^{2+} follows cytosolic Ca^{2+} transients or whether the change in matrix Ca^{2+} reflects a more time averaged change in cytosolic Ca^{2+}. Several studies [24, 25] report that mitochondrial Ca^{2+} cycles on a beat-to-beat basis. Others such as Miyata et al. [26] suggest that mitochondrial Ca^{2+} is in the range of 0.1–0.2 μM and increases to a higher steady state as cytosolic Ca^{2+} is raised by increasing beating frequency or by addition of isoproterenol. One key issue is whether a mitochondrial release mechanism exists with sufficient time resolution to extrude Ca^{2+} from the mitochondria on a beat-to-beat basis [27]. Calcium uptake into the mitochondria might exceed the normal mitochondrial Ca^{2+} efflux mechanisms especially during conditions of elevated Ca^{2+} (e.g., isoproterenol stimulation). Therefore, transient Ca^{2+} release via PTP opening might serve as a Ca^{2+}-release valve [21]. Experimental support for this hypothesis is sparse, likely due in part to the difficulty in measuring small differences in matrix Ca^{2+} in myocytes or cells in situ. Studies by Altschuld et al. showed that cyclosporin increased net $^{45}Ca^{2+}$ uptake and reduced efflux from myocytes, consistent with PTP inhibition as being a mechanism for mitochondrial Ca^{2+} efflux. Ichas et al. [23] also reported that Ca^{2+}-induced Ca^{2+} release from mitochondria was dependent on the PTP.

A transient opening of the PTP has been shown to occur in cultured hepatocytes and HM1C1 cells [20]. The PTP was measured in these cells by loading mitochondria with calcein-AM, using cobalt (Co^{2+}) to quench cytosolic calcein. This takes advantage of the lack of Co^{2+} entry into mitochondria. Over time, the calcein fluorescence in the mitochondria declined, and this decline was inhibited by cyclosporin, suggesting that the release of calcein from the mitochondria was through the PTP. Mitochondrial membrane potential was also measured and did not decline over the same time period. In this case, the opening of the PTP and release of calcein appeared to transiently occur in only a small percentage of mitochondria at any time, thereby preventing a macroscopic change in mitochondrial membrane potential [20]. These data provide strong evidence that transient PTP opening or PTP flickering occurs normally in cells without apparent detrimental consequences.

During ischemia-reperfusion, the PTP is overwhelmed with high mitochondrial Ca^{2+}, and its normal opening set point may also be altered by reactive oxygen species (ROS) generated at the start of reperfusion [28]. In the setting of ischemia-reperfusion, most mitochondria in the cell undergo PTP simultaneously, resulting in a cessation of ATP generation in the cell and increased generation of ROS, in contrast to the aforementioned case of a few mitochondria opening transiently without detriment to the cell's ability to make ATP. This scenario is consistent with the protection afforded by the cyclophilin D inhibitor cyclosporin in protocols of ischemia-reperfusion. Low pH inhibits the PTP, and low pH at the start of reperfusion is also reported to reduce ischemia-reperfusion injury [29]. It has also been suggested that postconditioning protects, at least in part, by maintaining a more acid intracellular pH during early reperfusion [30]. Furthermore, ischemic injury can be reduced by interventions that decrease the rise in cell Ca^{2+} during ischemia-reperfusion. In fact, there is considerable data to suggest that preconditioning and many cardioprotective drugs mediate protection at least in part by reducing Ca^{2+}

overload. For example, inhibitors of the plasma membrane Na^{+}-Ca^{2+} exchanger reduce ischemia-reperfusion injury and appear to do so by blunting the rise in cytosolic Ca^{2+} during ischemia and early reperfusion [31]. Preconditioning has also been shown to reduce the rise in Ca^{2+} during ischemia [32]. Further assessment of the composition of the PTP and potential pharmacological inhibition thereof will advance our understanding of PTP opening and our ability to design therapeutic interventions.

The following sections describe methods that can be employed to measure PTP opening viaCa^{2+} uptake and swelling in isolated mitochondria. The mitochondrial swelling assay measures sample absorbance following an addition of a high concentration of calcium. The mitochondria Ca^{2+} uptake assay measures extramitochondrial Ca^{2+} in response to repeated additions of Ca^{2+}.

2 Materials

2.1 Mitochondrial Swelling Assay

1. Isolated mitochondria (final concentration ~0.5 mg/mL).
2. Mitochondria EGTA-free buffer: 225 mM mannitol, 75 mM sucrose, 5 mM MOPS, 2 mM taurine, pH 7.25.
3. Swelling assay buffer: 120 mM KCl, 10 mM Tris–HCl, 5 mM MOPS, 5 mM KH_2PO_4, pH 7.4. Store at 4 °C. Bring to room temperature and stir to aerate before use.
4. Calcium chloride solution: Prepare a 1–10 mM stock in H_2O.
5. UV-Vis 96-well plate reader.
6. For a negative control, pore desensitizer cyclosporin A: 2 mM stock in DMSO, and prepare secondary stock of 100 μM.

2.2 Mitochondria Ca^{2+} Uptake Assay

1. Isolated mitochondria (stock ~2 mg/mL; final concentration ~0.25 mg/mL).
2. Ca^{2+} depletion buffer: 125 mM KCl, 20 mM HEPES, 15 mM NaCl, 5 mM $MgCl_2$, 1 mM K_2EDTA, 1 mM EGTA, 2 mM KH_2PO_4, 0.1 mM malate, pH 7.1. Store at 4 °C. Bring to room temperature and stir to aerate before use.
3. Ca^{2+} uptake assay buffer: 120 mM KCl, 5 mM MOPS, 5 mM KH_2PO_4, 10 mM glutamate, 5 mM malate, pH 7.4. Store at 4 °C. Bring to room temperature and stir to aerate before use. Add glutamate and malate fresh.
4. Calcium Green™-5N tetrapotassium salt solution (Molecular Probes, Carlsbad, CA): Prepare a 1–10 mM stock in H_2O.
5. Calcium chloride solution: Prepare a 1–10 mM stock in H_2O.
6. Fluorescence spectrometer with stirring capabilities (1 mL total volume) or UV-Vis 96-well plate reader (200 μL volume per well).

7. For a negative control, mitochondrial calcium uniporter inhibitor Ru360: 2 mM stock in DMSO, and prepare secondary stock of 300 μM.

3 Methods for Measuring PTP

3.1 Mitochondrial Swelling

PTP opening is commonly measured by following mitochondrial swelling. In the original study by Haworth and Hunter, they found that PTP opening results in a conformational change in isolated mitochondria which is accompanied by swelling that can be followed by measuring changes in light scattering. Thus, a common and convenient assay for PTP opening is to measure changes in light scattering by measuring absorbance following the addition of Ca^{2+} to mitochondria. Figure 1 shows an example trace of absorbance recorded from mitochondria following an addition of 250 μM Ca^{2+}. Mitochondrial swelling is observed as a decrease in absorbance, which is inhibited by cyclosporin A.

1. Wash the mitochondrial pellet gently 2× with mitochondria EGTA-free buffer to remove any EGTA remaining from the isolation.
2. Add mitochondria (0.1 mg/well) to a clear 96-well plate.
3. Add swelling assay buffer to bring total well volume to 200 μL.
4. Add cyclosporin A (final concentration 1 μM) to one well as a negative control.

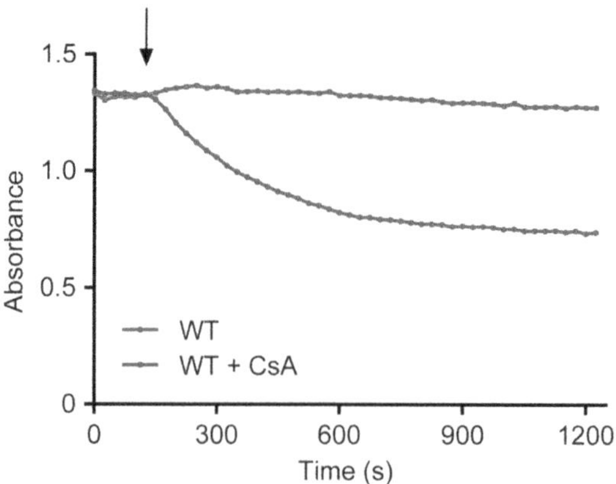

Fig. 1 Example traces of wild-type heart mitochondria, in the presence and absence of the PTP desensitizer cyclosporin A (CsA). The arrow marks the addition of 250 μM Ca^{2+}

5. Measure absorbance at 540 nm for ~5 min at 20 s intervals or until a steady baseline is achieved.
6. Induce swelling with the addition of 0–0.25 mM calcium chloride, and continue to monitor absorbance at 540 nm.

3.2 Mitochondrial Calcium Uptake Assay

PTP opening is also commonly measured by following the ability of mitochondria to take up and retain Ca^{2+}. In this assay, extramitochondrial Ca^{2+} levels are measured with a fluorescent dye, and Ca^{2+} is added to mitochondria in small increments. Fluorescence increases from baseline with each addition of calcium. As the mitochondria take up the added Ca^{2+}, the fluorescence decreases back to baseline until the mitochondria have taken up enough calcium for the PTP to open (i.e., fluorescence increases rapidly). Figure 2 depicts typical results obtained from a mitochondrial calcium uptake assay using a plate reader. As outlined in Subheading 2, either a fluorescence spectrometer with stirring capabilities (1 mL total volume) or a UV-Vis 96-well plate reader (200 μL volume per well) can be used for this assay. A spectrometer provides better kinetic resolution but requires more sample. A plate reader enables the measurement of multiple samples or conditions at the same time, at the expense of kinetic resolution.

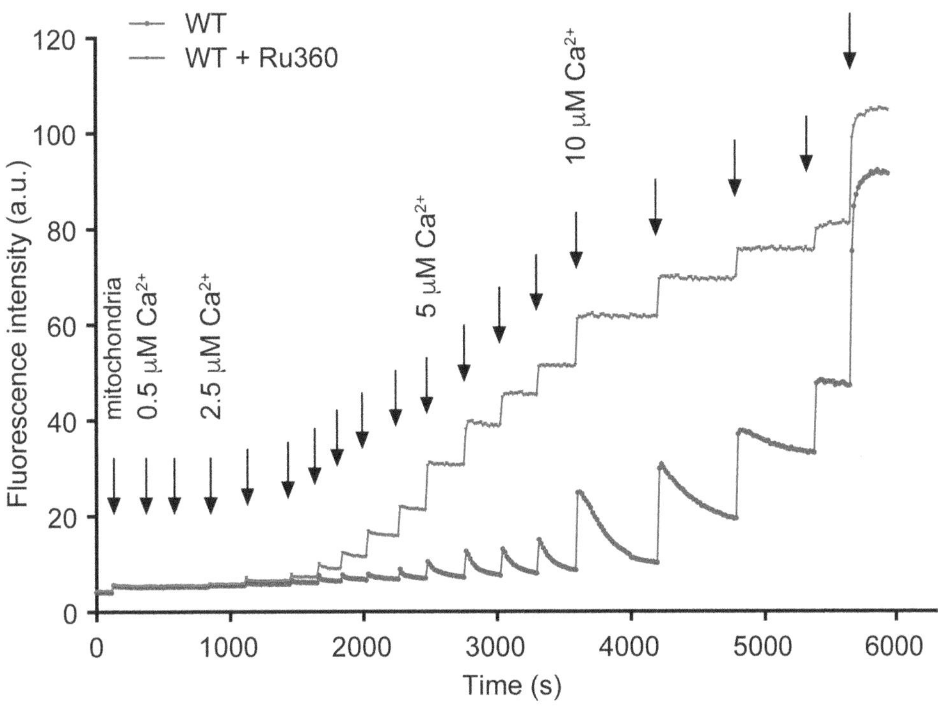

Fig. 2 Calcium uptake performed in a plate reader in isolated wild-type mouse brain mitochondria, with and without Ru360 (mitochondrial calcium uniporter inhibitor). Successive additions of Ca^{2+} are indicated by the arrows (successive unmarked arrows are repeated additions of the previous concentration). *See* text for details

1. Add Ca^{2+} uptake assay buffer containing 1 μM Calcium Green-5N to a stirred cuvette (1 mL volume) or wells of a 96-well plate (200 μL volume). Take a baseline background recording prior to adding mitochondria; Calcium Green-5N fluorescence is excited at 506 nm and emits at 532 nm.
2. To address concern about Ca^{2+} uptake during mitochondria isolation or if groups are being compared that may have basal differences in matrix Ca^{2+} levels, perform a Ca^{2+} depletion step. Incubate mitochondria in Ca^{2+} depletion buffer (0.3 mg/mL) for 5 min, and then spin mitochondria for 2 min at 11,000 × *g*. Resuspend mitochondria in Ca^{2+} uptake assay buffer.
3. For a negative control, add Ru360 (final concentration 3 μM) to inhibit mitochondrial Ca^{2+} uniporter activity (i.e., in one well of a plate).
4. Add mitochondria (0.25–0.5 mg/mL) to the cuvette or wells. Upon addition of the mitochondria, the Ca^{2+} in the buffer (there is always some Ca^{2+} contamination in the buffer) will decline because of Ca^{2+} uptake into the mitochondria. As discussed in Subheading 1, the mitochondria will reach a set point, and extramitochondrial Ca^{2+} will stabilize (*see* Fig. 2).
5. Add successive increments of Ca^{2+} (*see* Fig. 2). The appropriate concentration of Ca^{2+} depends on the source of mitochondria and should be adjusted such that multiple Ca^{2+} additions can be made before the mitochondria release their Ca^{2+}. Typically, additions of 10 μM Ca^{2+} (10 nmol/mL/0.25 mg/mL = 40 nmol Ca^{2+}/mg mitochondrial protein) are a good concentration to begin with, and if needed, the Ca^{2+} concentration should be adjusted to the mitochondria concentration. Initially, the mitochondria can sequester the initial additions of Ca^{2+}. As the Ca^{2+} levels rise to trigger PTP opening, the mitochondrial Ca^{2+} stores are released and lead to saturation of the fluorescent indicator.

References

1. Nicholls DG (1978) The regulation of extra-mitochondrial free calcium ion concentration by rat liver mitochondria. Biochem J 176 (2):463–474
2. McCormack JG, Halestrap AP, Denton RM (1990) Role of calcium ions in regulation of mammalian intramitochondrial metabolism. Physiol Rev 70(2):391–425
3. Denton RM, McCormack JG, Edgell NJ (1980) Role of calcium ions in the regulation of intramitochondrial metabolism. Effects of Na+, Mg2+ and ruthenium red on the Ca2+-stimulated oxidation of oxoglutarate and on pyruvate dehydrogenase activity in intact rat heart mitochondria. Biochem J 190 (1):107–117
4. McCormack JG, Denton RM (1980) Role of calcium ions in the regulation of intramitochondrial metabolism. Properties of the Ca2 +-sensitive dehydrogenases within intact uncoupled mitochondria from the white and brown adipose tissue of the rat. Biochem J 190(1):95–105
5. Territo PR, Mootha VK, French SA, Balaban RS (2000) Ca(2+) activation of heart mitochondrial oxidative phosphorylation: role of the F(0)/F(1)-ATPase. Am J Physiol Cell Physiol 278(2):C423–C435
6. Haworth RA, Hunter DR (1979) The Ca2+-induced membrane transition in mitochondria. II. Nature of the Ca2+ trigger site. Arch Biochem Biophys 195(2):460–467

7. Hunter DR, Haworth RA (1979) The Ca2+-induced membrane transition in mitochondria. I. The protective mechanisms. Arch Biochem Biophys 195(2):453–459
8. Hunter DR, Haworth RA (1979) The Ca2+-induced membrane transition in mitochondria. III. Transitional Ca2+ release. Arch Biochem Biophys 195(2):468–477
9. Griffiths EJ, Halestrap AP (1993) Protection by Cyclosporin A of ischemia/reperfusion-induced damage in isolated rat hearts. J Mol Cell Cardiol 25(12):1461–1469
10. Piot C, Croisille P, Staat P, Thibault H, Rioufol G, Mewton N, Elbelghiti R, Cung TT, Bonnefoy E, Angoulvant D, Macia C, Raczka F, Sportouch C, Gahide G, Finet G, Andre-Fouet X, Revel D, Kirkorian G, Monassier JP, Derumeaux G, Ovize M (2008) Effect of cyclosporine on reperfusion injury in acute myocardial infarction. N Engl J Med 359 (5):473–481
11. Cung TT, Morel O, Cayla G, Rioufol G, Garcia-Dorado D, Angoulvant D, Bonnefoy-Cudraz E, Guerin P, Elbaz M, Delarche N, Coste P, Vanzetto G, Metge M, Aupetit JF, Jouve B, Motreff P, Tron C, Labeque JN, Steg PG, Cottin Y, Range G, Clerc J, Claeys MJ, Coussement P, Prunier F, Moulin F, Roth O, Belle L, Dubois P, Barragan P, Gilard M, Piot C, Colin P, De Poli F, Morice MC, Ider O, Dubois-Rande JL, Unterseeh T, Le Breton H, Beard T, Blanchard D, Grollier G, Malquarti V, Staat P, Sudre A, Elmer E, Hansson MJ, Bergerot C, Boussaha I, Jossan C, Derumeaux G, Mewton N, Ovize M (2015) Cyclosporine before PCI in patients with acute myocardial infarction. N Engl J Med 373(11):1021–1031
12. Kokoszka JE, Waymire KG, Levy SE, Sligh JE, Cai J, Jones DP, MacGregor GR, Wallace DC (2004) The ADP/ATP translocator is not essential for the mitochondrial permeability transition pore. Nature 427(6973):461–465
13. Baines CP, Kaiser RA, Sheiko T, Craigen WJ, Molkentin JD (2007) Voltage-dependent anion channels are dispensable for mitochondrial-dependent cell death. Nat Cell Biol 9(5):550–555
14. Baines CP, Kaiser RA, Purcell NH, Blair NS, Osinska H, Hambleton MA, Brunskill EW, Sayen MR, Gottlieb RA, Dorn GW, Robbins J, Molkentin JD (2005) Loss of cyclophilin D reveals a critical role for mitochondrial permeability transition in cell death. Nature 434(7033):658–662
15. Leung AW, Halestrap AP (2008) Recent progress in elucidating the molecular mechanism of the mitochondrial permeability transition pore. Biochim Biophys Acta 1777(7–8):946–952
16. Leung AW, Varanyuwatana P, Halestrap AP (2008) The mitochondrial phosphate carrier interacts with cyclophilin D and may play a key role in the permeability transition. J Biol Chem 283(39):26312–26323
17. Gutierrez-Aguilar M, Douglas DL, Gibson AK, Domeier TL, Molkentin JD, Baines CP (2014) Genetic manipulation of the cardiac mitochondrial phosphate carrier does not affect permeability transition. J Mol Cell Cardiol 72:316–325
18. Kwong JQ, Davis J, Baines CP, Sargent MA, Karch J, Wang X, Huang T, Molkentin JD (2014) Genetic deletion of the mitochondrial phosphate carrier desensitizes the mitochondrial permeability transition pore and causes cardiomyopathy. Cell Death Differ 21 (8):1209–1217
19. Giorgio V, von Stockum S, Antoniel M, Fabbro A, Fogolari F, Forte M, Glick GD, Petronilli V, Zoratti M, Szabo I, Lippe G, Bernardi P (2013) Dimers of mitochondrial ATP synthase form the permeability transition pore. Proc Natl Acad Sci U S A 110(15):5887–5892
20. Petronilli V, Miotto G, Canton M, Brini M, Colonna R, Bernardi P, Di Lisa F (1999) Transient and long-lasting openings of the mitochondrial permeability transition pore can be monitored directly in intact cells by changes in mitochondrial calcein fluorescence. Biophys J 76(2):725–734
21. Bernardi P, Petronilli V (1996) The permeability transition pore as a mitochondrial calcium release channel: a critical appraisal. J Bioenerg Biomembr 28(2):131–138
22. Gunter TE, Pfeiffer DR (1990) Mechanisms by which mitochondria transport calcium. Am J Phys 258(5 Pt 1):C755–C786
23. Ichas F, Jouaville LS, Mazat JP (1997) Mitochondria are excitable organelles capable of generating and conveying electrical and calcium signals. Cell 89(7):1145–1153
24. Bell CJ, Bright NA, Rutter GA, Griffiths EJ (2006) ATP regulation in adult rat cardiomyocytes: time-resolved decoding of rapid mitochondrial calcium spiking imaged with targeted photoproteins. J Biol Chem 281 (38):28058–28067
25. Robert V, Gurlini P, Tosello V, Nagai T, Miyawaki A, Di Lisa F, Pozzan T (2001) Beat-to-beat oscillations of mitochondrial [Ca2+] in cardiac cells. EMBO J 20 (17):4998–5007
26. Miyamae M, Camacho SA, Weiner MW, Figueredo VM (1996) Attenuation of

postischemic reperfusion injury is related to prevention of [Ca2+]m overload in rat hearts. Am J Phys 271(5 Pt 2):H2145–H2153

27. O'Rourke B, Blatter LA (2009) Mitochondrial Ca2+ uptake: tortoise or hare? J Mol Cell Cardiol 46(6):767–774
28. Murphy E, Steenbergen C (2008) Mechanisms underlying acute protection from cardiac ischemia-reperfusion injury. Physiol Rev 88 (2):581–609
29. Inserte J, Barba I, Hernando V, Abellan A, Ruiz-Meana M, Rodriguez-Sinovas A, Garcia-Dorado D (2008) Effect of acidic reperfusion on prolongation of intracellular acidosis and myocardial salvage. Cardiovasc Res 77 (4):782–790
30. Cohen MV, Yang XM, Downey JM (2007) The pH hypothesis of postconditioning: staccato reperfusion reintroduces oxygen and perpetuates myocardial acidosis. Circulation 115 (14):1895–1903
31. Murphy E, Perlman M, London RE, Steenbergen C (1991) Amiloride delays the ischemia-induced rise in cytosolic free calcium. Circ Res 68(5):1250–1258
32. Steenbergen C, Perlman ME, London RE, Murphy E (1993) Mechanism of preconditioning. Ionic alterations. Circ Res 72(1):112–125

Chapter 12

Redox Equivalents and Mitochondrial Bioenergetics

James R. Roede, Young-Mi Go, and Dean P. Jones

Abstract

Mitochondrial energy metabolism depends upon high-flux and low-flux electron transfer pathways. The former provide the energy to support chemiosmotic coupling for oxidative phosphorylation. The latter provide mechanisms for signaling and control of mitochondrial functions. Few practical methods are available to measure rates of individual mitochondrial electron transfer reactions; however, a number of approaches are available to measure steady-state redox potentials (E_h) of donor/acceptor couples, and these can be used to gain insight into rate controlling reactions as well as mitochondrial bioenergetics. Redox changes within the respiratory electron transfer pathway are quantified by optical spectroscopy and measurement of changes in autofluorescence. Low-flux pathways involving thiol/disulfide redox couples are measured by redox Western blot and mass spectrometry-based redox proteomics. Together, the approaches provide the opportunity to develop integrated systems biology descriptions of mitochondrial redox signaling and control mechanisms.

Key words NADH, NADPH, NADH dehydrogenase, Ubiquinone, Cytochromes, Hydrogen peroxide, Glutathione, Thioredoxin-2, Redox Western blot, Redox proteomics, Peroxiredoxin

1 Introduction

Mitochondrial function depends upon oxidation-reduction (redox) processes with three areas especially relevant to contemporary mitochondrial research. The first is the use of energy from oxidation to support electrochemical coupling of oxidative phosphorylation [1]. The second involves a small fraction of the O_2 consumed by mitochondria which is converted to the so-called reactive oxygen species (ROS), superoxide anion radical (O_2^-), and H_2O_2 [2], with mitochondria being selectively vulnerable to oxidative damage [3, 4]. The third encompasses low-flux redox reactions which function in cell signaling and control [5]. Another chapter in this volume (Wieckowski) addresses mitochondrial ROS generation, so this will not be included in the present article.

The first section addresses measurement of steady-state redox potentials of the high-flux redox systems of the mitochondrial

Carlos M. Palmeira and António J. Moreno (eds.), *Mitochondrial Bioenergetics: Methods and Protocols*, Methods in Molecular Biology, vol. 1782, https://doi.org/10.1007/978-1-4939-7831-1_12,

respiratory chain. These methods were developed over 50 years ago and currently provide a convenient but underutilized approach to improve understanding of mitochondrial bioenergetics and respiratory control. As described by Britton Chance in 1957 [6], the mitochondrial electron transfer chain undergoes a "cushioning effect" in which the steady-state reduction of components changes with substrate conditions to maintain function of oxidative phosphorylation. Because the system functions in vivo under a non-equilibrium steady state, the steady-state redox potentials of the components within the chain are always relevant to accurate descriptions of respiratory function.

A second section describes use of redox Western blot methods to measure components of low-flux thiol/disulfide systems. These methods are relatively newly developed and provide the capability to selectively study mitochondrial function in intact cells and tissues due to the redox-sensitivities of specific mitochondrially localized proteins. Critically important enzymes within mitochondria are subject to redox regulation. Examples include NADH dehydrogenase which undergoes S-glutathionylation [7], apoptosis signal-regulating kinase-1 which is regulated by the redox state of thioredoxin-2 [8], and glutaredoxin-2 which is present as an inactive iron-sulfur complex until oxidatively activated [9].

A subsequent section describes a mass spectrometry-based method to measure fractional reduction of cysteines in specific proteins [10]. This is a highly versatile approach to profile redox-sensitive proteins which relies upon rapidly advancing technology. A final section provides some general comments concerning future needs for mitochondrial redox biology and briefly discusses related methods to study redox regulation by S-glutathionylation and S-nitrosylation. Combinations of immunoassay, mass spectrometry, and molecular manipulations provide means to define and discriminate these covalent modifications from other redox mechanisms in signaling and control.

1.1 Methodologic Limitations Due to Complexity of Mitochondria

Accurate quantification is essential to progress in redox biochemistry. However, several features of mitochondria have hindered accurate assessment of key reaction rates under relevant physiologic conditions. A first problem is that the tissue-specific regulation of mitochondria which occurs in vivo cannot readily be replicated in vitro. Because of the inability in in vitro studies to completely mimic pO_2, pCO_2, substrate supply, workload, and physiologic signaling as found in vivo, there is a possibility that key observations are misinterpreted. Furthermore, even if these conditions can be controlled, one is faced with the in vitro adaptations of cells and limitations concerning function of individual mitochondria within populations of mitochondria and cells. Mitochondria are not uniform in biochemical characteristics, and there is likelihood that spatial constraints within the cell further impact redox reactions at the individual mitochondrial level. Thus, the complexity of mitochondrial structure, function, and regulation requires that in vitro findings be validated in vivo and that "bottom-up" models of mitochondrial function be mirrored with "top-down" investigation.

Despite the limitations of in vitro studies, measurements of mitochondrial function at the cellular level can be more reflective of physiologic rates than studies of isolated mitochondria. Fractionation of cells and isolation of mitochondria result in perturbation of rate-control characteristics [11]. Isolated mitochondria invariably have some extent of physical damage, and even if this is considered minimal, there is an uncertainty about how to create a synthetic aqueous cytoplasm which appropriately reflects the intracellular environment. On the other hand, mitochondria which have been isolated from normal tissues contain proteins which are present at the correct in vivo level. Most tumor-derived cell lines have aberrant mitochondria, and mitochondrial characteristics often change dramatically when normal cells are grown in vitro. Thus, there is a need to study mitochondrial functions at multiple levels of organization. Importantly, the methods described in the present article can be applied at all levels of organization. Thus, an advantage of available quantitative redox biology methods is that they can provide a foundation for a comprehensive understanding of mitochondrial respiratory control.

1.2 Steady-State Reduction Potentials as Quantitative Descriptors in Redox Biology

As described elsewhere in this volume, quantitative information for specific biochemical processes can be obtained using isolated mitochondria and submitochondrial fractions. These preparations provide mechanistic detail that cannot be obtained with more intact systems. There is a critical need to integrate the information obtained from mitochondria and submitochondrial fractions with data obtained from cellular and in vivo models. A particular challenge lies in the difficulty of knowing whether kinetic principles or thermodynamic principles govern specific redox processes. Measurement of steady-state redox potential (E_h) values provides means to characterize operational characteristics of mitochondria without specific knowledge of reaction rates. E_h values are reduction potentials for electron acceptor/donor pairs relative to a standard hydrogen electrode [12]. By definition, the reduction potentials are given for an acceptor/donor pair, e.g., pyruvate/lactate. The biological and medical research literature more commonly uses the inverted expression of the donor/acceptor pair, e.g., lactate/pyruvate; while technically incorrect, common use requires that, regardless of how the couple is ordered, E_h expressions be interpreted as *reduction potentials*, unless explicitly stated otherwise. In the present article, we use donor/acceptor to refer to a redox couple and the term E_h with the acceptor to define the couple, e.g., for the GSH/GSSG couple, E_hGSSG refers to the reduction potential for the reaction GSSG $+2e^- + 2H^+ \leftrightarrow 2$ GSH.

For proteins, we use the protein abbreviation with the redox-sensitive thiols implied, e.g., for thioredoxin-2, E_hTrx2 refers to the reduction potential for the active site dithiol/disulfide couple.

In biological systems, E_h values are calculated using the Nernst equation, $E_h = E_o + RT/nF \ln([\text{acceptor}]/[\text{donor}])$, where E_o is the standard potential for the couple, R is the gas constant, T is the absolute temperature, n is the number of electrons transferred, and F is Faraday's constant. The difference in E_h values between two couples, ΔE_h, is related to Gibbs free energy for the electron transfer according to the equation $\Delta G = nF\Delta E_h$.

Changes in $\Delta\psi$ and ΔpH depend upon the ΔE_h for electron transfer, yet countless studies of $\Delta\psi$, proton leak, ROS production, etc. have been performed and interpreted without knowledge of E_h values for donors or ΔE_h for donor and acceptor couples under the conditions of assay. Because the validity of interpretation depends upon the energetics determined by the redox potentials of the respective upstream and downstream couples in the electron transfer pathway, there is a need to incorporate measurements redox potentials into bioenergetic descriptions.

1.3 Steady-State Redox Potentials of Mitochondrial Electron Transfer Components

In vivo, the mitochondrial electron transfer pathway operates under non-equilibrium steady-state conditions. The principles for measurement are derived from the historically important discovery of the mitochondrial cytochrome chain by David Keilin, in which oxidation-reduction reactions were characterized in terms of spectral changes in the cytochromes (Fig. 1) [13, 14]. While the rates of electron transfer through the entire pathway are conveniently measured in terms of O_2 consumption, measurement of specific donor pathways is more difficult. However, changes in steady-state potentials can be readily measured for many of the components because of their inherent spectral and fluorescent properties (Fig. 2) [15].

The materials needed to perform experiments include appropriate spectrophotometry or fluorometry instrumentation, appropriate means to calibrate the system for complete oxidation and reduction, and means to control O_2 and substrate availability. Spectral changes are measured as changes in light absorbance by a chemical at a specific wavelength defined by Beer's law, $A = \varepsilon Cl$, where A is the absorbance, ε is the extinction coefficient, C is the concentration, and l is the path length. ε is usually given as the mM extinction coefficient, and the path length is 1 cm in standard spectrophotometers. Studies of steady-state levels of oxidation of respiratory chain components in isolated mitochondria and cell suspensions can be obtained in 1 cm cuvettes or in small beakers [16] with appropriate positioning of light source and detector. With appropriate instrumentation, absorbance changes can be measured in perfused organs and adherent cells [17]. In all cases, light scatter can be limiting so that dual-wavelength spectrometry is used to minimize this problem [13, 17].

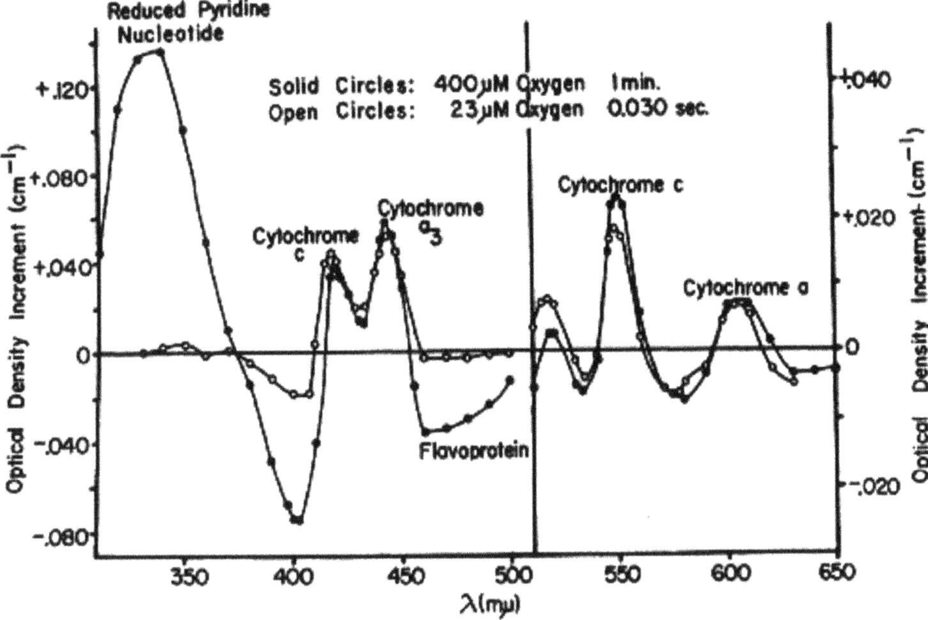

Fig. 1 Changes in steady-state redox levels of mitochondrial components are based upon absorbance characteristic as shown by this original study of yeast [13]. The reduced pyridine nucleotides, NADH and NADPH, absorb light at 320 nm. Oxidized flavoproteins absorb light at 455 in the 450–500 nm range. Cytochromes have distinct absorbance in the reduced forms. Reproduced with permission of the publisher

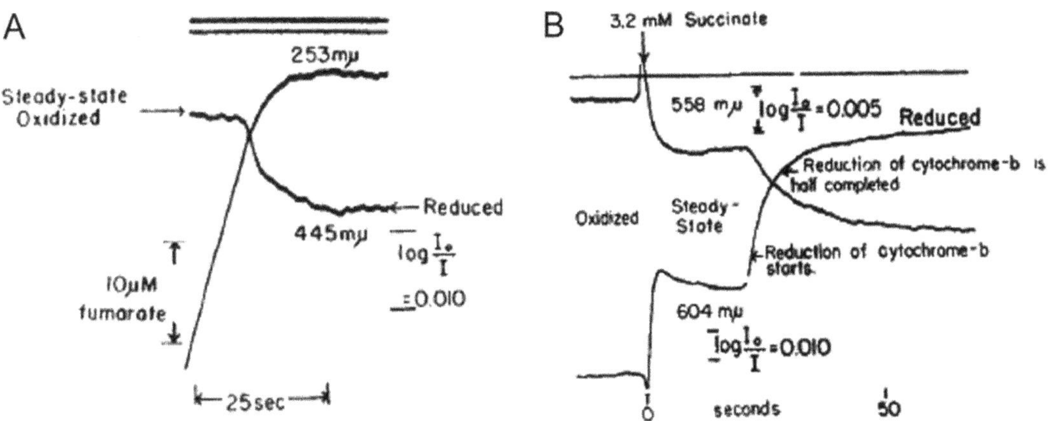

Fig. 2 Measurement of changes in steady-state levels of oxidation in the mitochondrial cytochrome chain. Changes in steady-state oxidation of (**a**) cytochrome a_3 and (**b**) cytochrome *b* are illustrated in this original study of Chance [15]. In **a**, the response of heart-muscle mitochondrial preparation to metabolism of added fumarate is shown. In **b**, changes in steady-state oxidation are observed in response to added succinate, with the cytochromes approaching complete reduction as O_2 becomes exhausted. Reproduced with permission of the publisher

Stock reagents needed for total oxidation are either 100 mM potassium ferricyanide (added at 5 μL/mL incubation) or 1 mM FCCP (carbonyl cyanide *p*-trifluoromethoxyphenylhydrazone) in ethanol (added at 1 μL/mL incubation). Reagent needed for total reduction is sodium dithionite. The latter must be maintained dry so that only a few grains can be added; if a solution is used, this is prepared fresh as a 1 M solution so that only 1 μL/mL incubation is needed to effect complete reduction. It should be noted that reduction by dithionite is slower with larger amounts of dithionite due to altered chemical reactions.

1.4 NADH/NAD⁺ Measurement

The NADH/NAD^+ couple provides a central electron carrier connecting oxidation of many food-derived metabolites and energy production. Enzyme-coupled and HPLC assays are available to measure NADH and NAD^+ concentrations [18, 19], and with selective solubilization of cell membranes, measurements of mitochondrial contents can be obtained [20]. However, a substantial fraction of the total NADH + NAD^+ pool is protein bound, so the measurements of the total amounts of NADH and NAD^+ in the mitochondria do not allow accurate estimate of $E_h NAD^+$. Studies of different dehydrogenase reactions in rat liver and hepatocytes showed that the reaction catalyzed by β-hydroxybutyrate dehydrogenase is near equilibrium so that the β-hydroxybutyrate/acetoacetate ratio can be used to calculate the matrix $E_h NAD^+$, which is about 318 mV [20]. Because E_o for NADH/NAD^+ is −337 mV, this means that free NAD^+ is present at a considerably higher concentration than free NADH. This characteristic means that the energetics of Complex I can be altered by the rates of the NAD^+-linked dehydrogenases.

Assay of β-hydroxybutyrate and acetoacetate can be performed with several standard approaches, including on a fluorescence plate reader using enzyme-coupled assays [19], by ^{1}H-NMR spectroscopy [21] or gas chromatography-mass spectrometry [22]. $E_h NAD^+$ is then estimated using the Nernst equation, the β-hydroxybutyrate and acetoacetate concentrations, an E_o value of −297 mV [20], and the assumption that the β-hydroxybutyrate dehydrogenase reaction is at equilibrium. It must be pointed out that it is not clear whether this reaction is at equilibrium in all mitochondria so this assumption will require validation for extension to other cell types and tissues.

Measures of relative changes in the NADH/NAD^+ couple can be readily obtained by absorbance and fluorescence methods. NADH has an absorbance at 340 nm (extinction coefficient, 6.2/mM/cm) which is not present in NAD^+. Although not directly useful for E_h determination, measurement with dual-wavelength spectroscopy can provide a sensitive means to measure relative changes. In this approach, absorbance at 375 nm is used as a reference to control for light scatter so that absorbance change can be used to provide a measure of the absolute change in NADH concentration in intact cells [23].

More sensitive detection of changes in NADH redox state is obtained by measuring fluorescence of NADH with excitation at 366 nm and emission at 450 nm [24, 25]. Protein-binding results in enhanced NADH fluorescence efficiency, and fractionation studies have shown that most of the NADH fluorescence is associated with mitochondria. Consequently, relative changes in fluorescence of NADH provide means to measure changes in redox state of this couple in intact cells and tissues. This approach has been used effectively to study redox control in cell culture, often described as measurement of "autofluorescence" because it does not require addition of fluorescence probes.

The flow of electrons from substrates to NAD^+ is regulated at many levels to balance utilization of carbohydrate, fat, and amino acid-derived precursors. Some of the key dehydrogenases are activated by Ca^{2+}, and several of the monocarboxylate, dicarboxylate and tricarboxylate substrates are exchanged by antiporters. Multiple fatty acid transporters control utilization of different chain lengths, and variations in amino acid transporters also contribute to rates of utilization of these substrates for mitochondrial respiration. The relatively limited range of substrates and conditions used for most studies of isolated mitochondria raises the possibility that key regulatory mechanisms related to supply of reducing equivalents to maintain $E_h NAD^+$ in different tissues remain to be discovered. To date, relatively limited efforts have been made to mimic physiologically relevant metabolic conditions in studies with isolated mitochondria.

1.5 Coenzyme Q and Complex III

Electron transfer from NADH to ubiquinone (CoQ) occurs through Complex I, a redox-driven proton pump. This is a large protein complex containing a covalently bound, redox-active flavin, as well as multiple iron-sulfur centers. The ΔE_h for the overall reaction is determined by the E_hCoQ relative to $E_h NAD^+$. CoQ exists as three redox forms, ubiquinone, ubiquinol, and ubisemiquinone. The redox characteristics of quinones have recently been reviewed [26]; CoQ is hydrophobic and mostly membranal; estimates of E_hCoQ are obtained using the Nernst equation with relative amounts of ubiquinol and ubiquinone obtained by HPLC [27]. Because CoQ is an intermediate between Complex I and Complex III and also substrate for Complex II and a number of other substrate-linked dehydrogenases (e.g., acyl CoA dehydrogenases), studies of Complex I function are subject to variable energetics due to steady-state reduction of the CoQ pool. In this way, changes in ΔE_h can alter the energetics for maintenance of $\Delta\psi$ independently of leak currents. Thus, there is a need to include ΔE_h measurements along with $\Delta\psi$ and O_2 consumption measurements.

Absorbance and fluorescence of the flavin in complex III change upon reduction, and these changes have also been used to measure redox in cells and intact tissues [28]. In contrast to the increases in absorbance and fluorescence of NADH upon reduction, flavoproteins have a decrease in absorbance and fluorescence upon reduction. Fluorescence is measured with excitation at 436 nm and emission at 570 nm, largely a measure of mitochondrial flavoproteins [29]. Because there are multiple flavoproteins which can contribute to the signals, there is a limitation to the mechanistic information which can be derived. On the other hand, ratiometric studies of the NADH fluorescence and flavoprotein fluorescence provide very sensitive indicators of tissue anoxia which can be used to visualize regional differences in oxidation due to pathophysiologic changes in vivo [24, 25]. With modern instrumentation, these approaches could provide improved capabilities for in vivo study of mitochondrial redox reactions.

1.6 Cytochromes

Complex III and Complex IV contain hemoproteins termed "cytochromes" which are readily detected by spectrophotometry. These were used as early as 1952 to measure mitochondrial redox changes in cells [15]. The measurements of the redox states of cytochromes depend upon the absorbance characteristics of the hemes in the cytochromes, which have more intense absorbance in the reduced forms (Fig. 1) [13]. In the visible range, the absorbance bands are termed α, β, and γ (Soret) absorbance bands. While the Soret band has the greatest extinction coefficients, it occurs at the shortest wavelength where there is the greatest light scatter in preparations containing mitochondria; consequently, measurements in this region can be more prone to error due to this light scatter. In addition, there is considerable overlap of signals from hemes *b* and *c*, and these can be further obscured by signals from myoglobin or hemoglobin [30]. Consequently, measurements in the Soret region are usually limited to cytochromes a_3. Cytochromes *a* and a_3 contain heme *a*, which is structurally distinct from heme *b*, present in hemoglobin, myoglobin, catalase, and *b*- and *c*-type cytochromes. Cytochromes *a* and a_3 also absorb light in the near-infrared region, and this has provided a means to measure oxidation in vivo [17]. Considerable spectral overlap occurs with the β bands, so most measurements are obtained with the α bands. Cytochromes b and c contain heme *b* (Fe-complexed protoporphyrin IX) but differ in absorbance because the heme is covalently bound in cytochromes *c* and c_1 but not in b_H or b_L. Other minor differences in absorbance maxima occur due to the interactions with amino acids in the vicinity of the heme. Addition of ligands like KCN, which bind to the O_2 site, can be used to discriminate cytochromes *a* and a_3. However, this is not useful for steady-state redox measurement because it blocks electron flow.

Complex III accepts two electrons from CoQ and transfers one to cytochrome c_1 and the other to the cytochrome b_H and b_L pair. The electron from cytochrome c_1 is then transferred to cytochrome *c*, while the electron from cytochrome *b* is transferred back to CoQ. The system is a proton pump, with energetics of proton pumping probably defined by steady-state potential of the donor ubiquinol/ ubisemiquinone couple and cytochrome c_1. As above, changes in the steady-state E_h values determine the energetics available for proton pumping. Consequently, interpretations of Complex III function in generation of $\Delta\mu_{H^+}$, ROS generation or support of ATP production, are limited by changes in steady-state E_h values which determine the available ΔE_h.

Steady-state reduction of cytochromes *b* is measured by dual-wavelength measurement with 575 as an isosbestic point. Absorbance maxima for the two forms are at 561 and 566 nm, but the signals overlap considerably so that redox changes of the mixture are most readily obtained by measuring absorbance at 562 nm relative to 575 nm. The extinction coefficient value for 562 nm minus 575 nm is 23/mM/cm. Fractional oxidation under steady-state conditions is obtained by use of an uncoupler to approximate 100% oxidation followed by addition of sodium dithionite to obtain 100% reduction. Because dithionite breaks down rapidly in solution, complete reduction is usually achieved by addition of a few grains of solid sodium dithionite. FCCP and CCCP (carbonyl cyanide *m*-chlorophenylhydrazone) are commonly used as uncouplers to obtain maximal oxidation, but this approach assumes that the only limitation to oxidation is the existence of the proton gradient. Because this assumption may not be valid, an alternative means to obtain maximal oxidation is to use a chemical oxidant such as 0.5 mM potassium ferricyanide [31]. In intact organs and some cell preparations, the fully reduced state can be approximated by removal of O_2 supply or addition of potassium cyanide [6, 15]; however, these treatments may not result in complete reduction due to endogenous respiratory control characteristics.

Cytochrome *c* has an absorbance maximum at 550 nm in the reduced form, while cytochrome c_1 has a maximum at 554 nm. The two are not easily resolved spectrally in cells or tissues; therefore, the two are measured together as cytochrome $c + c_1$. For these measures, dual-wavelength spectroscopy is used by measuring 550 nm minus 540 nm as an isosbestic point. The extinction coefficient is 19/mM/cm. This measurement is usually expressed relative to maximal oxidation and reduction as performed for cytochromes *b*.

Steady-state reduction of cytochromes *a* and a_3 is also frequently measured together as cytochrome $a + a_3$, using the wavelength pair 605 nm and 630 nm [32]. The extinction coefficient is 13.1/mM/cm.

1.7 Oxygen Consumption Rate

Changes in redox state of cytochromes can be used to measure O_2 consumption rate in closed systems where known amounts of O_2 are added [15, 23]. Typically, the system is allowed to consume all O_2 and become anoxic. Rapid addition of a known amount of O_2 then results in oxidation of the cytochromes, and the time required to consume the O_2 and become reduced allows calculation of O_2 consumption rate. Because the rate of O_2 consumption decreases when O_2 concentration reaches the low micromolar range, measurements are calculated from the difference in time to consume different amounts of added O_2. This approach has been directly validated relative to measurements with a Clark-type electrode [23]. With this approach, the duration of anoxia prior to addition of O_2 pulses affects the O_2 consumption rate [33], necessitating care in the timing of the measurements. A variation on this approach has also been used to measure rates of catalase activity in terms of the amount of O_2 produced and cytochrome oxidized following addition of H_2O_2 under anaerobic conditions [34].

1.8 NADPH/NADP⁺ Measurement

As indicated above, mitochondria contain low-flux redox systems that are qualitatively different from those which support oxidative phosphorylation and energy metabolism. Although originally described in terms of their role in protection against oxidative stress, accumulating evidence shows that these systems function in regulation of the redox states of cysteine (Cys) residues of proteins, controlling their structures and activities. Consequently, assays to measure fractional reduction of specific Cys residues of proteins are of considerable importance to understand mitochondrial signaling and control.

Important conceptual advances in understanding redox regulation of cell functions have come from redox proteomics methods, some of which are especially useful to measure mitochondrial redox systems. These systems are oxidized by the endogenous mitochondrial generation of O_2^- and H_2O_2. An early estimate that H_2O_2 production by mitochondria is about 1% of the total O_2 consumption rate [2] is frequently repeated, but accurate values for mitochondria in intact systems are not available. Measurements obtained with isolated mitochondrial preparations are likely to provide exaggerated values due to the presence of damage mitochondria and the analysis under supraphysiologic conditions of oxidizable substrates and pO_2. On the other hand, because there is also evidence for electron transfer bypass, in which NADH dehydrogenase generates O_2^-, the O_2^- is transported into the intermembrane space according to the electrochemical gradient, and the O_2^- donates the electron to cytochrome *c* [35]. The rate of this pathway is not known, so that there is a possibility that the estimated rate of 1% could also be an underestimate. Available methods to measure oxidant production by mitochondria under relevant in vivo conditions are difficult to calibrate so that better approaches are needed. Nonetheless, the data are accurate enough to allow the conclusion that only a small fraction of the total O_2 consumed by cells is converted by mitochondria to oxidants. Because thiol-dependent regulatory mechanisms are dependent upon these oxidants for oxidation, this low rate of oxidant production defines redox regulatory systems as "low-flux" systems when compared to mitochondrial respiration. The practical meaning is that minor disruption of normal high-flux electron transfer by the respiratory chain can produce oxidants at rates which are comparable to or higher than the rates which normally function in regulation. These issues have been previously reviewed and provide the basis for a definition of oxidative stress which includes the disruption of redox signaling and control pathways [36, 37].

While rates of electron transfer through the low-flux redox signaling and control pathways are not known, information about their function can be obtained from measurements of steady-state redox potentials. The thiol-dependent systems are maintained by electron transfer from NADPH. The NADPH/$NADP^+$ couple in mitochondria is maintained at relatively negative E_h value, about −415 mV in the liver [20]. As with the NADH/NAD^+ couple, direct measurement of total concentrations does not give a reliable measure of the potential because of extensive protein binding. Consequently, better estimates are obtained in the liver by measuring the glutamate, ammonia, and 2-oxoglutarate concentrations and calculating the E_h with the assumption that the reaction is at equilibrium. Alternatively, isocitrate, 2-oxoglutarate, and CO_2 can be used [20]. Little information is available concerning whether these reactions are at equilibrium in different tissues, so some extent of validation would be important in use of this approach.

2 Materials

2.1 GSH/GSSG Sample Buffer and Derivatization Reagents

1. 10% PCA/BA (10% perchloric acid/0.2 M boric acid) is prepared by adding 71 mL of 70% perchloric acid, 6.2 g boric acid (0.2 M) and adjusting total volume to 500 mL.
2. 10% PCA/BA with internal standard is made by adding 1.38 mg γ-glutamylglutamate (10 μM) to the solution above.
3. 20 mg/mL dansyl chloride solution prepared in 100% acetone.
4. 9.3 mg/mL iodoacetic acid solution prepared in water.
5. 1% digitonin (w/v) solution in water.
6. Chloroform.
7. Acetone.

2.2 . Mobile Phase Buffers for GSH/GSSG HPLC Determination

1. Solvent A: 80% methanol in water.
2. *Solvent B*: 64% methanol, 4 M sodium acetate buffer, pH 4.6.

2.3 SDS-Polyacrylamide Gel Electrophoresis

1. Running buffer (10×): 250 mM Tris–HCl, 1.92 M glycine, 1% SDS.
2. 30% acrylamide/bis solution 37.5:1 (2.6% C) (BioRad, Hercules, CA) and N,N,N,N′-tetramethylethylenediamine (TEMED) (Sigma).
3. Ammonium persulfate (APS): Prepare a 10% stock solution and store at −20 °C.
4. Precision plus protein standards (BioRad).
5. Gel casting stock solutions: 1.5 M Tris–HCl (pH 8.8) to be used for casting the resolving gel; 1.0 M Tris–HCl (pH 6.8) to be used in casting the stacking gel; 10% SDS.
6. Mini-Protean III cell (BioRad) and Power Pac 200 power source (BioRad).

2.4 Redox Western Blotting (Trx2, Prx3, TrxR2)

1. Towbin's transfer buffer: 25 mM Tris–HCl, 192 mM glycine, 20% methanol. Store at 4 °C.
2. Nitrocellulose membrane (BioRad) and extra thick filter paper (7.5 × 10 cm) (BioRad).
3. Trans-blot SD semidry transfer apparatus (BioRad) and Power Pac 200 power source (BioRad).
4. PBS with Tween-20 (PBS-T)/wash buffer: add 2 mL of Tween-20 (Sigma) to 1 L of 1× PBS.
5. Blocking buffer/antibody dilution buffer: 1:1 mixture of PBS-T and Li-Cor Blocking Buffer (Li-Cor, Lincoln, NE).
6. Primary antibody: rabbit antisera against Trx2; Prx3 mouse monoclonal antibody (Abcam, Cambridge, MA); TrxR2 rabbit polyclonal antibody (AbFrontier, Korea).
7. Secondary antibody: goat anti-rabbit or goat anti-mouse Alexafluor680 antibody (Invitrogen, Eugene, OR).
8. Li-Cor Odyssey infrared imager (Li-Cor).

2.5 Cell Lysis and Protein Alkylation: Trx2 Redox Western Blot

1. 10% trichloroacetic acid solution (Sigma).
2. 100% acetone (Sigma).
3. Phosphate-buffered saline (PBS) (Mediatech).
4. Disposable cell scraper (Fisher).
5. Alkylation buffer: 50 mM Tris–HCl (pH 8), 0.1% SDS, 15 mM 4-acetoamido-4′-maleimidylstilbene-2,2′-disulfonic acid (AMS) (Invitrogen).
6. Nonreducing sample loading buffer (5×): 300 mM Tris–HCl (pH 6.8), 50% glycerol, 1% SDS, 0.05% bromophenol blue.

2.6 Cell Lysis and Protein Alkylation: Prx3 Redox Western Blot

1. DMEM/F-12 (1:1) (with L-glutamine) cell culture medium (Mediatech, Manassas, VA) supplemented with 10% fetal bovine serum (Atlanta Biologicals, Atlanta, GA) and 1% penicillin/streptomycin (Hyclone, Logan, UT).

2. Paraquat (Sigma, St. Louis, MO) 10 mM stock solution in water. Prepare stock solution and use within 15 min of preparation.
3. Phosphate-buffered saline (Mediatech, Manassas, VA).
4. Alkylation buffer: 40 mM Hepes, 50 mM NaCl, 1 mM EGTA, protease inhibitor cocktail (Sigma), 100 mM N-ethylmaleimide (NEM). Add NEM to buffer immediately prior to use.
5. 25% CHAPS stock solution for cell lysis.
6. Disposable cell scraper (Fisher).
7. Nonreducing sample loading buffer (5×): 300 mM Tris–HCl (pH 6.8), 50% glycerol, 1% SDS, 0.05% bromophenol blue.

2.7 BIAM Modification and Pulldown: TrxR2 Redox Western Blot

1. Biotin iodoacetamide [*N*-(biotinoyl)-*N*′-(iodoacetyl)ethylenediamine, BIAM, Molecular Probes]: Dissolved in 50 mM Tris–HCl buffer (pH 6.8) to make a stock solution at 0.1 M concentration.
2. Iodoacetamide (IAM, Sigma-Aldrich Co.): Dissolve in Tris–HCl buffer at 5 mM concentration.
3. Lysis buffer A [50 mM Bis-Tris–HCl (pH 6.5), 0.5% Triton X-100, 0.5% deoxycholate, 0.1% SDS, 150 mM NaCl, 1 mM EDTA, leupeptin, aprotinin, and 0.1 mM PMSF] containing 10 μM BIAM.
4. Protein G Sepharose (Sigma-Aldrich, St. Louis, MO).

2.8 Cell Lysis and ICAT Reagents: Redox Proteomics

1. 10% TCA.
2. Denaturation buffer: 50 mM Tris–HCl, pH 8.5, 0.1% SDS.
3. ICAT assay kit (Applied Biosystems, Foster City, CA).

2.9 Mass Spectrometry-Based Redox Proteomics

1. Ultimate 3000 nanoHPLC system (Dionex) with a nanobore column (0.075 × 150 mm PepMap C18 100 Å, 3 μm, Dionex).
2. QSTAR XL Q-TOF mass spectrometer (Proxeon Biosystems).
3. ProteinPilot software (Applied Biosystems).

3 Methods

3.1 GSH/GSSG Measurement in Cultured Cells

Mitochondria contain two central systems controlling protein thiol/disulfide states, one dependent upon GSH and the other dependent upon a mitochondria-specific thioredoxin, thioredoxin-2 (Trx2). These systems are parallel, nonredundant systems [38] which are both reduced by NADPH-dependent mechanisms and oxidized by H_2O_2-dependent mechanisms. The GSH/GSSG couple is maintained by a splice variant of GSSG reductase which is targeted to mitochondria. Oxidation of GSH can occur by many enzymes, most notably glutathione peroxidase-1 (Gpx1) and glutathione peroxidase-4 (Gpx4) and a number of GSH transferases [39]. Glutaredoxin-2 also oxidizes GSH to GSSG in its reaction to remove GSH from glutathionylated proteins.

3.1.1 Sample Derivatization

1. Remove treatment media from cells and wash three times using cold PBS.
2. Add approximately 500 μL of 1% digitonin (w/v) to each plate/well of cells, and incubate for 5 min on ice to permeabilize.
3. Aspirate the digitonin solution from the cells and carefully wash the cells three times with cold PBS.
4. Apply approximately 500 μL of 10% PCA/BA directly to the permeabilized, adherent cells, scrape each well, and collect cell extracts in a 1.5 mL tube.
5. Samples are spun for 2 min in a microcentrifuge at approximately 14,000 × *g* to pellet protein.
6. An aliquot (300 μL) of each supernatant is transferred to a fresh microcentrifuge tube.
7. 9.3 mg/mL Iodoacetic acid solution (60 μL) is added to each tube and vortex to mix.
8. The pH is adjusted to 9.0 ± 0.2 with the KOH/tetraborate solution (approximately 220–250 μL).
9. After about 3 min to allow complete precipitation of potassium perchlorate, the pH of at least some of the samples should be checked to verify that they are in the correct range.
10. Incubate samples for 20 min at room temperature.
11. 300 μL of dansyl chloride (20 mg/mL acetone) is added, and the samples are mixed and placed in the dark at room temperature for 16–26 h. Dansylation of GSH is complete by 8 h, but GSSG has 2 amino groups that must be modified. The rate of the second dansylation is slower than for the first with the result that mono-dansyl derivatives of GSSG (eluting between the N-dansyl-S-carboxymethyl-GSH and bis-dansyl-GSSG) will be present if dansylation is incomplete.
12. After derivatization, chloroform (500 μL) is added to each tube to extract the unreacted dansyl chloride.
13. Samples are stored at 0–4° in the dark in the presence of both the perchlorate precipitate and the chloroform layer until assay by HPLC. Stability tests show that samples can be stored under these conditions for 12 months with little change in the amounts of GSH and GSSG derivatives.

3.1.2 HPLC Analysis of GSH/GSSG Redox

1. Samples are centrifuged for 2 min in a microcentrifuge prior to transfer of an aliquot of the upper (aqueous) layer to an autosampler.
2. Typical injection volume is 25 μL.
3. Separation is achieved on 3-aminopropyl columns (5 μm; 4.6 mm × 25 cm; Custom LC, Houston; or Supelcosil LC-NH2, Supelco, Bellefonte, PA).
4. Initial solvent conditions are 80% A, 20% B run at 1 mL/min for 10 min.
5. A linear gradient to 20% A, 80% B is run over the period from 10 to 30 min.
6. From 30 to 46 min, the conditions are maintained at 20% A, 80% B and returned to 80% A, 20% B from 46 to 48 min.
7. Equilibration time for the next run is 12 min.
8. Detection is obtained by fluorescence monitoring with band-pass filters, 305–395 nm excitation and 510–650 nm emission (Gilson Medical Electronics, Middleton, WI).
9. Quantification is obtained by integration relative to the internal standard, γ-glutamylglutamate.

3.2 Trx2 Redox Western Blot Analysis in Cultured Cells: Sample Preparation

Human Trx2 contains two Cys residues in the active site; the dithiol and disulfide forms are separated by nonreducing PAGE following modification of the protein with AMS, a reagent which increases the mass from 12 kD to 13 kD (1 kD/2 Cys) (Fig. 3). To obtain E_hTrx2, a redox Western blot method is used [40]. This approach has an advantage over the GSH/GSSG method in that study of proteins which are present only in mitochondria provides a

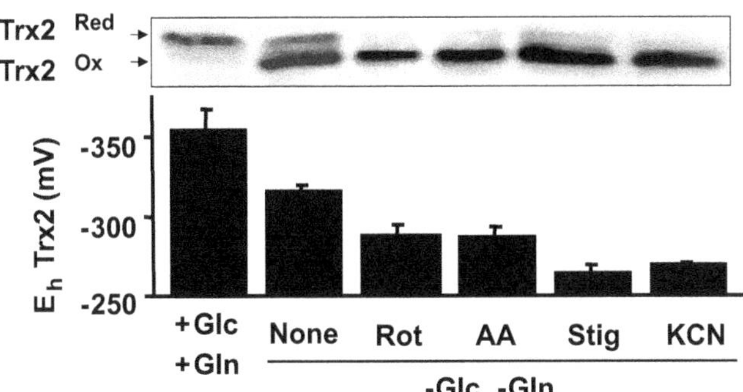

Fig. 3 Trx2 redox state in HT29 cells exposed to mitochondrial respiratory inhibitors in control media and glucose (Glc)-, glutamine (Gln)-free media. Cells were grown to 80% confluency and then cultured for 24 h with media as indicated. After 24 h, inhibitors were added at 5 μM (Rot, AA, Stig) or 0.5 mM (KCN) for 30 min prior to extraction and analysis by redox Western blotting. Separation was provided by reaction of samples with AMS, a thiol reagent which adds approximately 500 Da per thiol, thereby slowing mobility sufficiently to separate the reduced form from the disulfide form. All incubations in Glc,−Gln-free media were significantly different from respective +Glc,+Gln controls. $N = 5$. *Rot* rotenone, *AA* antimycin A, *Stig* stigmatellin

compartment-specific assay without subcellular fractionation. Redox Western blots have been developed which resolve reduced and oxidized forms of proteins based upon both changes in molecular mass and changes in charge. For the former, a thiol reagent with relatively large mass, such as AMS (4-acetoamido-4-′-maleimidylstilbene-2,2′-disulfonic acid), is used to increase the mass of the thiol-containing form of the protein. This allows separation by electrophoresis using SDS-PAGE and is described below.

3.2.1 Cell Lysis and Protein Alkylation

1. Assays are performed with cells grown in a 35-mm or 6-well culture plate (1–2 × 10^6 cells).
2. After experimental treatment, cells are washed with ice-cold PBS.
3. The cells are then treated with 1 mL of ice-cold TCA (10%), scraped, transferred to microcentrifuge tubes, incubated on ice for 30 min, and centrifuged at 12,000 × *g* for 10 min.
4. The supernatant is removed and the protein pellet is saved.
5. One milliliter of 100% acetone is added to the pellet. The tube is mixed, incubated on ice for 30 min, and centrifuged at 12,000 × *g* for 10 min.
6. The acetone is removed and the pellet (30–50 μg protein) is used for analysis.
7. After addition of 100 μL of lysis/derivatization buffer (50 mM Tris–HCl, pH 8.0, 0.1% SDS, 15 mM AMS), pellets are resuspended by sonication and incubated at room temperature in the dark for 3 h. The samples can be used for redox Western blotting at this step or be saved at –20° C for subsequent analysis.

3.3 Prx3 Redox Western Blot Analysis in Cultured Cells: Sample Preparation

Peroxiredoxins (Prx) are a class of thiol-dependent antioxidant proteins that exhibit peroxidase activity. These proteins utilize redox-active Cys residues in their active site to reduce hydrogen peroxide and other organic peroxides. Prx are classified into two distinct groups, 1-Cys and 2-Cys, based on the number of Cys residues involved in the reduction of peroxides. Mammals possess six isoforms, five of which are considered 2-Cys Prx (Prx1-5). Prx 6 is the sole member of the 1-Cys class. The reaction mechanism for Prx occurs in two steps, the first of which is common for both 1-Cys and 2-Cys Prx. First, peroxide reacts with the "peroxidatic" Cys residue in the active site resulting in the formation of a cysteine sulfenic acid. This newly formed sulfenic acid then reacts with the "resolving" Cys, resulting in the formation of a disulfide linkage. It should be noted that Prx are present as domain-swapped homodimers, where the "peroxidatic" Cys is located on one subunit and "resolving" Cys on the opposite subunit. Therefore, when a 2-Cys Prx becomes oxidized, the enzyme is locked in a dimer due to the newly formed disulfide bonds (Fig. 4). 2-Cys Prx utilize the

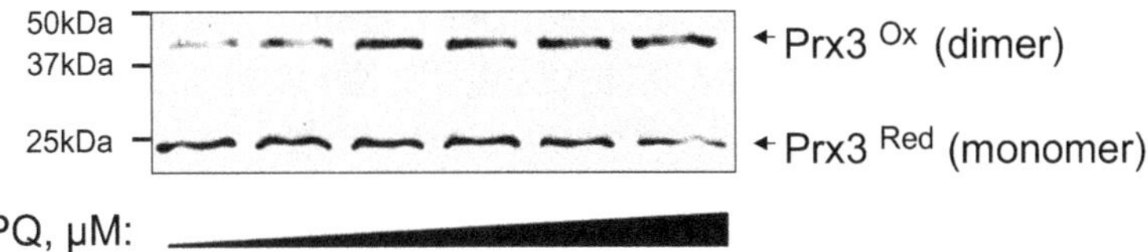

Fig. 4 Measurement of oxidation of Prx3 by redox Western blot analysis. SH-SY5Y human neuroblastoma cells were treated 24 h with increasing amounts of paraquat (0, 10, 25, 50, 75, 100 μM). Following extraction and treatment with NEM to prevent further oxidation, separation by SDS-PAGE followed by Western blotting reveals the oxidized (disulfide) form increases in response to PQ compared to the reduced, NEM-modified form (monomer)

thioredoxin/thioredoxin reductase system to reduce this disulfide and effectively recycle the enzyme back to its active form [41].

Prx3 and Prx5 are mitochondrial isoforms; however, Prx5 is an "atypical" 2-Cys Prx that does not form an intermolecular disulfide upon reaction with a peroxide because it functions as a monomer. Based on the fact that Prx form stable disulfides resulting in a "locked" dimerized state, Cox et al. have exploited this concept in order to measure redox changes in Prx3 [42]. This method involves alkylation of cellular proteins with N-ethylmaleimide (NEM) followed by SDS-PAGE and standard immunoblotting techniques.

3.3.1 Cell Lysis and Protein Alkylation

1. Approximately 1–2 × 10^6 SH-SY5Y neuroblastoma cells are plated into each well of a 6-well plate containing 2 mL of media and allowed to adhere overnight.
2. Medium is aspirated from each well and replaced with treatment media containing desired concentrations of paraquat, and cells are incubated for 24 h.
3. At the completion of the treatment period, the medium is aspirated from each well, and wells are gently washed three times with 2 mL of cold PBS.
4. After washing approximately 100 μL of alkylation buffer containing 100 mM NEM is added to each well and allowed to incubate 10–15 min at room temperature.
5. Add approximately 4 μL of 25% CHAPS solution to each well (1% final concentration) to lyse cells.
6. Using a cell scraper, scrape cells and debris from each well, and collect cell extracts in 1.5 mL tubes and place on ice.
7. Using a benchtop sonicator, sonicate each tube briefly to ensure complete cell lysis and protein extraction. Keep cells on ice.
8. Pellet debris and insoluble protein by centrifugation at 17,000 × *g* for 5 min on a benchtop microcentrifuge. Place samples back on ice.
9. Assay protein content of each sample using BCA protein assay (Thermo Scientific, Rockford, IL).

10. Add 5× nonreducing sample buffer to approximately 20–25 μg of protein. Place samples on a heating block at 95 °C for 5 min.

3.4 TrxR2 Redox Western Blot Analysis: Sample Preparation

Trx system is composed of Trx, thioredoxin reductase (TrxR), NADPH, and Trx peroxidases/peroxiredoxins. Trx reduces protein disulfides directly and serves as a reductant for the peroxiredoxins [43]. Oxidized form of Trx is reduced by catalytic activity of TrxR using an electron from reduced NADPH [44]. TrxR2 is an isoform of TrxRs including *E. coli* TrxR and human TrxR1. The catalytic site of TrxR is -Cys-Val-Asn-Val-Gly-Cys- and located in the FAD domain of enzymes [45]. Furthermore, TrxR2 has a C-terminal selenocysteine residue that is required for catalytic activity but is not part of the conserved active site [46]. Since TrxRs are known to reduce oxidized Trx, alterations in TrxR activity may regulate Trx activity. TrxR2 is localized in mitochondria, while TrxR1 is predominantly found in cytosol and nucleus [47]. In this chapter, we describe an assay for measuring mitochondrial TrxR2 redox state using BIAM-labeling technique (Fig. 5). The thiol-reactive biotin iodoacetamide and biotin maleimide derivatives can be used for BIAM-labeling technique described in this chapter. This method is based on the procedure of Kim et al. [48], with modifications to measurement of redox states of proteins containing thiols in catalytic site such as TrxR and redox factor-1 (Ref-1). The procedures are described below.

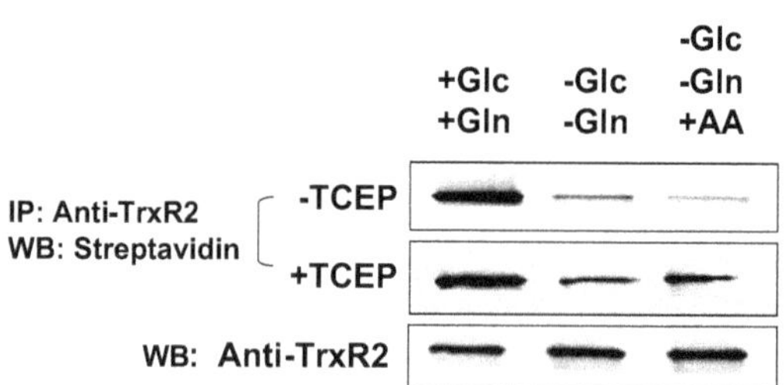

Fig. 5 Semiquantitative analysis of fractional reduction of thioredoxin reductase-2 (TrxR2) by BIAM-blot. Cells were incubated 24 h with +Glc,+Gln, –Glc,–Gln, or –Glc,–Gln and antimycin A (30 min with 5 μM; +AA). Aliquots of cell lysates were treated with the biotinylated iodoacetamide reagent, BIAM, and parallel aliquots were reduced with TCEP and then reacted with BIAM. Following immunoprecipitation with anti-TrxR2, samples were separated by SDS-PAGE, blotted and probed with fluorescently labeled streptavidin. Controls for recovery following immunoprecipitation were performed by Western blotting with anti-TrxR2 and showed similar recovery. Although these methods are reproducible, the limiting conditions of BIAM labeling selected to maximize detection of reactive thiols in the presence of less reactive thiols do not allow strict quantification

3.4.1 Cell Culture and Affinity Purification

1. Cells (1–2 × 10^7 in 10 cm plate) after treatment [e.g., glucose (Glc)- and glutamine (Gln)-deficient media] were washed with cold PBS and fractionated, and then nuclear fraction was lysed with 1 mL of lysis buffer A [50 mM Bis-Tris-HCl (pH 6.5), 0.5% Triton X-100, 0.5% deoxycholate, 0.1% SDS, 150 mM NaCl, 1 mM EDTA, leupeptin, aprotinin, and 0.1 mM PMSF] containing 10 μM BIAM. As a control, cells that were not stimulated with treatment (e.g., Glc- and Gln-deficient media) were lysed and labeled with BIAM.
2. After incubation for 10 min at 37 °C in the dark, the labeling reaction was stopped by adding IAM to 5 mM. TrxR2 in the reaction mixtures was precipitated with the use of rabbit antibodies to TrxR2 and Protein G Sepharose (40 μL per sample). (Optional: Streptavidin-agarose can be used instead of TrxR2 antibody to immunoprecipitate all BIAM-labeled proteins.)
3. Immunocomplex of BIAM-labeled TrxR2 (use of TrxR2 antibody) or other BIAM-labeled proteins (use of streptavidin-agarose, 40 μL) will be washed with 1 mL of ice-cold lysis buffer A three times.
4. Add 40 μL of 2× gel loading buffer to each sample; heat at 95 °C for 10 min.

3.5 SDS-PAGE: Trx2, Prx3, and TrxR2 Redox Western Analysis

1. These instructions assume the use of a BioRad Mini-Protean II or III gel electrophoresis system. Clean each glass plate first with water then 100% methanol prior to gel casting.
2. For Trx2 and Prx3: prepare a 1.5 mm thick, 15% resolving gel by mixing 2.3 mL water, 5 mL 30% acrylamide solution, 2.5 mL 1.5 M Tris–HCl (pH 8.8), 100 μL 10% SDS, 100 μL APS, and 4 μL TEMED in a 15 mL tube. Invert tube to mix, pour gel solution between glass plates, and allow gel to polymerize for approximately 45–60 min at room temperature. Note: be sure to leave sufficient room for a stacking gel. Also, fill the remainder of the glass plate with water to ensure that the gel polymerizes with a clean, straight edge.
3. For TrxR2: prepare a 1.5 mm thick, 10% resolving gel by mixing 4 mL water, 3.3 mL 30% acrylamide solution, 2.5 mL 1.5 M Tris–HCl (pH 8.8), 100 μL 10% SDS, 100 μL APS, and 4 μL TEMED in a 15 mL tube. Invert tube to mix, pour gel solution between glass plates, and allow gel to polymerize for approximately 45–60 min at room temperature. Note: be sure to leave sufficient room for a stacking gel. Also, fill the remainder of the glass plate with water to ensure that the gel polymerizes with a clean, straight edge.
4. After polymerization of resolving gel, prepare and pour the 5% stacking gel by mixing the following in a 15 mL tube: 2.7 mL water, 0.67 mL 30% acrylamide solution, 0.5 mL 1.0 M

Tris–HCl (pH 6.8), 40 μL 10% SDS, 40 μL APS, and 4 μL TEMED. Pour off the water layer on the resolving gel and add stacking gel to the top of the glass plates. Place in desired comb (10 or 15 well) and allow to polymerize for 15–30 min.

5. Assemble the gel apparatus.
6. Prepare 500 mL of running buffer by adding 50 mL 10× running buffer to 450 mL water. Then fill inner gel chamber and add remainder of buffer to outer chamber. Allow the assembly to sit for a few min to ensure that there are now leaks. If no leaks are discovered, then samples may be loaded.
7. Load samples and the precision plus protein standards to individual wells.
8. Place lid on gel apparatus and connect to the Power Pac 200. Run the gel at 150 V for approximately 60–80 min or until the dye front runs off the gel.

3.6 Western Blotting: Trx2, Prx3, and TrxR2 Redox Western Analysis

1. After proteins are separated on the SDS-PAGE gel, proteins need to be transferred to a nitrocellulose (or PVDF) membrane for immunoblotting. The procedure described here assumes the use of a BioRad semidry transfer apparatus. A "wet" transfer can also be conducted; however, this procedure will not be described here.
2. Preincubate two filter papers and an appropriately cut nitrocellulose membrane in cold transfer buffer for approximately 10–15 min.
3. After SDS-PAGE, discard the running buffer and carefully pry apart the glass plates to free the gel.
4. Place a filter paper saturated with transfer buffer on the semidry transfer apparatus, and use a serological pipette to roll out any trapped air bubbles. Next place the membrane on the filter paper and roll out any air bubbles. Place the gel on top of the membrane and orient it to your liking. Again, carefully roll out any air bubbles. Place the last filter paper on top the gel and roll out any air bubbles. Note: it is important to not introduce any air bubbles in the filter paper-membrane-gel-filter paper sandwich. These bubbles will prevent proper protein transfer and produce poor quality blots.
5. After the filter paper sandwich is made, assemble the transfer apparatus, and connect it to the Power Pac 200. Run the transfer at 20 V for 60 min. Note: if you cast a thinner gel, run the transfer for slightly shorter time period, i.e., 0.75 mm gels for 45 min at 20 V.
6. After the transfer is complete, place the membrane in 10–20 mL of blocking buffer, and incubate on a shaker table for at least 30 min at room temperature.
7. Prepare primary antibody solution by adding 4 μL of peroxiredoxin 3 antibody to 10 mL of fresh blocking buffer (1:2500

dilution); 10 μL of Trx2 or TrxR2 antibody to 10 mL of fresh blocking buffer (1:1000 dilution).

8. Discard blocking solution and add the primary antibody solution to the membrane and incubate on a shaker table overnight at 4 °C.
9. After primary antibody incubation, pour used antibody solution into a clean 15 mL tube and save for another use.
10. Wash membrane 3 times for 15 min each in 10–20 mL of PBS-T. Discard wash buffer after every wash.
11. For Trx2 and TrxR2: Prepare secondary antibody solution immediately prior to use by adding 2 μL of goat anti-rabbit Alexa Fluor 680 secondary antibody to 10 mL of blocking buffer (1:5000 dilution). Protect this solution from light to avoid photobleaching of the fluorophore.
12. For Prx3: Prepare secondary antibody solution immediately prior to use by adding 2 μL of goat anti-mouse Alexa Fluor 680 secondary antibody to 10 mL of blocking buffer (1:5000 dilution). Protect this solution from light to avoid photobleaching of the fluorophore.
13. Add secondary antibody solution to the membrane, and incubate 45 min in the dark at room temperature on a shaker table.
14. Wash membrane three times for 15 min each in 10–20 mL of PBS-T in the dark. Discard wash buffer after every wash.
15. After the final washes, carefully place the membrane between two Kimwipes, and allow the membrane to dry in the dark.
16. Once the membrane is dry, scan it on a Li-Cor Odyssey infrared scanner at 700 nm. Adjust the intensity, contrast, and brightness of the scan to optimize your result.

3.7 Mass Spectrometry-Based Redox Proteomics

To measure the fractional reduction of mitochondrial proteins, we have used the isotope-coded affinity tag reagent (Applied Biosystems, CA) with a sequential treatment procedure designed to yield the ratio of reduced/oxidized forms of specific tryptic peptides, as measured by tandem mass spectrometry (Fig. 6) [49]. Data from this approach show that many proteins have Cys residues which are partially oxidized under steady-state conditions. Consequently, application of the mass spectrometry-based redox proteomics method can be expected to considerable improve understanding of redox signaling and control. Importantly, this approach allows detection of mitochondrial proteins even without mitochondrial isolation so that the oxidation of mitochondrial proteins can be determined within the context of other cellular compartments.

The methods used for redox proteomic analyses of mitochondrial proteins are based upon those used for nuclear proteins [10]. Analyses can be performed in three different ways, by studying

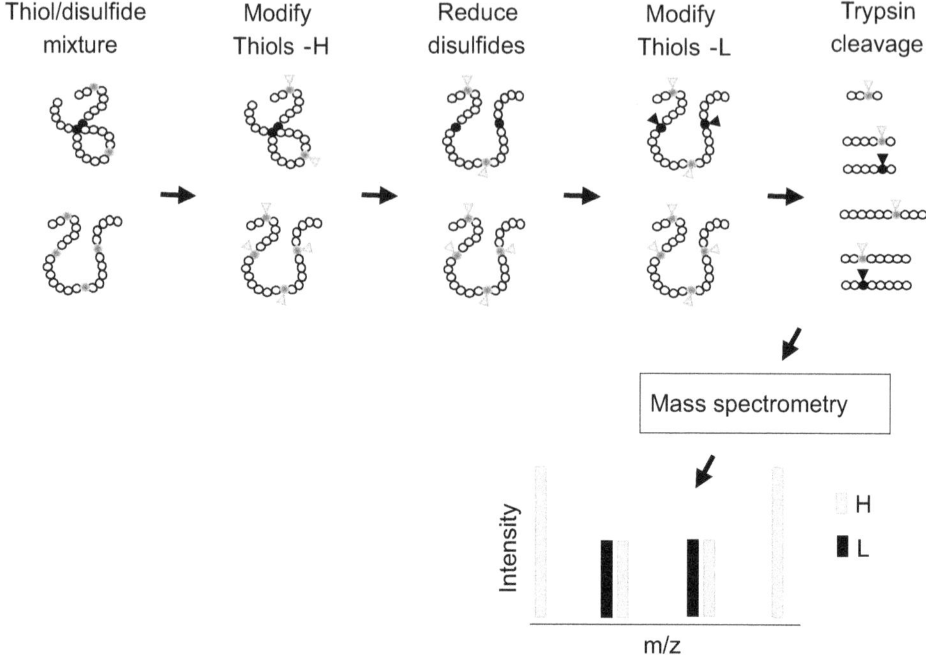

Fig. 6 Mass spectrometry-based analysis of fractional reduction of protein using ICAT reagents. Proteins are extracted and treated with the heavy ICAT reagent (H) to label thiols. Following removal of excess reagent, samples are treated with TCEP to reduce disulfides. The newly formed thiols are modified by treatment with the light reagent (L). Following tryptic digestion, analysis by LC-MS/MS allows calculation of fractional reduction from the H:L ratio

isolated mitochondria, by using digitonin to permeabilize cells and remove contaminating cytoplasm, or by directly analyzing cells and selecting mitochondrial proteins from the larger list of cytoplasmic proteins and proteins from other subcellular compartments.

3.7.1 Cell Lysis and Protein Collection

1. Cells [mouse aortic endothelial cells (MAEC), 1–2 × 10^6 in 3.5 cm plate] after treatment [e.g., reduced (−150 mV) or oxidized (0 mV) extracellular E_hCySS] were washed with cold PBS.
2. Add 1 mL 10% TCA to cells, scrape, and collect lysates in a clean 1.5 mL tube. Precipitation of white protein pellets will be observed.
3. Place on ice for 30 min and spin for 10 min at 16,100 × *g* (4 °C).
4. Remove supernatant, add 1 mL acetone to protein pellet, vortex, and spin again at 16,100 × *g* for 10 min (4 °C).
5. Remove supernatant and air-dry pellet for 1–2 min.

6. Add 100 μL denaturation buffer (50 mM Tris–HCl, pH 8.5, 0.1% SDS) to pellet, and resuspend pellet by sonication for 2 s on ice.
7. Perform protein assay.
8. Transfer 150 μg protein in denaturation buffer, and add more denaturation buffer if needed to make 80 μL as total sample volume.

3.7.2 ICAT Labeling

1. Add 150 μg protein in 80 μL denaturation buffer prepared above to 20 μL of heavy ICAT reagent (H).
2. Vortex and snap-centrifuge.
3. Incubate for 60 min at 37 °C.
4. Add 10 μL 100% TCA to sample and vortex.
5. Add additional 400 μL of 10% TCA to "4" and transfer all to a clean 1.5 mL tube.
6. Place on ice for 30 min.
7. Centrifuge at 16,100 × *g* for 10 min (4 °C).
8. Remove supernatant.
9. Wash protein pellet with 500 μL, 100% acetone.
10. Centrifuge at 16,100 × *g* for 10 min.
11. Remove supernatant.
12. Add 80 μL denaturation buffer to pellet.
13. Add 2 μL TCEP to sample and sonicate for 2 s on ice.
14. Incubate for 20 min at 37 °C.
15. Add 20 μL light ICAT reagent (L) reagent to samples.
16. Vortex and snap-centrifuge.
17. Incubate for 60 min at 37 °C.

3.7.3 Trypsinization of ICAT-Labeled Proteins

1. Dissolve trypsin included in ICAT assay kit in 200 μL of MilliQ-H_2O.
2. Add 200 μL of trypsin to a sample labeled with H- and L-ICAT reagents above.
3. Vortex and snap-centrifuge.
4. Incubate 12–16 h at 37 °C.
5. Vortex and snap-centrifuge.

3.7.4 Cation Exchange

1. Place sample in 3 mL tube.
2. Add 2 mL of Cation Exchange Buffer (CEB)-Load to sample (1 drop/s).
3. Vortex and snap-centrifuge.

4. Check pH; it should be around 2.5–3.3—if not add more CEB-Load until pH is correct.
5. Assemble cartridge holder with cation exchange cartridge.
6. Equilibrate cartridge by injecting 2 mL CEB-Load (1 drop/s).
7. Inject sample onto cartridge slowly (1 drop/s).
8. Wash with 1 mL CEB-Load (1 drop/s).
9. Elute peptides with 500 μL CEB-Elute (1 drop/s).
10. Collect flow-through.

3.7.5 Cleaning and Storing Cation Exchange Cartridge

1. Inject 1 mL CEB to clean cartridge (1 drop/s).
2. Inject 2 mL CEB before storing cartridge at 4 °C.

3.7.6 Purifying Peptides and Cleaving Biotin

1. Insert avidin cartridge into cartridge holder.
2. Equilibrate cartridge by injecting 2 mL of Affinity Buffer (AB)-Elute (1 drop/s) and discard waste.
3. Inject 2 mL AB-Load (1 drop/s) and discard waste.
4. Neutralize samples by adding 500 μL of AB, vortex, and snap-centrifuge.
5. Check pH; pH should be at 7—if not add more AB until it is correct.
6. Vortex and snap-centrifuge.
7. Slowly inject sample onto cartridge (1 drop/s).
8. Collect flow-through; this fraction contains unlabeled fragments and can be analyzed at a later time if necessary.
9. Inject an additional 500 μL of AB-Load and collect in the same tube (or a fresh tube) as **step 8**.
10. Inject 1 mL of AB (1 drop/s)-first wash and discard flow-through.
11. Inject 1 mL of AB (1 drop/s)-second wash and collect first 500 μL but discard the latter 500 μL.
12. Inject 1 mL of MilliQ-H_2O and discard flow-through.

3.7.7 Eluting ICAT-Labeled Samples

1. Inject 800 μL AB-Elute (1 drop/s).
2. Discard initial 50 μL.
3. Collect the remaining 750 μL in tube.
4. Vortex and snap-centrifuge.

3.7.8 Cleaning and Storing Avidin Cartridge

1. Inject 2 mL AB (1 drop/s) to clean avidin cartridge.
2. Inject 2 mL AB and store at 4 °C.

3.7.9 Cleaving Biotin

1. Evaporate samples in speed vacuum.
2. In another tube mix cleaving agents A (95 μL) and B (5 μL) in a ratio of 95:5.
3. Add 95 μL of cleaving reagent solution to each sample.
4. Vortex and snap-centrifuge.
5. Incubate for 2 h at 37 °C.
6. Vortex and snap-centrifuge.
7. Evaporate in speed vacuum.
8. Send off for mass spectrometry analysis.

3.7.10 Mass Spectrometry Analysis of Redox ICAT

An Ultimate 3000 nanoHPLC system (Dionex) with a nanobore column (0.075 × 150 mm PepMap C18 100 Å, 3 μm, Dionex) is used with the LC eluent being directly sprayed into a QSTAR XL system using a nanospray source from Proxeon Biosystems. The data from each salt cut is combined and processed by ProteinPilot software (Applied Biosystems). All quantification is performed by the ProteinPilot V2.0.1 software using the Swiss-Prot database. Quantification for proteins of interest is manually validated by examination of the raw data.

3.7.11 Calculation of Protein Redox State

Table 1 shows an example of measuring protein reduction/oxidation state affected by changes in extracellular redox conditions [49]. The ratios of reduced (H) to oxidized (L) thiols analyzed by mass spectrometry enable to calculate protein redox state, e.g., % oxidation = [L/(H + L)] × 100. Thirty mitochondrial proteins were sorted out from the original data [49] to calculate redox states of mitochondrial proteins using this approach.

4 Notes Regarding Redox Analyses

4.1 GSH/GSSG HPLC Analysis

1. To facilitate simultaneous measurement of CySS (typically >40 μM) and GSSG (typically <200 nM), two detectors with different sensitivity settings are used in series. Fluorometric detectors with monochromators set at 335 nm for excitation and 515 nm emission can also be used with equivalent results, but sensitivity is substantially less because the narrower bandwidths limit the intensity of both excitation and emission light.
2. To calculate the E_hGSSG, use the Nernst equation as follows: $E_h = E_o + RT/nF\ ln\ ([GSSG]/[GSH]^2)$, where E_o for this equation is equal to −276 mV.

4.2 Trx2 Redox Western Blot

1. An alternative redox western approach is adapted from the original method of Holmgren and Fagerstedt [50] for *E. coli* Trx in which iodoacetate is used to add negative charges to

Table 1

Redox ICAT results of mitochondrial proteins in mouse aortic endothelial cells treated with extracellular E_hCySS of −150 mV (reduced) or 0 mV (oxidized)

Accession	Name	−150 mV H:L	0 mV H:L	−150 mV % oxidation	0 mV % oxidation	Δ% oxidation
P63038	60 kDa heat shock protein, mitochondrial precursor	1.9	1.1	34.4	46.9	12.6
Q9CR21	Acyl carrier protein, mitochondrial precursor	3.9	2.9	20.4	25.7	5.3
P40124	Adenylyl cyclase-associated protein 1	2.3	1.3	30.0	42.9	12.9
P48962	ADP/ATP translocase 1	4.7	3.5	17.5	22.3	4.8
P51881	ADP/ATP translocase 2	2.5	2.3	28.5	30.5	2.0
P62331	ADP-ribosylation factor 6	1.5	1.2	40.6	45.5	4.8
P47738	Aldehyde dehydrogenase, mitochondrial precursor	4.9	3.1	16.8	24.4	7.6
P05202	Aspartate aminotransferase, mitochondrial precursor	2.9	2.2	25.6	30.9	5.3
Q9DCX2	ATP synthase D chain, mitochondrial	4.1	2.7	19.6	26.7	7.1
Q03265	ATP synthase subunit alpha, mitochondrial precursor	3.8	4.0	20.9	20.1	0.8
Q61753	D-3-phosphoglycerate dehydrogenase	2.9	2.2	25.7	31.5	5.8
Q9EQ06	Dehydrogenase/reductase SDR family member 8 precursor	2.9	1.8	25.7	35.1	9.4
O08749	Dihydrolipoyl dehydrogenase, mitochondrial precursor	2.0	1.5	33.3	39.7	6.4
Q91YQ5	Dolichyl-diphosphooligosaccharide-protein glycosyltransferase 67 kDa subunit precursor	3.7	0.8	21.3	54.1	32.8
Q99LC5	Electron transfer flavoprotein subunit alpha, mitochondrial precursor	1.8	1.7	35.2	37.0	1.9
P26443	Glutamate dehydrogenase 1, mitochondrial precursor	2.5	1.8	28.5	36.1	7.6
P54071	Isocitrate dehydrogenase [NADP], mitochondrial precursor	3.5	1.2	22.4	44.7	22.2
P08249	Malate dehydrogenase, mitochondrial precursor	3.2	2.3	23.8	30.1	6.3

Q791V5	Mitochondrial carrier homolog 2	2.3	3.1	30.1	24.4	5.7
P52503	NADH dehydrogenase [ubiquinone] iron-sulfur protein 6, mitochondrial precursor	2.5	1.9	28.7	34.5	5.8
Q9DCN2	NADH-cytochrome b5 reductase 3	2.8	2.1	26.5	32.1	5.5
Q922Q4	Pyrroline-5-carboxylate reductase 2	2.4	2.0	29.4	33.5	4.1
P52480	Pyruvate kinase isozymes M1/M2	3.6	2.0	21.6	33.8	12.2
Q8K2B3	Succinate dehydrogenase [ubiquinone] flavoprotein subunit, mitochondrial precursor	3.2	2.3	23.8	30.7	6.9
Q9R112	Sulfide:quinone oxidoreductase, mitochondrial precursor	2.7	1.9	27.0	35.0	8.1
Q9CZ13	Ubiquinol-cytochrome-c reductase complex core protein 1, mitochondrial precursor	4.5	3.3	18.3	23.4	5.1
Q9Z1Q9	Valyl-tRNA synthetase	1.2	0.9	45.2	51.7	6.5
Q60932	Voltage-dependent anion-selective channel protein 1	3.9	2.8	20.5	26.4	5.9
Q60930	Voltage-dependent anion-selective channel protein 2	4.1	1.2	19.7	45.2	25.6
Q60931	Voltage-dependent anion-selective channel protein 3	2.6	1.1	27.6	47.3	19.6

thiols so that separation can be obtained under native gel conditions.

2. Trx2 resolves at approximately 10–15 kDa, and care should be taken to prevent these lower mass proteins from running off of the gel.
3. Care must be taken to avoid loading too much protein because overloading interferes with separation of reduced and oxidized Trx2.

4.3 Prx3 Redox Western Blot

1. The amount of protein that you load for each blot may be different depending on the cell type that is used. It is wise to optimize this condition to achieve the best result.
2. The mouse monoclonal antibody from Abcam is used in this current protocol; however, other antibodies against peroxiredoxin 3 exist and can be used if blotting conditions are optimized.
3. It is very important to make sure that there are no air bubbles present in the filter paper-gel sandwich during the protein transfer step. These bubbles can obscure bands of interest in ruin an entire experiment.
4. If one chooses to utilize PVDF membrane instead of nitrocellulose, remember to activate the PVDF by wetting with 100% methanol prior to placing in transfer buffer. Also, *do not* wet nitrocellulose with 100% methanol prior to use.
5. The semidry transfer technique has been chosen because it is far less cumbersome and easier to perform compared to the wet transfer technique. Wet transfer can be done if the investigator so wishes.
6. As a negative control for peroxiredoxin 3 oxidation, simply take an aliquot of your alkylated cell extract, and add reducing sample buffer (containing DTT or β-mercaptoethanol). This will result in only one band (approx. 25 kDa) on the blot corresponding to a completely reduced protein.

5 Comments and Perspectives

Electron transfer reactions between some of the components within the respiratory chain are very rapid, presenting a false impression that redox processes within the mitochondrion occur under near equilibrium conditions. In reality, the overall process is a non-equilibrium state, and partial rate control occurs at many steps. Similarly, proteins are synthesized with the thiol of Cys residues in the reduced state, presenting the impression that thiols are maintained in that form. However, the reality is that a fraction of the Cys residues are readily oxidized, and the steady-state generation of

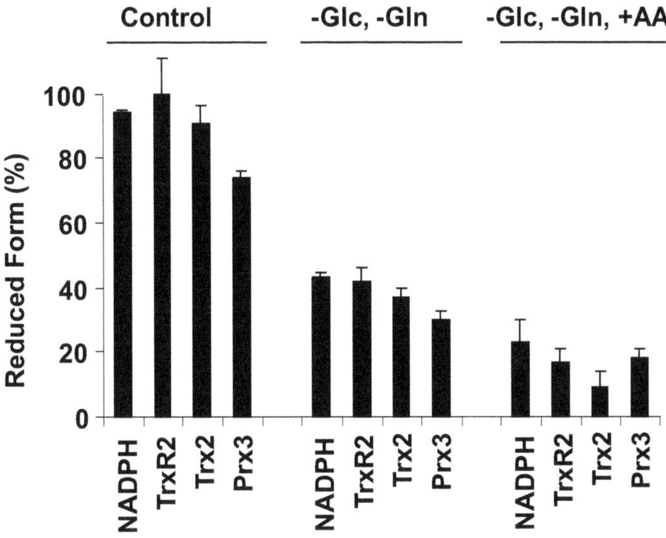

Fig. 7 Measurement of steady-state reduction of Trx2 by redox Western blotting and TrxR2 and Prx3 by BIAM blotting shows that mitochondrial thiol redox systems exist under non-equilibrium conditions in cells. The reduced fraction of TrxR2, Trx2, and Prx3 was decreased both by –Glc,–Gln media and by addition of respiratory substrates. Total cell NADPH was measured by HPLC and is shown for comparison. Results are representative of four experiments

oxidants within mitochondria is sufficient to maintain many proteins in a partially oxidized state. An example is shown in Fig. 7 in which steady-state reduction of proteins in the pathway from NADH through TrxR2, Trx2, and Prx3 was examined under conditions which caused oxidation of NADPH. Results show that the entire pathway functions in a non-equilibrium steady state. Of critical importance, the reactions which are not at equilibrium (within the electron transfer chain and among the protein thiols) are of most interest in terms of control of mitochondrial functions, and these are the most difficult to study because of the methodological challenges to accurately trap the steady-state values.

Finally, an unresolved complexity lies in the relatively large number of modifications which thiols undergo under relevant biologic conditions. Irreversible modification occurs by reaction with the lipid oxidation product, 4-hydroxynonenal [51], and other reactive electrophiles. In addition, subsets of proteins are physiologically regulated by glutathionylation and nitrosylation [52]. The assays described above do not discriminate the contributions of these other modifications. Consequently, accurate descriptions of mitochondrial function will require more global approaches which capture the contributions of each rate controlling step.

Acknowledgments

Supported by NIH grants ES009047, ES011195, and ES012870.

References

1. Mitchell P (1979) Keilin's respiratory chain concept and its chemiosmotic consequences. Science 206:1148–1159
2. Chance B, Sies H, Boveris A (1979) Hydroperoxide metabolism in mammalian organs. Physiol Rev 59:527–605
3. Meredith MJ, Reed DJ (1982) Status of the mitochondrial pool of glutathione in the isolated hepatocyte. J Biol Chem 257:3747–3753
4. Wallace DC (1999) Mitochondrial diseases in man and mouse. Science 283:1482–1488
5. Jones DP (2006) Disruption of mitochondrial redox circuitry in oxidative stress. Chem Biol Interact 163:38–53
6. Chance B (1957) Cellular oxygen requirements. Fed Proc 16:671–680
7. Taylor ER, Hurrell F, Shannon RJ, Lin TK, Hirst J, Murphy MP (2003) Reversible glutathionylation of complex I increases mitochondrial superoxide formation. J Biol Chem 278:19603–19610
8. Zhang R, Al-Lamki R, Bai L et al (2004) Thioredoxin-2 inhibits mitochondria-located ASK1-mediated apoptosis in a JNK-independent manner. Circ Res 94:1483–1491
9. Lillig CH, Berndt C, Vergnolle O et al (2005) Characterization of human glutaredoxin 2 as iron-sulfur protein: a possible role as redox sensor. Proc Natl Acad Sci U S A 102:8168–8173
10. Go YM, Pohl J, Jones DP (2009) Quantification of redox conditions in the nucleus. Methods Mol Biol 464:303–317
11. Jones DP (1984) Effect of mitochondrial clustering on O2 supply in hepatocytes. Am J Phys 247:C83–C89
12. Schafer FQ, Buettner GR (2001) Redox environment of the cell as viewed through the redox state of the glutathione disulfide/glutathione couple. Free Radic Biol Med 30:1191–1212
13. Chance B (1954) Spectrophotometry of intracellular respiratory pigments. Science 120:767–775
14. Keilin D (1966) The history of cell respiration and cytochrome. Cambridge University Press, Cambridge
15. Chance B (1952) Spectra and reaction kinetics of respiratory pigments of homogenized and intact cells. Nature 169:215–221
16. Jones DP, Thor H, Andersson B, Orrenius S (1978) Detoxification reactions in isolated hepatocytes. Role of glutathione peroxidase, catalase, and formaldehyde dehydrogenase in reactions relating to N-demethylation by the cytochrome P-450 system. J Biol Chem 253:6031–6037
17. Tamura M, Hazeki O, Nioka S, Chance B (1989) In vivo study of tissue oxygen metabolism using optical and nuclear magnetic resonance spectroscopies. Annu Rev Physiol 51:813–834
18. Jones DP (1981) Determination of pyridine dinucleotides in cell extracts by high-performance liquid chromatography. J Chromatogr 225:446–449
19. Williamson JR, Corkey BE (1969) Assays of intermediates of the citric acid cycle and related components by fluorometric enzyme methods. Methods Enzymol 13:434–513
20. Sies H (1982) Nicotinamide nucleotide compartmentation. In: Sies H (ed) Metabolic compartmentation. Academic Press, London, pp 205–231
21. Kirwan GM, Coffey VG, Niere JO, Hawley JA, Adams MJ (2009) Spectroscopic correlation analysis of NMR-based metabonomics in exercise science. Anal Chim Acta 652:173–179
22. Beylot M, Beaufrère B, Normand S, Riou JP, Cohen R, Momex R (1986) Determination of human ketone body kinetics using stable-isotope labelled tracers. Diabetologia 29:90–96
23. Jones DP, Kennedy FG (1982) Intracellular oxygen supply during hypoxia. Am J Phys 243:C247–C253
24. Chance B, Schoener B (1962) Correlation of oxidation-reduction changes of intracellular reduced pyridine nucleotide and changes in electroencephalogram of the rat in anoxia. Nature 195:956–958

25. Chance B, Cohen P, Jobsis F, Schoener B (1962) Intracellular oxidation-reduction states in vivo. Science 137:499–508
26. Song Y, Buettner GR (2010) Thermodynamic and kinetic considerations for the reaction of semiquinone radicals to form superoxide and hydrogen peroxide. Free Radic Biol Med 49 (6):919–962
27. Yamamoto Y, Ubiquinol YS (2002) Ubiquinone ratio as a marker of oxidative stress. Methods Mol Biol 186:241–246
28. Matsubara M, Ranji M, Leshnower BG et al (2010) In vivo fluorometric assessment of cyclosporine on mitochondrial function during myocardial ischemia and reperfusion. Ann Thorac Surg 89:1532–1537
29. Scholz R, Thurman RG, Williamson JR, Chance B, Bucher T (1969) Flavin and pyridine nucleotide oxidation-reduction changes in perfused rat liver. I. Anoxia and subcellular localization of fluorescent flavoproteins. J Biol Chem 244:2317–2324
30. Tamura M, Oshino N, Chance B, Silver IA (1978) Optical measurements of intracellular oxygen concentration of rat heart in vitro. Arch Biochem Biophys 191:8–22
31. Estabrook RW (1961) Studies of oxidative phosphorylation with potassium ferricyanide as electron acceptor. J Biol Chem 236:3051–3057
32. Jones DP, Orrenius S, Mason HS (1979) Hemoprotein quantitation in isolated hepatocytes. Biochim Biophys Acta 576:17–29
33. Aw TY, Andersson BS, Jones DP (1987) Suppression of mitochondrial respiratory function after short-term anoxia. Am J Phys 252: C362–C368
34. Jones DP (1982) Intracellular catalase function: analysis of the catalatic activity by product formation in isolated liver cells. Arch Biochem Biophys 214:806–814
35. Guidot DM, Repine JE, Kitlowski AD et al (1995) Mitochondrial respiration scavenges extramitochondrial superoxide anion via a nonenzymatic mechanism. J Clin Invest 96:1131–1136
36. Jones DP (2006) Redefining oxidative stress. Antioxid Redox Signal 8:1865–1879
37. Sies H, Jones DP (2007) Oxidative stress. In: Fink G (ed) Encyclopedia of stress, 2nd edn. Elsevier, Amsterdam, pp 45–48
38. Zhang H, Go YM, Jones DP (2007) Mitochondrial thioredoxin-2/peroxiredoxin-3 system functions in parallel with mitochondrial GSH system in protection against oxidative stress. Arch Biochem Biophys 465:119–126
39. Jones DP (2008) Radical-free biology of oxidative stress. Am J Physiol Cell Physiol 295: C849–C868
40. Halvey PJ, Watson WH, Hansen JM, Go YM, Samali A, Jones DP (2005) Compartmental oxidation of thiol-disulphide redox couples during epidermal growth factor signalling. Biochem J 386:215–219
41. Wood ZA, Schroder E, Robin Harris J, Poole LB (2003) Structure, mechanism and regulation of peroxiredoxins. Trends Biochem Sci 28:32–40
42. Cox AG, Winterbourn CC, Hampton MB (2010) Measuring the redox state of cellular peroxiredoxins by immunoblotting. Methods Enzymol 474:51–66
43. Padgett CM, Whorton AR (1995) S-nitrosoglutathione reversibly inhibits GAPDH by S-nitrosylation. Am J Phys 269: C739–C749
44. Stadtman TC (2002) Discoveries of vitamin B12 and selenium enzymes. Annu Rev Biochem 71:1–16
45. Gasdaska PY, Gasdaska JR, Cochran S, Powis G (1995) Cloning and sequencing of a human thioredoxin reductase. FEBS Lett 373:5–9
46. Mustacich D, Powis G (2000) Thioredoxin reductase. Biochem J 346(Pt 1):1–8
47. Soini Y, Kahlos K, Napankangas U et al (2001) Widespread expression of thioredoxin and thioredoxin reductase in non-small cell lung carcinoma. Clin Cancer Res 7:1750–1757
48. Kim JR, Lee SM, Cho SH et al (2004) Oxidation of thioredoxin reductase in HeLa cells stimulated with tumor necrosis factor-alpha. FEBS Lett 567:189–196
49. Go YM, Park H, Koval M et al (2010) A key role for mitochondria in endothelial signaling by plasma cysteine/cystine redox potential. Free Radic Biol Med 48:275–283
50. Holmgren A, Fagerstedt M (1982) The in vivo distribution of oxidized and reduced thioredoxin in Escherichia coli. J Biol Chem 257:6926–6930
51. Roede JR, Jones DP (2010) Reactive species and mitochondrial dysfunction: mechanistic significance of 4-hydroxynonenal. Environ Mol Mutagen 51:380–390
52. Requejo R, Chouchani ET, Hurd TR, Menger KE, Hampton MB, Murphy MP (2010) Measuring mitochondrial protein thiol redox state. Methods Enzymol 474:123–147

Chapter 13

Mitochondrial Bioenergetics by ^{13}C–NMR Isotopomer Analysis

Rui A. Carvalho and Ivana Jarak

Abstract

Metabolic reprogramming has been associated to a plethora of diseases, and there has been increased demand for methodologies able to determine the metabolic alterations that characterize the pathological states and help developing metabolically centered therapies. In this chapter, methodologies for monitoring TCA cycle turnover and its interaction with pyruvate cycling and anaplerotic reactions will be presented. These methodologies are based in the application of stable ^{13}C isotope "tracers"/substrates and ^{13}C–NMR isotopomer analysis of metabolic intermediates. These methodologies can be applied at several organizational levels, ranging from isolated organelles and organs to whole organisms/humans. For the sake of simplicity, only very simple and well-defined models will be presented, including isolated heart mitochondria and isolated perfused hearts and livers.

Key words NMR isotopomer analysis, Mitochondria, TCA cycle, ^{13}C isotope tracers, Intermediary metabolism

1 Mitochondrial Metabolism

The word *metabolism* has its origin in the Greek metabole which means "change." Therefore, in a physiological context, metabolism is a term used to characterize the set of biochemical reactions able to convert specific molecules into different ones for either energy production or synthesis of biomolecules required for structural and proliferation purposes. In particular, the production of energy, supported by the oxidative metabolism, assumes special importance since it is vital for cell sustainability. The tricarboxylic acid cycle (TCA) is vital in oxidative metabolism but also in biosynthesis; thus its functioning under normal physiological conditions is subject to a very elaborate set of regulation mechanisms, in which disturbance is frequently associated with pathological states that are characteristic of many important diseases. Being able to monitor its activity constitutes a major asset toward understanding the origin of the

Carlos M. Palmeira and António J. Moreno (eds.), *Mitochondrial Bioenergetics: Methods and Protocols*, Methods in Molecular Biology, vol. 1782, https://doi.org/10.1007/978-1-4939-7831-1_13,

disturbance and elaborating possible treatment strategies to revert the metabolic disturbance and hopefully the pathology.

Mathematical models describing the oxidation of substrates in the TCA cycle, based in the use of stable ^{13}C isotopically enriched tracers and ^{1}H–NMR (**P-II**) plus ^{13}C–NMR (**P-III**) detection of ^{13}C isotope incorporation in metabolic intermediates, have long been available [1–9]. Nevertheless, such tools are frequently avoided by researchers due to their overall complexity, and in many circumstances simpler and more straightforward approaches can be adopted and still be able to provide the much needed metabolic information.

There is a need to understand how the TCA cycle functions and how its functioning induces specific ^{13}C enrichment patterns in its metabolic intermediates [1, 2]. The two schemes presented in Fig. 1 represent the TCA cycle and the pyruvate cycling pathways and some of the possible ^{13}C labeling patterns of glutamate or glucose. The two schemes presented will be used to discuss metabolism in the heart (a), a catabolically oriented organ, and the liver (b), an organ responsible for numerous biosynthetic processes (e.g., production of glucose from gluconeogenic sources for glucose homeostasis).

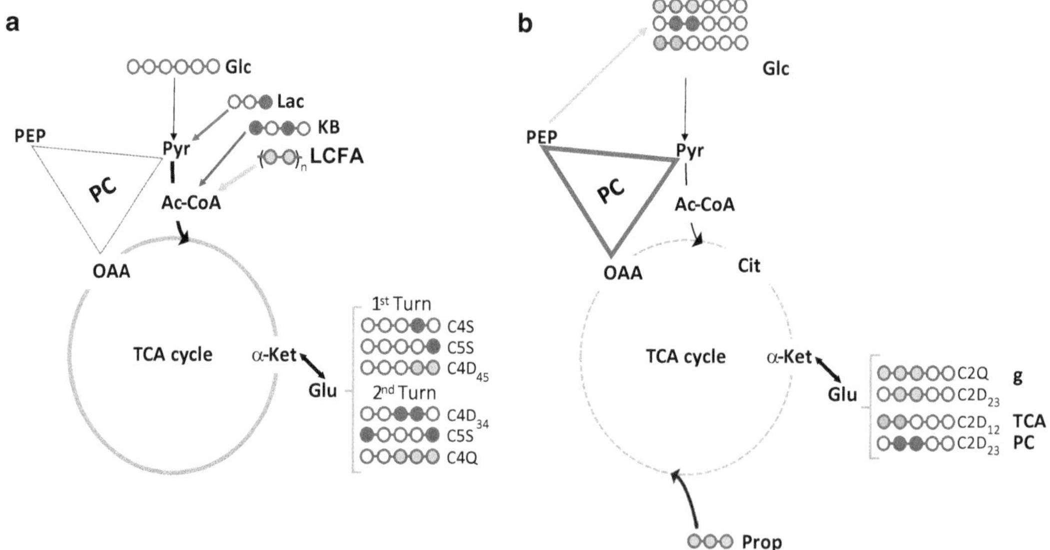

Fig. 1 Schematic representations of the TCA cycle plus the pyruvate-cycling pathways. In the heart (**a**) there is a predominance of cataplerotic activity, and the V_{TCA} is much higher than the activity of the PC pathway. Substrate competition analysis is possible by providing the heart with a set of substrates with complimentary ^{13}C enrichment patterns that generate different acetyl-CoA isotopomers and could be distinguished at the level of glutamate, through a substrate competition analysis (see Fig. 2 for details). In the liver (**b**) the PC and gluconeogenic (g) pathways are frequently very active and have fluxes higher than V_{TCA}. Providing [U-^{13}C] propionate as an anaplerotic substrate to probe gluconeogenic pathways, it is possible to detect the prevalence of PC and TCA activities by analysis of the C2 resonance of glucose, since their activity leads to the appearance of distinct glucose isotopomers

2 Substrate Competition Analysis in the Heart by ^{13}C–NMR Isotopomer Analysis

Scheme **a** in Fig. 1 portrays the activity of the TCA cycle in the heart. The origin of the acetyl-CoA is variable and dependent on substrate availability and also, to a high degree, on the organs' preference. Under physiological conditions, a healthy heart will use long-chain fatty acids (LCFA) as its most important source of acetyl-CoA, but other substrates including ketone bodies (-β-hydroxybutyrate and acetoacetate), lactate, pyruvate, and glucose can also feed the acetyl-CoA pool. There have been numerous studies [10–13] reporting changes in substrate preferences by the heart due to pathological conditions (e.g., left ventricular hypertrophy), and a simple methodology to determine the contributions up to four classes of substrates has been developed [10]. This methodology uses complimentary ^{13}C labeling patterns of the four major classes of compounds, distinguishable at the acetyl-CoA and α-ketoglutarate/glutamate levels. Fig. 1 (scheme **a**) and Fig. 2 depict this process: LCFA are provided uniformly ^{13}C–enriched ([U-^{13}C]LCFA) and produce [1,2-$^{13}C_2$]acetyl-CoA by β-oxidation; ketone bodies are provided with simultaneous enrichment in carbons 1 and 3 ([1,3-$^{13}C_2$]KB) and generate [1-^{13}C] acetyl-CoA; lactate and pyruvate, treated as a single class of

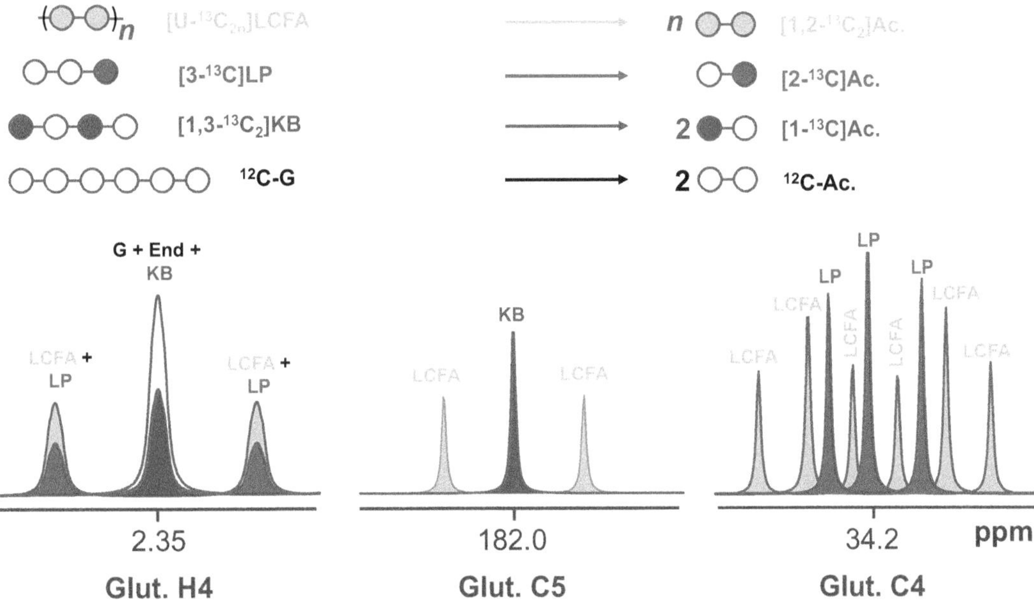

Fig. 2 Substrate competition in the heart. Through a choice of complimentary ^{13}C–enriched classes of substrates, distinct ^{13}C–acetyl moieties are generated in the acetyl-CoA pool that can be distinguished by a ^{13}C–NMR isotopomer analysis of glutamate (see equations in text), using both fractional enrichments of glutamate carbons (C2F or C3F plus C4F) and isotopomer information data ($C4D_{34}$, C4Q, C5S, and C5D). LCFA, long-chain fatty acids; KB, ketone bodies (α-hydroxybutyrate and acetoacetate); LP, lactate and pyruvate; G, glucose; End, unenriched endogenous sources; Ac, acetyl from acetyl-CoA

compounds (LP) due to their fast interconversion through lactate dehydrogenase (LDH) activity, are provided with enrichment in carbon 3 ([3-^{13}C]LP) that turns into carbon 2 of acetyl once pyruvate is oxidized to acetyl-CoA by pyruvate dehydrogenase (PDH), thus generating [2-^{13}C]acetyl-CoA; finally glucose is administered unenriched (at natural abundance ^{13}C enrichment levels) and generates unenriched acetyl-CoA. This exact choice takes into consideration the complementarity of the acetyl-CoA enrichment patterns, to allow distinction of oxidative sources, and also the usual substrate preferences. Consumption of higher quantities of ^{13}C enriched classes of compounds contributes to improved sensitivity by ^{13}C–NMR and more accurate ^{13}C–NMR isotopomer analysis. The switch of patterns among substrates is possible under conditions where significant alterations in substrate preferences could be anticipated.

This substrate competition analysis (**P-V**) can be performed in isolated hearts using the Langendorff perfusion (**P-I**), allowing the observation of shifts in substrate oxidation associated with various heart diseases. Figure 3 shows the expansions of the glutamate C4 resonances derived from the ^{13}C–NMR spectra (**P-III**) of a PCA extract (**P-II**) of a healthy control heart (Ctrl) and the extract of a heart with left ventricular hypertrophy (LVH). We could appreciate

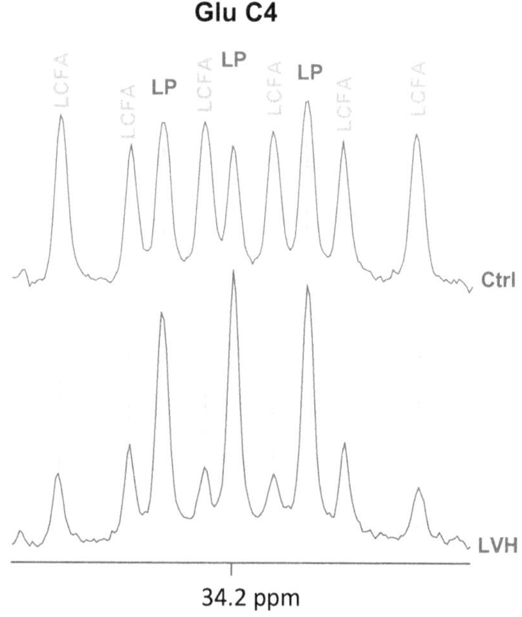

Fig. 3 Expanded glutamate C4 (Glu C4) resonances from the ^{13}C–NMR spectra (11.7 T) of the extracts of a healthy rat heart (Ctrl) and a rat heart with left ventricular hypertrophy (LVH). The resonances due to long-chain fatty acids (LCFA) are considerably diminished in the heart with LVH, while the ones due to lactate plus pyruvate (LP) are increased

from a simple observation of the two expansions that a dramatic shift in substrate preference toward a significant reduction in consumption of long-chain fatty acids (LCFA) and increase in consumption of lactate plus pyruvate occurs due to the LVH condition.

3 TCA Cycle Kinetics by ^{13}C–NMR Isotopomer Analysis

TCA cycle kinetics (**P-VI**) is another topic of great interest when comparing the oxidative capacity of the heart. The rate of incorporation of ^{13}C in TCA cycle intermediates is ultimately dependent on the rate of oxidation [1, 2, 5, 14–16] of the tissue and could be used to monitor TCA flux (V_{TCA}). In these studies a single ^{13}C–enriched substrate is used, with a labeling pattern that allows enrichment of acetyl-CoA in the carbon 2 of the acetyl moiety ([2-^{13}C] acetyl-CoA) or eventually in carbons 1 and 2 ([1,2-$^{13}C_2$]acetyl-CoA). The importance of having the carbon 2 enriched is due to the fact that this carbon is not immediately oxidized in the TCA cycle and remains in its metabolic intermediates throughout several turns of the cycle. This allows kinetic analysis by measuring rates of incorporation, in contrast with enrichment in carbon 1 that is immediately eliminated in the second turn of the cycle. A good tracer for such kinetic analysis could be [3-^{13}C]pyruvate and/or [3-^{13}C]lactate. The [2-^{13}C]acetyl-CoA they generate feeds the TCA cycle, and as the cycle turns over, the relative contributions of multiply ^{13}C–enriched isotopomers increase. This is the case for TCA cycle intermediates and for metabolic intermediates that exchange with their pools, like glutamate. Studies of TCA cycle kinetics have been performed in the heart [5, 14–16] and the influences of cytosolic redox potential (altered by changes in lactate/pyruvate) and Ca^{2+} concentration investigated [16].

Figure 4 shows the ^{13}C–NMR spectra (**P-III**) of the PCA extracts (**P-II**) from hearts perfused with a mixture of [3-^{13}C]lactate and [3-^{13}C]pyruvate in ratios that varied from 10:1 (a) to 1:10 (b). This major shift in lactate/pyruvate ratios causes a significant alteration in the cytosolic redox by reduction of NADH/H^+ levels due to pyruvate utilization by LDH when the ratio of the two metabolites is inverted (1:10) relatively to what is considered a physiological condition, the 10:1 ratio. This oxidation of the cytosol causes a massive alteration in the sizes of the metabolic intermediates, moving from a "conventional" situation where glutamate dominates the spectrum (Fig. 4a) to a situation where other metabolic intermediates, including some from the TCA cycle, appear at levels similar to glutamate or even higher (Fig. 4b). This observation demonstrates the sensitivity of intermediate pools to the functioning of shuttles that are responsible for moving those intermediates as a function of the availability of cytosolic NADH/H^+.

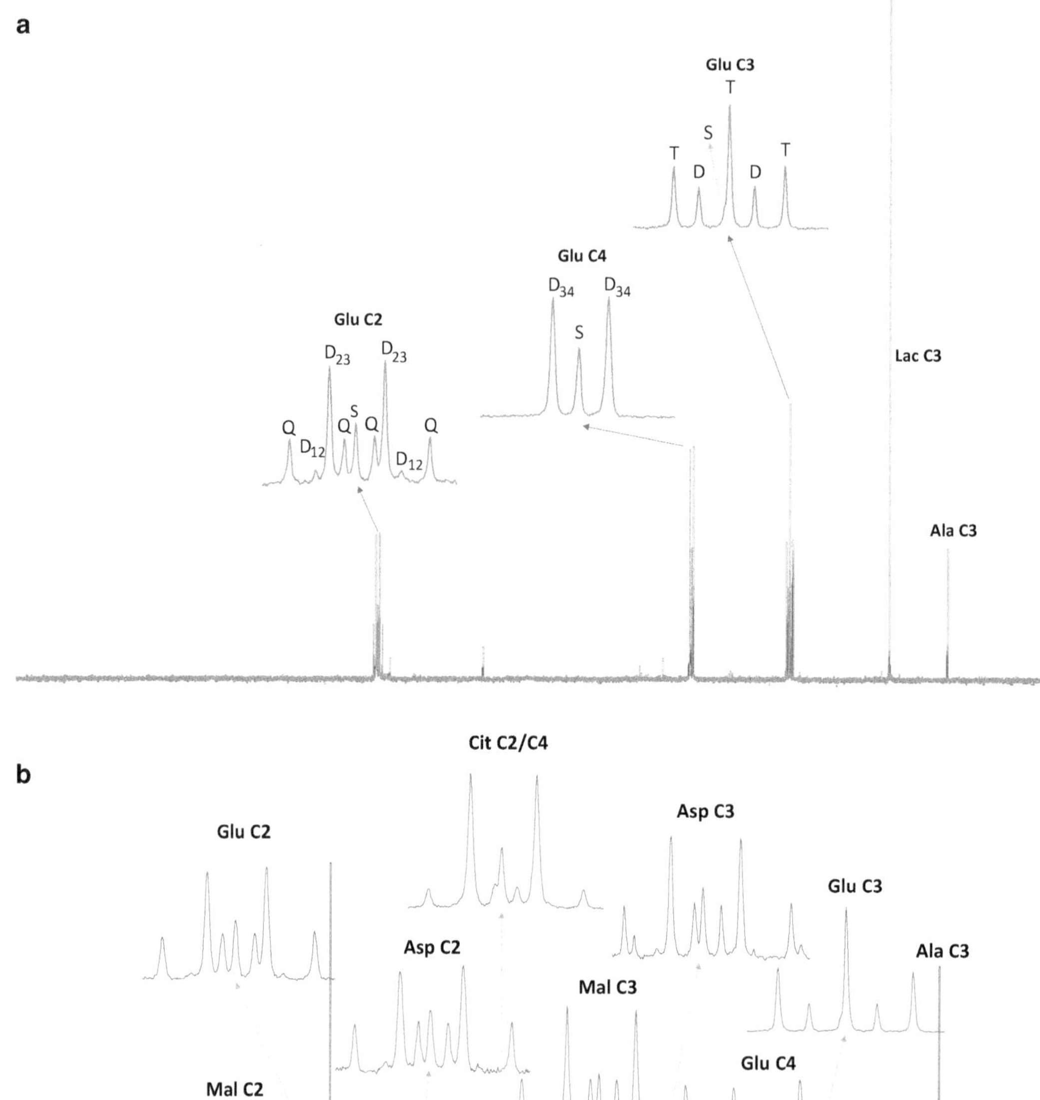

Fig. 4 11.7 T ^{13}C–NMR spectra of extracts from rat hearts perfused with 0.4 mM [3-^{13}C]pyruvate plus 4.0 mM [3-^{13}C]lactate (**a**) or 4.0 mM [3-^{13}C]pyruvate plus 0.4 mM [3-^{13}C]lactate (**b**). Expansions in A refer to glutamate carbons C2 (Glu-C2), C3 (Glu-C3), and C4 (Glu-C4). Aside from glutamate carbons (Glu-C2, Glu-C3, and Glu-C4), expansions in B also show (left to right) citrate C3 (Cit-C3), malate C2 (Mal-C2), aspartate C2 (Asp-C2), citrate C2 and C4 (Cit-C2C4), malate C3 (Mal-C3), and aspartate C3 (Asp-C3)

The levels of ^{13}C enrichment are also affected by the cytosolic redox; however, a more robust interference in the flux of the TCA cycle is given by the changes in Ca^{2+} [16–18]. A higher concentration of extracellularly available Ca^{2+}, provided in the perfusion medium, causes increases in intracellular Ca^{2+} in myocytes and subsequent raises in heart rate. The result is a faster incorporation of ^{13}C in each of the TCA cycle intermediates and glutamate as well. The exchange between TCA cycle intermediates and glutamate is very dependent on the activity of transaminases. An inhibition of this exchange has dramatic effects in the appearance of ^{13}C enrichment in glutamate and interferes with the flux measurements that could be made from the analysis of intermediates like glutamate [15–18]. This is easily monitored by the appearance of signals due to multiply-enriched isotopomers at earlier times of perfusion [15].

To analyze kinetic issues at an organelle level, it is possible to isolate mitochondria from tissues, namely, heart tissue (**P-VII**), and incubate the isolated mitochondria in the presence of the enriched ^{13}C substrate(s). Figure 5a shows the ^{13}C–NMR spectrum from the PCA extract of the heart mitochondria incubated with [3-^{13}C] pyruvate (**P-VIII**). In the expansions, the resonances due to glutamate carbons 2, 3, and 4 show the multiplet patterns due to the occurrence of multiply ^{13}C–enriched glutamate isotopomers. In Fig. 5b the Glu C2 and Glu C4 resonances appear expanded for two experimental conditions, control and in the presence of aminooxyacetate (AOA), a nonspecific transaminase inhibitor. The presence of the AOA causes massive alterations in the rates of appearance of ^{13}C enrichment in the two carbons, consistent with an inhibition of the enrichment of the glutamate pool.

4 Gluconeogenesis, TCA Cycle, and Pyruvate Cycling in the Isolated Perfused Liver by ^{13}C–NMR Isotopomer Analysis

The isolated liver can be perfused (**P-IX**) in the presence of gluconeogenic tracers with specific ^{13}C enrichment patterns that allow the monitoring of the intricate interplay between gluconeogenesis (g), TCA cycle flux (V_{TCA}), and pyruvate cycling (PC). Figure 6 shows the expanded ^{13}C resonance of carbon 2 (C2) from the glucose derivative monoacetone glucose. The observed multiplets allow the characterization of the metabolic behavior of the liver in terms of its intermediary metabolism. The choice of [U-^{13}C]propionate as gluconeogenic tracer allows distinction of direct flux from this tracer to glucose (linear pathway, quartet) from the indirect pathways originated from TCA cycle (V_{TCA}, doublet 12) and pyruvate cycling (PC, doublet 23) interferences.

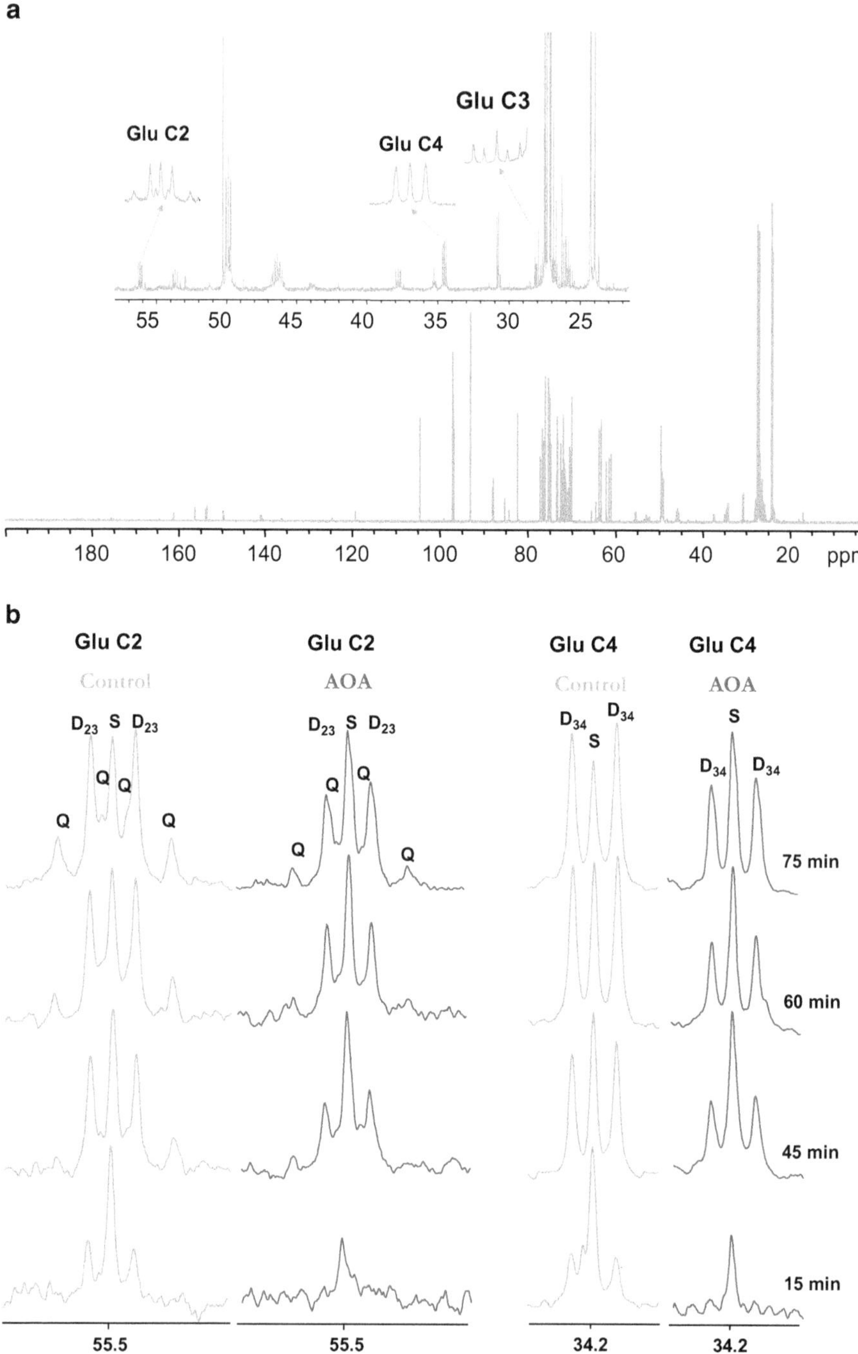

Fig. 5 (**a**) 11.7 T ^{13}C–NMR spectrum from the PCA extract of mitochondria incubated with [3-^{13}C]pyruvate. The resonances due to glutamate C2, C3, and C4 are expanded to demonstrate the appearance of multiplets due to multiply ^{13}C–enriched glutamate isotopomers. (**b**) Expansions of the resonances due to glutamate C2 and C4 carbons in the absence (Control) and presence (AOA) of the nonspecific transaminase inhibitor aminooxyacetate. The presence of AOA alters the dynamics of the enrichment of glutamate carbons, delaying the appearance of ^{13}C enrichment in the glutamate pool

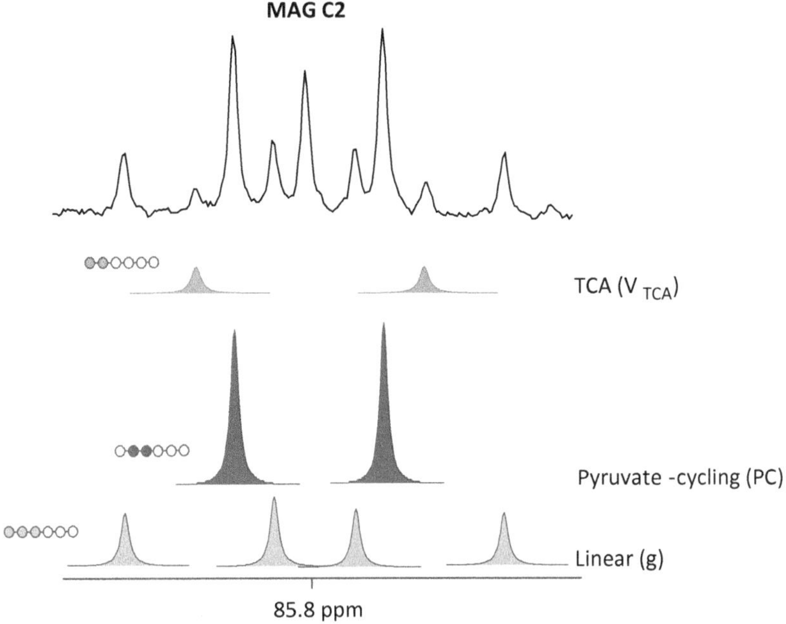

Fig. 6 Expanded resonance of monoacetone glucose (MAG) C2 from a 14.1 T ^{13}C–NMR spectrum of monoacetone glucose derived from glucose produced by an isolated liver being perfused with [U-^{13}C] propionate. The set of multiplets in MAG C2 is associated with specific metabolic activities in the hepatocytes: the quartet 123 (Linear – g), due to the glucose isotopomer enriched in carbons 1, 2, and 3 simultaneously, denotes the direct pathway from propionate to glucose; the doublet 12 (PC), due to glucose isotopomers enriched in carbons 1 and 2 simultaneously, is overemphasized relative to the quartet by pyruvate cycling; the doublet 23 (TCA; V_{TCA}), due to glucose enriched in carbons 2 and 3 simultaneously, is the result of TCA cycle turnover, and the higher the V_{TCA}, relative to g and PC, the more prominent this doublet becomes (*see* also Fig. 1b)

5 Experimental Protocols

5.1 Protocol I: Isolated Heart Perfusion

Hearts from mice or rats can be perfused after removal from the animal using the Langendorff or retrograde heart perfusion.

1. After general anesthesia, mice or rat hearts are rapidly removed and placed in ice-cold perfusion medium, the aorta is immediately cannulated, and hearts are subject to a perfusate column height of 100 cm H_2O. The excess tissue and fat can be removed while the heart rate stabilizes. The modified Krebs-Henseleit buffer, containing (in mM) 119.2 NaCl, 4.7 KCl, 1.25 $CaCl_2$, 1.2 $MgSO_4$, and 25 $NaHCO_3$, is continuously bubbled with 95% O_2–5% CO_2 and in most studies is used without recirculation to avoid perfusate alteration.
2. Several metabolic analyses can be performed using ^{13}C–enriched substrates as a supplement of the KH bicarbonate buffer (e.g., substrate competition analysis and kinetic analysis).

3. In substrate competition measurements by ^{13}C–NMR isotopomer analysis (**P-V**), four distinct sources of acetyl-CoA can be distinguished in a single experiment. Most frequently the mixture chosen contains [U-^{13}C]LCFA (0.22 mM), [1,3-^{13}C]-β-hydroxybutyrate (0.12 mM) plus [1,3-^{13}C]acetoacetate (0.09 mM), [3-^{13}C]lactate (1.2 mM) plus [3-^{13}C]pyruvate (0.12 mM), and glucose (5.5 mM) at natural abundance ^{13}C enrichment levels (~1.1%). These four varieties of substrates originate acetyl-CoA with alternate ^{13}C labeling patterns which are distinguishable by analysis of the ^{13}C labeling patterns of glutamate carbons C4 and C5. This substrate competition analysis does not require metabolic and isotopic steady state to be reached. Simple equations correlate the ^{13}C multiplets in glutamate carbons with the composition of the acetyl-CoA pool at any time of the perfusion.
4. In evaluations of Krebs cycle kinetics, most frequently only one substrate is provided with ^{13}C enrichment. The choice of the substrate is made so that the acetyl-CoA generated becomes enriched in carbon 2 of the acetyl moiety. This will enrich carbon 4 of glutamate on a first turn of the cycle, and that labeling will appear in glutamate carbons C2 or C3 on subsequent turns. The appearance of multiply-enriched glutamate isotopomers will thus be a measure of Krebs cycle turnover and could be correlated with corresponding measures of oxygen consumption. The choice of singly labeled precursors capable of generating [2-^{13}C]acetyl-CoA is also crucial from a sensitivity point of view. If the enrichment is to be followed ex vivo, having the heart being perfused inside the magnet, then a smaller ^{13}C–NMR multiplicity will be favorable for detection by direct ^{13}C–NMR, and better time resolution could be achieved in the analysis.
5. The entire all-glass perfusion system is jacketed and maintained at 37 °C throughout the entire duration of the heart perfusion.
6. The perfusion system can have several buffer chambers, but typically two chambers are chosen. In one of the chambers, there is perfusion buffer alone or supplemented with unenriched substrates, used for initial washing and stabilization procedures, and a second chamber containing the ^{13}C–enriched substrates which is used for metabolic studies by ^{13}C–NMR isotopomer analysis. Of course such chambers can also be used to follow up the effects on metabolism of the administration of hormones and or metabolic inhibitors.
7. By the end of the perfusion protocol, the perfused heart is in most cases freeze-clamped using nitrogen precooled aluminum tongs. The obtained tissue is ready for extraction protocols, namely, PCA extraction as described above.

5.2 Protocol II: Perchloric Acid Extraction

Perchloric acid extraction for preparation of tissue extracts for high-resolution NMR analysis:

1. Deep-freeze tissue is pulverized into a fine powder using nitrogen precooled mortar and pestle.
2. Two volumes of ice-cold perchloric acid (PCA) 7% are added for each gram of wet tissue; upon addition of PCA, vortex and allow tissue to melt—keep tissue in ice. Perform vigorous vortex several times for sample homogenization.
3. Centrifuge the homogenate at 1,077 × g, at 4 °C, for 15 min. The supernatant (aqueous phase) containing water-soluble metabolites is recovered and the pellet containing membranes, cell debris, and other metabolites discarded.
4. Neutralize the supernatant with KOH (pH 6.9–7.0) keeping the solution in ice. This ensures a more significant precipitation of the salt being formed, potassium perchlorate ($KClO_4$).
5. Centrifuge at 1,077 × g, at 4 °C, for 15 min to remove precipitated salt. Resulting supernatant is subject to lyophilization to concentrate intermediate metabolites.
6. In the PCA extract, all water-soluble metabolites can be found including amino acids (e.g., glutamate, glutamine, aspartate, and alanine), ketoacids (e.g., pyruvate, oxaloacetate, α-ketoglutarate), lactate, and glucose.
7. The obtained extract still contains significant amount of salts. This could constitute a major problem when acquiring high-resolution NMR spectra. Two further steps can help improve the spectra quality. First, upon dissolution of the extract in ^2H_2O, allow the sample to stay at 4 °C. This will cause further precipitation of salt (mostly $KClO_4$), and sample viscosity will be significantly reduced. Second, add a considerable amount of a chelating agent, namely, ethylenediaminetetraacetic acid (disodium salt), to complex most cations in solution which otherwise will cause significant line broadening in the NMR spectrum.
8. While performing this extraction procedure, special care has to be taken while adjusting the pH with KOH (**step 4**). This adjustment to pH ~7.0 has to be made slowly since an incorrect adjustment of the pH can lead to destruction of the sample upon lyophilization if pH is left considerably low or to an excessive increment in solution salt content due to the need to add further PCA if pH keeps going above 7.0 by excessive addition of KOH. Several solutions of KOH must be prepared in order to allow the smallest addition of KOH solution but at the same time be as correct as possible in pH adjustment. In the initial steps, a saturated solution of KOH should be used, but as pH approaches the target value, KOH solutions 0.5 or even

0.1 M should be used. At all times try to keep the added volume to a minimum since an increment in sample solution will consequently increase the amount of salt which is dissolved and is not removed by the second centrifugation step. Still associated with pH, misadjustment is another problem which is frequently encountered when running NMR spectra from tissue extracts. If the pH on the dissolve lyophilized sample approaches the pK_a value of the amino group of amino acids like glutamate, then due to chemical exchange phenomena, the resonances of those molecules suffer considerable line broadening. This effect can be avoided by dissolving the tissue extract using ^2H_2O containing a significant buffer capacity, given, for example, by a phosphate buffer.

5.3 Protocol III: Acquisition of High-Resolution ^{1}H–NMR Spectra

1. While this might sound a simple matter, and in fact it is, one frequently encounters incorrect procedures in an NMR laboratory setting which must be avoided if the data is intended to have any meaning. The first worry should always be to have the "best sample" if possible. This implies careful sample handling to avoid several types of interferences: (a) avoid high salt content in samples whenever possible; (b) use adequate (see specifications by the NMR spectrometer provider) sample heights, and keep them constant in all analysis (this simple caution speeds up considerably NMR analysis by avoiding the need for significant NMR field adjustments on sample change); (c) use high-quality NMR tubes; and (d) avoid by all means the presence of air bubbles in the sample—this causes major field inhomogeneities and abnormal resonance broadening.
2. Most one-dimensional (1D) proton NMR spectra used in metabolic studies are acquired with the intent of quantifying metabolites in solution. In order to be quantitative, fully relaxed spectra need to be obtained. This implies correct knowledge about the relaxation behavior of all compounds in the solution being analyzed. This does not necessarily imply that a full relaxation (T_1/T_2) analysis has to be made for each sample to be run, but in the beginning of any study, an estimation of those parameters should be made to warrant that spectra will be acquired quantitatively, a major requirement for correct evaluation of metabolite levels.
3. The simplest 1D NMR experiment possesses three essential stages:
   (a) A first stage which comprises a delay time, which needs to be adjusted in accordance with the relaxation behavior of the sample. This stage could also include a period of solvent saturation in case the solvent resonance is much more intense than the resonances of the metabolites being

analyzed, frequently the situation when running aqueous extracts/samples.

(b) A second stage which is simply the excitation period and implies the application of a radiofrequency pulse for a given duration and power. This again has to be adjusted according to sample specifications; high salt content leads to significant increases in pulse duration, and pulse calibrations should be made prior to quantitative spectra acquisition.

(c) A third stage which consists in free-induction-decay (FID) acquisition—the duration of this period, also called acquisition time, is chosen to provide enough signal digitization to derive well-resolved spectra. The total duration of the three stages is frequently referred to as the experimental "repetition time" (RT).

4. Adequate signal-to-noise ratio is required for accurate determinations of metabolite concentrations. Increment in signal to noise is achieved by increasing the number of transients accumulated for each FID. In the case of tissue extracts, there are some metabolites which are frequently abundant such as glutamate and lactate and for which there is not much need for transient accumulation for signal averaging. However, for metabolites less abundant like many of the Krebs cycle intermediates, there is frequently the need for performing signal averaging, and total acquisition times can easily be 30 min to several hours, depending on the amount of tissue material available. The repetition time used between transients is a function of the relaxation behavior as stated above as well as a function of the pulse width being used. Increased signal-to-noise ratios are possible maintaining the quantitative character of the spectrum by lowering considerably the excitation pulse length from a maximum of 90°. Frequently used parameters for tissue PCA extracts include a radiofrequency excitation pulse of 30° and a RT of 15 s, from which 9 s is the relaxation delay; 3 s is the pre-saturation period for solvent suppression, using a saturation pulse with the lowest possible saturation power to avoid saturation of nearby resonances; and finally 3 s is the time for FID acquisition. This acquisition period needs to be optimized to provide enough spectral resolution and is intimately associated with parameters like signal linewidths (the sharper the signals, the longer the acquisition time for providing enough points defining the resonance) and the spectral widths.

5. In summary, a quantitative NMR spectrum requires long enough repetition time to ensure full relaxation and good signal-to-noise ratio to allow accurate estimates of metabolite concentrations by resonance deconvolution and comparison with a well-defined internal standard.

5.4 Protocol IV: Acquisition of High-Resolution ^{13}C–NMR Spectra

1. All aspects mentioned above apply to ^{13}C–NMR in the same way as described for ^{1}H–NMR. However, while using ^{1}H–NMR one can be absolutely quantitative, when using ^{13}C–NMR spectra, several extra considerations need to be taken into account in order to obtain "acceptable" quantitative information, namely, nuclear Overhauser effects (NOE) and signal saturation by oversampling.
2. ^{13}C–NMR spectra are almost always acquired with ^{1}H broadband decoupling to remove scalar coupling from directly coupled protons and from protons attached to neighboring carbons. The need for broadband decoupling arises from the need to reduce spectral complexity and, most importantly, to improve considerably the signal to noise of ^{13}C–NMR spectra. This improvement in signal to noise can be further enhanced by allowing for NOE to build up during the delay before radiofrequency excitation. While this NOE improves signal to noise and allows an improvement on detection limits, it confuses quantification procedures due to the fact that NOE effects are not identical for all aliphatic carbons and are essentially absent in quaternary carbons. The only possibility for being quantitative resides in those circumstances in finding the NOE enhancement factors so that normalization of signal intensities can be made after spectrum acquisition and quantification.
3. The problem of oversampling is always present when acquiring ^{13}C–NMR spectra of tissue extracts. This is the result of a significant difference between the relaxation behaviors of aliphatic carbons relative to quaternaries. Differences in relaxation times of one order of magnitude are frequently found, and allowing full relaxation of quaternary carbons would essentially render the experiment obsolete due to total lack of signal to noise. Most frequently enough time is given to allow full relaxation to all aliphatic carbons of a given metabolite and quantitative comparisons can be made between them, but such time is too small to allow full relaxation of carboxyl or carbonyl carbons. The practical result is a ^{13}C–NMR spectrum in which the aliphatic carbons appear much more intense than the quaternaries and direct comparison of their areas is totally impossible.
4. A typical ^{13}C–NMR spectrum of a tissue extract uses a repetition time between transients of 3 s and a radiofrequency pulse width equivalent to 45°. These parameters ensure an almost complete relaxation of aliphatic carbons, and, providing that adequate NOE normalization factors are obtained for such samples, the areas derived from the ^{13}C–NMR spectrum can be used as a measure of ^{13}C enrichment.

5. Within each carbon resonance, the comparison among multiplets is possible since the changes in relaxation behavior of the carbon nuclei caused by other enriched carbons are considerably low. This comparison is critical for the ^{13}C–NMR isotopomer analysis that uses relative contributions of resonance multiplets to derive isotopomer data for metabolic studies.

5.5 Protocol V: Substrate Competition by ^{13}C–NMR Isotopomer Analysis

1. ^{13}C–NMR spectra of tissue extracts acquired under conditions described in Protocol IV can be deconvoluted using several NMR processing software packages. The overall concept is to derive from each NMR spectrum the ratios of ^{13}C–NMR multiplets for each resonance of the metabolic intermediate under analysis, which is frequently glutamate for sensitivity reasons.
2. Resonances of glutamate carbons C3, C4, and C5 are deconvoluted and relative values (% of total resonance) for C3S, C3D, and C3T (of resonance C3); for C4S, C4D34, C4D45, and C4Q (of resonance C4); and for C5S and C5D (of resonance C5) determined. Some of these values are used in simple equations derived to determine the relative fractions of acetyl-CoA isotopomers that form the acetyl-CoA pool.
3. The equations are as follows: (a) fraction of acetyl-CoA with acetyl enriched in both carbons, $F_{C3} = C4Q^*(C4F/C3F) = \%$ [U-^{13}C]LCFA; (b) fraction of acetyl-CoA with acetyl enriched in carbon, $F_{C1} = C5S/C5D^*F_{C3} = \%$ [1,3-$^{13}C_2$]KB (-β-hydroxybutyrate + acetoacetate); (c) $F_{C2} = C4D34^*(C4F/C3F) = \%$ [3-^{13}C]LP (lactate + pyruvate); and (d) $F_{C0} = 1 - (F_{C3} + F_{C2} + F_{C1}) = \%$ glucose + unenriched endogenous sources.
4. With the above set of equations, the % contributions of each major class of substrates are determined. Its conversion into absolute contributions only requires the determination of the use of one of the substrates. The schematic representation of Fig. 2 is very intuitive and allows understanding the whole concept. In circumstances where the labeling patterns of the substrates are intentionally altered, the procedure is identical, but the $F_{Ci\ (i\ =\ 0,1,2,3)}$ fractions will report distinct classes of substrates. For example, if LCFA are enriched in odd carbons and KB are uniformly enriched, the F_{C3} fraction will report KB contribution, while F_{C1} will refer to LCFA contribution.

5.6 Protocol VI: Kinetic ^{13}C–NMR Isotopomer Analysis of Isolated Perfused Heart

1. Hearts can be perfused (**P-I**) for distinct periods in order to allow measuring the rates of ^{13}C incorporation into metabolic intermediates. By the end of a given time period (3, 6, 9, 12, 15, 20, 30, 45 min), hearts are freeze-clamped and PCA extracted for ^{1}H- and ^{13}C–NMR acquisition and ^{13}C–NMR isotopomer analysis.

2. With time the evolution of multiplets from each of the glutamate carbon resonances is determined and curves of their evolution drawn. This TCA flux analysis uses both the fractional enrichment at each carbon position (C1F, C2F, C3F, C4F, and C5F) and the multiplets of each resonance to account for isotopomer information (C1S, C1D; C2S, C2D12, C2D23, C2Q; C3S, C3D, C3T; C4S, C4D34, C4D45, C4Q; C5S, C5D). The rates of appearance of ^{13}C enrichment in carbons 2 and 3 of glutamate are important keys of the oxidative flux through the TCA cycle, and curves of evolution of ^{13}C enrichment of carbons 2 (C2F) and 4 (C3F) or 3 (C3F) and 4 (C4F) are very elucidative of the metabolic behavior of the heart and allow the monitoring of metabolic reprogramming.
3. Many factors can influence the rate of appearance of ^{13}C in glutamate. The energy requirements to sustain heart rate are necessarily a major factor, but other issues like changes in cytosolic redox and presence of transaminase inhibitors tailor the C4F versus C3F/C2F enrichment curves in predictable ways. The rates of appearance of 13C in carbons 2 and 3 of glutamate are very sensitive to TCA flux (V_{TCA}) and intermediate pool sizes.

5.7 Protocol VII: Isolation of Heart Mitochondria

Several protocols for heart mitochondria isolation are available in literature [19, 20, 21]. A very simple procedure to obtain high-quality heart mitochondrial preparations is the following:

1. Rats are sacrificed by cervical dislocation and the hearts immediately excised and minced finely in an ice-cold isolation medium containing 250 mM sucrose, 1 mM EGTA, 10 mM Hepes–KOH (pH 7.4), and 0.1% defatted BSA.
2. Minced blood-free tissue is then resuspended in 40 mL of isolation medium containing 1 mg protease Type VIII (Sigma No. P-5390) per milligram of tissue and homogenized with a tightly fitted homogenizer (Teflon/glass pestle). The suspension is incubated for 1 min (4 °C) and then rehomogenized.
3. The homogenate is then centrifuged at 10,000 × *g* for 10 min (Sorvall RC-5C, Plus, SS 34 rotor, 4 °C). The supernatant fluid is decanted, and the pellet, essentially devoid of protease, is gently homogenized to its original volume with a loose-fitting homogenizer.
4. The suspension is centrifuged at 500 × *g* for 10 min, and the resulting supernatant is centrifuged at 10,000 × *g* for 10 min.
5. The pellet is resuspended using a paint brush and repelleted twice at 10,000 × *g* for 10 min. EGTA and defatted BSA should be omitted from the final washing medium.

6. Mitochondrial protein content can be determined by the biuret method calibrated with BSA.
7. Oxygen consumption of isolated heart mitochondria can be monitored polarographically with a Clark oxygen electrode connected to a suitable recorder in a 2 mL thermostated water-jacketed closed chamber with magnetic stirring, at 25 °C. Mitochondria are suspended at a concentration of 1 mg/mL in the respiratory medium (100 mM KCl, 50 mM sucrose, 10 mM Tris, 30 mM EGTA, 1 mM KH_2PO_4, pH 7,4). State IV respiration is measured in the presence of 8 mM succinate (plus 4 mM rotenone). ADP (600 nmoles), oligomycin (2 mg), and FCCP (0.5 mM) are added to induce, respectively, state III respiration, inhibition of state III respiration, and uncoupled respiration. Respiratory control ratio is calculated as the ratio between state III and state IV respiration.

5.8 Protocol VIII: Incubation of Isolated Mitochondria with [3-^{13}C]Pyruvate

Isolated healthy mitochondria (**P-VI**) can be subsequently incubated in the presence of [3-^{13}C]pyruvate to follow TCA cycle kinetics in the presence and absence of the nonspecific transaminase inhibitor aminooxyacetate (AOA). There is a need for a mixture of specific metabolic intermediates in the incubation medium to allow the appearance of ^{13}C enrichment in glutamate.

1. 30 mL of mitochondrial incubation medium (in mM – 120 KCl, 30 D-glucose, 25 $NaHCO_3$, 20 KH_2PO_4, 5 ATP, 5 ADP, 2 $MgCl_2$, 0.2 malate, 0.113 glutamate, 0.02 aspartate, 0.02 α-ketoglutarate, 8.0 [3-^{13}C]pyruvate; 0.25% BSA fatty acids free; pH 7.4) is added to a water-jacketed chamber kept at a temperature of 37 °C. A saturated atmosphere of carbon is kept with constant stirring.
2. Upon temperature stabilization, hexokinase (15 U/mL of medium) is added to the incubation medium, followed by mitochondria addition.
3. The first aliquot of medium with mitochondria (5 mL) is removed immediately after the homogenization and corresponds to time zero. Other aliquots are removed every 15 min (15, 30, 45, 65, 75) to follow kinetic incorporation of ^{13}C in glutamate. To each aliquot, 270 μL of concentrated (70%) $HClO_4$ is added to warrant a final concentration of ~3.6%. This stops metabolism by denaturation of all protein in the sample.
4. PCA extraction was performed as described in Protocol II. The lyophilized extract is dissolved in D_2O and ^{1}H- and ^{13}C–NMR spectra acquired as described in Protocols III and IV, respectively.

5. Incubation is made in the presence or absence of the transaminase inhibitor in order to follow exchange between the TCA intermediates and the pool of glutamate.
6. The incubation medium was prepared from Krebs-Henseleit (KH) medium (in mM – 120 KCl, 25 $NaHCO_3$, 20 KH_2PO_4, 2 $MgCl_2$) by addition of ATP, ADP, D-glucose, BSA, [3-^{13}C] pyruvate, and aliquots of previously prepared stock solutions of malate, glutamate, aspartate, and α-ketoglutarate immediately before the addition of mitochondria.

5.9 Protocol IX: Perfusion of Isolated Livers

1. Prior to the final experimental procedure, rats are fasted for 24 h. Rats are then anesthetized via intraperitoneal injection of ketamine/chlorpromazine (2 mL/kg of body weight). The liver is exposed and the portal vein immediately cannulated. The hepatic vein and inferior vena cava are dissected, and non-recirculating KH buffer is pumped through the portal vein at >20 mL/min, as described previously [22].
2. The liver is then carefully removed and suspended in a container, which was partially submerged in perfusion medium maintained at 37 °C by a controlled water bath circulating around the container [23].
3. The anatomical absence of a gallbladder in the rat makes this perfusion system unique for the evaluation of the metabolic response of the liver to the ^{13}C–enriched substrates given through the perfusion buffer. After 10 min of washout with modified KH bicarbonate buffer (ionic composition in mM – 143 Na^+, 128.1 Cl^-, 4.7 K^+, 1.2 Ca^{2+}, 1.2 Mg^{2+}, SO_4^{2-}, 25 HCO_3^-; pH 7.4), the livers are perfused for an additional period of 30 min in the same conditions but with ^{13}C–enriched KH bicarbonate buffer.
4. At the end of perfusion, livers are immediately freeze-clamped using liquid-nitrogen-precooled aluminum tongs.

References

1. Chance EM, Seeholzer SH, Kobayashi K, Williamson JR (1983) Mathematical analysis of isotope labeling in the citric acid cycle with applications to ^{13}C NMR studies in perfused rat hearts. J Biol Chem 258:13785–13794
2. Malloy CR, Sherry AD, Jeffrey FM (1987) Carbon flux through citric acid cycle pathways in perfused heart by ^{13}C NMR spectroscopy. FEBS Lett 212(1):58–62
3. Malloy CR, Sherry AD, Jeffrey FM (1990) Analysis of tricarboxylic acid cycle of the heart using ^{13}C isotope isomers. Am J Phys 259(3 Pt 2):H987–H995
4. Weiss RG, Gloth ST, Kalil-Filho R, Chacko VP, Stern MD, Gerstenblith G (1992) Indexing tricarboxylic acid cycle flux in intact hearts by carbon-13 nuclear magnetic resonance. Circ Res 70:392–408
5. Yu X, White LT, Doumen C, Damico LA, LaNoue KF, Alpert NM, Lewandowski ED (1995) Kinetic analysis of dynamic 13C NMR spectra: metabolic flux, regulation, and compartmentation in hearts. Biophys J 69 (5):2090–2102
6. Mason GF, Gruetter R, Rothman DL, Behar KL, Shulman RG, Novotny EJ (1995) Simultaneous determination of the rates of the TCA

cycle, glucose utilization, alpha-ketoglutarate/ glutamate exchange, and glutamine synthesis in human brain by NMR. J Cereb Blood Flow Metab 15:12–25

7. Gruetter R, Seaquist ER, Ugurbil K (2001) A mathematical model of compartmentalized neurotransmitter metabolism in the human brain. Am J Physiol Endocrinol Metab 281: E100–E112
8. Henry P-G, Öz G, Provencher S, Gruetter R (2003) Toward dynamic isotopomer analysis in the rat brain in vivo: automatic quantitation of 13C NMR spectra using LCModel. NMR Biomed 16:400–412
9. Des Rosiers C, Lloyd S, Comte B, Chatham JC (2004) A critical perspective of the use of ^{13}C-isotopomer analysis by GCMS and NMR as applied to cardiac metabolism. Metab Eng 6 (1):44–58
10. Malloy CR, Thompson JR, Jeffrey FM, Sherry AD (1990) Contribution of exogenous substrates to acetyl coenzyme a: measurement by ^{13}C NMR under non-steady-state conditions. Biochemistry 29(29):6756–6761
11. Jessen ME, Kovarik TE, Jeffrey FM, Sherry AD, Storey CJ, Chao RY, Ring WS, Malloy CR (1993) Effects of amino acids on substrate selection, anaplerosis, and left ventricular function in the ischemic reperfused rat heart. J Clin Invest 92(2):831–839
12. Carvalho RA, Sousa RP, Cadete VJ, Lopaschuk GD, Palmeira CM, Bjork JA, Wallace KB (2010) Metabolic remodeling associated with subchronic doxorubicin cardiomyopathy. Toxicology 270:92–98
13. Ragavan M, Kirpich A, Fu X, Burgess SC, McIntyre LM, Merritt ME (2017) A comprehensive analysis of myocardial substrate preference emphasizes the need for a synchronized fluxomic/metabolomic research design. Am J Physiol Heart Circ Physiol 312(6): H1215–H1223
14. Jeffrey FM, Reshetov A, Storey CJ, Carvalho RA, Sherry AD, Malloy CR (1999) Use of a single ^{13}C NMR resonance of glutamate for measuring oxygen consumption in tissue. Am J Phys 277:E1111–E1121
15. Carvalho RA, Zhao P, Wiegers CB, Jeffrey FM, Malloy CR, Sherry AD (2001) TCA cycle kinetics in the rat heart by analysis of ^{13}C isotopomers using indirect ^{1}H{^{13}C}detection. Am J Physiol Heart Circ Physiol 281: H1413–H1421
16. Carvalho RA, Rodrigues TB, Zhao P, Jeffrey FM, Malloy CR, Sherry AD (2004) A ^{13}C isotopomer kinetic analysis of cardiac metabolism: influence of altered cytosolic redox and $[Ca^{2+}]_o$. Am J Physiol Heart Circ Physiol 287(2): H889–H895
17. Lewandowski ED, Yu X, LaNoue KF, White LT, Doumen C, O'Donnell JM (1997) Altered metabolite exchange between subcellular compartments in intact postischemic rabbit hearts. Circ Res 81:165–175
18. O'Donnell JM, Doumen C, LaNoue KF, White LT, Yu X, Alpert NM, Lewandowski ED (1998) Dehydrogenase regulation of metabolite oxidation and efflux from mitochondria in intact hearts. Am J Phys 274(2 Pt 2):H467–H476
19. Jin ES, Jones JG, Merritt M, Burgess SC, Malloy CR, Sherry AD (2004) Glucose production, gluconeogenesis, and hepatic tricarboxylic acid cycle fluxes measured by nuclear magnetic resonance analysis of a single glucose derivative. Anal Biochem 327 (2):149–155
20. Darley-Usmar VM, Rickwood D, Wilson MT (1987) Mitochondria - a practical approach. IRL Press Limited, England
21. Oliveira PJ, Rolo AP, Sardão VA, Coxito PM, Palmeira CM, Moreno AJ (2001) Carvedilol in heart mitochondria: protonophore or opener of the mitochondrial K(ATP) channels? Life Sci 69(2):123–132
22. Nunes PM, Jones JG, Rolo AP, Palmeira CM, Carvalho RA (2011) Ursodeoxycholic acid treatment of hepatic steatosis: a ^{13}C-NMR metabolic study. NMR Biomed 24 (9):1145–1158
23. Bartosek I, Guaitani A, Miller LL (1973) Isolated liver perfusion and its applications. Raven Press, New York

Chapter 14

Computational Modeling of Mitochondrial Function from a Systems Biology Perspective

Sonia Cortassa, Steven J. Sollott, and Miguel A. Aon

Abstract

The advent of "big data" in biology (e.g., genomics, proteomics, metabolomics), holding the promise to reveal the nature of the formidable complexity in cellular and organ makeup and function, has highlighted the compelling need for analytical and integrative computational methods to interpret and make sense of the patterns and changes in those complex networks. Computational models need to be built on sound physicochemical mechanistic principles in order to integrate, interpret, and simulate high-throughput experimental data. Energy transduction processes have been traditionally studied with thermodynamic, kinetic, or thermo-kinetic models, with the latter proving superior to understand the control and regulation of mitochondrial energy metabolism and its interactions with cytoplasmic and other cellular compartments. In this work, we survey the methods to be followed to build a computational model of mitochondrial energetics in isolation or integrated into a network of cellular processes. We describe the use of analytical tools such as elementary flux modes, linear optimization of metabolic models, and control analysis, to help refine our grasp of biologically meaningful behaviors and model reliability. The use of these tools should improve the design, building, and interpretation of steady-state behaviors of computational models while assessing validation criteria and paving the way to prediction.

Key words Mitochondrial energetics, Ordinary differential equations (ODEs), Metabolic control and dynamic analyses, Elementary flux modes, Linear optimization of metabolic models

1 Introduction

Systems biology comprises integrative approaches that aim at describing and characterizing extensive cellular or organ composition changes in gene expression at the RNA, protein, or metabolite levels. Combining high-throughput technologies with a variety of quantitative methods, including bioinformatics analyses and computational modeling, systems biology attempts to gain a comprehensive understanding of systemic changes in these biological properties [1, 2]. Computational approaches include thermodynamic, kinetic, and stoichiometric models which have been succinctly described in the previous edition of this volume [3].

Carlos M. Palmeira and António J. Moreno (eds.), *Mitochondrial Bioenergetics: Methods and Protocols*, Methods in Molecular Biology, vol. 1782, https://doi.org/10.1007/978-1-4939-7831-1_14,

Stoichiometric models constitute the basis of elementary flux mode analysis, introduced by Schuster and Fell [4], and of flux balance analysis (FBA), also known as constraint-based metabolic modeling, as developed by Palsson and coworkers [5–8]. The linear optimization tools from FBA can be used to adjust the values of the maximal rate of processes to estimate the steady-state behavior of detailed kinetic models [9].

Our modeling approach is modular, with each module accounting for the known or hypothetical kinetic scheme available for a process, e.g., the reaction catalyzed by citrate synthase or the transport of Ca^{2+} through the mitochondrial Ca^{2+} uniporter [10–13]. According to the modular approach of computational modeling, a module represents an edge connecting at least two nodes, in analogy with the topological view of networks, conceived as a collection of edges and nodes exhibiting large-scale organization [14, 15]. Because the kinetic schemes considered are based on experimental studies, our modular approach can incorporate functional and regulatory information of reconstructed networks, a decisively important feature for analyzing the dynamics of complex physiological processes and their response to specific stimuli. In this regard, a realistic representation of transient dynamic behavior, other than steady-state fluxes [5, 16, 17], requires detailed kinetic modeling including mechanistic regulatory information, such as flux regulation by substrate or product inhibition/activation. Kinetic models can also accommodate the nonlinear behavior imparted by feedback and feed-forward regulatory mechanisms [9].

The modular approach of computational modeling also allows "zooming in or out" to a network of processes, thus enabling the modeler to choose the desired level of detail to consider. For instance, the tricarboxylic acid (TCA) cycle, or the glycolytic pathway, may be described as a set of reactions (8 or 10 reactions, respectively) or in a single aggregated step. Likewise, a biochemical reaction may be written down as an aggregate single step or in detail considering the individual elementary kinetic steps of the catalytic cycle [18]. Newly emergent spatiotemporal organizational properties, which could not have been anticipated from the properties of the isolated modules, may appear because of interaction when the modules are linked into a unified ensemble.

Using a specific model of mitochondrial function, in this chapter, we succinctly describe the modular approach as applied to bioenergetic studies, with a special emphasis on elementary flux mode (EFM) analysis, linear optimization of kinetic models, and control analysis at the steady state. Other applications related to time-dependent, nonlinear behavior such as oscillations, wave propagation, and chaos are developed elsewhere [19–21].

2 Materials

This section has been extensively developed in the chapter published previously [3], which we briefly recapitulate in the following while expanding with greater detail on the new material introduced herein.

1. Basic biochemical information about the metabolic pathways and transport processes that are being modeled. This information can be obtained in biochemistry or physiology textbooks and some biochemical databases [22–24].
2. Repository databases of models are a source for building our own model, constituting a good starting point for a beginner in the field of computational modeling. The models in these databases are stored according to a series of standards in language, syntax, and annotations (e.g., SBML, CellML) and are publicly available for download [25].
3. Experimental data needed to introduce realistic constraints to each of the modules' kinetics, usually obtained under in vitro conditions (*see* below), and the assembled model, are also accessible. Metabolomics data is an important dataset, now available under different experimental and physiological conditions. The latter can be "untargeted," which informs only about relative changes in metabolites' levels or concentrations, or "targeted," providing actual concentrations in case the appropriate internal standards are included in the samples analysis.
   Metabolite concentrations (in molar units) are required to calculate the fluxes through the network as inputs of the rate expressions from the kinetic model.

Alternatively, a few representative metabolites can be quantified by different methods. The metabolite concentration values are used to calibrate the area of the mass spectrometric [MS] peaks to determine the concentrations of other metabolites present in the MS profiles as follows:

$$\mathrm{Mass}_{\mathrm{metabolite}} = \left(\frac{1}{\mathrm{Mean\ Area}_{\mathrm{std}} / \mathrm{Mass}_{\mathrm{std}}} \right) \times \mathrm{Area}_{\mathrm{metabolite}} \quad (1)$$

Mean $\mathrm{Area}_{\mathrm{std}}$ can be determined from the area under the peak of the corresponding molecular ion in the mass spectrum. The $\mathrm{Mass}_{\mathrm{std}}$ was calculated from the number of moles measured enzymatically in the same samples that were analyzed by MS, multiplied by the mass of the molecular ion. The number of moles of metabolite in the sample was calculated according to:

$$\text{Metabolite (mol)} = \left({}^{\text{Mass}_{\text{metabolite}}}\!/\!_{\text{MW molecular ion}}\right) \quad (2)$$

where MW stands for molecular weight.

In this quantitation, the underlying assumption is that the area under the peak is proportional to the mass of the molecular ion impacting the MS detector. The metabolite concentrations calculated were used as input data for the model to optimize maximal velocity (V_{max}) values as described before [9]. After optimization, the model can be utilized to simulate the metabolite concentration profile.

4. Computational tools to work with the modules, such as MATLAB® or Wolfram's Mathematica, or any piece of software designed to solve systems of ordinary differential equations (ODEs) by numerical integration, are required. These tools may be used in the modular analysis, while the model is built, or with the assembled model. Some useful computational modeling packages have been developed for software such as MATLAB. Among them the graphical package MatCont is able to simulate time-dependent behavior and calculate the stability and the parameter sensitivity of the model [26].
5. Elementary flux mode (EFM) analysis requires the software Metatool (currently, version 5.1 for MATLAB) freely available for academic use (available at http://pinguin.biologie.uni-jena.de/bioinformatik/networks/metatool/metatool5.1/metatool5.1.html) [27, 28]. This software enables the computation of structural properties of biochemical reaction networks while informing about the system's capacity, as designed, to achieve a steady state or if there may be missing components.
6. Conceptual tools of metabolic control analysis (MCA). These concepts were masterfully explained by David Fell in his book: *Understanding the Control of Metabolism* [17]. Herein, in the Methods section, we will succinctly explain the main tools of MCA as needed to apply MCA to study quantitatively metabolic networks.

3 Methods

1. The molecular and supramolecular (molecular complexes) level of organization is generally where bioenergetic processes take place. Thus, modelers should consider these levels to choose the right variables and processes with the appropriate relaxation times. Ion transport processes, such as channels and pumps, occur on very fast time scales (on the order of milliseconds), whereas metabolic reactions and organized behavior of

metabolic networks have much longer relaxations times (on the order of seconds to minutes). Although computational modeling can consider spatial in addition to time coordinates, for simplicity we restrict the scope of this article to ODEs that only account for time as independent variable.

2. To study the model's steady-state behavior, it is relevant to know whether the model can evolve to a steady state. This is achieved by calculating the EFMs [4], which compute both the null-space matrix (whose existence means that the system can achieve a steady state) and the explicit EFMs the system exhibits at the steady state.
3. The choice of the kinetic expressions describing the rates of the set of processes relevant for the model. According to our modular approach, the rate expression that best represents the known mechanism underlying the behavior of a specific process will be chosen. Examples of kinetic models of transport include the **Goldman-Hodgkin-Katz** voltage equation [29, 30] or **Fick's laws** of diffusion for non-charged molecules following a concentration gradient in a homogeneous medium [31].
4. The first step is to represent the kinetic behavior of a particular module as a function of the variables participating in the rate eq. [11, 12, 19, 29, 32]. At this point it is critical to choose the right set of parameters, which in the example of the case of an enzyme-catalyzed reaction process following Michaelis-Menten kinetics corresponds to the K_M values for each of the substrates, products, or dissociation constants for effectors taking part in the reaction kinetics of the enzyme and the V_{max} value of the enzyme at saturating levels of all substrates and/or effectors. Those parameter values can be obtained from experimental data, if available [22]. These values may need adjustments since the conditions in which they have been measured may not correspond to physiological ones (*see* Subheading 4).
5. The availability of data from metabolomics which could be used to calculate metabolite concentrations has led to the development of a new strategy to adjust V_{max} values using either unique solutions or optimization of metabolic fluxes [9]. Because the strategy involves the use of the whole integrated model of the metabolic network under study, it will be developed more in depth in **step 12**.
6. The examination of the sensitivity of the curve relating the rate of a given process to its substrates, products, and effectors will provide information about ways to modify a module behavior in the fully assembled model, when the simulated output differs from experimental data. It also provides information about mechanisms through which its activity might be modulated

$$V_{IDH} = \frac{k_{cat}^{IDH}\,E_T^{IDH}}{\left(1+\frac{[H^+]}{k_{h,1}}+\frac{k_{h,2}}{[H^+]}\right)+\frac{\left(\frac{K_M^{ISOC}}{[ISOC]}\right)^{ni}}{\left(1+\frac{[ADP]_m}{K_{ADP}^{a}}\right)\left(1+\frac{[Ca^{2+}]}{KCa}\right)}+\left(\frac{K_M^{NAD}}{[NAD]}\right)\left(1+\frac{[NADH]}{K_{i,NADH}}\right)+\frac{\left(\frac{K_M^{ISOC}}{[ISOC]}\right)^{ni}\left(\frac{K_M^{NAD}}{[NAD]}\right)\left(1+\frac{[NADH]}{K_{i,NADH}}\right)}{\left(1+\frac{[ADP]_m}{K_{ADP}^{a}}\right)\left(1+\frac{[Ca^{2+}]}{KCa}\right)}}$$

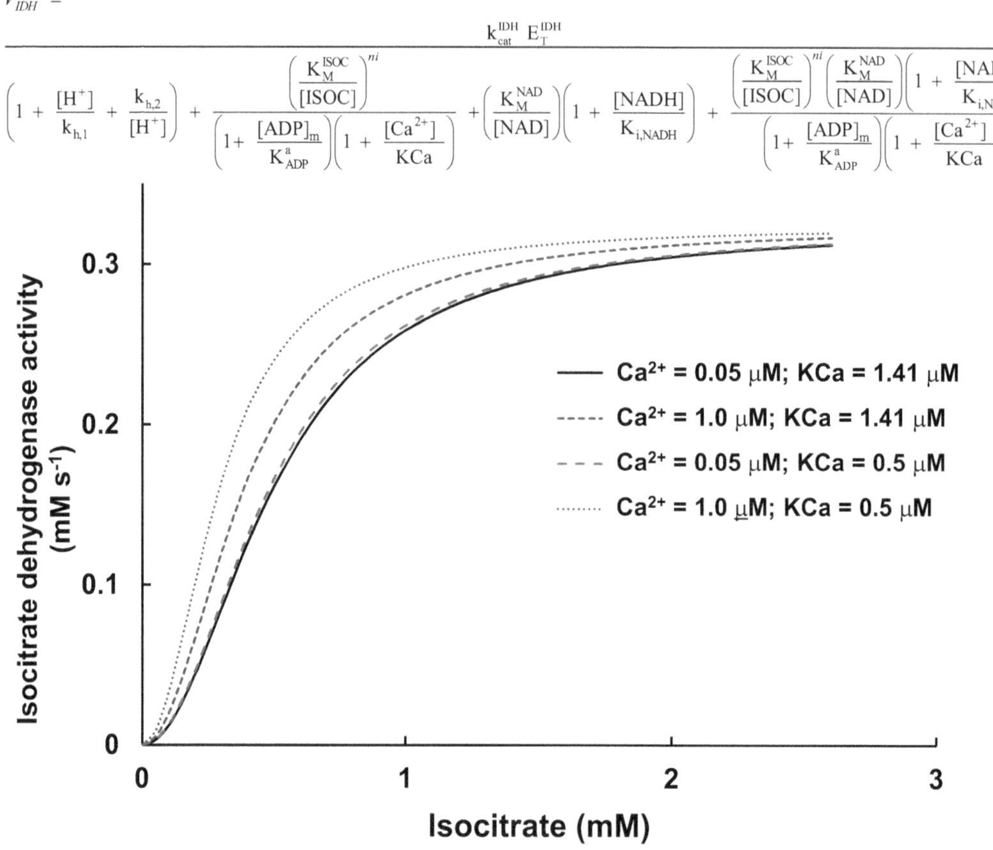

Fig. 1 Sensitivity of the rate of isocitrate dehydrogenase as a function of isocitrate, the level of Ca^{2+}, and value of the Ca^{2+} activation constant. The activity of isocitrate dehydrogenase was studied as a function of its substrate, isocitrate, at different levels of Ca^{2+} activation. To achieve a good reproduction of the stimulatory effect of Ca^{2+} as has been reported in [37], the Ca^{2+} association constant, KCa (*see* equation on top of the plot), was decreased to 0.5 μM with respect to the original level reported in [11] (Reproduced from Cortassa et al. [11] by copyright permission of the Biophysical Society)

in vivo. As an illustration of this point, Fig. 1 shows the plots of the rate of isocitrate dehydrogenase as a function of isocitrate at two levels of Ca^{2+} (enzyme activator) and for different values of the activation constant for Ca^{2+}, KCa. By decreasing KCa we were able to simulate in ventricular myocytes the stimulation of the TCA cycle flux upon changes in workload, within physiological Ca^{2+} levels in the mitochondrial matrix, and at different stimulation frequencies [12]. The sensitivity of the rate equation should be compared with experimental data, and the parameters may be adjusted until a good agreement (qualitative, quantitative, or both) can be attained.

7. The modules may be assembled once their individual behavior is proved satisfactory according to the criteria 1–6 above [11]. The next step is to decide which of the variables

participating in the equations will be state variables (i.e., whose values describe the mathematical state of a dynamical system at any time point) and which ones will be adjustable parameters. For each state variable, an ODE will be written according to the following general form:

$$\frac{\mathrm{dM}}{\mathrm{d}t} = +V_{\text{production}} \pm V_{\text{transport}} - V_{\text{consumption}} \tag{3}$$

where M represents an intermediary (metabolite) in the system, or in a compartment thereof; $V_{\text{production}}$, the sum of the rates of all processes contributing M; $V_{\text{consumption}}$, the sum of the rates of all processes consuming M; and $V_{\text{transport}}$, the rates of the process that carry M inside or outside the compartment or system being modeled. Worthy of notice at this stage is that the rate parameters in ODEs may be subjected to scaling factors to account for volume or buffering effects. In the case that several compartments are considered, the compartment's volume ratio must be used to scale the rates if processes occur in different compartments.

8. To perform a simulation, we need to consider the format with which to write the differential equations. The best advice here is to take a sample model from the chosen simulation package and modify it to include the specific ODEs from your own model. Then, parameter files should be assembled and linked or included, as well as a set of initial values of the state variables (initial conditions).
9. To choose the integration algorithm to be used to simulate the model's behavior, the "stiffness" of the model needs to be considered, i.e., if the time scales in which different processes attain steady state are very different (more than two orders in the time units used), usually milliseconds for channel gating events and seconds for mitochondrial energy transduction processes. If some rapid kinetic process is being accounted for, such as fast ion transport mechanisms, then "stiff" solvers may be required such as ode15, ode23, or ode113 in MATLAB®.
10. Simulate the behavior of the assembled model as a function of time, until it reaches a steady state, i.e., where the derivatives of the state variables are less than 10^{-6}–10^{-8} in relative terms ($\Delta x/x$).
11. At this stage, the modeler should start comparing the model output with experimental data (e.g., steady-state values of variables or fluxes). Such comparison will lead to the fine-tuning of parameters, and the criteria utilized to do the parameter adjustment will need to be made explicit (*see* **Note 1** on this subject).

To improve the understanding of the model and provide additional validation through a comparison with experimental data, we will now introduce some analytical tools mainly designed to study steady-state behavior:

12. Elementary flux modes (EFMs): An elementary flux is defined as the minimal set of enzyme processes that could operate at steady state with the irreversible reactions proceeding in the thermodynamically permissible direction. If there is a blockade in any of the participating reactions, all flux through that mode will cease. Figure 2 shows the EFMs occurring in the mitochondrial bioenergetic network that we are utilizing to illustrate the analytical tools in this article. EFMs can additionally inform us about network robustness and flexibility, two related concepts. Flexibility refers to the capacity of a network to "rewire" its topology (i.e., the way it is connected) to function even under conditions in which one or more steps are blocked, whereas robustness denotes the network's ability to keep its function despite rewiring. Generally, the higher the number of EFMs, the higher the network's flexibility and robustness, at least under steady-state function. The calculation of EFMs employs software developed to operate in either MATLAB or Octane environments (Metatool version 5.1) and requires a text file with the declaration of the metabolites (internal and external) and the reactions relating those metabolites with a specific text format. The Metatool routine can be called within MATLAB, and the software produces a series of matrices that contain all the formal information of the metabolic network under analysis and its functioning under steady-state conditions. That information includes the stoichiometric matrix, null-space matrix, and EFMs.
13. Linear optimization of V_{max} values from metabolomics data to accurately represent model behavior under specific physiological conditions.

The method involves replacing the concentrations of metabolites in the rate expressions from the kinetic model to solve and optimize the model at the steady state for the V_{max} values of all the metabolic steps. Model simulations are run to corroborate the metabolites' steady-state level. V_{max} is the parameter to be optimized because it is the most sensitive one to gene expression modulation and posttranslational modifications. Indirectly, V_{max} integrates gene expression and posttranslational information thus accounting for mass-energy transformations and signaling [2, 10]. Since all rate expressions linearly depend on V_{max}, linear algebraic methods can be used.

The V_{max} optimization method is based on the general equation:

$$dx/dt = S*v + b \tag{4}$$

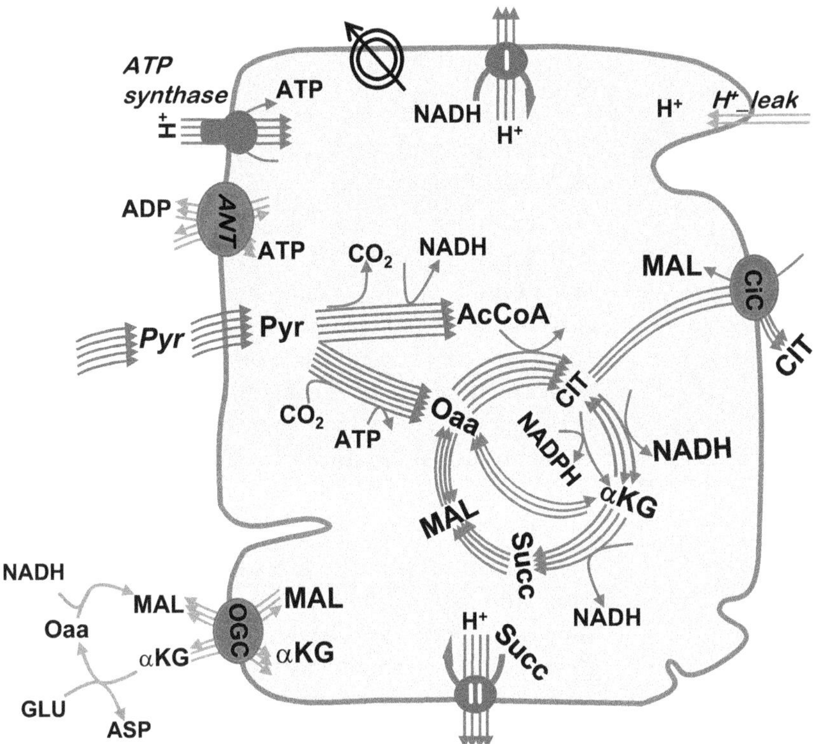

Fig. 2 Schematic representation of the elementary flux modes (EFMs) exhibited by the mitochondrial energetic model. Depicted are the mitochondrial electrochemical and metabolic pathways, encompassing oxidative phosphorylation (OxPhos) and matrix-based processes, along with the interactions accounted for by the model. The tricarboxylic (TCA) cycle in the mitochondrial matrix is fed by pyruvate (Pyr), supplied by glycolysis. In the mitochondrial matrix, Pyr can be oxidized, first, to AcCoA by pyruvate dehydrogenase [36], a highly regulated enzymatic complex by NADH/NAD^+, Ca^{2+}, and AcCoA/CoA. TCA cycle oxidizes AcCoA to CO_2 producing NADH and $FADH_2$ which provide the redox driving force for OxPhos. Alternatively, Pyr can undergo carboxylation to contribute to the replenishment of the TCA cycle which is drained by the malic enzyme and glutamate dehydrogenase (omitted in the scheme for simplicity but considered in the EFMs analysis). Additionally, NADH can be imported from the cytosol via the malate-aspartate shuttle. EFMs are represented by arrows of which their number is proportional to the number of EFMs or its nearest fraction (i.e., each arrow corresponding to 100 EFMs) possible through the step considered and the indicated direction. For the sake of simplification, the arrows leading to consumption/production of co-substrates/cofactors are indicated only once, but they have the same number of modes as the reaction to which they belong. Since EFMs denote fluxes with a definite direction, double arrows in the scheme mean that the EFMs from reactions or transport steps can go in either direction. For the EFM calculations, cytoplasmic MAL, αKG, NADH, and Oaa were considered external variables, thus not participating in the EFMs calculated. Key to symbols: $\Delta\Psi_m$ is represented by the concentric circles with an arrow across located at the inner mitochondrial membrane

with dx/dt representing the vector of derivatives of the state variables (intermediary metabolite concentrations); S, the stoichiometric matrix; and b, the vector of transport and demand processes, (e.g., biosynthesis). Lastly, v stands for the rates' vector expressed as the rate equation in the model with known kinetic parameters except for V_{max} which is subjected to optimization.

Optimization of the linear algebraic system is performed with the Simplex or an equivalent algorithm (e.g., implemented in the linprog optimization method in MATLAB). Different objective functions may be used as optimization criteria such as maximization of ATP synthesis fluxes or minimization of redox consumption.

Firstly, the function "linprog" from MATLAB is utilized with an algorithm for optimization (e.g., "simplex" or "interior point") of V_{max} values. Secondly, within the volume of possible solutions, a maximal V_{max} of the first step (glucose transport) is chosen, and computer model simulations are performed choosing those solutions that correspond to the vertices of the volume of solutions that satisfy the metabolite concentrations and the set of ODEs that relate them (*see* Fig. 3 and its legend).

When the chosen model can simulate the initial input of metabolites profile at steady state, we can also use it to determine the rate or flux through each individual step of the metabolic network. The set of metabolic fluxes determined in this way corresponds to the fluxome [9].

Linear algebraic functions from MATLAB are utilized for matrix calculations as required by the method employed for calculating the main rate-controlling and rate-regulatory steps of the metabolic network, as described next.

14. Metabolic control analysis (MCA) of the optimized model.

A main benefit of having a realistic representation of a biological/bioenergetic system is the ability to calculate how the metabolic network is controlled and regulated. Using the tools of MCA, it is possible to attain this aim. MCA was originally developed to understand how the dynamics of a metabolic network depends on the activities of the individual processes participating in such a network (reviewed in [17]). Figure 4 shows the results of applying MCA to mitochondrial respiration, utilizing specific inhibitors targeting the processes, e.g., respiration, whose control is under study. Reder [33] developed a generalized linear algebraic method to analyze the sensitivity of metabolic systems to perturbations triggered by either a change in the internal state of the system or by the environment. Due to its generality, it can be applied to metabolic networks of any topology. We implemented Reder's method to analyze the behavior of a mitochondrial model isolated or integrated to ventricular cardiomyocyte function [34].

To quantify control in a metabolic network, the following coefficients are used: flux control coefficient, $C_{E_k}^{J_i}$; metabolite concentration control coefficient, $C_{E_k}^{M_i}$; and the elasticity coefficient, $\varepsilon_{S_k}^{v_i}$, that are, respectively, defined as follows:

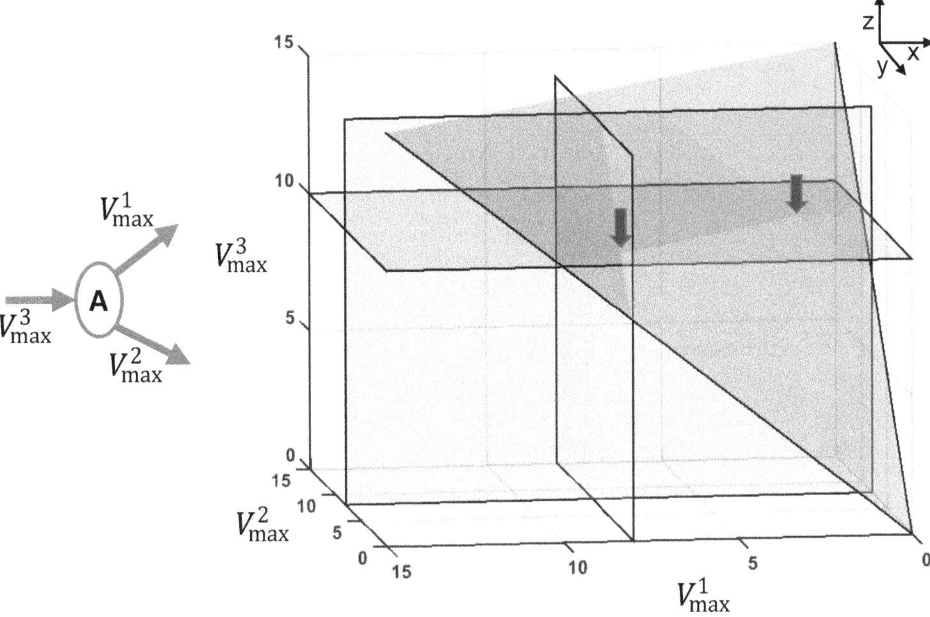

Fig. 3 Solution space in a hypothetical branched metabolic network with three unknown fluxes. The network displayed on the left comprises three fluxes (unknowns) and a single metabolite (equation) thus underdetermined, meaning that its solution is not unique but a solution space in 2D, i.e., a surface of solutions (purple and orange planes). According to the procedure proposed to find the V_{max} values (*see* Subheading 3, **step 13** and **Note 1** for details), the solutions chosen (identified by dark arrows) fulfill two conditions: (a) $V_{max} > 0$ for all enzymes in the network and (b) belongs to the solution space of the network (orange pentagon contained between the absolute V_{max} values whose boundaries are given by the yellow, green, and teal surfaces corresponding to V^1_{max}, V^2_{max}, and V^3_{max}, respectively) (Redrawn from Price et al. [38]).

$$C_{E_k}^{J_i} = \frac{\frac{\partial J_i}{J_i}}{\frac{\partial E_k}{E_k}}, \quad C_{E_k}^{M_i} = \frac{\frac{\partial M_i}{M_i}}{\frac{\partial E_k}{E_k}}, \quad \varepsilon_{S_k}^{v_i} = \frac{\frac{\partial v_i}{v_i}}{\frac{\partial S_k}{S_k}} \tag{5}$$

These expressions indicate the relative change in a flux (J_i) through reaction i, or metabolite concentration (M_i), of the metabolic network under study upon a relative change in the enzyme activity (E_k) of reaction k. The elasticity coefficient computes the relative change in the enzyme activity of reaction i, upon a relative change of intermediate S_k which has a direct impact on the activity. Control and elasticity coefficients differ in that the former reflects global, systemic properties of the network, and as such if a control coefficient changes, in response to a stimulus or perturbation, other coefficients will also be modified. On the other hand, elasticity coefficients reflect local properties within a network of reactions or processes and thus depend only on the concentrations of substrates, products, and rate-modifying effectors.

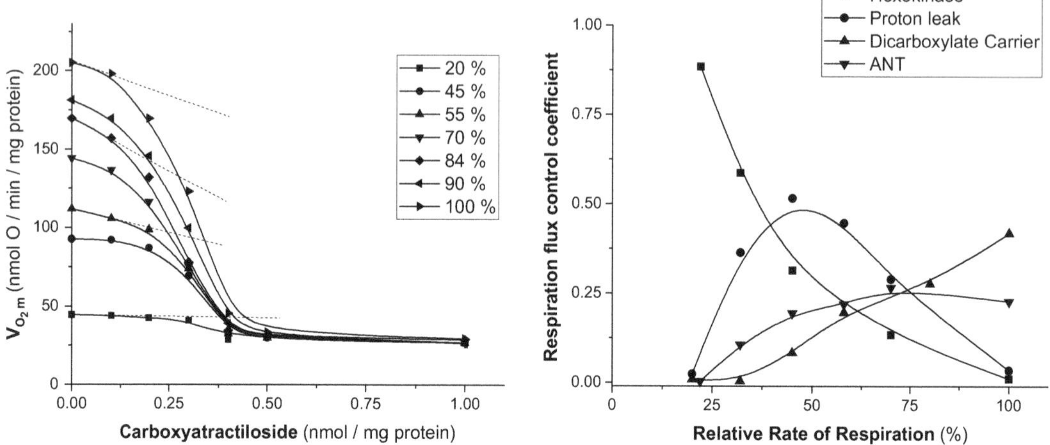

Fig. 4 Metabolic control analysis of respiration in rat liver mitochondria. The left panel depicts the determination of the ANT control coefficient performed with the inhibitor titration method utilizing carboxyatractyloside at various levels of respiratory activity in the presence of succinate and malate. Respiration was modulated by the amount of hexokinase added together with glucose and ATP. The higher the amount of hexokinase, the larger the supply of ADP to stimulate respiration. The right panel summarizes the results of the experiments with inhibitor titrations of the dicarboxylate carrier using phenylsuccinate; hexokinase, assessed by adding small amounts of hexokinase and recording the increase in hexokinase activity; and the proton leak, quantified after successive additions of proton uncoupler FCCP. The changes elicited in the respiration levels are analogous to the transitions from state 4 (low respiration, low ADP supply) to state 3 (high respiration, high ADP supply) (Redrawn from Groen et al. [39])

The departure point of the generalized linear algebraic method is the *stoichiometric matrix*, obtained from the set of ODEs describing the model. The stoichiometric matrix defines the structural relationships between the processes/reactions and the intermediates from the metabolic network. The information present in the stoichiometric matrix is independent from both the enzyme kinetics and the parameters that rule the dynamic behavior of the network. The *elasticity matrix*, defined by the dependence of each process/reaction on the intermediates (e.g., ions or metabolites) included in the model, is the second piece of information required to perform control analysis. The elasticity matrix comprises the quantification of the rates of individual processes through the derivatives with respect to each possible intermediate/effector. By applying matrix algebra, the corresponding control coefficients matrices are obtained. The regulation and control in the network are quantified by both types of matrix. The network regulation exerted by internal or external factors can be quantified by the *response coefficient* [35].

The following matrix relationships were used in the computation of flux and metabolite concentration control coefficients:

$$C = Id_r - D_x vL(N_r\, D_x v\, L)^{-1} N_r \tag{6}$$

$$\Gamma = -L(N_r\, D_x v\, L)^{-1} N_r \tag{7}$$

with C and Γ referring to the matrices of flux- and metabolite-concentration control coefficients, respectively; Id_r, the identity matrix of dimension *r* or the number of processes in the network under study; $D_x v$ the elasticity matrix; N_r the reduced stoichiometric matrix; and L, the link matrix that relates the reduced- to the full-stoichiometric matrix of the system.

4 Notes

1. Regarding the issue of adjusting model parameters, the main "inconvenience" is apparent when different sources of data render very dissimilar values. Different organisms or tissue sources from which an enzymatic activity of interest has been isolated and/or determined, or different assay conditions, can provide widely different results (e.g., V_{max}). The V_{max} of a process is the most likely parameter to be adjusted in the scenario that an inconsistency in the dynamic behavior of the assembled model is detected. This is the underlying rationale to develop an algorithmic method to optimize and resolve the steady-state values of all V_{max} parameters in the model. It is usually the case that a model of a metabolic network contains more processes (reactions, transport steps) than metabolites leading to underdetermination, meaning that the system admits multiple (rather than unique) solutions, thus the need to apply optimization criteria to algebraically solve the system while restricting the size of the solution space (i.e., the set of V_{max} values for all the processes participating in the network). Figure 3 shows an example for a hypothetical network in which there is a single intermediary metabolite and three processes. The solutions are bound by the maximal V_{max} determined that correspond to the planes parallel to the *x-y* (teal), *x-z* (green), or *y-z* (yellow) origin of coordinates. The solution space of the system is contained in the purple plane, bound by its intersections with the maximal rate planes. The set of maximum V_{max} values that are not zero are indicated by the thick dark arrows that point to the vertices of the solution space (denoted by an orange polygon).
2. MCA, as applied to computational models of mitochondrial energetics, may assist researchers to interpret and understand the physiological meaning of the information provided by high-throughput technologies such as genomics, proteomics, or metabolomics. Figure 5 shows the flux- (FCC) and

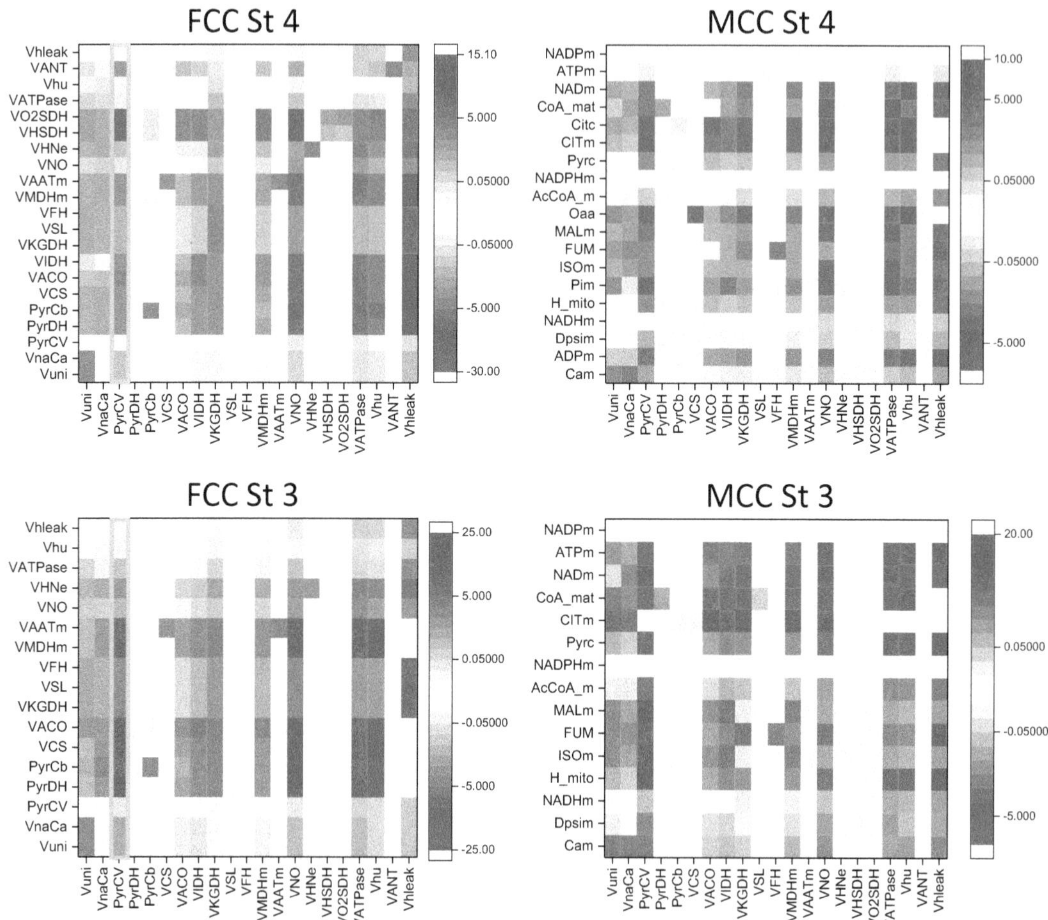

Fig. 5 Metabolic control analysis of the mitochondrial computational model. Flux control coefficients (FCC) and metabolite concentration control coefficients (MCC) at steady state under state 4 (St 4) or state 3 (St 3) respiration. The nonlinear scale of values for the control coefficients ranges from negative (blue shades) to positive (red shades) and is shown next to each panel. Only those fluxes and metabolites (in rows) for which the summation theorem fulfills the expected result of 1.0 (for FCC) or zero (for MCC), respectively, are displayed. Key to flux and metabolite acronyms: *m* at the end of the acronym means the enzyme activity or metabolite is located in the mitochondrial matrix or inner membrane; Vuni, Ca^{2+} uniporter; VnaCa, Na^+/Ca^+ exchanger; PyrCV, pyruvate carrier; PyrDH, pyruvate dehydrogenase complex; PyrCb, pyruvate carboxylase; VCS, citrate synthase; VACO, aconitase; VIDH, isocitrate dehydrogenase; VKGDH, α-ketoglutarate dehydrogenase; VSL, succinyl CoA lyase; VFH, fumarate hydratase; $VMDH_m$, mitochondrial malate dehydrogenase; $VAAT_m$, mitochondrial amino aspartate transferase; VNO, NADH-driven respiratory electron transport; VHNe, NADH-driven respiratory proton transport; VHSHD, succinate-driven respiratory proton transport; VO2SDH, succinate dehydrogenase-driven electron transport; VATPase, ATP synthase; Vhu, proton transport through ATP synthase; VANT, adenine nucleotide translocator; V_{hleak}, proton leak across inner mitochondrial membrane, Ca_m; mitochondrial free Ca^{2+}; $Dpsi_m$, mitochondrial inner membrane potential; H_mito, matrix H^+; ISO_m, mitochondrial isocitrate; FUM, fumarate; MAL_m, mitochondrial malate; $AcCoA\_{m}$, mitochondrial AcCoA; CITm,c mitochondrial, cytosolic (respectively) citrate; CoA_mat; matrix CoA

metabolite-control (MCC) coefficients of a mitochondrial model (schematized in Fig. 2) under both state 4 and 3 respirations. The heat map representation enables the visualization of processes that exert more positive (red) or negative (blue) control. The heat map layout also highlights the distributed nature of the control in distinction to the outdated concept of "rate-limiting" step. For example, the rate of pyruvate transport (PyrCV, x-axis, outlined in green) exerts a large control over most fluxes through the TCA cycle under state 3 respiration (bottom left), but the control is shared with the activity of ATP synthase and NADH-driven respiration. The control of pyruvate transport on, e.g., the TCA cycle flux, is negative under state 4 but positive under state 3 respiration. Processes that are highly regulated, e.g., pyruvate dehydrogenase (PyrDH), do not exert a significant control on mitochondrial energetics [36].

Metabolites' concentration is controlled by the same steps that control the flux through the pathway to which the metabolite belongs and negatively controlled by the processes that consume such metabolite.

In summary, we described the modular approach of kinetic modeling emphasizing the use of EFM analysis, linear optimization of kinetic models, and control analysis at the steady state, to help refine our grasp of biologically meaningful behaviors and model reliability. The use of these tools should help improve the design, building, and interpretation of the steady-state behavior of computational models and the assessment of validation criteria while paving the way to prediction.

Acknowledgments

This work was supported by the Intramural Research Program of the National Institutes of Health, National Institute on Aging.

References

1. Winslow RL, Cortassa S, Greenstein JL (2005) Using models of the myocyte for functional interpretation of cardiac proteomic data. J Physiol 563(Pt 1):73–81
2. Aon MA (2014) Complex systems biology of networks: the riddle and the challenge. In: Aon MA, Saks V, Schlattner U (eds) Systems biology of metabolic and signaling networks: energy, mass and information transfer, Springer series in biophysics, vol 16. Springer-Verlag, Berlin, Berlin, pp 19–35. https://doi.org/10.1007/978-3-642-38505-6_2
3. Cortassa S, Aon MA (2012) Computational modeling of mitochondrial function. Methods Mol Biol 810:311–326. https://doi.org/10.1007/978-1-61779-382-0_19
4. Schuster S, Dandekar T, Fell DA (1999) Detection of elementary flux modes in biochemical networks: a promising tool for pathway analysis and metabolic engineering. Trends Biotechnol 17(2):53–60
5. Savinell JM, Palsson BO (1992) Network analysis of intermediary metabolism using linear optimization. I. Development of mathematical formalism. J Theor Biol 154(4):421–454

6. Vo TD, Palsson BO (2007) Building the power house: recent advances in mitochondrial studies through proteomics and systems biology. Am J Physiol Cell Physiol 292(1):C164–C177
7. Orth JD, Thiele I, Palsson BO (2010) What is flux balance analysis? Nat Biotechnol 28 (3):245–248. https://doi.org/10.1038/nbt.1614
8. Thiele I, Price ND, Vo TD, Palsson BO (2005) Candidate metabolic network states in human mitochondria. Impact of diabetes, ischemia, and diet. J Biol Chem 280 (12):11683–11695. https://doi.org/10.1074/jbc.M409072200
9. Cortassa S, Caceres V, Bell LN, O'Rourke B, Paolocci N, Aon MA (2015) From metabolomics to fluxomics: a computational procedure to translate metabolite profiles into metabolic fluxes. Biophys J 108(1):163–172. https://doi.org/10.1016/j.bpj.2014.11.1857
10. Cortassa S, Aon MA, Iglesias AA, Aon JC, Lloyd D (2012) An Introduction to Metabolic and Cellular Engineering, 2nd Edition edn. World Scientific Publishers, Singapore
11. Cortassa S, Aon MA, Marban E, Winslow RL, O'Rourke B (2003) An integrated model of cardiac mitochondrial energy metabolism and calcium dynamics. Biophys J 84(4):2734–2755
12. Cortassa S, Aon MA, O'Rourke B, Jacques R, Tseng HJ, Marban E, Winslow RL (2006) A computational model integrating electrophysiology, contraction, and mitochondrial bioenergetics in the ventricular myocyte. Biophys J 91 (4):1564–1589
13. Magnus G, Keizer J (1997) Minimal model of beta-cell mitochondrial Ca2+ handling. Am J Phys 273(2 Pt 1):C717–C733
14. Jeong H, Tombor B, Albert R, Oltvai ZN, Barabasi AL (2000) The large-scale organization of metabolic networks. Nature 407 (6804):651–654
15. Almaas E, Kovacs B, Vicsek T, Oltvai ZN, Barabasi AL (2004) Global organization of metabolic fluxes in the bacterium Escherichia Coli. Nature 427(6977):839–843
16. Cortassa S, Aon JC, Aon MA (1995) Fluxes of carbon, phosphorylation, and redox intermediates during growth of Saccharomyces cerevisiae on different carbon sources. Biotechnol Bioeng 47(2):193–208
17. Fell DA (1996) Understanding the control of metabolism. Frontiers in Metabolism. Portland Press, London
18. Hill TL, Chay TR (1979) Theoretical methods for study of kinetics of models of the mitochondrial respiratory chain. Proc Natl Acad Sci U S A 76(7):3203–3207
19. Cortassa S, Aon MA, Winslow RL, O'Rourke B (2004) A mitochondrial oscillator dependent on reactive oxygen species. Biophys J 87 (3):2060–2073. https://doi.org/10.1529/biophysj.104.041749
20. Kurz FT, Kembro JM, Flesia AG, Armoundas AA, Cortassa S, Aon MA, Lloyd D (2017) Network dynamics: quantitative analysis of complex behavior in metabolism, organelles, and cells, from experiments to models and back. Wiley Interdiscip Rev Syst Biol Med 9 (1). https://doi.org/10.1002/wsbm.1352
21. Zhou L, Aon MA, Almas T, Cortassa S, Winslow RL, O'Rourke B (2010) A reaction-diffusion model of ROS-induced ROS release in a mitochondrial network. PLoS Comput Biol 6(1):e1000657. https://doi.org/10.1371/journal.pcbi.1000657
22. Barthelmes J, Ebeling C, Chang A, Schomburg I, Schomburg D (2007) BRENDA, AMENDA and FRENDA: the enzyme information system in 2007. Nucleic Acids Res 35(Database issue):D511–D514
23. Kanehisa M, Furumichi M, Tanabe M, Sato Y, Morishima K (2017) KEGG: new perspectives on genomes, pathways, diseases and drugs. Nucleic Acids Res 45(D1):D353–D361. https://doi.org/10.1093/nar/gkw1092
24. Nelson DL, Cox MM (2013) Lehninger principles of biochemistry, 6th edn. W. H. Freeman and Company, New York
25. Nickerson D, Stevens C, Halstead M, Hunter P, Nielsen P (2006) Toward a curated CellML model repository. Conf Proc IEEE Eng Med Biol Soc 1:4237–4240
26. Dhooge A, Govaerts W, Kuznetsov YA, Meijer HGE, Sautois B (2008) New features of the software MatCont for bifurcation analysis of dynamical systems. Math Comput Model Dyn Syst 14(2):147–175
27. Schuster S, von Kamp A, Pachkov M (2007) Understanding the roadmap of metabolism by pathway analysis. Methods Mol Biol 358:199–226. https://doi.org/10.1007/978-1-59745-244-1_12
28. von Kamp A, Schuster S (2006) Metatool 5.0: fast and flexible elementary modes analysis. Bioinformatics 22(15):1930–1931. https://doi.org/10.1093/bioinformatics/btl267
29. Gunn RB, Curran PF (1971) Membrane potentials and ion permeability in a cation exchange membrane. Biophys J 11 (7):559–571

30. Hille B (2001) Ion channels of excitable membranes, 3rd edn. Sinauer, Sunderland, MA
31. Crank J (1975) The mathematics of diffusion, 2d edn. Clarendon Press, Oxford, England
32. Segel IH (1975) Enzyme kinetics: behavior and analysis of rapid equilibrium and steady state enzyme systems. Wiley, New York
33. Reder C (1988) Metabolic control theory: a structural approach. J Theor Biol 135 (2):175–201
34. Cortassa S, O'Rourke B, Winslow RL, Aon MA (2009) Control and regulation of mitochondrial energetics in an integrated model of cardiomyocyte function. Biophys J 96 (6):2466–2478
35. Ainscow EK, Brand MD (1999) The responses of rat hepatocytes to glucagon and adrenaline. Application of quantified elasticity analysis. Eur J Biochem 265(3):1043–1055
36. Cortassa S, Sollott SJ, Aon MA (2017) Substrate selection and its impact on mitochondrial respiration and redox. Molecular Basis for Mitochondrial Signalling Springer International Publishing:349–375. https://doi.org/10.1007/978-3-319-55539-3
37. Rutter GA, Denton RM (1988) Regulation of NAD+−linked isocitrate dehydrogenase and 2-oxoglutarate dehydrogenase by Ca2+ ions within toluene-permeabilized rat heart mitochondria. Interactions with regulation by adenine nucleotides and NADH/NAD+ ratios. Biochem J 252(1):181–189
38. Price ND, Schellenberger J, Palsson BO (2004) Uniform sampling of steady-state flux spaces: means to design experiments and to interpret enzymopathies. Biophys J 87 (4):2172–2186. https://doi.org/10.1529/biophysj.104.043000
39. Groen AK, Wanders RJ, Westerhoff HV, van der Meer R, Tager JM (1982) Quantification of the contribution of various steps to the control of mitochondrial respiration. J Biol Chem 257 (6):2754–2757

Chapter 15

Monitoring the Mitochondrial Dynamics in Mammalian Cells

Luca Simula and Silvia Campello

Abstract

Mitochondria exist in a dynamic state inside mammalian cells. They undergo processes of fusion and fission to adjust their shape according to the different cell needs. Different proteins tightly regulate these dynamics: Opa-1 and Mitofusin-1 and Mitofusin-2 are the main profusion proteins, while Drp1 and its different receptors (Mff, Fis1, MiD49, MiD51) regulate mitochondrial fission. The dynamic nature of the mitochondrial network has become evident and detectable, thanks to recent advances in live imaging video microscopy and to the availability of mitochondria-tagged fluorescent proteins. High-resolution confocal reconstruction of mitochondria over time allows researchers to visualize mitochondria shape changes in living cells, under different experimental conditions. Moreover, in recent years, different techniques in living cells have been developed to study the process of mitochondria fusion in more details. Among them are fluorescence recovery after photobleaching (FRAP) of mitochondria-tagged GFP (mtGFP), use of photoactivatable mtGFP, polyethylene glycol (PEG)-based fusion of mtGFP and mtRFP cells, and Renilla luciferase assay (for population studies). In addition, in combination with imaging, the analysis of the expression levels of the different mitochondria-shaping proteins, along with that of their activation status, represents a powerful tool to investigate potential modulations of the mitochondrial network. Here, we review this aspect and then mention a number of techniques, with particular attention to their relative protocols.

Key words Mitochondrial dynamics, Fusion, Fission, Microscopy, Live cells

1 Introduction

The morphology of the mitochondrial network is highly plastic in mammalian cells [1]. The shape of these organelles dynamically shifts between two extremes: (1) a long, elongated, and tubular mitochondrial network or (2) a high number of small grain-shaped mitochondria. These organelles are continuously subjected to a dynamic equilibrium between these two ends, thanks to processes of fusion and fission, which adjust the shape of mitochondria

Electronic supplementary material: The online version of this chapter (https://doi.org/10.1007/978-1-4939-7831-1_15) contains supplementary material, which is available to authorized users.

Carlos M. Palmeira and António J. Moreno (eds.), *Mitochondrial Bioenergetics: Methods and Protocols*,
Methods in Molecular Biology, vol. 1782, https://doi.org/10.1007/978-1-4939-7831-1_15,

according to (patho)physiological cell needs. For example, an elongated network sustains oxidative phosphorylation (OXPHOS) for an optimal ATP production and maintains the overall quality of the mitochondrial mass [2], while mitochondria fragmentation frequently occurs concomitantly with a metabolic switch from OXPHOS toward glycolysis [3]. Fragmented, often damaged, mitochondria can be isolated from the healthy network by the cell and addressed to degradation through mitophagy [4]. Moreover, fragmentation of the mitochondrial network is a physiological step required during different cellular processes, such as apoptosis induction [5], cell migration [6, 7], and cell proliferation [8]. In addition, while a fused network promotes OXPHOS-dependent ATP production and the assembly of electron transport chain (ETC) complexes, a fragmented network reduces ETC activity, thus increasing ROS production as respiration by-products [8, 9].

Several mitochondria-shaping proteins tightly control the shape of the mitochondrial network. The activity of the GTPase proteins Mitofusin-1 (Mfn1) and Mitofusin-2 (Mfn2) regulates the fusion of the outer mitochondrial membrane (OMM) [10]. Mfns can be posttranslationally ubiquitinated by different E3 ubiquitin ligases, which can alternatively promote their stabilization (and thus mitochondrial elongation) or proteasome-mediated degradation (and thus mitochondrial fission). For example, while PINK/Parkin-dependent ubiquitination promotes Mfns degradation and mitochondrial fission [11], other E3 ligases can promote Mfns activity by adding stabilizing ubiquitin chains [12]. The fusion of the inner mitochondrial membrane (IMM) is regulated by the activity of the Optic Atrophy-1 (Opa1) protein [13]. Opa-1 is normally present in two different forms: the Opa1-long isoforms (Opa1-L1 and Opa1-L2), which promote fusion of IMM, and the Opa1-short isoforms (Opa1-S3, Opa1-S4, and Opa1-S5), soluble forms unable to mediate fusion and obtained by Opa1-L proteolytic cleavage by the proteases Yme1L and Oma1 [14]. Besides regulating the IMM fusion, the formation of complexes between Opa1-L and Opa1-S isoforms controls the width of the mitochondrial *cristae*, where cytochrome C (cytC) and ETC complexes are localized. Thus, Opa1-L-to-Opa1-S conversion is an important step that controls cytochrome C (cytC) release from mitochondria, upon apoptotic conditions [15]. In addition, cytosolic MSTO1 has been recently described as an additional regulator of mitochondria fusion process [16].

Fragmentation of the mitochondrial network is mainly driven by the activity of the GTPase protein Dynamin-Related Protein-1 (Drp1) [17]. Drp1 is recruited from the cytosol to OMM upon different stimuli. Once at the organelle, it encircles mitochondria, severing their membranes through GTP hydrolysis [18]. Different proteins help Drp1-driven mitochondrial fission, acting directly as Drp1 receptors on OMM, such as Fis1 [19] and Mff [20], or as ancillary proteins, such as MiD49 and MiD51 [21]. Drp1 activity is posttranslationally regulated by different modifications: phosphorylation on Serine637 or Serine693 prevents the Drp1 translocation to OMM and thus mitochondrial fission, while phosphorylation on Serine616 is strictly required for an optimal Drp1 activity [17, 22]. In addition, Drp1 SUMOylation enhances Drp1 activity [23], similarly to MARCH5-dependent ubiquitination [24]. Moreover, also Drp1 O-GlcNAcylation activates Drp1 in different cell models [17]. Very recently, different noncanonical proteins have been described to mediate mitochondrial fission, such as the cytoplasmic kinase leucine-rich repeat kinase 2 (LRRK2) [25] and dynamin-2 (Dyn-2) [26].

In this chapter, we will review some of the methodologies currently used to investigate the shape of the mitochondrial network in mammalian cells, by directly visualizing mitochondria fission and/or fusion events or by indirectly investigating the activity of the mitochondria-shaping proteins.

2 Visualization of the Mitochondrial Network by 3D Live Imaging Using Mitochondria-Tagged Fluorescent Proteins

The best way to investigate the dynamic nature of the mitochondrial network in cells is probably live imaging confocal microscopy; this follows the mitochondrial network dynamic changes by using mitochondria-tagged fluorescent proteins (e.g., mitochondria-tagged GFP) or fusion constructs (between a mitochondrial protein and a fluorescent one, e.g., TOM20-GFP) overexpressed in living cells. In recent decades, giant steps have been made in this field, thanks to this technique, and impressive and more and more detailed new results are still achieved with the sophisticated last-generation microscopes, such as the super-resolution or the correlative microscope. Nowadays, countless constructs are commercially available, expressing variably colored fluorescent proteins (i.e., GFP, RFP, YFP) fused with specific sequences targeting their localization in the outer (OMM) or the inner mitochondrial membrane (IMM) or inside the mitochondrial matrix [27, 28]. Post-acquisition high-resolution 3D reconstruction of the mitochondrial network during time allows investigating changes in the shape of the cell mitochondria, in both a qualitative and quantitative way. Indeed, the microscope slice acquisition along the sample z-axis (z-stack acquisition) allows a 3D reconstruction of the mitochondrial network, while time-lapse video imaging provides the visualization of changes of mitochondrial morphology over time. Obviously, given the tight link between the shape of mitochondria and the metabolic and health status of the cell, the sample must be handled with extreme caution to avoid perturbations of the organelle function and health that could lead to artefactual considerations on their shape and vice versa.

2.1 Materials

1. Cells transfected (preferentially stably but also transiently) with a mitochondrial-targeted fluorescence protein plasmid, e.g., with mitochondria-tagged GFP (mtGFP), kept at 37 °C and 5% CO_2 after transfection, in appropriate medium.
2. Glass-based imaging chambers for video microscopy (or glass coverslip). Please note that for transient transfection of adherent cells, transfection has to be performed on cells plated on glass coverslips (suitable for imaging) at least 24 h before imaging.
3. Video recording live imaging confocal laser scanning microscope system with dedicated heating system to control temperature and CO_2 during video acquisition.
4. Medium for imaging (suited to the cells of interest), pre-warmed at 37 °C and 5% CO_2 before acquisition.

2.2 Methods

1. The imaging medium (the choice of the medium depends on the cell type chosen for analysis; or it can also be DPBS pH 7) has to be pre-warmed and CO_2-equilibrated at 37 °C and 5% CO_2 for 30–60 min. Meanwhile, preheat the microscope stage holder at 37 °C and 5% CO_2.
2. Grow adherent cells (stably transfected with mtGFP) in a glass-based imaging chamber up to 70% confluency in appropriate medium for 24 h before performing the assay. For cells in suspension, such as T cells, coat the imaging chamber by incubating it for 1 h at RT with 10 μg/mL fibronectin before adding cells. Then, let the cells adhere to the chamber, for at least 30 min in the incubator. In all cell cases, immediately before starting the microscope acquisition, add pre-warmed medium to the imaging chamber containing them.
3. Transfer cells inside the microscope stage holder, stably set at 37 °C and 5% CO_2. It is very important, at this stage, that any temperature oscillation is avoided. Moreover, it is possible to create a thin mineral oil layer above the imaging medium, in order to prevent medium evaporation.
4. Choose the appropriate laser configuration and start the acquisition avoiding high laser intensity that could bleach the sample during the time-lapse z-stack acquisition, especially for long-time acquisitions. For acquisition and microscope parameter details and suggestions, *see* ref. 29. For this kind of acquisitions, the microscope setup needs to have a good autofocus device (in order to keep the focus during the time lapse) and a sensible z-stack holder enabling precise submicron movements.
5. Usually, a good 3D reconstruction of the mitochondrial network can be obtained from z-stack acquisitions of the whole cell volume by 0.4 μm steps. Once set all the parameters acquire a series of z-stack for the desired time span. Live imaging

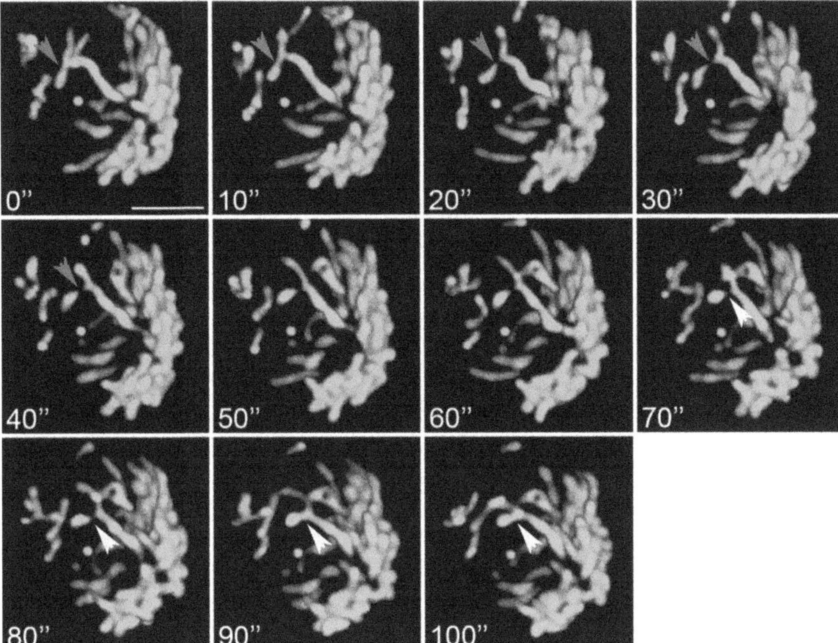

Fig. 1 Time-lapse confocal video microscopy of mtYFP-expressing Jurkat cell. Representative time-lapse images of confocal 3D reconstruction of the mitochondrial network in a single mtYFP-expressing Jurkat cell. Acquisitions have been performed by time-lapse video microscopy, by exciting with a 488 nm laser and performing a 0.4 μm z-stack scanning every 10 s. The red arrowhead indicates the fragmentation of a small mitochondrion from the rest of the network, while the yellow arrowhead indicates its subsequent fusion. Scale bar 10 μm

confocal video recording on cells expressing mitochondrial markers can be performed for up to 24 h. Representative images of the mitochondrial dynamics by time-lapse video confocal recording in resting Jurkat cells transfected with an mtYFP-expressing plasmid are reported in Fig. 1 (*see* also Supplementary Movie 1).

2.3 Notes

1. The use of stable cell lines expressing mitochondria-tagged fluorescent proteins (such as mtGFP or mtYFP) or of primary cells isolated from mice carrying a mitochondria-tagged fluorescent protein [30] is preferable, if possible, when compared to transiently transfected cell lines. This is obviously due to the stressing conditions caused by a transient transfection, which could lead per se to alterations of mitochondrial morphology.
2. Since the resolution of an image is inversely proportional to the wavelength used for acquisition, the lower wavelength of the green laser excitation (488 nm) offers a better image resolution and definition of cell structures compared to the use of a red laser (560 or 630 nm). Thus, given the small size of single mitochondria and the difficulties to visualize fission or fusion

events that normally occur within the range of few micrometers, we recommend the use of green laser-based fluorescent proteins, such as GFP or YFP. This point is of the highest importance when analyzing mitochondria in small cells, such as T lymphocytes. However, it should also be noted that the higher energy transferred by a short-wavelength laser to the samples can damage cells more than a long-wavelength laser, particularly for prolonged exposure times.

3. Some mitochondrial dyes, small molecules able to cross mitochondrial membranes and accumulate inside these organelles, such as JC-1 and TMRE (tetramethylrhodamine ethyl ester), are highly dependent on an intact mitochondrial membrane potential ($\Delta\psi$) to be loaded within the organelles; indeed, they are used as $\Delta\psi$ indicator [31]. On the contrary, others, such as MitoTracker FM dyes, are routinely used to stain whole mitochondria, since their capability to permeate the mitochondrial membranes is much less dependent on $\Delta\psi$ [32] and since some of them are retained inside the organelles also after cell fixation. However, given that all these dyes exploit potential differences across the mitochondrial membranes to specifically localize into these organelles, small alterations in the mitochondrial functionality have the impact to alter their signal distribution. Therefore, we preferentially recommend the use of $\Delta\psi$-independent markers, such as mitochondria-tagged proteins.

4. Alternative to live imaging analysis, an immunofluorescence analysis on fixed cells with antibodies against mitochondrial proteins (such as TOM20) can be used. In this case, we suggest the use of 0.4% Nonidet P-40 to optimally permeabilize mitochondrial membranes and antibodies against OMM-targeted proteins (such as TOM20) to better visualize the mitochondrial network morphology.

5. In order to perform a quantitative analysis of mitochondrial morphology, it is possible to use specific software such as the free-available ImageJ "mitochondrial morphology" macro [33]. Representative images of the 3D reconstruction of the mitochondrial morphology (z-stack confocal microscopy acquisitions) of fixed mtYFP-expressing HeLa and Jurkat cells are reported in Fig. 2a, b. By ImageJ macro, after manually setting a threshold for the mitochondria signal intensity, you can measure average circularity, perimeter, area, and minor axis of mitochondria for each cell (Fig. 3a, b). Fragmented mitochondria will display a higher circularity than resting "normal" cells; by contrast, completely elongated mitochondria will have a higher area/perimeter ratio. Therefore, the index (area/perimeter)/circularity is a bona fide indicator of mitochondria shape. In addition, the index (area/perimeter)/minor axis will

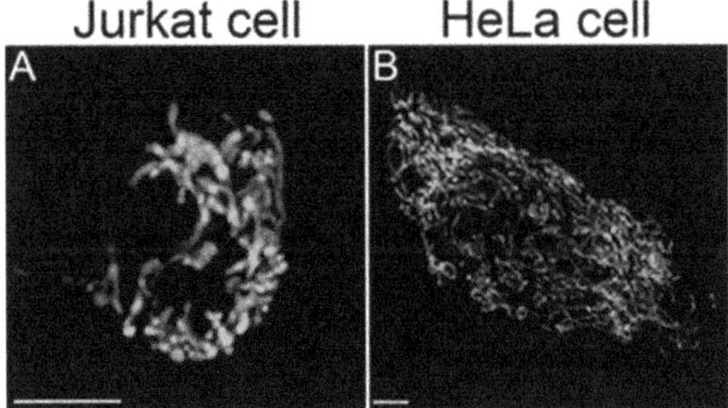

Fig. 2 Representative morphology of the mitochondrial network in Jurkat and HeLa cells. Confocal z-stack acquisition and 3D reconstruction of the mitochondrial network of Jurkat (**a**) and HeLa (**b**) single cells after immunofluorescence with anti-TOM20 antibody (0.4 μm z-stack steps). Scale bar 10 μm

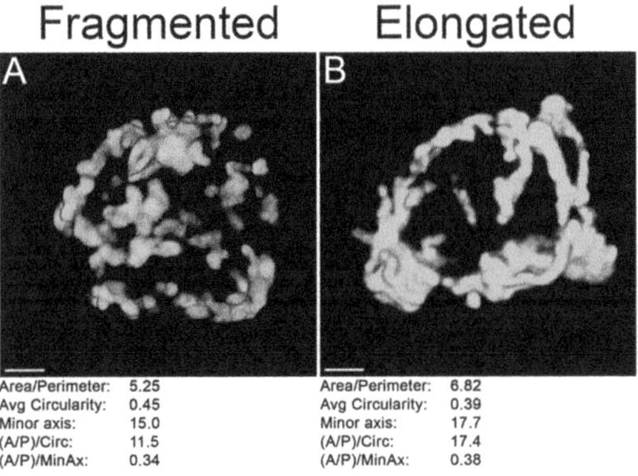

Fig. 3 Images of a representative fragmented or elongated mitochondrial network in Jurkat cells. Confocal 3D reconstruction of the mitochondrial network in Jurkat single cell, after immunofluorescence staining with anti-TOM20 antibody (0.4 μm z-stack steps). (**a**) Cell with fragmented mitochondria. (**b**) Cell with a clear elongated network. Several parameters, measured by applying the ImageJ mitochondrial morphology macro, are reported below the figures (see text for details). Scale bar 2 μm

indicate if the increase in the area/perimeter ratio is the consequence of a real mitochondria elongation or of swollen mitochondria. We recommend to use appropriate controls for each mitochondrial morphology the first time, in order to set and validate all parameters.

6. The morphology of the mitochondrial network in transfected cells might be altered by the overexpressed proteins. Thus, when performing for the first time a similar experiment, transfected cell mitochondria morphology should be compared to the morphology of non-transfected cells, in which the organelles have been labelled by immune fluorescence (IF) against a mitochondrial marker (such as TOM20).

7. Several drugs can be used to specifically modulate the activity of mitochondria-shaping proteins in mammalian cells, thus specifically altering the morphology of the mitochondrial network. One of the most used is the small Drp1 inhibitor mdivi-1 [34]. This drug is able to inhibit Drp1 oligomerization on OMM and prevent its GTP hydrolysis, thus inhibiting mitochondrial fission [18]. Mdivi-1 best concentration is in the range of 10–50 μM. It is preferable to pretreat the cells with this drug 1–2 h before starting the live imaging recording. Since a prolonged exposure to mdivi-1 is toxic [35], the researcher should avoid experiments longer than 6–8 h. Of note, it should be mentioned that very recently mdivi-1 has been suggested as a modulator of the activity of electron transport chain (ETC) complexes and reactive oxygen species (ROS) production in a Drp1-independent way [36, 37]. Two other recently discovered drugs, likely affecting the mitochondrial dynamics, are p110 and hydrazone M1. P110 blocks Drp1 interaction with its receptor Fis1, thus partially preventing Drp1 recruitment to mitochondria [38]. M1 is another inducer of mitochondrial elongation, and its activity strictly requires the profusion protein Opa1 and at least one Mitofusin protein (either Mfn1 or Mfn2) [39]. However, its inhibition mechanism is still not completely understood. Until now, no drugs are known that can promote mitochondrial fission, or inhibit fusion, in mammalian cells.

In addition, the morphology and dynamism of the mitochondrial network can be modulated by altering the cytoskeleton assembly, such as by using nocodazole [40] or downregulating the kinesin motors [41, 42]. The same phenotype can be obtained by pharmacologically impairing the mitochondrial functionality, as in the case of the mitochondrial membrane potential ($\Delta\psi$) dissipation with FCCP [43]. Any drugs can be directly added to the cells inside the imaging chamber during the microscope acquisition, by carefully removing half of the chamber medium and replacing it with new (heated) medium containing twice the desired final concentration of the drug. Obviously, during this step, the imaging chamber must be handled with extreme caution to avoid losing the focus plane.

3 Fluorescence Recovery After Photobleaching (FRAP)

Fluorescence recovery after photobleaching (FRAP) is a method that can be applied during live imaging time-lapse acquisitions to investigate the dynamical movements of cellular structures over time. After photobleaching fluorescent molecules in an area of interest (referred to as "bleached area"), the researcher follows, and acquires, the fluorescence recovery in the bleached area over time. This recovery is due to movements of fluorescent proteins or organelles from the surrounding non-bleached areas within the bleached one. A simple diffusion of mitochondria-tagged fluorescent proteins moving along the mitochondria from regions just adjacent to the photobleached one indicates that the organelles are fused, corresponds to a good fluorescence recovery, and occurs within few minutes. The fluorescent recovery due to mitochondrial dynamics indicates active fusion machinery and processes and can take up to 1 h [29]. On the contrary, a fragmented (or fragmenting) mitochondrial network will not be able to recover the fluorescence signal, even in hours.

3.1 Materials

The material necessary for this kind of experiments, performed in cells (preferentially stably) transfected with mtRFP or mtGFP plasmids, is the same of Subheading 1.

3.2 Methods

1. Repeat **steps 1–4** from Methods section 2.2 (Visualization of the mitochondrial network by z-stack live imaging acquisition, by using mitochondria-tagged fluorescent proteins).
2. Choose a suitable cell with correct shape and check for a mitochondrial morphology typical of a resting cell by fast-scanning acquisition. Select a region of interest (ROI) inside the cell where to perform the FRAP. Within the ROI, aim at including 30–40% of the total mitochondrial network, spanning 5–20% of the total cell area to appreciate the fluorescence recovery from the unbleached zones (60–70% of mitochondria), in an appropriate time span. In addition, choose a ROI possibly in the center of the cell, where mitochondria are uniformly distributed and they can reach the bleached zone from the surrounding unbleached ones all around.
3. Perform photobleaching by using the appropriate laser (depending on the mitochondria-tagged fluorescent protein of choice, with the GFP being highly indicated) at its maximum intensity. It is very important to have a complete bleaching of the ROI. This can be achieved by performing multiple laser irradiation (always at maximum intensity), in a short time window (but prolonged laser irradiation can damage mitochondria).
4. Set the total imaging time and the time interval between each acquisition. These parameters should be optimized and strongly depend on cell type. As starting condition, it is possible to acquire up to 1 h after photobleaching with a time interval of 0.5–60 s between each acquisition. Of course, the longer is the recovery acquisition time, the higher has to be the frequency of the acquisitions (also up to 5–10 min). Qualitative visualization of fluorescence recovery can be performed by

direct measuring the total and mean fluorescent signal in the ROI, starting time 0 after photobleaching onward. For a more quantitative analysis of the fluorescence recovery in the ROI, we recommend to refer to reference [44].

3.3 Notes

1. For FRAP experiments, it is important to choose a bright and photo-stable fluorescent dye under excitation with a low laser power, but capable to be as well strongly and irreversibly bleached under excitation with an intense one.
2. *See* also **Notes 1**, **2**, **6**, and 7 of section 2.3.
3. If using the mtRFP as a fluorescent marker, it might be necessary to use all available lasers together with the RFP-specific 543 nm to achieve a complete photobleaching of the signal.
4. It is recommended to start the acquisition, for a short time, before starting the photobleaching, in order to have a control of the original signal. The FRAP effect will be also clearer, in this way. Nowadays, the most common setup software have FRAP-dedicated preset applications that greatly facilitate the researcher.

4 Fusion Assay with Photoactivatable Fluorescent Proteins

Another assay to visualize and investigate mitochondrial fusion events in living cells, by confocal time-lapse video microscopy, is based on the use of mitochondrial matrix-tagged photoactivatable fluorescent GFP (mito-PAGFP) probes [45]. This engineered version of the classical fluorescent protein GFP emits poorly when excited with a 488 nm green laser, but its emission with the same laser can be "switched on" (100-fold increase) by a previous photoactivation with the 400 nm blue laser. Similar to the FRAP, this technique allows visualization of the movements of the photoactivated molecules, and thus of mitochondria fusion events, in real time by video microscopy. In mito-PAGFP-expressing cells, a brief excitation with the 400 nm laser allows the "photoactivation" of this marker only in the region of interest (ROI). Then, by acquiring with the 488 nm laser, it is possible to follow in real time the redistribution in the mitochondrial network of photoactivated GFP, diffusing from the activated ROI toward the adjacent non-activated mitochondria regions, this due to organelle fusion. Besides a direct visualization of fusion events, this assay also allows a quantification of the fusion process rate.

4.1 Materials

The material necessary for this kind of experiments, performed in cells (preferentially stably) transfected with mitochondrial matrix-tagged photoactivatable GFP (mito-PAGFP), is the same of Subheading 1.

4.2 Methods

1. Repeat **steps 1–3** from Methods section 2.2.
2. Adjust the power of the green laser in order to obtain a low and barely detectable signal from the "non-photoactivated" GFP that depicts the entire mitochondrial network. This will allow visualization of all mitochondria also after photobleaching.
3. Select a region of interest (ROI) to photoactivate, including several mitochondria (visualized as described in **step 2**). The ROI should be relatively small (around 5 μm in diameter), when compared to the whole network, and should be located in the perinuclear area with high density of mitochondria, thus allowing fusion events in all directions.
4. Capture a few pre-activation images with the 488 nm blue laser, and perform photoactivation in the ROI with a brief 400 nm laser pulse. The duration and the number of pulses must be optimized according to the cell type, in order to obtain an efficient result (a strong 488-excited GFP signal). Also in this case, long and iterated 400 nm laser pulses can lead to mitochondria damage and consequent mitochondrial fission, thus preventing fusion events.
5. Acquire different images with 488 nm blue laser at different time points after photobleaching (one image has to be taken immediately after photoactivation). "Photoactivated" GFP will rapidly move in a few minutes from ROI toward adjacent non-activated areas by simple diffusion inside the mitochondrial matrix of interconnected mitochondria (this leads also to a drop in GFP signal). Further redistribution of the GFP signal to non-activated and non-interconnected mitochondria upon the next 15–45 min will be the outcome of new fusion events among the organelles. Therefore, a time-lapse video recording, with image acquisitions every 5–10 min (or every min, if single fusion events must be studied), for up to 1 h should be sufficient for satisfactory results.
6. Effective mitochondrial fusion can be inferred by measuring with analysis software the redistribution rate of mito-PAGFP from photoactivated mitochondria to non-activated areas. One should quantify the pixel intensities in both activated and non-activated areas of the cell. The values (normalized to the pre-activation intensity, set as 1) can be plotted as a function of time. A tenfold post-activation increase in the ROI signal should be reached, in order to clearly observe an increase in pixel intensities also in non-activated areas (just to minimize the "noise" of nonspecific fluorescence).

4.3 Notes

1. For obvious reasons, it would be extremely difficult to recognize and distinguish the mito-PAGFP transfected cells; thus, a double transfection with an RFP plasmid allows visualizing

transfected cells. We recommend a 3:1 ratio of mito-PAGFP: RFP plasmids to be sure that RFP-transfected cells also contain mito-PAGFP. It would be more appropriate co-transfecting the cells with a mitochondria-tagged RFP (mtRFP) construct; this will allow to obtain and directly compare the morphology of the whole mitochondrial network with the "photoactivated" GFP diffusion from the ROI. This will also allow "correcting" the PAGFP signal (constantly performing a PAGFP:RFP ratio) for possible fusion-unrelated events, such as the loss of focus of the image or the photobleaching verifiable over time.

2. *See* also **Notes 6** and 7 of section 2.3.

5 PEG-Based Mitochondria Fusion Assay

Although in vivo imaging of mitochondrial dynamism allows a direct visualization of fusion events, this method does not distinguish between a complete organelle fusion, with exchange of mitochondrial matrix material, and the formation of simple mitochondrial junctions or partial fusion of OMM, without involvement of the IMM. In order to address this point, years ago it has been developed an assay based on the capability of heterokaryon cells (derived from mitochondrial matrix-tagged GFP- or RFP-overexpressing cells) to fuse their mitochondria, thus combining green and red fluorescence in a yellow signal related to fused mitochondria. This signal can be detected by either live imaging or after cell fixation [28]. The use of mitochondrial matrix-targeted fluorescent proteins ensures that fusion events observed involve both the OMM and the IMM, thus indicating that the whole mitochondrial content (i.e., including mtDNA) has been mixed.

5.1 Materials

Cells stably transfected with mtRFP or mtGFP plasmids, kept at 37 °C and 5% CO_2 in appropriate medium (with 10% fetal bovine or calf serum).

Reagents required	Cycloheximide (CHX)
	PEG (polyethylene glycol) 1500
	Fibronectin-coated glass coverslips
	MEM (Minimum Essential Medium, without serum)

5.2 Methods

1. Co-culture mtGFP- and mtRFP-expressing cells at 1:1 ratio on glass coverslips for 16–40 h before performing the assay.
2. Transfer the glass coverslips in a 35 mm culture dish, and wait until you reach confluency up to 70–100%.
3. Wash the cells with MEM without serum and add 50% w/v PEG 1500 for 45–60 s.

4. Wash extensively with MEM, 10% serum to finally leave the cells rest in pre-warmed culture medium. Heterokaryon cell formation should be observed in 3–4 h.
5. Visualize the results by live imaging microscopy, or fix the cells with 4% paraformaldehyde 20 min at RT for proceeding with immunostaining. Fused mitochondria generated from mtGFP- and mtRFP-derived organelles become yellow over time.

5.3 Notes

1. Stably mtGFP-transfected cells can be preventively incubated with 1 μM trichostatin-A for 24–48 h to increase GFP expression. In this case, it is recommended to remove trichostatin-A at least 12–24 h before performing PEG-mediated cell fusion, to avoid trichostatin-A undesired effects, such as cell cycle arrest and/or apoptosis.
2. In order to inhibit in the transfected cells the protein translation after PEG-mediated cell fusion, it is recommended to use CHX. This prevents de novo synthesis of the constructs after heterokaryon formation, which could induce formation of complementation proteins that do not depend on mitochondrial fusion. 50–100 μg/mL CHX can be added to the cells 30 min before performing the assay and kept in the medium thereafter.
3. The effect of drugs affecting the activity of the mitochondria-shaping proteins, or the mitochondrial functionality (*see* **Note** 7 of section 2.3), can be tested by adding them to the medium after the PEG treatment, thus avoiding interference with cell fusion processes.

6 Renilla Luciferase Complementation Assay

The PEG-based mitochondria fusion assay allows efficient detection of fusion events in vivo in single cells. More recently, it has been developed a novel assay based on the complementation of two inactive fragments of a luciferase protein, which is functional only upon fragment association. This alternative assay allows detecting fusion events on a cell population scale [46]. Briefly, the cells can be transfected exclusively with one of the two following plasmids. The first plasmid contains the N-terminal of Venus-YFP fluorescent protein, a leucine zipper helix, and the N-terminal of a FLAG- or HA-tagged *Renilla* luciferase protein. The second plasmid contains the C-terminal of Venus-YFP fluorescent protein, a leucine zipper helix, and the C-terminal of a FLAG- or HA-tagged *Renilla* luciferase protein. These constructs are, respectively, termed N-MitoVZL and C-MitoLZV. Then, N-MitoVZL cells and C-MitoLZV cells are co-cultured and treated with PEG to promote cell fusion. In the heterokaryon cells derived, the mitochondria fuse together and N-MitoVZL and C-MitoLZV can fuse together, thanks to the interaction between the leucine zipper helixes. This leads to complementation of N- and C-terminal of both Venus-YFP and *Renilla* luciferase proteins. While reconstituted fluorescent YFP allows visualization of fusion events, the activity of the reconstituted *Renilla* luciferase allows population-based analysis of fusion events based on light photon emission in a plate reader.

6.1 Materials

Cells transfected, preferentially stably but also transiently, with N-MitoVZL or C-MitoLZV plasmids, kept at 37 °C and 5% CO_2 in appropriate medium.

Luminometer

Reagents required	Cycloheximide (CHX) PEG (polyethylene glycol) 1500 5 mM EDTA (ethylenediaminetetraacetic acid) Dulbecco's phosphate-buffered saline (DPBS) Renilla luciferase assay system

6.2 Methods

1. Co-culture 4×10^5 N-MitoVZL- and C-MitoLZV-expressing cells at 1:1 ratio in 12-well plate for 18 h before performing the assay. As in the point 1 of the section 5.2, it is possible also in this case, and in the same way, to check for the correct localization of the recombinant constructs.
2. Add 50% PEG for 1 min, and immediately wash four times with complete medium (with 10% serum).
3. Culture the cells for up to 16 h, and then collect them in DPBS, 5 mM EDTA.
4. Proceed with the cell lysis and the preparation of the luciferase assay detection as indicated in any commercial kits for this assay, and measure the luminescence signal by luminometer. Read luminescence over a period of 10 s. To set the negative background luminescence signal (RLU = relative light unit), read the luminescence recovered from N-MitoVZL- and C-MitoLZV-expressing co-cultured cells *without* PEG 1500 addition. Fusion events can be normally observed starting from 30 min after PEG addition and peaking at 2–3 h (optimization is required in agreement with the cell type used). Alternatively, analyze in vivo mitochondrial fusion events by visualizing reconstituted Venus-YFP (see the protocol described in Subheading 1).

6.3 Notes

1. See the procedure indicated in the **Note 2** of section 5.3 ("PEG-based mitochondria fusion assay"), in order to inhibit by using CHX, also in this case, protein translation after PEG-mediated cell fusion.
2. The effect of drugs affecting the activity of mitochondria-shaping proteins, or mitochondrial functionality (*see* **Note 4**

of section 2.3), can be tested by adding them to the medium after the PEG treatment, thus avoiding interference with cell fusion processes.

7 Evaluating by Western Blot the Levels of the Mitochondria-Shaping Proteins and their Activity

Different mitochondria-shaping proteins tightly regulate the dynamism of the mitochondrial network. The Western blot analysis of the expression levels of these mitochondria-shaping proteins is a valuable tool, in combination with direct imaging of mitochondrial network, to investigate the potential dynamic capacity of these organelles. When available, antibodies against specific posttranslational modifications in these proteins (such as Drp1 phosphorylation sites) may help in assessing their activity. While the general and well-known Western blot technique can be easily performed to visualize most of these shaping proteins and their modifications, a specific protocol has to be followed to discriminate the different Opa1 isoforms by Western blot. In this section, we will describe this specific protocol.

7.1 Materials

RIPA buffer	(50 mM Tris–HCl pH 8, 150 mM NaCl, 1% Nonidet-P40, 0.5% sodium deoxycholate, 0.1% SDS, stored in dark at 4 °C).
Tris-acetate buffer	(2.5 mM Tricine, 2.5 mM Tris base, 0.005% SDS, pH 8.2)

7.2 Methods

1. Collect at least one million confluent adherent cells, or four to five million for smaller cells, such as lymphocytes.
2. Prepare the samples as for a common Western blot assay. We just recommend the use of RIPA buffer to lysate the cells, since it strongly solubilizes membrane-bound proteins, so achieving an optimal isolation of mitochondrial membrane-bound proteins. Run samples in 3–8% acrylamide gel in Tris-acetate buffer, which allows a better separation of the Opa-1 isoforms.
3. Follow the canonical Western blot procedure for developing the blot and visualizing the Opa1 bands [47], which should appear as reported in Fig. 4. For a better interpretation of the results, consider that the expression of eight Opa-1 splice variants and their proteolytic processing leads to the formation of two different Opa-1 long isoforms with a transmembrane (TM) domain inserted in the IMM (called L1 and L2). Two different cleavage sites (S1 and S2) are present in these long isoforms, which can be differentially cleaved by specific proteases, such as Yme1l and Oma1, thus generating three soluble

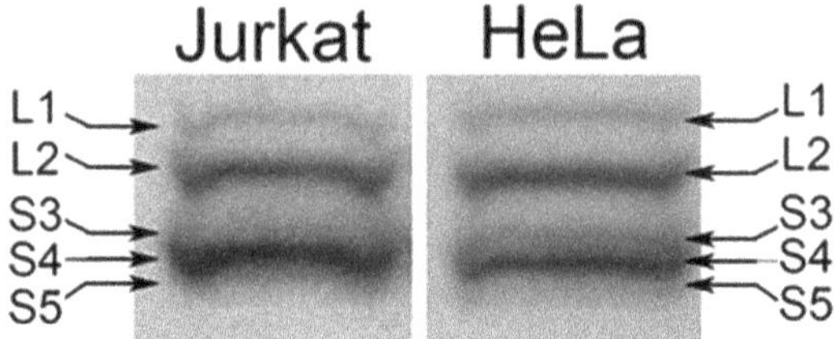

Fig. 4 Western blot analysis of the Opa-1 isoforms in Jurkat and HeLa cells. Opa-1 isoform (80–100 kDa) expression levels, in Jurkat and HeLa cells, are detected by Western blot analysis. Arrows indicate the five different isoforms: Opa-1-long-1 (L1), Opa-1-long-2 (L2), Opa-1-short-3 (S3), Opa-1-short-4 (S4), and Opa-1-short-5 (S5)

Opa-1 short isoforms resident in the IMS (called S3, S4, and S5) [14].

7.3 Note

1. Under physiological conditions, Opa-1 monomers interact among them giving rise to high molecular weight oligomers (up to 250 kDa), which are required to control *cristae* width [15, 48]. These oligomers can be isolated by cross-linking sulfhydryl groups before performing cell lysis and Western blot. If interested on Opa-1 oligomers, thus, instead of its isoforms, one should then pretreat cells with 1 mM BMH (bismaleimidohexane) for 20 min at 37 °C to cross-link proteins [48] and, then, wash twice with 1% β-mercaptoethanol (in DPBS) to quench residual BME. Subsequent protein extraction can be performed by using BME-supplemented RIPA buffer. Opa-1 oligomers can be extracted also from isolated mitochondria [49].

Acknowledgments

This work was funded by the Italian Ministry of Health (GR-2011-02351643) and IG-19826 by AIRC to SC and by grants from Fondazione Roma and FISM.

Movie Caption

Supplementary Movie 1. Time-lapse confocal video microscopy of a mtYFP-expressing Jurkat cell.

Confocal 3D reconstruction of the mitochondrial network in a mtYFP-expressing Jurkat cell over time. Acquisitions have been performed with a 488 nm laser, by 0.4 μm z-stack steps, each 10 s (MP4 269 kb)

References

1. Detmer SA, Chan DC (2007) Functions and dysfunctions of mitochondrial dynamics. Nat Rev Mol Cell Biol 8(11):870–879
2. Cogliati S, Frezza C, Soriano ME, Varanita T, Quintana-Cabrera R, Corrado M, Cipolat S, Costa V, Casarin A, Gomes LC, Perales-Clemente E, Salviati L, Fernandez-Silva P, Enriquez JA, Scorrano L (2013) Mitochondrial cristae shape determines respiratory chain supercomplexes assembly and respiratory efficiency. Cell 155(1):160–171
3. Prieto J, León M, Ponsoda X, Sendra R, Bort R, Ferrer-Lorente R, Raya A, López-García C, Torres J (2016) Early ERK1/2 activation promotes DRP1-dependent mitochondrial fission necessary for cell reprogramming. Nat Commun 7:11124
4. Twig G, Elorza A, Molina AJ, Mohamed H, Wikstrom JD, Walzer G, Stiles L, Haigh SE, Katz S, Las G, Alroy J, Wu M, Py BF, Yuan J, Deeney JT, Corkey BE, Shirihai OS (2008) Fission and selective fusion govern mitochondrial segregation and elimination by autophagy. EMBO J 27(2):433–446
5. Karbowski M, Lee YJ, Gaume B, Jeong SY, Frank S, Nechushtan A, Santel A, Fuller M, Smith CL, Youle RJ (2002) Spatial and temporal association of Bax with mitochondrial fission sites, Drp1, and Mfn2 during apoptosis. J Cell Biol 159(6):931–938
6. Campello S, Lacalle RA, Bettella M, Mañes S, Scorrano L, Viola A (2006) Orchestration of lymphocyte chemotaxis by mitochondrial dynamics. J Exp Med 203(13):2879–2886
7. Zhao J, Zhang J, Yu M, Xie Y, Huang Y, Wolff DW, Abel PW, Tu Y (2013) Mitochondrial dynamics regulates migration and invasion of breast cancer cells. Oncogene 32 (40):4814–4824
8. Qian W, Choi S, Gibson GA, Watkins SC, Bakkenist CJ, Van Houten B (2012) Mitochondrial hyperfusion induced by loss of the fission protein Drp1 causes ATM-dependent G2/M arrest and aneuploidy through DNA replication stress. J Cell Sci 125(Pt 23):5745–5757
9. Chen H, Chomyn A, Chan DC (2005) Disruption of fusion results in mitochondrial heterogeneity and dysfunction. J Biol Chem 280 (28):26185–26192
10. Eura Y, Ishihara N, Yokota S, Mihara K (2003) Two mitofusin proteins, mammalian homologues of FZO, with distinct functions are both required for mitochondrial fusion. J Biochem 134(3):333–344
11. Gegg ME, Cooper JM, Chau KY, Rojo M, Schapira AH, Taanman JW (2010) Mitofusin 1 and mitofusin 2 are ubiquitinated in a PINK1/parkin-dependent manner upon induction of mitophagy. Hum Mol Genet 19 (24):4861–4870
12. Yue W, Chen Z, Liu H, Yan C, Chen M, Feng D, Wu H, Du L, Wang Y, Liu J, Huang X, Xia L, Liu L, Wang X, Jin H, Wang J, Song Z, Hao X, Chen Q (2014) A small natural molecule promotes mitochondrial fusion through inhibition of the deubiquitinase USP30. Cell Res 24(4):482–496
13. Cipolat S, Martins de Brito O, Dal Zilio B, Scorrano L (2004) OPA1 requires mitofusin 1 to promote mitochondrial fusion. Proc Natl Acad Sci U S A 101(45):15927–15932
14. Anand R, Wai T, Baker MJ, Kladt N, Schauss AC, Rugarli E, Langer T (2014) The i-AAA protease YME1L and OMA1 cleave OPA1 to balance mitochondrial fusion and fission. J Cell Biol 204(6):919–929
15. Frezza C, Cipolat S, Martins de Brito O, Micaroni M, Beznoussenko GV, Rudka T, Bartoli D, Polishuck RS, Danial NN, De Strooper B, Scorrano L (2006) OPA1 controls apoptotic cristae remodeling independently from mitochondrial fusion. Cell 126 (1):177–189
16. Gal A, Balicza P, Weaver D, Naghdi S, Joseph SK, Várnai P, Gyuris T, Horváth A, Nagy L, Seifert EL, Molnar MJ, Hajnóczky G (2017) MSTO1 is a cytoplasmic pro-mitochondrial fusion protein, whose mutation induces myopathy and ataxia in humans. EMBO Mol Med 9 (7):967–984
17. Otera H, Ishihara N, Mihara K (2013) New insights into the function and regulation of mitochondrial fission. Biochim Biophys Acta 1833(5):1256–1268
18. Smirnova E, Griparic L, Shurland DL, van der Bliek AM (2001) Dynamin-related protein Drp1 is required for mitochondrial division in mammalian cells. Mol Biol Cell 12 (8):2245–2256
19. James DI, Parone PA, Mattenberger Y, Martinou JC (2003) hFis1, a novel component of the mammalian mitochondrial fission machinery. J Biol Chem 278(38):36373–36379
20. Otera H, Wang C, Cleland MM, Setoguchi K, Yokota S, Youle RJ, Mihara K (2010) Mff is an essential factor for mitochondrial recruitment of Drp1 during mitochondrial fission in mammalian cells. J Cell Biol 191(6):1141–1158

21. Palmer CS, Osellame LD, Laine D, Koutsopoulos OS, Frazier AE, Ryan MT (2011) MiD49 and MiD51, new components of the mitochondrial fission machinery. EMBO Rep 12(6):565–573
22. Chou CH, Lin CC, Yang MC, Wei CC, Liao HD, Lin RC, Tu WY, Kao TC, Hsu CM, Cheng JT, Chou AK, Lee CI, Loh JK, Howng SL, Hong YR (2012) GSK3beta-mediated Drp1 phosphorylation induced elongated mitochondrial morphology against oxidative stress. PLoS One 7(11):e49112
23. Zunino R, Schauss A, Rippstein P, Andrade-Navarro M, McBride HM (2007) The SUMO protease SENP5 is required to maintain mitochondrial morphology and function. J Cell Sci 120(Pt 7):1178–1188
24. Karbowski M, Neutzner A, Youle RJ (2007) The mitochondrial E3 ubiquitin ligase MARCH5 is required for Drp1 dependent mitochondrial division. J Cell Biol 178 (1):71–84
25. Wang X, Yan MH, Fujioka H, Liu J, Wilson-Delfosse A, Chen SG, Perry G, Casadesus G, Zhu X (2012) LRRK2 regulates mitochondrial dynamics and function through direct interaction with DLP1. Hum Mol Genet 21 (9):1931–1944
26. Lee JE, Westrate LM, Wu H, Page C, Voeltz GK (2016) Multiple dynamin family members collaborate to drive mitochondrial division. Nature 540(7631):139–143
27. Rizzuto R, Brini M, De Giorgi F, Rossi R, Heim R, Tsien RY, Pozzan T (1996) Double labelling of subcellular structures with organelle-targeted GFP mutants in vivo. Curr Biol 6(2):183–188
28. Legros F, Lombès A, Frachon P, Rojo M (2002) Mitochondrial fusion in human cells is efficient, requires the inner membrane potential, and is mediated by mitofusins. Mol Biol Cell 13(12):4343–4354
29. Mitra K, Lippincott-Schwartz J (2010) Analysis of mitochondrial dynamics and functions using imaging approaches. Curr Protoc Cell Biol Chapter 4 Unit 4.25.1–21
30. Sterky FH, Lee S, Wibom R, Olson L, Larsson NG (2011) Impaired mitochondrial transport and Parkin-independent degeneration of respiratory chain-deficient dopamine neurons in vivo. Proc Natl Acad Sci U S A 108 (31):12937–12942
31. Scaduto RC, Grotyohann LW (1999) Measurement of mitochondrial membrane potential using fluorescent rhodamine derivatives. Biophys J 76(1 Pt 1):469–477
32. Poot M, Zhang YZ, Krämer JA, Wells KS, Jones LJ, Hanzel DK, Lugade AG, Singer VL, Haugland RP (1996) Analysis of mitochondrial morphology and function with novel fixable fluorescent stains. J Histochem Cytochem 44 (12):1363–1372
33. Dagda RK, Cherra SJ, Kulich SM, Tandon A, Park D, Chu CT (2009) Loss of PINK1 function promotes mitophagy through effects on oxidative stress and mitochondrial fission. J Biol Chem 284(20):13843–13855
34. Cassidy-Stone A, Chipuk JE, Ingerman E, Song C, Yoo C, Kuwana T, Kurth MJ, Shaw JT, Hinshaw JE, Green DR, Nunnari J (2008) Chemical inhibition of the mitochondrial division dynamin reveals its role in Bax/Bak-dependent mitochondrial outer membrane permeabilization. Dev Cell 14(2):193–204
35. Rosdah AA, Holien JK, Delbridge LM, Dusting GJ, Lim SY (2016) Mitochondrial fission - a drug target for cytoprotection or cytodestruction? Pharmacol Res Perspect 4(3):e00235
36. Kushnareva Y, Andreyev AY, Kuwana T, Newmeyer DD (2012) Bax activation initiates the assembly of a multimeric catalyst that facilitates Bax pore formation in mitochondrial outer membranes. PLoS Biol 10(9):e1001394
37. Bordt EA, Clerc P, Roelofs BA, Saladino AJ, Tretter L, Adam-Vizi V, Cherok E, Khalil A, Yadava N, Ge SX, Francis TC, Kennedy NW, Picton LK, Kumar T, Uppuluri S, Miller AM, Itoh K, Karbowski M, Sesaki H, Hill RB, Polster BM (2017) The putative Drp1 inhibitor mdivi-1 is a reversible mitochondrial complex I inhibitor that modulates reactive oxygen species. Dev Cell 40(6):583–594.e6
38. Qi X, Qvit N, Su YC, Mochly-Rosen D (2013) A novel Drp1 inhibitor diminishes aberrant mitochondrial fission and neurotoxicity. J Cell Sci 126(Pt 3):789–802
39. Wang D, Wang J, Bonamy GM, Meeusen S, Brusch RG, Turk C, Yang P, Schultz PG (2012) A small molecule promotes mitochondrial fusion in mammalian cells. Angew Chem Int Ed Engl 51(37):9302–9305
40. Woods LC, Berbusse GW, Naylor K (2016) Microtubules are essential for mitochondrial dynamics-fission, fusion, and motility-in Dictyostelium discoideum. Front Cell Dev Biol 4:19
41. Tanaka Y, Kanai Y, Okada Y, Nonaka S, Takeda S, Harada A, Hirokawa N (1998) Targeted disruption of mouse conventional kinesin heavy chain, kif5B, results in abnormal perinuclear clustering of mitochondria. Cell 93(7):1147–1158

42. da Silva AF, Mariotti FR, Máximo V, Campello S (2014) Mitochondria dynamism: of shape, transport and cell migration. Cell Mol Life Sci 71(12):2313–2324
43. Cereghetti GM, Costa V, Scorrano L (2010) Inhibition of Drp1-dependent mitochondrial fragmentation and apoptosis by a polypeptide antagonist of calcineurin. Cell Death Differ 17 (11):1785–1794
44. Goodwin JS, Kenworthy AK (2005) Photobleaching approaches to investigate diffusional mobility and trafficking of Ras in living cells. Methods 37(2):154–164
45. Karbowski M, Cleland MM, Roelofs BA (2014) Photoactivatable green fluorescent protein-based visualization and quantification of mitochondrial fusion and mitochondrial network complexity in living cells. Methods Enzymol 547:57–73
46. Huang H, Choi SY, Frohman MA (2010) A quantitative assay for mitochondrial fusion using Renilla luciferase complementation. Mitochondrion 10(5):559–566
47. Mahmood T, Yang PC (2012) Western blot: technique, theory, and trouble shooting. N Am J Med Sci 4(9):429–434
48. Patten DA, Wong J, Khacho M, Soubannier V, Mailloux RJ, Pilon-Larose K, MacLaurin JG, Park DS, McBride HM, Trinkle-Mulcahy L, Harper ME, Germain M, Slack RS (2014) OPA1-dependent cristae modulation is essential for cellular adaptation to metabolic demand. EMBO J 33(22):2676–2691
49. Wei MC, Lindsten T, Mootha VK, Weiler S, Gross A, Ashiya M, Thompson CB, Korsmeyer SJ (2000) tBID, a membrane-targeted death ligand, oligomerizes BAK to release cytochrome c. Genes Dev 14(16):2060–2071

Chapter 16

Plate-Based Measurement of Superoxide and Hydrogen Peroxide Production by Isolated Mitochondria

Hoi-Shan Wong, Pierre-Axel Monternier, Adam L. Orr, and Martin D. Brand

Abstract

Superoxide and hydrogen peroxide produced by mitochondria play important roles in various physiological and pathological processes. This chapter describes a plate-based method to measure rates of superoxide and/or hydrogen peroxide production at specific sites in isolated mitochondria.

Key words Mitochondria, Superoxide, Hydrogen peroxide, Amplex UltraRed

1 Introduction

Mitochondrial superoxide ($O_2^{\cdot -}$) and hydrogen peroxide (H_2O_2) are proposed to be involved in various physiological and pathological processes [1–6]. To date 11 sites that produce $O_2^{\cdot -}$ and/or H_2O_2 have been identified in mitochondria isolated from rat skeletal muscle [7–14]. It is hypothesized that physiological and pathological processes are differentially influenced by the origin and magnitude of mitochondrial $O_2^{\cdot -}/H_2O_2$ production [15, 16]. Reliable and robust measurement of $O_2^{\cdot -}/H_2O_2$ production from different mitochondrial sites is now possible. In this chapter we describe in detail a 96-well microplate-based method that provides a high-throughput assay to measure site-specific $O_2^{\cdot -}/H_2O_2$ production by isolated mitochondria. It measures the rates of $O_2^{\cdot -}/H_2O_2$ production by (1) the quinone-binding site of mitochondrial complex I during reverse electron transport (site I_Q) [8], (2) the flavin site of mitochondrial complex II (site II_F) [9], (3) the outer quinone-binding site of mitochondrial complex III (site III_{Qo}) [10], (4) the flavin site of complex I plus upstream dehydrogenases (site I_F plus DH) [11], (5) the flavin site of the pyruvate dehydrogenase complex (site P_F) [11], and (6) the quinone-binding site of mitochondrial glycerol phosphate dehydrogenase (site G_Q) [12]. All assays measure the maximum rate from the designated

Carlos M. Palmeira and António J. Moreno (eds.), *Mitochondrial Bioenergetics: Methods and Protocols*, Methods in Molecular Biology, vol. 1782, https://doi.org/10.1007/978-1-4939-7831-1_16,

site, except for the site III_{Qo} assay, which measures the submaximal but less condition-dependent rate at high QH_2/Q ratio [10]. Compared to serial cuvette-based assays, this plate-based assay has much higher throughput, making it suitable for screening compound libraries for their effects on mitochondrial H_2O_2 production. It allows parallel comparisons of rates from different sites under different assay conditions or using mitochondria from different sources, such as comparing a disease model to a control. Although we routinely use it with mitochondria isolated from rat skeletal muscle, it is readily adaptable to mitochondria from different tissues or species.

2 Principles

The detection of $O_2^{\cdot -}/H_2O_2$ uses the fluorometric probe Amplex® UltraRed, an improved derivative of Amplex® Red [17]. Amplex UltraRed is a fluorogenic substrate for horseradish peroxidase (HRP), which catalyzes the reaction of Amplex UltraRed with H_2O_2 in a 1:1 stoichiometric ratio with high specificity to produce a resorufin product, Amplex® UltroxRed, a brightly fluorescent and strongly absorbing reaction product (excitation 568 nm/emission 581 nm). The long wavelength spectra of Amplex UltroxRed allow it to be used to detect H_2O_2 with little interference from the blue and green autofluorescence found commonly in biological samples. The assay's specificity for H_2O_2 is provided by the active site of the horseradish peroxidase, avoiding most of the potential specificity artifacts associated with purely chemical indicators such as rhodamine derivatives, dihydroethidium, or MitoSOX. Amplex UltroxRed has a higher fluorescence and a lower pK_a value than the reaction products for similar fluorogenic substrates such as Amplex Red, giving Amplex UltraRed utility across a broader physiological pH range.

In each of the following assays, H_2O_2 generated during mitochondrial substrate oxidation is released into the assay medium, where excess exogenous horseradish peroxidase catalyzes its reaction with Amplex UltraRed to form Amplex UltroxRed. The production of $O_2^{\cdot -}$ is also captured by the addition of excess exogenous superoxide dismutase (SOD) to convert $O_2^{\cdot -}$ to H_2O_2 in the assay medium. Omission of exogenous SOD can be used to assess whether $O_2^{\cdot -}$ production to the extramitochondrial space contributes a large proportion of the total signal [18]. Note that $O_2^{\cdot -}$ produced in the mitochondrial matrix is transformed by endogenous superoxide dismutase to H_2O_2, which then diffuses across the mitochondrial inner membrane into the assay medium [18].

The focus of this chapter is to provide a protocol to assess the rates of $O_2^{\cdot -}/H_2O_2$ production by several different sites in isolated mitochondria. For more insight into the mechanisms of $O_2^{\cdot -}/H_2O_2$ production from these sites, *see* [7].

3 Materials

Equipment:

Microplate reader	Equipped with fluorescence detection module (top reading). We use a BMG Labtech microplate reader, model PHERAstar FS
Humidified incubator	37 °C
Water bath	37 °C
Assay plates	96-Well microplates (black; clear or solid bottom)
Multichannel pipette	

Reagents:

Base medium (KHE)	120 mM KCl, 5 mM HEPES, 1 mM EGTA, 1 μg/mL oligomycin, adjusted to pH 7.4 at 37 °C using KOH
Assay medium (KHEB)	KHE plus 0.3% w/v bovine serum albumin (BSA), adjusted to pH 7.4 at 37 °C using KOH (*see* **Note 1**)
Standard	H_2O_2 solution (30% w/w, equivalent to 8 M) (*see* **Note 2**)
Stock solutions	Amplex® UltraRed (10 mM in DMSO) (*see* **Notes 3** and **4**)
	HRP (1000 U/mL in KHE) (*see* **Note 5**)
	SOD (5000 U/mL in KHE) (*see* **Note 5**)
	Substrates (Table 1):
	L-Succinate, L-malate, L-glutamate, pyruvate (free acids, dissolved in KHE, adjusted to pH 7.4 at 37 °C using KOH), L-carnitine, and *sn*-glycerol-3-phosphate (dissolved in KHE) (*see* **Note 6**)
	Inhibitors (Table 2):
	Oligomycin, rotenone, antimycin A, myxothiazol (dissolved in 96% v/v ethanol), and L-malonate (free acid, dissolved in KHE, adjusted to pH 7.4 at 37 °C using KOH) (*see* **Note 6**)

Table 1
Stock and working concentrations of substrates used in the measurement of rates of $O_2^{\cdot -}/H_2O_2$ production by specific sites in isolated mitochondria

Substrate[a]	Stock concentration (mM)	Working concentration (mM)
L-Succinate	1000	5 (or 0.5)
L-Malate	1000	5
L-Glutamate	1000	5
Pyruvate	500	2.5
sn-Glycerol-3-phosphate	5000	25
L-Carnitine	1000	5

[a]Dissolved in KHE

Table 2
Stock and working concentrations of inhibitors used in the measurement of rates of $O_2^{\cdot -}/H_2O_2$ production by specific sites in isolated mitochondria

Inhibitor[a]	Stock concentration (mM)	Working concentration (mM)
Oligomycin	0.2 mg/mL	1 μg/mL
Rotenone	4.0	0.004
Antimycin A	2.5	0.0025
Myxothiazol	2.0	0.002
L-Malonate	200	2.0

[a]Dissolved in 96% v/v ethanol, except for L-malonate (dissolved in KHE)

4 Methods

(a) Plate reader setup

Detection wavelength:	Excitation 540 nm/emission 590 nm (*see* **Note 7**)
Reading height (focal height)	8 mm (adjust for optimal sensitivity according to the volume of the final reaction mixture, the thickness of the bottom of microplates, etc.)
Temperature (plate heating)	37 °C

(continued)

Measurement sensitivity (gain)	Usually automatically adjusted by the software bundled with the microplate reader (depending on the intensity of the measured signal, the gain of the measurement can be adjusted accordingly)
Number of cycles	30
Measurement interval	0.1 s (the time period after a well of the microplate is moved to the measurement position before the measurement begins)
Plate shaking	Agitation before each measurement cycle

(b) Reaction mixture

The following protocol describes assays in a total volume of 200 μL per microplate well. We generally run three replicate wells and average the results for each condition. The maximum rates (capacities) of $O_2^{\cdot-}/H_2O_2$ production by specific mitochondrial sites are measured. The concentrations of substrates and inhibitors of the mitochondrial electron transport chain (usually saturating) are chosen to induce optimal reduction of the targeted redox center for each assay (and thereby maximum rates of $O_2^{\cdot-}/H_2O_2$ production). For site III_{Qo} the standard assay uses a high succinate concentration, to give a submaximal but less condition-dependent rate.

For all assays described in this protocol, the common base of the reaction mixture contains:

- HRP (5 Units/mL).
- SOD (25 U/mL).
- Amplex UltraRed (0.025 mM).

  In KHEB; specific substrate-inhibitor mixes for assessing rates of $O_2^{\cdot-}/H_2O_2$ production from particular sites are described in Table 3.

  Note that a group of wells, each containing a different site-specific inhibitor(s) of the mitochondrial electron transport chain, is included in each of the assays. This serves to identify the "background" rate of change of fluorescence.

(c) Procedures

- An H_2O_2 standard curve (at concentrations ranging from 0 to 0.5 mM) is prepared by diluting the appropriate amount of H_2O_2 stock solution in KHEB. Final H_2O_2 concentrations in the microplate are 100-fold lower (i.e., 0–5 μM) (*see* **Note 8**).
- 100 μL of reaction mixture is loaded into each well of a 96-well microplate.

Table 3
Substrate-inhibitor mixes for the measurement of rates of $O_2^{\cdot -}/H_2O_2$ production by specific sites in isolated mitochondria

Assay	Substrate(s) and inhibitor(s)	Site-specific inhibitor(s)
Site I_F plus DH[a] [11]	L-Glutamate, L-malate, rotenone	No specific inhibitor available (omit L-glutamate and L-malate)
Site I_Q[b] [8, 16]	L-Succinate (5 mM)	Rotenone
Site II_F [9]	L-Succinate (0.5 mM), rotenone, antimycin A, myxothiazol	L-Malonate
Site III_{Qo} [10]	L-Succinate (5 mM), rotenone, antimycin A	Myxothiazol
Site P_F[c] [11]	Pyruvate, L-carnitine, rotenone	3-Methyl-2-oxobutanoate [11] (or omit pyruvate)
Site G_Q [12, 18]	*sn*-Glycerol-3-phosphate, rotenone, L-malonate, myxothiazol, antimycin A	iGP-1 or iGP-5 [18] (or omit *sn*-glycerol-3-phosphate)

[a]Site I_F alone has a very low maximum capacity for $O_2^{\cdot -}/H_2O_2$ production and is best assayed in more sensitive cuvette-based assays [11]
[b]Site I_Q could also be assayed using glycerol 3-phosphate as substrate if the contribution from site G_Q is corrected for
[c]Superoxide/H_2O_2 production by other oxoacid dehydrogenase complexes could be assayed in a similar way using the appropriate substrates [11, 14]

- Aliquots of 2 μL of vehicle/site-specific inhibitors/H_2O_2 standard/test compound(s) are added to the appropriate wells of the 96-well microplate.
- The 96-well microplate is then pre-warmed by incubating it in a humidified incubator for 10 min at 37 °C.
- Immediately before **the next step**, freshly isolated mitochondria are suspended in pre-warmed (37 °C) KHEB at a final concentration of 0.4 mg protein/mL.
- The reaction is initiated by the addition of 100 μL of these suspended mitochondria (0.4 mg protein/mL) to each well using a multichannel pipette. The working concentration of mitochondria in each well is 0.2 mg protein/mL. H_2O_2 production is monitored by measuring fluorescence changes (ex540/em590) for 30 cycles (approximately 35 min) at 37 °C. The timing of mitochondrial addition is critical for the accuracy of the measurement. Therefore, the time between mitochondrial addition and the beginning of the measurement should be as small and consistent as practicable.

Figure 1 shows a typical plate layout.

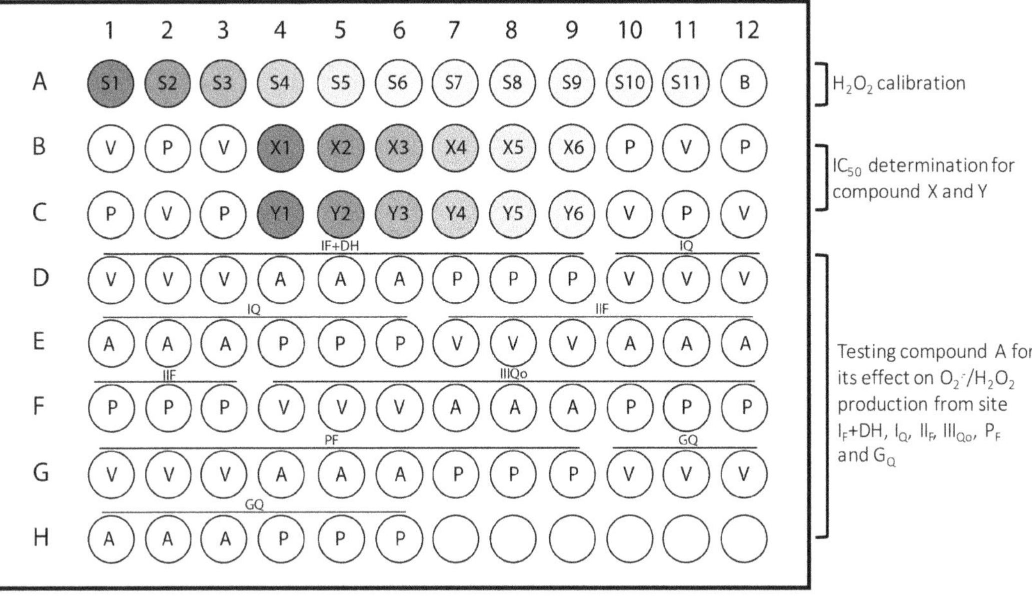

Keys:
S1 to S11 – H_2O_2 standards; B – blank (no H_2O_2); V – vehicle; P – specific inhibitor of a particular assay; X1 to X6 – compound X at different concentrations; Y1 to Y6 – compound Y at different concentrations; A – compound A

Fig. 1 A typical plate layout for H_2O_2 calibration (Row A), IC_{50} determination for compounds X and Y (Rows B and C), and examination of the specificity of a test compound (compound A) (Rows D–H), with appropriate positive controls

5 Data Processing

(a) Plot an H_2O_2 calibration curve

An H_2O_2 calibration curve is prepared by plotting the measured Amplex UltroxRed fluorescence against the concentrations of H_2O_2 standards (Fig. 2) (*see* **Note 8**).

(b) *Perform regression analysis of the H_2O_2 calibration curve*

A linear regression is used to establish the equation that best describes the linear relationship between measured fluorescence intensity and the concentrations of H_2O_2 standards (using the least-squares approach). The relationship is described by the equation of the line, $y = ax + b$, where "a" is the linear regression slope of the line and "b" is its intercept with the y-axis. The equations for calculating a and b are:

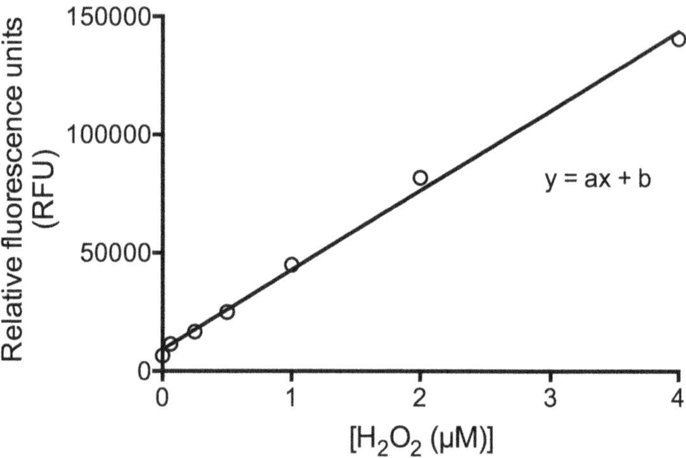

Fig. 2 An H_2O_2 standard curve. Data are means of 3 wells in a single plate

$$a = \frac{\sum_{i=1}^{n} \{(x_i - \bar{x})(y_i - \bar{y})\}}{\sum_{i=1}^{n} (x_i - \bar{x})^2}$$

and

$$b = \bar{y} - a\bar{x}$$

Software supplied with the instrument or packages such as Microsoft Excel® can be used to perform the regression analysis.

(c) *Use the* H_2O_2 *calibration function to calculate rates for test samples*

The calibration equation (the linear regression slope and the intercept) can be used to calculate the amount of H_2O_2 produced by each of the samples. This requires each sample to be analyzed under the same conditions as the calibration standards (Fig. 3a and b).

(d) *Plot the results*

Figure 2b shows time-dependent changes in H_2O_2 produced by mitochondria supplied with a substrate-inhibitor mix for $O_2^{\cdot -}/H_2O_2$ production by a specific site. The time course can be divided into three phases: (1) an early phase, (2) a pseudo-linear phase, and (3) a slowing phase.

- A nonlinear relationship of H_2O_2 production from mitochondria with time happens during the early phase of the measurement. The nonlinearity may be the result of

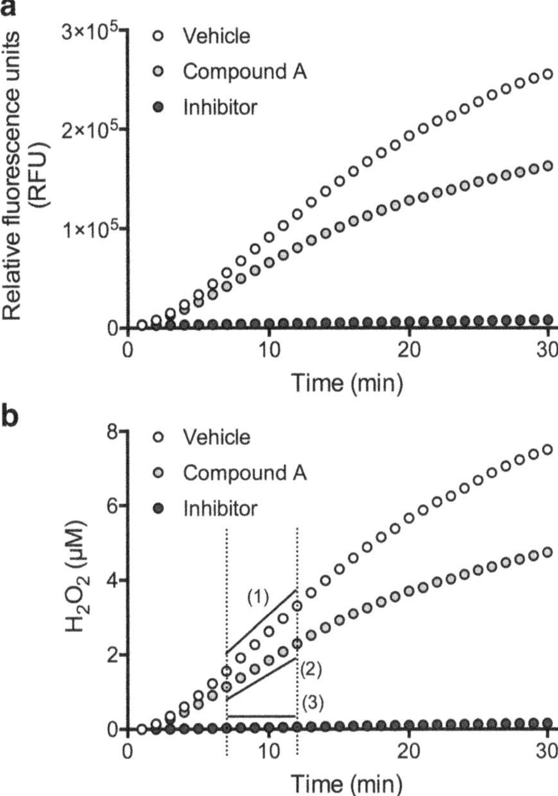

Fig. 3 Time-dependent changes of (**a**) relative fluorescence (RFU) and (**b**) calculated H_2O_2 produced during mitochondrial substrate oxidation. Compound A is an arbitrary example of a suppressor of $O_2^{\cdot -}/H_2O_2$ production by the site being investigated. Inhibitor is an inhibitor of electron transport at the relevant site. Data are for site I_Q (mean of 3 wells in a single plate). Regions 1, 2, and 3 denote pseudo-linear slopes used for calculating rates

heating up of the reaction mixture and/or the reactivation of enzyme complexes upon exposure to respiratory substrates.

- The nonlinear phase is followed by a pseudo-linear region of the curve. This phase represents the steady state after the accelerating reaction is completed and before the reaction starts to slow.
- The slowing phase is not well understood but may involve product accumulation (e.g., fumarate or oxaloacetate) and instability of mitochondria or reaction components.

One should optimize the protocol to maximize the linearity of the time-dependent H_2O_2 production by pre-heating the substrate-inhibitor mix and mitochondria before initiation of H_2O_2 production.

(e) *Estimate the rate of H_2O_2 production*

The rates of H_2O_2 production by mitochondria are estimated from the slopes of pseudo-linear regions of the curves plotting H_2O_2 content against time (Regions 1, 2, and 3 of Fig. 3b). The site-specific H_2O_2 production is then defined by the rate differences in the absence and presence of site-specific inhibitors (i.e., the difference in the rates of H_2O_2 production of the vehicle-incubated group and inhibitor-incubated group in Fig. 3b). The rates of H_2O_2 production are then expressed in units of μmol/min/mg mitochondrial protein.

(f) *Assessing the efficacy of potential suppressors of site-specific $O_2^{\cdot -}$/ H_2O_2 production in isolated mitochondria*

This microplate-based protocol can be used to examine potencies of effectors of $O_2^{\cdot -}/H_2O_2$ production, such as electron transport inhibitors [19] or site-specific suppressors [15, 16]. Potencies can be assessed by determining the concentrations of test suppressors that lead to 50% inhibition of H_2O_2 production (the IC_{50}).

- *Normalize rates of H_2O_2 production*

  The rates of H_2O_2 production of mitochondria incubated with test suppressors are normalized to that of negative (usually vehicle, scaled as 100%) and positive controls (potent inhibitors of specific mitochondrial sites, scaled as 0%). A concentration-response curve can then be generated by plotting the percent rate of H_2O_2 production of mitochondria with different concentrations of the test suppressors (Fig. 4a).

- *Generate a log (suppressor)-response curve*

  For convenience of presentation and calculation, the *x*-axis of the concentration-response curve is transformed into the logarithm of the concentrations of the test suppressors to generate a sigmoidal log (suppressor)-response curve (Fig. 4b). This transformation allows optimal visualization of nonlinear curve fits in the later steps for the estimation of IC_{50} of test suppressors.

- *Fit a sigmoidal curve*

  Using a standard software package such as GraphPad Prism, a nonlinear curve for (1) a standard sigmoidal dose response (known as the three-parameter logistic) or (2) a sigmoidal dose response (variable slope, known as the four-parameter logistic) is fitted to the data points (percent rate

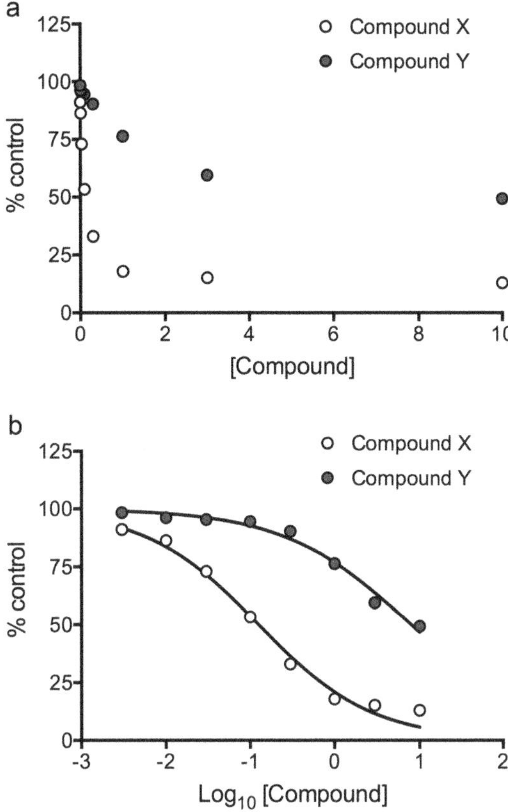

Fig. 4 Dose responses of site-specific H_2O_2 production to a site-specific suppressor of $O_2^{\cdot -}/H_2O_2$ production plotted on (**a**) linear and (**b**) log scales, fitted using the variable slope approach. Compounds X and Y are arbitrary examples of site-specific suppressors

of H_2O_2 production at different concentration of test suppressors). The standard sigmoidal dose response assumes a standard "steepness" (i.e., a Hill slope of ~1) of the curve and fits only the top, bottom, and log IC_{50} of data points, whereas the sigmoidal dose response (variable slope) fits the line based on all data points and allows some flexibility of the "steepness" of the curve. When examining uncharacterized suppressors, we recommend use of the variable slope approach to decrease possible misinterpretations.

- *Calculate IC_{50} values of test suppressors*
  Using a suitable software package, such as Excel or Prism, the IC_{50} values of test suppressors can be calculated from the concentrations of suppressors required to decrease the rate of H_2O_2 production by 50% (Fig. 4b).

6 Notes

1. The detection of H_2O_2 by Amplex UltraRed works best between pH 6.0 and 7.5.
2. H_2O_2 stock solutions decompose with time, particularly when diluted, so keep stocks concentrated, and replace them regularly (every 2 weeks).
3. Amplex UltraRed reagent is sensitive to air and light; keep all vials containing the reagent tightly capped and wrapped in foil when not in use, and prepare all necessary solutions promptly after opening vials.
4. Amplex UltraRed stock solution (in DMSO) may be stored for future use. The stock solution should be kept in the dark with desiccant at −20 °C. When stored properly, this solution is stable for at least 6 months.
5. HRP and SOD should be titrated from time to time to ensure they are in excess and not limiting signal production. HRP has a pH optimum near 6.0. Buffers that contain Tris-HCl or have pH over 8.0 contribute to increased background and may result in decreased signal-to-noise ratios.
6. Stock solutions of substrates and inhibitors are recommended to be stored as single-use aliquots at −20 °C. Pyruvate should be prepared fresh each day.
7. Optimal wavelength settings may vary slightly between instruments. If excitation at 530 nm results in signal saturation when the emission is read at 590 nm, one may lower the excitation wavelength to 490–525 nm, although it is better to lower the amount of mitochondria in the assay.
8. The range of concentrations of the standards should cover the full range of concentrations likely to be found in test samples.

References

1. Lenaz G (2001) The mitochondrial production of reactive oxygen species: mechanisms and implications in human pathology. IUBMB Life 52:159–164. https://doi.org/10.1080/15216540152845957
2. Bergamini CM, Gambetti S, Dondi A, Cervellati C (2004) Oxygen, reactive oxygen species and tissue damage. Curr Pharm Des 10:1611–1626
3. Le Bras M, Clément M-V, Pervaiz S, Brenner C (2005) Reactive oxygen species and the mitochondrial signaling pathway of cell death. Histol Histopathol 20:205–219. https://doi.org/10.14670/HH-20.205
4. Starkov AA (2008) The role of mitochondria in reactive oxygen species metabolism and signaling. Ann N Y Acad Sci 1147:37–52. https://doi.org/10.1196/annals.1427.015
5. Hamanaka RB, Chandel NS (2010) Mitochondrial reactive oxygen species regulate cellular signaling and dictate biological outcomes. Trends Biochem Sci 35:505–513. https://doi.org/10.1016/j.tibs.2010.04.002
6. Sena LA, Chandel NS (2012) Physiological roles of mitochondrial reactive oxygen species. Mol Cell 48:158–167. https://doi.org/10.1016/j.molcel.2012.09.025

7. Brand MD (2016) Mitochondrial generation of superoxide and hydrogen peroxide as the source of mitochondrial redox signaling. Free Radic Biol Med 100:14–31. https://doi.org/10.1016/j.freeradbiomed.2016.04.001
8. Treberg JR, Quinlan CL, Brand MD (2011) Evidence for two sites of superoxide production by mitochondrial NADH-ubiquinone oxidoreductase (complex I). J Biol Chem 286:27103–27110. https://doi.org/10.1074/jbc.M111.252502
9. Quinlan CL, Orr AL, Perevoshchikova IV, Treberg JR, Ackrell BA, Brand MD (2012) Mitochondrial complex II can generate reactive oxygen species at high rates in both the forward and reverse reactions. J Biol Chem 287:27255–27264. https://doi.org/10.1074/jbc.M112.374629
10. Quinlan CL, Gerencser AA, Treberg JR, Brand MD (2011) The mechanism of superoxide production by the antimycin-inhibited mitochondrial Q-cycle. J Biol Chem 286:31361–31372. https://doi.org/10.1074/jbc.M111.267898
11. Quinlan CL, Goncalves RLS, Hey-Mogensen M, Yadava N, Bunik VI, Brand MD (2014) The 2-oxoacid dehydrogenase complexes in mitochondria can produce superoxide/hydrogen peroxide at much higher rates than complex I. J Biol Chem 289:8312–8325. https://doi.org/10.1074/jbc.M113.545301
12. Orr AL, Quinlan CL, Perevoshchikova IV, Brand MD (2012) A refined analysis of superoxide production by mitochondrial *sn*-glycerol 3-phosphate dehydrogenase. J Biol Chem 287:42921–42935. https://doi.org/10.1074/jbc.M112.397828
13. Hey-Mogensen M, Goncalves RLS, Orr AL, Brand MD (2014) Production of superoxide/H_2O_2 by dihydroorotate dehydrogenase in rat skeletal muscle mitochondria. Free Radic Biol Med 72:149–155. https://doi.org/10.1016/j.freeradbiomed.2014.04.007
14. Goncalves RLS, Bunik VI, Brand MD (2016) Production of superoxide/hydrogen peroxide by the mitochondrial 2-oxoadipate dehydrogenase complex. Free Radic Biol Med 91:247–255. https://doi.org/10.1016/j.freeradbiomed.2015.12.020
15. Orr AL, Vargas L, Turk CN, Baaten JE, Matzen JT, Dardov VJ, Attle SJ, Li J, Quackenbush DC, Goncalves RLS, Perevoshchikova IV, Petrassi HM, Meeusen SL, Ainscow EK, Brand MD (2015) Suppressors of superoxide production from mitochondrial complex III. Nat Chem Biol 11:834–836. https://doi.org/10.1038/nchembio.1910
16. Brand MD, Goncalves RLS, Orr AL, Vargas L, Gerencser AA, Borch Jensen M, Wang YT, Melov S, Turk CN, Matzen JT, Dardov VJ, Petrassi HM, Meeusen SL, Perevoshchikova IV, Jasper H, Brookes PS, Ainscow EK (2016) Suppressors of superoxide-H_2O_2 production at site I_Q of mitochondrial complex I protect against stem cell hyperplasia and ischemia-reperfusion injury. Cell Metab 24:582–592. https://doi.org/10.1016/j.cmet.2016.08.012
17. Quinlan CL, Perevoschikova IV, Goncalves RLS, Hey-Mogensen M, Brand MD (2013) The determination and analysis of site-specific rates of mitochondrial reactive oxygen species production. Methods Enzymol 526:189–217
18. St-Pierre J, Buckingham JA, Roebuck SJ, Brand MD (2002) Topology of superoxide production from different sites in the mitochondrial electron transport chain. J Biol Chem 277:44784–44790
19. Orr AL, Ashok D, Sarantos MR, Ng R, Shi T, Gerencser AA, Hughes RE, Brand MD (2014) Novel inhibitors of mitochondrial *sn*-glycerol 3-phosphate dehydrogenase. PLoS One 9:e89938. https://doi.org/10.1371/journal.pone.0089938

Chapter 17

Plate-Based Measurement of Respiration by Isolated Mitochondria

Shona A. Mookerjee, Casey L. Quinlan, Hoi-Shan Wong, Pratiksha Dighe, and Martin D. Brand

Abstract

Measuring respiration rate can be a powerful way to assess energetic function in isolated mitochondria. Current, plate-based methods have several advantages over older, suspension-based systems, including greater throughput and the requirement of only μg quantities of material. In this chapter, we describe a plate-based method for measuring oxygen consumption by isolated adherent mitochondria.

Key words Mitochondrial respiration, Oxygen consumption, Respiratory control ratio, Electron flow

1 Introduction

The rate of mitochondrial oxidative phosphorylation is an important measure of cellular energetic status, dynamics, and capacity. Though the ATP that results is the primary currency of cellular energetics, it is difficult to measure the rate of ATP production directly. However, mitochondrial oxygen consumption is an indirect measurement of oxidative phosphorylation that is easily monitored by different instruments [1–4]. It is also relatively simple to establish experimental conditions that shift the control of respiration between substrate oxidation, oxidative phosphorylation, and proton leak, allowing the rate of respiration to be used to report flux through different sets of reactions within the linked network [5].

The current state of the art for these measurements is the use of fluorescent probes that quench under high O_2 tension. The fluorescent signal increases as O_2 tension declines, as with mitochondrial respiration. Robust measurement of respiration rates can be achieved using these probes either in solid state (e.g., Agilent Seahorse applications) or suspension (e.g., Luxcel MitoXpress) forms

Carlos M. Palmeira and António J. Moreno (eds.), *Mitochondrial Bioenergetics: Methods and Protocols*, Methods in Molecular Biology, vol. 1782, https://doi.org/10.1007/978-1-4939-7831-1_17,

to monitor mitochondrial O_2 consumption. However, apparent rates of O_2 consumption by biological samples can be affected by O_2 diffusion into and out of the measurement chamber, which need to be accounted for as part of any analysis of biological O_2 consumption (e.g., [6]).

Isolated mitochondria provide a useful experimental system for analysis, as they can be recovered from cultured cells and most tissues from animals at any age. Despite some limitations (e.g., relatively nonphysiological system, possible artifacts of isolation), they have the powerful advantage of allowing extensive control over the experimental conditions to enable elucidation of the molecular mechanisms underlying biological processes and pathologies. Here, we present protocols for isolating mitochondria from rat skeletal muscle tissue, preparing mitochondria for analysis using an Agilent Seahorse XF24 Analyzer, and running XF24 assays of mitochondrial respiratory control and electron flow. With relatively minor modifications, these protocols can also be used for mitochondria from many other sources.

2 Materials

2.1 Mitochondrial Isolation from Rat Skeletal Muscle

1. *Animals*—Wistar rats, aged between 6 and 10 weeks.
2. *Equipment*—Sharp scissors and herb mincer (Oxo Good Grips® Herb Mincer); cutting board, Polytron homogenizer; four 50 mL Falcon tubes; eleven 50 mL plastic centrifuge tubes; one 1.5 mL Eppendorf tube, two square pieces of muslin or cheese cloth; paintbrush (round tip, size 0); full-size refrigerated centrifuge (we use a Beckman Coulter Avanti J-26 XPI centrifuge).
3. *Solutions*—Chappell-Perry medium 1 (CP1): 100 mM KCl, 50 mM Tris–HCl (pH 7.2 at 20 °C), and 2 mM EGTA; Chappell-Perry medium 2 (CP2): 100 mM KCl, 50 mM Tris–HCl (pH 7.4 at 4 °C), 2 mM EGTA, 0.5% w/v BSA, 5 mM $MgCl_2$, 1 mM ATP, and 250 units/100 mL Protease Type VIII.
4. *Protein assay*—Biuret reagent or similar protein measurement system for determination of mitochondrial yield and concentration.

2.2 Mitochondrial Oxygen Consumption

1. *Sample support*—Seahorse XF24 cell culture microplates (Agilent Technologies, Santa Clara, CA).
2. *Assay medium*—Mitochondrial assay solution (MAS-1 medium): 2 mM HEPES, 10 mM KH_2PO_4, 1 mM EGTA, 70 mM sucrose, 220 mM mannitol, 5 mM $MgCl_2$, 0.2% w/v

fatty-acid-free bovine serum albumin (Sigma A3803), pH 7.4 at 37 °C.

3. Measurement probe—XF24 sensor cartridges (Agilent Technologies, Santa Clara, CA).
4. Seahorse XF24 extracellular flux analyzer (Agilent Technologies, Santa Clara, CA).

2.3 Compound Preparation

1. Substrate stock solutions (prepare in H_2O; dilute to working concentrations in MAS-1 on day of experiment): pyruvate 1 M; malate 1 M, ADP 0.5 M, ascorbate 1 M, tetramethyl-*p*-phenylenediamine (TMPD) 0.01 M. Pyruvate stock should be made fresh on day of experiment; succinate, malate, and ADP may be stored in single-use aliquots at −20 °C (*see* **Note 1**).
2. Inhibitor stock solutions (in EtOH, dilute to working concentrations in MAS-1): oligomycin 2 mg/mL, FCCP 10 mM, rotenone 10 mM, antimycin A or myxothiazol 10 mM (*see* **Note 2**).

3 Methods

3.1 Isolating Mitochondria from Rat Skeletal Muscle

The following protocol is based on methods described by Ashour et al. [7] and Letellier et al. [8].

1. Euthanize one or two rats with CO_2 gas exposure for 2–3 min in a closed chamber. Perform cervical dislocation to ensure that the animal is not alive.
2. Dissect out skeletal muscle by cutting around hind limbs with sharp scissors and peeling back skin from the fascia layer. Quickly retrieve as much muscle as possible, including some from the animal's dorsal area (carefully avoiding scapular brown fat deposits), and put it immediately into a beaker containing chilled CP1 (~100–200 mL) kept on ice.
3. Discard the CP1 after dissection is completed, and weigh the tissue; yields are typically 10–15 g/rat. After weighing, return the tissue immediately to fresh chilled CP1 (~100–200 mL) kept on ice.
4. Place tissue on a precooled cutting board (we use an acrylic cutting board). Trim fat and connective tissue away from the muscle. Finely chop the remaining tissue with sharp scissors, and mince it thoroughly with an herb mincer until the tissue forms a homogeneous-looking cohesive mass. If tissue becomes sticky, add small volumes of chilled CP1 medium until tissue mass holds together but does not stick to blades or preparation board. Mincing can be done with razor blades, but we use a dedicated OXO brand herb mincer with multiple

circular rolling blades in parallel. Place the minced tissue into ~300 mL fresh CP1 kept on ice.

5. Mix the minced tissue in CP1, and allow it to settle for 3–4 min on ice, until a visible layer of clear or mostly clear CP1 develops at the top of the beaker. Then, carefully decant this layer, pouring off any fragments of connective tissue.
6. Add the rinsed tissue to ice-cold CP2, making sure to use 40 mL CP2 for every 4–5 g tissue (*see* **Note 3**). Stir gently for 5 min on ice to allow protein digestion (carefully monitoring protease exposure—total time for **steps 6–8** should not exceed 20 min).
7. Divide the minced tissue into four Falcon tubes (50 mL), and homogenize it further with a Polytron for 5–10 s at low speed (1300 rpm) (*see* **Note 4**). Repeat three times for each tube.
8. Return the homogenized tissue to CP2 kept on ice, and stir for another 5 min.
9. Divide the tissue between four plastic 50 mL centrifuge tubes, and spin at 500 × *g* for 11 min (4 °C).
10. Filter the supernatant through two layers of muslin, transfer to four clean 50 mL centrifuge tubes, and spin at 10,000 × *g* for 11 min (4 °C).
11. Carefully decant the supernatant, leaving the mitochondrial pellet at the bottom. Add a few mL of ice-cold CP1 for each pellet, and using a paintbrush, carefully brush the brown mitochondrial pellet free from the pellet mass and into the CP1, leaving the red blood cell (RBC) pellet on the tube wall. Resuspend the mitochondrial pellet completely using a 1 mL pipette. Pool all the mitochondrial suspensions into one 50 mL centrifuge tube. Increase the volume to 35–40 mL total with CP1, and spin at 10,000 × *g* for 11 min (4 °C).
12. Resuspend the pellet again in CP1 as in **step 11**, avoiding the RBC pellet. Transfer the mitochondrial suspension to a fresh 50 mL centrifuge tube, and spin at 500 × *g* for 6 min (4 °C).
13. Transfer the supernatant to a clean 50 mL centrifuge tube. Spin at 10,000 × *g* for 11 min (4 °C).
14. Resuspend the final pellet in approximately 0.8–1 mL CP1; transfer the mitochondrial suspension to a 1.5 mL Eppendorf tube, and keep it on ice at all times (*see* **Note 5**).
15. Determine the mitochondrial protein concentration (e.g., using the biuret assay). This isolation procedure yields typically 20–30 mg mitochondrial protein from 10 g of rat muscle.

3.2 Preparing the XF Assay Cartridge

1. *Equilibrate the cartridge*—Open the XF24 sensor cartridge package, and remove the cartridge from the base, taking care not to touch the sensor probes to any surface. Add 1 mL

Seahorse XF Calibrant to each well of the cartridge base. Replace sensor cartridge. Incubate cartridge at 37 °C for a minimum of 1 h (maximum 48 h as recommended by the manufacturer).

2. *Preparation of working stocks for making port additions*—Typical port addition volumes range between 50 and 75 μL and starting volumes range between 400 and 700 μL. Dilute substrate and inhibitor stocks into MAS-1 medium (*see* **Note 6**) to make ~10× working stocks so that equal volumes are added with each injection. Final concentrations after each addition should be calculated using the final volume that results from that addition. For example, for adding 2 mg/mL oligomycin to a final concentration of 2 μg/mL in the assay well with the first addition, the (final concentration × final volume)/stock concentration is (2 μg/mL) (450 μL)/2000 μL = 0.45 μL. This should be diluted into MAS-1 medium for a final volume of 50 μL, or 50–0.45 = 49.55 μL. If oligomycin is added in the second addition, the calculation would be (2 μg/mL) (500 μL)/2000 μL = 0.5 μL, plus 49.5 μL MAS-1, per port addition. Make a master mix of $n + 1$ port additions per plate for each compound. Alternatively, injection volumes may vary if preparation of exactly 10× concentrations is desired (*see* **Notes 7** and **8**).
3. When cartridge is equilibrated and port additions are ready to add, remove cartridge from 37 °C incubator, and rest on a heat-insulated surface (e.g., polystyrene panel or thermostable Gel-Pak warmed to 37 °C) to set up the port additions (*see* **Note 9**).
4. Distribute each working stock into the appropriate port (A, B, C, or D) of the XF sensor cartridge. Add the same volume of MAS-1 medium to the ports of the temperature control wells (A1, B4, C3, D6) and to the appropriate ports of any vehicle-control experimental wells (*see* **Note 10**).
5. Following distribution of experimental additions to the ports, return the cartridge to 37 °C for 5 min to allow rewarming.
6. During the rewarming in **step 5**, program the XF Analyzer according to the experimental design.
7. Follow calibration instructions to load the XF assay cartridge into the XF Analyzer, and begin the calibration process.

3.3 Setting Up a Mitochondrial Sample Plate for the XF Assay

The XF Analyzer calibration cycle default time is approximately 25 min; this is an ideal amount of time to prepare the XF24 microplates with the isolated mitochondria. It is important to set up the sample plate as close to the assay start time as possible.

1. No coating or other preparation is required for the XF cell culture microplates prior to seeding with mitochondria. Open the package containing the assay microplate, and remove the lid.
2. Following Subheading 3.2, **step** 7, resuspend isolated mitochondria in MAS-1 medium (final concentration will be based upon the amount of mitochondria required per well as determined by the substrate utilized to support mitochondrial respiration; *see* **step 3** and **Note 11**).
3. Pipet 25 μL of the mitochondrial suspension into the center of each experimental well for a final distribution of the mitochondrial protein per well. For the temperature control wells, pipet 25 μL MAS-1 medium into the center of each well (*see* **Note 12**). To obtain optimal mitochondrial oxygen consumption rates for analysis, adjustment of the amount of mitochondria for each well may be required when using different combinations of substrates to support mitochondrial respiration. For mitochondria isolated from rat skeletal muscle, 5–8 μg of mitochondria are recommended for complex I-linked substrates (i.e., pyruvate plus malate; glutamate plus malate, etc.), whereas 3–5 μg of mitochondria are sufficient to obtain a good signal when oxidizing succinate as the substrate [9–12].
4. Centrifuge sample plate in a centrifuge equipped with a swinging-bucket rotor (sized for standard microplates) at 4 °C for 20 min at 2000 × *g*, programmed with slow acceleration and deceleration steps.
5. Remove sample plate from centrifuge, and place on a heat-insulated surface. Carefully add 375 μL MAS-1 at 37 °C to each microplate well, for 400 μL total, taking care not to disturb the adherent mitochondrial layer.
6. Using a standard microscope, visually inspect a representative number of wells at various positions on the plate to confirm consistency of plating and distribution of mitochondria on the bottom surface.
7. Rest sample plate at 37 °C until calibration cycle is complete.
8. Remove lid from assay plate, and follow instrument prompts to replace the calibration plate with the sample plate in the XF Analyzer when calibration cycle is complete. The equilibration step may be omitted to shorten the assay; however, if drifts in baseline level (O_2 or pH) measurements are observed, an equilibration step is recommended to minimize this drift.

3.4 Respiratory Control Assay (see Notes 13–15)

A typical measurement cycle in the XF Analyzer consists of a programmed mix, followed by a wait step, followed by a longer measurement period. For experiments with 3–8 μg mitochondrial protein/well, a 1-min mix, 1-min wait, and 3-min measurement are appropriate. One to five measurement cycles are typical for each segment of the experiment. The following experiment contains three measurements per segment. Assay medium: MAS-1 containing 10 mM succinate, 2 μM rotenone.

Under "protocol," program the XF Analyzer to run the following steps:

Calibrate.

Equilibrate.

*Mix 1 min.

Wait 1 min.

Measure 3 min.

Repeat from * twice more for three measurement cycles total.

Inject Port A (4 mM ADP).

*Mix 1 min.

Wait 1 min.

Measure 3 min.

Repeat from * twice more for three measurement cycles total.

Inject Port B (2 μg/mL oligomycin).

*Mix 1 min.

Wait 1 min.

Measure 3 min.

Repeat from * twice more for three measurement cycles total.

Inject Port C (4 μM FCCP).

*Mix 1 min.

Wait 1 min.

Measure 3 min.

Repeat from * twice more for three measurement cycles total.

Inject Port D (2.5 μM antimycin A).

*Mix 1 min.

Wait 1 min.

Measure 3 min.

Repeat from * twice more for three measurements total.

3.5 Electron Flow Assay (see Note 16)

Initial medium: MAS-1 assay medium containing 10 mM pyruvate, 2 mM malate, and 4 μM FCCP. Under "protocol," program the XF Analyzer to run the following steps:

Calibrate.

Equilibrate.

*Mix 1 min.

Wait 1 min.

Measure 3 min.

Repeat from * twice more for three measurements total.

Inject Port A (4 μM rotenone).

*Mix 1 min.

Wait 1 min.

Measure 3 min.

Repeat from * twice more for three measurements total.

Inject Port B (10 mM succinate).

*Mix 1 min.

Wait 1 min.

Measure 3 min.

Repeat from * twice more for three measurements total.

Inject Port C (2.5 μM antimycin A).

*Mix 1 min.

Wait 1 min.

Measure 3 min.

Repeat from * twice more for three measurements total.

Inject Port D (10 mM ascorbate, 100 μM TMPD).

*Mix 1 min.

Wait 1 min.

Measure 3 min.

Repeat from * twice more for three measurements total.

4 Notes

1. Substrate stock solutions, like the assay medium, should be adjusted to pH 7.4 at 37 °C.
2. EtOH concentration should not exceed 1% (v/v) in final assay volume.
3. The ratio of tissue to CP2 medium is critically important because it sets the appropriate conditions for protease-mediated muscle digestion. Too much digestion (i.e., less tissue or more CP2) will result in degradation of tissue and poor mitochondrial quality, whereas too little digestion (i.e., more tissue or less CP2) will result in a low final yield.
4. Polytron instrument parameters (e.g., homogenization speed, blade sharpness) will affect mitochondrial yield and quality: too

rapid homogenization will damage mitochondria, but too slow a speed will lower the yield. Optimum homogenization speed for mitochondrial isolation as described here should be empirically determined for each instrument.

5. Strengths and limitations of the "crude" mitochondrial preparation. The mitochondrial isolation protocol described here results in a mixed population of organelles, mostly mitochondria, but perhaps also containing endoplasmic reticulum, lysosomes, peroxisomes, and other microsomal bodies [13]. This isolation protocol preserves a relatively high proportion of intact mitochondria and therefore is preferred for the analysis of whole mitochondrial function. This relative impurity makes it possible (or likely, in mitochondrial preparations from, e.g., liver) for non-mitochondrial oxygen consumption to occur. The use of mitochondrial respiratory chain inhibitors (e.g., rotenone, antimycin A, and myxothiazol) allows differentiation between mitochondrial and non-mitochondrial oxygen consumption rate (OCR) and is therefore a crucial final addition to all experiments to enable subtraction of the non-mitochondrial rate within each well when accurate measurement of mitochondrial respiration is desired. Higher degrees of mitochondrial purity are possible but typically at the cost of decreased mitochondrial function.
6. Principles of assay medium composition. Mitochondria in different tissues exhibit different morphologies, membrane compositions, and substrate preferences. Accordingly, the composition of the medium used for mitochondrial assays should be considered. The mitochondrial assay solution (MAS-1) used for these experiments is a lightly buffered, high osmotic-strength, low ionic-strength medium that maintains skeletal muscle mitochondria in a relatively condensed state. HEPES and phosphate both provide pH buffering, and the phosphate also provides a source of Pi for ATP synthesis from ADP. Chelation of Ca^{2+} by EGTA helps to prevent mitochondrial damage caused by Ca^{2+} accumulation. Sucrose and mannitol maintain isotonicity with respect to the mitochondrial matrix (for discussion of sucrose versus mannitol and effects on mitochondrial respiration, *see*, e.g., [14, 15], though other strategies also exist, e.g., [16]). Mg^{2+} lowers K^+ permeability, buffers the concentrations of free adenine nucleotides, and supports the activity of kinases in experiments where they are added [17]. Finally, 0.2% BSA helps to maintain mitochondrial coupling by sequestering fatty acids remaining or generated after the isolation [18].
7. Addition of hydrophobic compounds via injection ports may alter the final amount of compound in the well, as they can remain stuck to the side wall of the port. The influence of this

effect on experimental outcome is probably small when a saturating amount is all that is needed, but for compound titrations, e.g., for determining EC_{50} or IC_{50} values, the effect of compound sticking and loss may be higher.

8. Use of colored compounds either in the assay medium (e.g., phenol red) or in an injection port can influence the apparent rates recorded by the Seahorse instrument by interfering with the fluorescence measurement. It is recommended to perform a pilot experiment to test this effect, by comparing different mitochondria-free wells, some with assay medium alone, some with assay medium plus compound, or the same wells before and after compound addition from an injection port. Traces for both OCR and extracellular (extramitochondrial) acidification rate (ECAR) should be examined (and corrected if necessary) for the effects of colored compounds on these measurements.
9. Sample randomization across the plate. To ensure that results are not contaminated by artifacts of well positioning, care should be taken to randomize both technical replicates within a plate and samples between replicate plates.
10. Number of technical replicates. For the 24-well format, a minimum of three technical replicates of each sample per plate is recommended to increase measurement accuracy. For pilot experiments involving titration or other design applications, fewer replicates may be preferred if reliable results can still be obtained.
11. Optimization of mitochondrial amount. The amount of mitochondria to add per well for optimal rates depends on the tissue source, mitochondrial quality, substrate conditions, and assay parameters. To empirically determine this amount, mitochondria should be titrated as in [4]. Examine the O_2 raw data traces for the range of mitochondrial amounts that confer high rates of respiration without severely depleting O_2 tension in the measurement well.
12. It is important to pipet the mitochondrial sample directly into the center of the well bottom when preparing the assay plate, to avoid measurement artifacts caused by uneven mitochondrial distribution in the well following centrifugation.
13. Principles of the respiratory control ratio (RCR) assay. Each portion of the RCR experiment measures specific aspects of overall mitochondrial activity, which together allow assessment of this activity and reveal the activities that are altered by experimental intervention [4, 5]. The basal respiration rate (classically defined as state 2 or state 4) is driven by proton leak. The addition of ADP activates ATP production through the F_1F_O ATP synthase (state 3). Oligomycin inhibits the ATP synthase and decreases this respiration to the rate that is

required to oppose mitochondrial proton leak (state 4_o, mimicking the depletion of ADP that defines state 4). FCCP addition drives maximal uncoupled respiration (state 3_u). Antimycin A inhibits mitochondrial respiration. The non-mitochondrial OCR after antimycin A addition is assumed to be constant during the experiment and is therefore subtracted from the entire trace to yield mitochondrial respiration rate. These rates are explained in detail elsewhere [5]. From here, proton leak rate (state 4_o), coupling efficiency (% mitochondrial respiration that is phosphorylating; 100 × (state 3 − state 4_o)/state 3), and respiratory control ratio (state 3/state 4) can be calculated. Using the protocols described here, we typically record RCR values >6 when mitochondria are given glutamate plus malate as substrates and ~4–5 when mitochondria are given succinate. These values are good indicators of overall mitochondrial quality, as the numerator and denominator each reflect multiple important aspects of total mitochondrial behavior.

14. Calculating ATP production rates. Rates of mitochondrial oxygen consumption under basal conditions directly reflect the rates of ATP production by mitochondrial oxidative phosphorylation. OCR can therefore be converted to rate of ATP production by oxidative phosphorylation [5], using the coupling efficiency multiplied by the appropriate modern value of the P/O ratio for a particular substrate [19].

15. Normalization. Raw measured rates (e.g., of oxygen consumption) are properly expressed as pmol [O]/min/well or pmol O_2/min/well. Normalization to protein content per well to yield rates in pmol/min/μg protein is straightforward, as the (crude) mitochondrial protein content is known. However, if desired (e.g., comparison between two samples is complicated by differing proportions of mitochondria to other cellular material), normalization to other quantifying measurements (e.g., amount of a particular protein as measured by western blot, or amount of enzyme activity) is also valid. For plate-based phenotypic screening or other applications where the relative effect of a compound addition on OCR is desired, normalizing to a control rate (e.g., basal rate before addition, or rate after vehicle control addition) is useful to decrease error between replicates and reveal small, true effects. The goal of the analysis should therefore be carefully considered when choosing a normalization strategy.

16. Principles of the electron flow assay. This assay reveals the electron flow (and resulting O_2 consumption) through sequential respiratory chain complexes as follows: At the assay start, pyruvate and malate drive electron entry via NADH oxidation at complex I through complexes III and IV. This

rate is not controlled by ATP turnover, as there is no ADP present and FCCP is added to uncouple pmf dissipation from phosphorylation. The addition of rotenone inhibits electron transfer to ubiquinone, inhibiting further respiration by electron entry at complex I. Electron flow is reengaged by addition of succinate, whose oxidation yields electrons that directly reduce the ubiquinone pool and reveals electron flow through complexes II, III, and IV. This flow is then inhibited by the complex III inhibitor antimycin A. Electron flow through complex IV is then reengaged by simultaneous additions of ascorbate and TMPD, which is subsequently inhibited by addition of the complex IV inhibitor azide.

References

1. Warburg O (1923) Versuche an überlebendem Carcinom-Gewebe (Methoden). Biochem Zeitschr 142:317–333
2. Warburg O (1924) Verbesserte Methode zur Messung der Atmung und Glykolyse. Biochem Z 152:51–63
3. Wu M, Neilson A, Swift AL, Moran R, Tamagnine J, Parslow D, Armistead S, Lemire K, Orrell J, Teich J, Chomicz S, Ferrick DA (2006) Multiparameter metabolic analysis reveals a close link between attenuated mitochondrial bioenergetic function and enhanced glycolysis dependency in human tumor cells. Am J Physiol Cell Physiol 292(1): C125–C136. https://doi.org/10.1152/ajpcell.00247.2006
4. Rogers GW, Brand MD, Petrosyan S, Ashok D, Elorza AA, Ferrick DA, Murphy AN (2011) High throughput microplate respiratory measurements using minimal quantities of isolated mitochondria. PLoS One 6(7):e21746. https://doi.org/10.1371/journal.pone.0021746.t001
5. Brand MD, Nicholls DG (2011) Assessing mitochondrial dysfunction in cells. Biochem J 435(2):297–312. https://doi.org/10.1113/expphysiol.2006.034330
6. Gerencser AA, Neilson A, Choi SW, Edman U, Yadava N, Oh RJ, Ferrick DA, Nicholls DG, Brand MD (2009) Quantitative microplate-based respirometry with correction for oxygen diffusion. Anal Chem 81(16):6868–6878. https://doi.org/10.1021/ac900881z
7. Ashour B, Hansford RG (1983) Effect of fatty acids and ketones on the activity of pyruvate dehydrogenase in skeletal-muscle mitochondria. Biochem J 214:725–736
8. Letellier T, Malgat M, Mazat JP (1993) Control of oxidative phosphorylation in rat muscle mitochondria: implications for mitochondrial myopathies. Biochim Biophys Acta 1141:58–64
9. Orr AL, Ashok D, Sarantos MR, Ng R, Shi T, Gerencser AA, Hughes RE, Brand MD (2014) Novel inhibitors of mitochondrial sn-glycerol 3-phosphate dehydrogenase. PLoS One 9(2): e89938. https://doi.org/10.1371/journal.pone.0089938
10. Orr AL, Ashok D, Sarantos MR, Shi T, Hughes RE, Brand MD (2013) Inhibitors of ROS production by the ubiquinone-binding site of mitochondrial complex I identified by chemical screening. Free Radic Biol Med 65:1047–1059. https://doi.org/10.1016/j.freeradbiomed.2013.08.170
11. Orr AL, Vargas L, Turk CN, Baaten JE, Matzen JT, Dardov VJ, Attle SJ, Li J, Quackenbush DC, Goncalves RL, Perevoshchikova IV, Petrassi HM, Meeusen SL, Ainscow EK, Brand MD (2015) Suppressors of superoxide production from mitochondrial complex III. Nat Chem Biol 11(11):834–836. https://doi.org/10.1038/nchembio.1910
12. Hey-Mogensen M, Goncalves RL, Orr AL, Brand MD (2014) Production of superoxide/H_2O_2 by dihydroorotate dehydrogenase in rat skeletal muscle mitochondria. Free Radic Biol Med 72:149–155. https://doi.org/10.1016/j.freeradbiomed.2014.04.007
13. Wieckowski MR, Giorgi C, Lebiedzinska M, Duszynski J, Pinton P (2009) Isolation of mitochondria-associated membranes and mitochondria from animal tissues and cells. Nat Protoc 4(11):1582–1590. https://doi.org/10.1038/nprot.2009.151
14. Siess EA (1983) Influence of isolation media on the preservation of mitochondrial

functions. Hoppe Seylers Z Physiol Chem 364:279–289

15. Whipps DE, Halestrap AP (1984) Rat liver mitochondria prepared in mannitol demonstrate increased mitochondrial volumes compared with mitochondria prepared in sucrose media. Biochem J 221:147–152
16. Corcelli A, Saponetti MS, Zaccagnino P, Lopalco P, Mastrodonato M, Liquori GE, Lorusso M (2010) Mitochondria isolated in nearly isotonic KCl buffer: focus on cardiolipin and organelle morphology. Biochim Biophys Acta 1798(3):681–687. https://doi.org/10.1016/j.bbamem.2010.01.005
17. Mildvan AS (1987) Role of magnesium and other divalent cations in ATP-utilizing enzymes. Magnesium 6(1):28–33
18. Panov AV, Vavilin VA, Lyakhovich VV, Brooks BR, Bonkovsky HL (2010) Effect of bovine serum albumin on mitochondrial respiration in the brain and liver of mice and rats. Bull Exp Biol Med 149:187–190
19. Mookerjee SA, Gerencser AA, Nicholls DG, Brand MD (2017) Quantifying intracellular rates of glycolytic and oxidative ATP production and consumption using extracellular flux measurements. J Biol Chem 292(17):7189–7207. https://doi.org/10.1074/jbc.M116.774471

Chapter 18

Detection of Iron Depletion- and Hypoxia-Induced Mitophagy in Mammalian Cells

Shun-ichi Yamashita and Tomotake Kanki

Abstract

Mitochondrial autophagy or mitophagy is a process that selectively degrades mitochondria via autophagy. It is believed that mitophagy degrades damaged or unnecessary mitochondria and is important for maintaining mitochondrial homeostasis. To date, it is known that several stimuli can induce mitophagy. However, some of these stimuli (including iron depletion, hypoxia, and nitrogen starvation) induce mild mitophagy, which is difficult to detect by measuring the decrease in mitochondrial mass. Recently, we have successfully detected mitophagy induced under these conditions using mito-Keima as a reporter. In this chapter, we describe the protocols for induction and detection of iron depletion- and hypoxia-induced mitophagy using the mito-Keima-expressing cells.

Key words Mitochondria, Autophagy, Mitophagy, Hypoxia, Iron depletion, Deferiprone, Keima

1 Introduction

Mitochondria are essential organelles that participate in important cellular events and numerous metabolic processes including ATP production, calcium buffering, apoptosis regulation, fatty acid β-oxidation, and heme and iron–sulfur cluster biosynthesis. Cellular ATP is primarily produced via oxidative phosphorylation by complexes present in the inner mitochondrial membrane. During this process, electrons are transported to oxygen via the electron transport chain, which results in the formation of a proton gradient across the inner mitochondrial membrane, termed the membrane potential. The membrane potential is utilized as a driving force for ATP synthase present in the inner mitochondrial membrane. Although most electrons pass through the oxidative phosphorylation complexes, some of them occasionally leak out. The leaked electrons eventually produce reactive oxygen species (ROS). Therefore, mitochondria (especially the oxidative phosphorylation complexes and mitochondrial DNA) are constitutively exposed to ROS

Carlos M. Palmeira and António J. Moreno (eds.), *Mitochondrial Bioenergetics: Methods and Protocols*, Methods in Molecular Biology, vol. 1782, https://doi.org/10.1007/978-1-4939-7831-1_18,

and eventually get disordered. It is believed that the damaged mitochondria are selectively degraded by mitochondrial autophagy (hereafter referred to as mitophagy), which serves to maintain the mitochondrial function and eliminate ROS-induced toxicity [1, 2].

In several cellular processes, iron plays an important role as an enzyme cofactor. In contrast, free iron leads to oxidative stress due to the generation of free radicals via the Fenton reaction. Thus, the intracellular level of iron is optimally maintained by regulating its uptake, export, and sequestration. Intracellular iron is primarily utilized in mitochondria where heme and iron–sulfur clusters are synthesized [3, 4]. Defects in iron homeostasis can lead to abnormal levels of heme and iron–sulfur clusters and confer pathological phenotypes such as anemia and hemochromatosis [5, 6]. The transferrin (Tf)–transferrin receptor (TfR) pathway mediates the uptake and reduction of serum iron. Subsequently, the reduced iron is recruited into the mitochondrial matrix and utilized there [7, 8].

Ferritin, an iron-storage protein, is present in both the cytoplasm and mitochondria. It sequesters iron under iron-sufficient conditions. Under iron-depletion conditions, in addition to the Tf–TfR pathway, iron is obtained from the autophagic degradation of cytoplasmic ferritin and supplied for the biosynthesis of heme and iron–sulfur clusters in mitochondria [9]. Interestingly, in addition to the autophagic degradation of cytoplasmic ferritin, mitophagy is also induced due to long-term treatment with an iron-chelating agent [10]. Iron depletion-induced mitophagy may have two potential physiological roles. The first role is the autophagic degradation of mitochondrial ferritin for supplying iron during iron-depletion conditions. The second role is when the iron–sulfur cluster depletion leads to nonfunctional oxidative phosphorylation complexes, resulting in the degradation of defective mitochondria. Further studies are required for elucidating the molecular mechanisms and physiological relevance of mitophagy under iron-depletion conditions.

During the last decade, several mitophagic pathways have been reported in mammalian cells including the PINK1/Parkin-dependent [11, 12], cellular differentiation-related [13–16], hypoxia-induced [17–20], and iron depletion-induced [10, 20] pathways. The PINK1/Parkin-dependent mitophagy has been most commonly studied because this type of mitophagy can be easily induced through the loss of membrane potential in cultured mammalian cells. During this process, most mitochondria are degraded by autophagy within 48 h. Therefore, mitophagy can be assessed by a decrease in the mitochondrial proteins or mitochondrial DNA using immunoblotting, immunofluorescence microscopy, or quantitative polymerase chain reaction (PCR) [21–24]. Similarly, mitochondria degradation during the differentiation of erythroid cells can also be detected by a decrease in the mitochondrial proteins [13, 14]. In contrast to these mitochondria degradation pathways, the flux through iron depletion- or hypoxia-induced mitophagy is relatively low and thus difficult to detect by monitoring the decrease in mitochondrial mass.

Recently, several groups (including ours) have reported that low mitophagic flux can be detected using the mitochondria-targeted fluorescent protein Keima (hereafter referred to as mito-Keima) [17, 25, 26]. Keima is an atypical fluorescent protein and possesses two useful features for the detection of mitophagy [25, 27, 28]. The first feature is that it is resistant to lysosomal proteases; thus, the delivery of mito-Keima into the lysosomal lumen can be monitored during mitophagy. The other feature is that it has a bimodal excitation spectrum with peaks at 440 and 586 nm under neutral and acidic conditions, respectively. This phenomenon is due to the conformational changes in the chromophore, in response to environmental pH [25]. Thus, by using the mito-Keima system, we can detect the mitochondria that are delivered into the lysosomes (as fluorescent signals) by selectively exciting them at 586 nm [17]. These features allow us to sensitively detect low levels of mitophagy, such as iron depletion- and hypoxia-induced mitophagy (Fig. 1).

2 Materials

2.1 Mammalian Cells (See Note 1)

1. HeLa cells.
2. SH-SY5Y cells.
3. Mouse embryonic fibroblast (MEF) cells.
4. Platinum-E (Plat-E) retroviral packaging cells, Ecotropic (Cell Biolabs, Inc.).

2.2 Media for Cell Culturing

1. Dulbecco's Modified Eagle's Medium (DMEM) with high glucose (Wako).
2. Fetal bovine serum (FBS; GIBCO).

2.3 Plasmids and cDNA

1. pMXs-Puro (Cell Biolabs, Inc.).
2. pMT-mKeima-Red (MBL).
3. pcDNA3.1-Hygro (Invitrogen).
4. First-strand cDNA from mouse brain (BioChain).

2.4 Reagents

2.4.1 For Generation of the Mito-Keima-Expressing Cells

1. Opti-MEM reduced serum medium (GIBCO).
2. FuGENE HD transfection reagent (Promega).
3. Puromycin (Wako).
4. Polybrene (Santa Cruz Biotechnology).

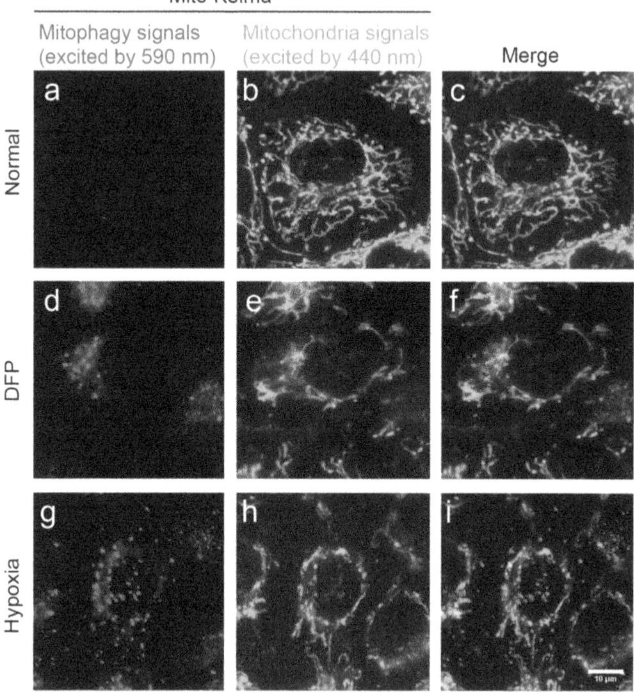
A HeLa
Mito-Keima
Mitophagy signals (excited by 590 nm)
Mitochondria signals (excited by 440 nm)
Merge
a
b
c
d
e
f
g
h
i
Normal
DFP
Hypoxia
10 μm

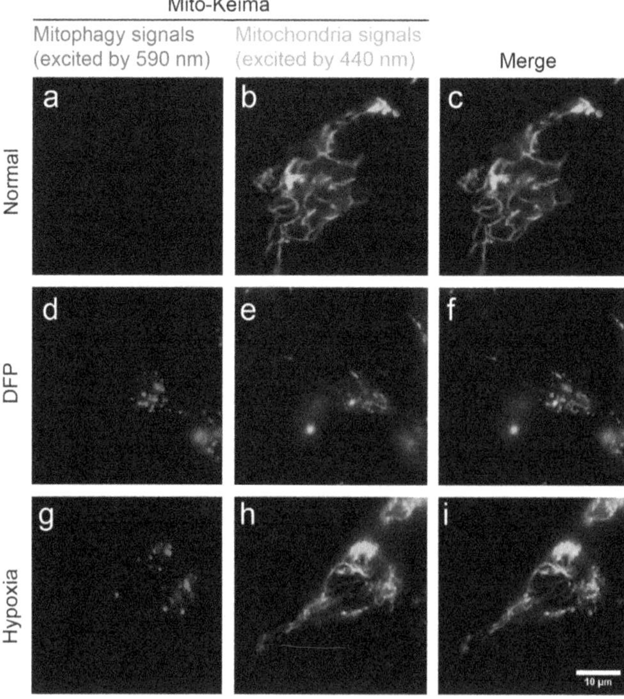
B SH-SY5Y
Mito-Keima
Mitophagy signals (excited by 590 nm)
Mitochondria signals (excited by 440 nm)
Merge
a
b
c
d
e
f
g
h
i
Normal
DFP
Hypoxia
10 μm

2.4.2 For the Induction of Iron Depletion-Induced Mitophagy

1. 3-Hydroxy-1,2-dimethyl-4(1H)-pyridone [Deferiprone, DFP; Wako].

3 Methods

3.1 Generation of Stable Cell Lines Expressing Mito-Keima (See Note 2)

3.1.1 Construction of the Retroviral Vector for Mito-Keima Expression (pMXs-Puro-Mito-Keima)

1. Amplify the mito-Keima cDNA from pMT-mKeima-Red (template) by performing PCR with the following primers: Fw, 5′-CGGGATCCGCCACCATGCTGAGCCTG-3′; Rv, 5′-GGAATTCTTAACCGAGCAAAGAGTGGCGTG-3′.
2. Ligate the PCR fragment into the BamHI–EcoRI restriction site of the vector pMXs-Puro.

3.1.2 Construction of the mCAT1 Expression Vector (pcDNA3.1-Hygro-mCAT1)

1. Amplify the cDNA encoding for the mouse cationic amino acid transporter (mCAT1) gene from the mouse first-strand cDNA pool by performing PCR with the following primers: Fw, 5′-CTAGCTAGCGCCACCATGGGCTGCAAAAACCTGCTCGG-3′; Rv, 5′-CGGGATCCTCATTTGCACTGGTCCAAGTTGCTGTCAGG-3′.
2. Ligate the PCR fragment into the NheI–BamHI restriction site of the vector pcDNA3.1-Hygro.

3.1.3 Preparation of the Retroviral Supernatant

1. Plate the Plat-E cells at a density of 3×10^5 cells per well in a 6-well culture plate in 2 mL DMEM supplemented with 10% FBS, and culture overnight.
2. Add 3 μg pMXs-Puro-mito-Keima plasmid into a 1.5-mL microcentrifuge tube, along with 150 μL Opti-MEM. Next, add 9 μL FuGENE HD, mix well by pipetting gently, and incubate for 5 min at room temperature. The following process should be performed in a Class II biological safety cabinet.
3. Add the plasmid–FuGENE HD mixture to the Plat-E cells precultured in 6-well plates, and culture for 24 h.
4. Replace the medium with 2 mL DMEM supplemented with 10% FBS, and incubate for another 24 h.

Fig. 1 Detecting iron depletion- and hypoxia-induced mitophagy. (**A**) HeLa and (**B**) SH-SY5Y stable cell lines expressing mito-Keima were cultured under DFP (an iron chelating agent) treatment, hypoxic, or normal conditions. The mito-Keima signals observed after excitation at 590 nm indicate mitochondria delivery into the lysosomes via mitophagy (a, d, and g indicated by red). The mito-Keima signals observed after excitation at 440 nm indicate mitochondria present in the cytoplasm (b, e, and h indicated by green). Merged views are shown in c, f, and i. Scale bar, 10 μm

5. Collect the supernatant (which contains the retroviruses), and filter it with a 0.45-μm syringe filter to remove cell debris.
6. Add polybrene at a final concentration of 8 μg/mL. This results in the generation of a retrovirus harboring the mito-Keima gene (mito-Keima retroviral solution). This retroviral solution can be stored at −80 °C for several months. However, we recommend using this solution immediately to prevent the reduction of viral titer.

3.1.4 Infecting Host Cells with the Mito-Keima Retroviral Solution

1. In this section, we discuss the protocol for generating a mito-Keima-expressing cell line using an ecotropic retroviral infection system. Since the HeLa and SH-SY5Y cells do not express the receptor for an ecotropic retrovirus, mCAT1 must be expressed in these cells prior to retroviral infection. Cells derived from mice or rats, including MEFs, intrinsically express mCAT1.
2. Plate the HeLa or SH-SY5Y cells at a density of 5×10^4 cells per well in a 12-well plate in 1 mL DMEM supplemented with 10% FBS, and culture overnight. Prepare the pcDNA3.1-Hygro-mCAT1–FuGENE HD mixture (obtained by mixing 1 μg vector and 3 μL FuGENE HD in 50 μL Opti-MEM; *see* Subheading 3.1.3), and incubate for 5 min at room temperature. Add the vector–FuGENE HD mixture to the cells pre-cultured in 12-well plates, and culture for 24 h.
3. Plate the host cells (MEFs or pcDNA3.1-Hygro-mCAT1-transfected HeLa or SH-SY5Y cells) at a density of 1×10^5 cells per well in a 12-well culture plate in 1 mL DMEM supplemented with 10% FBS, and culture overnight.
4. Replace the medium with the mito-Keima retroviral solution, and culture for 24 h.
5. Replace the retroviral solution with 1 mL DMEM supplemented with 1 μg/mL puromycin, and culture for >3 days to exclude the uninfected cells.
6. Maintain the mito-Keima-expressing cells in DMEM containing 1 μg/mL puromycin for subsequent use in Subheading 3.1.5.

3.1.5 Cloning of Cells Expressing Mito-Keima at a Moderate Level

1. Plate the cells expressing mito-Keima (*see* Subheading 3.1.4) at a density of 2 cells per mL in a 96-well culture plate with 0.2 mL DMEM supplemented with 10% FBS per each well, and culture for 1–2 weeks.
2. Transfer the cells grown from a single colony at a density of 1×10^4 cells per well into a 96-well glass-bottom plate, and culture overnight.

3. Observe the cells using a fluorescence microscope with the filter set for detecting Keima (*see* Subheading 3.2.1), and select the clone showing moderate Keima expression. If the cells are cultured under normal conditions and not grown to a high confluence, typical tubular mitochondria will be observed after excitation at 440 nm. No/faint signals will be observed after excitation at 590 nm (Figs. 1A, B).

3.2 Induction and Measurement of Mitophagy

3.2.1 Iron Depletion-Induced Mitophagy

1. Plate the mito-Keima-expressing cells at a density of 1×10^4 cells per well into a 96-well glass-bottom plate in 100 μL DMEM supplemented with 10% FBS, and culture overnight (*see* **Note 3**).
2. Replace the medium with 100 μL DMEM supplemented with 10% FBS (containing 1 mM DFP), and culture the cells for 24 h.
3. Observe the cells using a fluorescence microscope (Fig. 1). Mito-Keima can be observed under a fluorescence microscope with a specific filter set for detecting the two excitation peaks of Keima (*see* **Note 4**). The mito-Keima signals post-excitation at 590 or 440 nm indicate the mitochondria delivered into the lysosomes by mitophagy (seen as punctate structures) or the mitochondria present in the cytoplasm (seen as long or short tubular structures), respectively.

3.2.2 Hypoxia-Induced Mitophagy (See Note 5)

1. Plate the stable cell line expressing mito-Keima (from Subheading 3.1.5) at a density of 1×10^4 cells per well into a 96-well glass-bottom plate in 100 μL DMEM supplemented with 10% FBS, and culture overnight (*see* **Note 3**).
2. Transfer the plate to a hypoxic chamber, and culture for 24 h (*see* **Note 6**).
3. Observe the cells directly using a fluorescence microscope (*see* Subheading 3.2.1, Fig. 1).

3.3 Estimating the Extent of Mitophagy

1. The extent of mitophagy induced can be estimated by counting the number of punctate structures observed after excitation at 590 nm.
2. Here, we show the method for measuring mitophagy using the MetaMorph software.
3. Set the threshold values, and run the region measurements program to count the number of mitophagy dots per cell.
4. Calculate the percentage of cells with over 15 mitophagy dots as mitophagy-positive, after analyzing at least 50 cells.

4 Notes

1. We have tested HeLa, SH-SY5Y, and MEF cells using this method. In theory, most mammalian cultured cells can be utilized. Recently, mito-Keima was used for the detection of mitophagy in mice [29]. In the mito-Keima-expressing transgenic mice, mitophagy can be detected in most cell types using mito-Keima as a reporter.
2. The cells transiently expressing mito-Keima can also be used for examining mitophagy. However, when mito-Keima shows extremely strong or weak cellular expression, mitophagy may not be detected with a good signal-to-noise ratio. The cells showing high mito-Keima expression may occasionally give false-positive signals. Hence, we strongly recommend generating stable cell lines expressing mito-Keima and to specifically clone the cells showing moderate mito-Keima expression.
3. We regularly use the 96-well glass-bottom plates for mito-Keima analysis. In case of the mito-Keima-expressing HeLa cells, the cells are plated onto the 96-well glass-bottom plates at a density of 1×10^4 cells per well and cultured overnight, until they reach 70–80% confluency. Different sizes of glass-bottom dishes/plates can be used, but cellular confluency should be adjusted accordingly.
4. Mito-Keima is observed using a fluorescence microscope equipped with a specific filter set for detecting the two excitation peaks of Keima. Mito-Keima localizes to the mitochondrial matrix where the neutral environment is excited using the 430 ± 24-nm excitation filter (Chroma), whereas mito-Keima delivered into the lysosomal lumen is excited using the 560 ± 40-nm excitation filter (Chroma). Both excitation peaks are detected using the 624 ± 20-nm emission filter (Chroma) and a dichroic mirror for Texas Red (Semrock).
5. To induce hypoxic conditions, the cells were cultured under conditions of 1% O_2 and 5% CO_2 in a multigas incubator (Astec corporation: APM-30D).
6. This method using mito-Keima is very sensitive and thus useful for detecting the Parkin-dependent and Parkin-independent mitophagy, which includes hypoxia- and iron depletion-induced mitophagy [10, 17].

Acknowledgment

This work was supported in part by the Japan Society for the Promotion of Science KAKENHI [Grant numbers 17H03671 (T.K.), 16H01198 (T.K.), 16H01384 (T.K.), and 17 K15088 (S.Y.)], Yujin Memorial Grant (Niigata University School of Medicine) (T.K.), and Takeda Science Foundation (S.Y. and T.K.).

References

1. Twig G, Elorza A, Molina AJ, Mohamed H, Wikstrom JD, Walzer G, Stiles L, Haigh SE, Katz S, Las G, Alroy J, Wu M, Py BF, Yuan J, Deeney JT, Corkey BE, Shirihai OS (2008) Fission and selective fusion govern mitochondrial segregation and elimination by autophagy. EMBO J 27:433–446
2. Wallace DC (2005) A mitochondrial paradigm of metabolic and degenerative diseases, aging, and cancer: a dawn for evolutionary medicine. Annu Rev Genet 39:359–407
3. Ajioka RS, Phillips JD, Kushner JP (2006) Biosynthesis of heme in mammals. Biochim Biophys Acta 1763:723–736
4. Lill R (2009) Function and biogenesis of iron-sulphur proteins. Nature 460:831–838
5. Sheftel A, Stehling O, Lill R (2010) Iron-sulfur proteins in health and disease. Trends Endocrinol Metab 21:302–314
6. Ye H, Rouault TA (2010) Human iron-sulfur cluster assembly, cellular iron homeostasis, and disease. Biochemistry 49:4945–4956
7. Anderson GJ, Vulpe CD (2009) Mammalian iron transport. Cell Mol Life Sci 66:3241–3261
8. Schultz IJ, Chen C, Paw BH, Hamza I (2010) Iron and porphyrin trafficking in heme biogenesis. J Biol Chem 285:26753–26759
9. Asano T, Komatsu M, Yamaguchi-Iwai Y, Ishikawa F, Mizushima N, Iwai K (2011) Distinct mechanisms of ferritin delivery to lysosomes in iron-depleted and iron-replete cells. Mol Cell Biol 31:2040–2052
10. Allen GF, Toth R, James J, Ganley IG (2013) Loss of iron triggers PINK1/Parkin-independent mitophagy. EMBO Rep 14:1127–1135
11. Geisler S, Holmstrom KM, Skujat D, Fiesel FC, Rothfuss OC, Kahle PJ, Springer W (2010) PINK1/Parkin-mediated mitophagy is dependent on VDAC1 and p62/SQSTM1. Nat Cell Biol 12:119–131
12. Matsuda N, Sato S, Shiba K, Okatsu K, Saisho K, Gautier CA, Sou YS, Saiki S, Kawajiri S, Sato F, Kimura M, Komatsu M, Hattori N, Tanaka K (2010) PINK1 stabilized by mitochondrial depolarization recruits Parkin to damaged mitochondria and activates latent Parkin for mitophagy. J Cell Biol 189:211–221
13. Novak I, Kirkin V, McEwan DG, Zhang J, Wild P, Rozenknop A, Rogov V, Lohr F, Popovic D, Occhipinti A, Reichert AS, Terzic J, Dotsch V, Ney PA, Dikic I (2010) Nix is a selective autophagy receptor for mitochondrial clearance. EMBO Rep 11:45–51
14. Sandoval H, Thiagarajan P, Dasgupta SK, Schumacher A, Prchal JT, Chen M, Wang J (2008) Essential role for Nix in autophagic maturation of erythroid cells. Nature 454:232–235
15. Schweers RL, Zhang J, Randall MS, Loyd MR, Li W, Dorsey FC, Kundu M, Opferman JT, Cleveland JL, Miller JL, Ney PA (2007) NIX is required for programmed mitochondrial clearance during reticulocyte maturation. Proc Natl Acad Sci U S A 104:19500–19505
16. Zhang Y, Goldman S, Baerga R, Zhao Y, Komatsu M, Jin S (2009) Adipose-specific deletion of autophagy-related gene 7 (atg7) in mice reveals a role in adipogenesis. Proc Natl Acad Sci U S A 106:19860–19865
17. Hirota Y, Yamashita S, Kurihara Y, Jin X, Aihara M, Saigusa T, Kang D, Kanki T (2015) Mitophagy is primarily due to alternative autophagy and requires the MAPK1 and MAPK14 signaling pathways. Autophagy 11:332–343
18. Liu L, Feng D, Chen G, Chen M, Zheng Q, Song P, Ma Q, Zhu C, Wang R, Qi W, Huang L, Xue P, Li B, Wang X, Jin H, Wang J, Yang F, Liu P, Zhu Y, Sui S, Chen Q (2012) Mitochondrial outer-membrane protein FUNDC1 mediates hypoxia-induced mitophagy in mammalian cells. Nat Cell Biol 14:177–185
19. Zhang H, Bosch-Marce M, Shimoda LA, Tan YS, Baek JH, Wesley JB, Gonzalez FJ, Semenza GL (2008) Mitochondrial autophagy is an

HIF-1-dependent adaptive metabolic response to hypoxia. J Biol Chem 283:10892–10903

20. Yamashita SI, Jin X, Furukawa K, Hamasaki M, Nezu A, Otera H, Saigusa T, Yoshimori T, Sakai Y, Mihara K, Kanki T (2016) Mitochondrial division occurs concurrently with autophagosome formation but independently of Drp1 during mitophagy. J Cell Biol 215:649–665
21. Narendra D, Kane LA, Hauser DN, Fearnley IM, Youle RJ (2010) p62/SQSTM1 is required for Parkin-induced mitochondrial clustering but not mitophagy; VDAC1 is dispensable for both. Autophagy 6:1090–1106
22. Narendra D, Tanaka A, Suen DF, Youle RJ (2008) Parkin is recruited selectively to impaired mitochondria and promotes their autophagy. J Cell Biol 183:795–803
23. Vives-Bauza C, Zhou C, Huang Y, Cui M, de Vries RL, Kim J, May J, Tocilescu MA, Liu W, Ko HS, Magrane J, Moore DJ, Dawson VL, Grailhe R, Dawson TM, Li C, Tieu K, Przedborski S (2010) PINK1-dependent recruitment of Parkin to mitochondria in mitophagy. Proc Natl Acad Sci U S A 107:378–383
24. Lazarou M, Sliter DA, Kane LA, Sarraf SA, Wang C, Burman JL, Sideris DP, Fogel AI, Youle RJ (2015) The ubiquitin kinase PINK1 recruits autophagy receptors to induce mitophagy. Nature 524:309–314
25. Katayama H, Kogure T, Mizushima N, Yoshimori T, Miyawaki A (2011) A sensitive and quantitative technique for detecting autophagic events based on lysosomal delivery. Chem Biol 18:1042–1052
26. Kageyama Y, Hoshijima M, Seo K, Bedja D, Sysa-Shah P, Andrabi SA, Chen W, Hoke A, Dawson VL, Dawson TM, Gabrielson K, Kass DA, Iijima M, Sesaki H (2014) Parkin-independent mitophagy requires Drp1 and maintains the integrity of mammalian heart and brain. EMBO J 33:2798–2813
27. Kogure T, Karasawa S, Araki T, Saito K, Kinjo M, Miyawaki A (2006) A fluorescent variant of a protein from the stony coral Montipora facilitates dual-color single-laser fluorescence cross-correlation spectroscopy. Nat Biotechnol 24:577–581
28. Violot S, Carpentier P, Blanchoin L, Bourgeois D (2009) Reverse pH-dependence of chromophore protonation explains the large Stokes shift of the red fluorescent protein mKeima. J Am Chem Soc 131:10356–10357
29. Sun N, Yun J, Liu J, Malide D, Liu C, Rovira II, Holmstrom KM, Fergusson MM, Yoo YH, Combs CA, Finkel T (2015) Measuring in vivo mitophagy. Mol Cell 60:685–696

Chapter 19

The Importance of Calcium Ions for Determining Mitochondrial Glycerol-3-Phosphate Dehydrogenase Activity When Measuring Uncoupling Protein 1 (UCP1) Function in Mitochondria Isolated from Brown Adipose Tissue

Kieran J. Clarke and Richard K. Porter

Abstract

Glycerol-3-phosphate is an excellent substrate for FAD-linked mitochondrial glycerol-3-phosphate dehydrogenase (mGPDH) in brown adipose tissue mitochondria and is regularly used as the primary substrate to measure oxygen consumption and reactive oxygen consumption by these mitochondria. mGPDH converts cytosolic glycerol-3-phosphate to dihydroxyacetone phosphate, feeding electrons directly from the cytosolic side of the mitochondrial inner membrane to the CoQ-pool within the inner membrane. mGPDH activity is allosterically activated by calcium, and when calcium chelators are present in the mitochondrial preparation medium and/or experimental incubation medium, calcium must be added to insure maximal mGPDH activity. It was demonstrated that in isolated brown adipose tissue mitochondria (1) mGPDH enzyme activity is maximal at free calcium ion concentrations in the 350 nM–1 μM range, (2) that ROS production also peaks in the 10–100 nM range in the presence of a UCP1 inhibitory ligand (GDP) but wanes with further increasing calcium concentration, and (3) that oxygen consumption rates peak in the 10–100 nM range with rates being maintained at higher calcium concentrations. This article provides easy-to-follow protocols to facilitate the measurement of mGPDH-dependent UCP1 activity in the presence of calcium for isolated brown adipose tissue mitochondria.

Key words Glycerol-3-phosphate dehydrogenase, Uncoupling protein 1, Brown adipose tissue, Mitochondria, Calcium

1 Introduction

The FAD-linked mitochondrial glycerol-3-phosphate dehydrogenase (mGPDH, EC) is a fascinating enzyme, the physiological importance of which is still an area of intensive research [1–3]. mGPDH is an integral membrane enzyme accessible from the outer side of the mitochondrial inner membrane. It consists of a single 74-kDa subunit that converts cytosolic glycerol-3-phosphate

Carlos M. Palmeira and António J. Moreno (eds.), *Mitochondrial Bioenergetics: Methods and Protocols*, Methods in Molecular Biology, vol. 1782, https://doi.org/10.1007/978-1-4939-7831-1_19,

to dihydroxyacetone phosphate. Electrons from this reaction are channelled via FAD to the coenzyme Q (CoQ)-pool within the mitochondrial inner membrane. Glycerol-3-phosphate is formed by cytoplasmic glycerol-3-phosphate dehydrogenase from dihydroxyacetone phosphate and $NADH_2$ [3]. Thus glycerol-3-phosphate acts as a substrate for oxidative phosphorylation providing electron from the cytosolic side of the mitochondrial inner membrane which are ultimately used to reduce oxygen as cytochrome oxidase [3]. mGPDH is also a direct source of reactive oxygen species [2, 4–7].

The most important endogenous activator of mGPDH is calcium [8–11]. mGPDH has a calcium binding site which the canonical EF-hand motif [12] predicted to be on the cytosolic/ intermembrane space side of the mitochondria. The binding of Ca^{2+} decreases the Km of mGPDH for its substrates [10].

Possible metabolic roles for mGPDH include (1) reoxidation of cytosolic $NADH_2$ under conditions of high glycolysis; (2) a more hard-wired electron delivery system, bypassing complex 1, to oxidise..... to oxidize cytosolic $NADH_2$ for oxidative phosphorylation without having to transfer reducing equivalents across the mitochondrial inner membrane via transporters to generate matrix $NADH_2$; (3) regulation of cytosolic glycerol-3-phosphate, a metabolite that links glycolysis and lipogenesis to oxidative phosphorylation; and (4) regulation of thermogenesis and/or metabolic efficiency [1, 3].

In mammals, expression of mGPDH is regulated predominantly at the transcriptional level by triiodothyronine [13]. mGPDH also demonstrates a high degree of differential tissue expression [14]. The lowest levels of mGPDH are found in the heart and liver whereas the highest levels are found in the muscle, brain, and brown adipose tissue (BAT) [2, 3]. Interestingly, mGPDH has a significant level of expression in β-pancreatic cells, testis, and placenta [3, 15–17].

BAT is activated in response to cold exposure as a result of the release of noradrenaline by the sympathetic nervous system resulting in what is termed non-shivering thermogenesis [18, 19]. The noradrenaline signal in BAT is mediated via the β_3-, and α_2- adrenoreceptors [20]. Noradrenaline can also activate α_1-receptors on brown adipocytes which leads to a rapid increase in the intracellular Ca^{2+} levels through release from intracellular stores and plasma membrane fluxes [21].

An increase in mitochondrial Ca^{2+} levels can have a wide range of downstream consequences, one of which is the full activation of intramitochondrial metabolite dehydrogenases, which require Ca^{2+} for full activation [11], and one such dehydrogenase accessible from the outer side of the inner membrane of brown adipose tissue mitochondria is mGPDH. The mitochondrial inner membrane of brown adipose tissue also contains mitochondrial uncoupling protein 1 (UCP1) [18, 19, 22, 23]. UCP1 acts to transfer protons across the mitochondrial inner membrane, thus facilitating a catalyzed proton leak and increased catabolism which defines the molecular basis of non-shivering thermogenesis [18, 19]. A convenient way to assay the function of UCP1 in isolated brown adipose tissue mitochondria is to measure oxygen consumption and/or reactive oxygen species production in the presence and absence of a ligand that can access and inhibit UCP1, namely, purine nucleotide, e.g., GDP [18, 19, 22, 24]. This article provides easy-to-follow protocols to facilitate the measurement of mGPDH-dependent UCP1 activity in the presence of calcium for isolated brown adipose tissue mitochondria.

2 Materials

All solutions were prepared using ultrapure water from a Millipore Elix Advantage ten Water Purification System (resistivity >5 MΩ cm) and analytical grade reagents. All reagents were stored at room temperature (unless indicated otherwise). We adhered to waste disposal regulations rigorously.

2.1 Preparation for Assay of Mitochondrial Glycerol-3-Phosphate Dehydrogenase (mGPDH)

1. Our assay medium contains 25 mM potassium phosphate, 5 mM $MgCl_2$, 2.5 mg/mL bovine serum albumin, 2 μg/mL antimycin A, 2 μg/mL rotenone, 2 mM KCN, 50 μM 2,6-dichloroindophenolate hydrate, pH 7.2 (*see* **Note 1**).

2.2 Preparative Steps for Isolation of Mitochondria from Brown Adipose Tissue

1. Isolation STE buffer: At least a day before isolation of mitochondria, weigh out 85.75 g sucrose, 121.14 g Tris, and 0.38 g EGTA. Add 900 mL water. Mix and adjust the pH to 7.4 with concentrated HCl. Make the isolation medium up to 1 L with water, and store at 4 °C (*see* **Note 2**). 200 mL of the isolation was set aside for inclusion of defatted BSA (2%, w/v) with the pH adjusted accordingly (*see* **Note 3**).
2. Isolation apparatus: Place a Dounce/Potter homogenizer and pestles (covered in tinfoil) on ice (*see* **Note 4**).
3. Brown adipose tissue: Five young male/female Wistar rats c.12 weeks old/190 g (*see* **Note 5**).

1. The basic incubation medium for brown adipose tissue mitochondria is predominantly an ionic medium containing 120 mM KCl, 5 mM Hepes, 1 mM EGTA, and 0.1% (w/v) defatted bovine serum albumin, pH 7.4 (with KOH).
2. The Oroboros Respirometer was used to measure oxygen consumption by brown adipose tissue mitochondria (*see* **Note 6**).
3. Stocks of the following solutions are required to give final concentrations of 1 μg/mL oligomycin, 5 μM atractyloside,

2.3 Incubation Media and Apparatus for Measurement of Oxygen Consumption and Reactive Oxygen Species Production by Brown Adipose Tissue Mitochondria

and 10 mM glycerol-3-phosphate in the incubation medium for oxygen consumption rate measurements by non-phosphorylating thymus mitochondria.

4. Amplex Red conversion to Resorufin was used to detect reactive oxygen species production by mitochondria (*see* **Note** 7).
5. Stocks of the following solutions are required to final concentrations of 1 μg/mL oligomycin, 5 μM atractyloside, 10 mM glycerol-3-phosphate, 5 μM Amplex Red, 10 U/mL horseradish peroxidase, and 30 U/mL superoxide dismutase for measurements of reactive oxygen species production by non-phosphorylating thymus mitochondria.
6. The UCP1 inhibitor GDP was added to give final a concentration of 1 mM.

3 Methods

3.1 Glycerol-3-Phosphate Dehydrogenase Activity Assay

1. The activity of glycerol-3-phosphate dehydrogenase was determined by monitoring the reduction of 2,6-dichloroindophenolate at 600 nm and is based on the original assay by Dawson and Thorne [25].
2. Absorbance changes were monitored on a Shimadzu UVmini-1240 UV-Vis spectrophotometer at 30 °C.
3. Mitochondria (25 μg) were incubated in assay buffer containing 25 mM potassium phosphate, 5 mM $MgCl_2$, 2.5 mg/mL bovine serum albumin, 2 μg/mL antimycin A, 2 μg/mL rotenone, 2 mM KCN, 50 μM 2,6-dichloroindophenolate, pH 7.2.
4. Baseline activity was monitored for 3 min prior to initiation of the reaction. The reaction was initiated by the addition of glycerol-3-phosphate (20 mM final). The rate of reduction of 2,6-dichloroindophenolate was monitored for 3 min.
5. Calcium chloride was added to effect a series of final added concentrations of 0, 0.1, 0.2. 0.5, 0.75, and 1 mM.
6. The free calcium ion concentrations were calculated according to the online algorithm referenced from Schoenmakers et al. [26] (*see* **Note 8**).
7. Figure 1 demonstrates the effect of calcium on glycerol-3-phosphate dehydrogenase activity in brown adipose tissue.

3.2 Isolation of Mitochondria from Brown Adipose Tissue (BAT)

1. Rats were euthanized by carbon dioxide asphyxiation (*see* **Note** 9) in accordance with strict guidelines on animal welfare as outlined in Directive 2010/63/EU and S.I No. 543 2012.
2. BAT was removed from the interscapular region of the rat between the head and the shoulder blades and placed into a pre-weighed 50 mL beaker containing 30 mL ice-cold isolation medium.

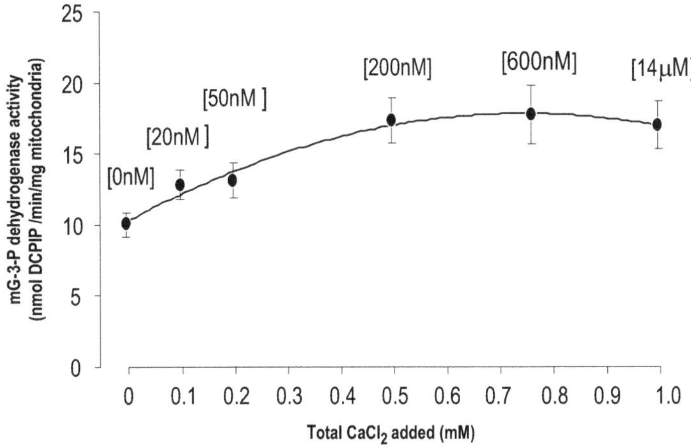

Fig. 1 Mitochondrial glycerol-3-phosphate dehydrogenase (mGPDH) activity in brown adipose tissue mitochondria increases in the presence of increasing amounts of Ca^{2+} ions. The activity of mGPDH was determined by monitoring the reduction of 2,6-dichloroindophenolate at 600 nm. Absorbance changes were monitored on a Shimadzu UVmini-1240 UV-Vis spectrophotometer at 30 °C. Mitochondria (25 μg) were incubated in 3 ml assay buffer containing 25 mM potassium phosphate, 5 mM $MgCl_2$, 2.5 mg/mL bovine serum albumin, 2 μg/mL antimycin A, 2 μg/mL rotenone, 2 mM KCN, 1 mM EGTA, 50 μM 2,6-dichloroindophenolate, pH 7.2. Baseline activity was monitored for 3 min. The reaction was initiated by the addition of 10 mM glycerol-3-phosphate and increasing concentrations of $CaCl_2$. Values in square brackets are the calculated free calcium ion concentrations based on the algorithm of Schoenmakers et al. [26]. Data is expressed as mean ± S.E.M. for at least three experiments and each experiment performed in triplicate. A one-way ANOVA analysis comparing enzymic rates at free calcium concentrations of zero with those at 200 nM gives 1.8-fold increase with a p-value of 0.0146. These data are consistent with data in the literature for the concentration of calcium required for activation of mGPDH and for the degree of activation achieved by calcium [2]. GDP(1 mM) has no effect on mGPDH enzymic activity

3. BAT mitochondria were prepared by the method of Scarpace et al. [27] (*see* **Note 10**).
4. The BAT was chopped carefully in the beaker and poured into a Potter homogenizer tube. The tissue was then homogenized by hand with four passes using a pestle of 0.26 in. (loose) clearance, followed by homogenization by hand with six passes using a pestle of 0.12 in. (tight) clearance.
5. The homogenate was then filtered through four layers of muslin cloth, and the filtrate was centrifuged at 8600 × *g* for 10 min at 4 °C.
6. The supernatant was discarded, and the sides of the centrifuge tube were wiped with tissue to remove any fat deposits.
7. The pellet was resuspended in approximately 20 mL STE buffer containing 2% (w/v) defatted BSA and centrifuged at 750 × *g* for 10 min at 4 °C.
8. The pellet was discarded, and the supernatant was centrifuged at 8600 × *g* for 10 min at 4 °C.

9. The pellet was then resuspended in STE buffer and centrifuged as above (*see* **Note 11**).
10. The resulting mitochondrial pellet was resuspended gently and thoroughly with 0.05 mL of STE buffer per gram of original tissue and used within 6–8 h of isolation (*see* **Note 12**). Any unused mitochondria were stored at −20 °C.

3.3 Measurement of Mitochondrial Reactive Oxygen Species (H_2O_2) Generation

1. A Perkin Elmer LS 55 fluorometer with excitation set at 570 ± 8 nm and emission at 585 ± 4 nm was used to detect fluorescence of Resorufin.
2. H_2O_2 was detected by Amplex Red conversion to Resorufin, essentially as described by Dlasková et al. [24] (*see* **Note 13**).
3. After addition of 100 μg mitochondria/mL incubation medium, final concentrations of 1 μg/mL oligomycin, 5 μM atractyloside, 10 mM glycerol-3-phosphate, 5 μM Amplex Red, 10 U/mL horseradish peroxidase, and 30 U/mL superoxide dismutase were added to detect reactive oxygen species production (Fig. 5).
4. GDP (1 mM final concentration), an inhibitor proton leak due to UCP1 and which increases reactive oxygen species production, was added to mitochondrial suspensions in fluorimeter cuvettes.
5. Figure 2 demonstrates the effect of calcium on glycerol-3-phosphate fuelled reactive oxygen species production by brown adipose tissue mitochondria in the presence and absence of GDP, an inhibitor of UCP1 function.

3.4 Measurement of Oxygen Consumption Rate

1. The Oroboros Respirometer was used to measure oxygen consumption by brown adipose tissue mitochondria. The respirometer contains 2 × 2 mL glass sealable chambers.
2. Oxygen consumption rates were determined essentially as described in Dlasková et al. [24].
3. After addition of 100 μg mitochondria/mL incubation medium, final concentrations of 1 μg/mL oligomycin, 5 μM atractyloside, and 10 mM glycerol-3-phosphate were added for oxygen consumption rate measurements by non-phosphorylating mitochondria (*see* **Note 14**).
4. Non-phosphorylating brown adipose mitochondrial oxygen consumption rates were measured when a stable signal was established (within a minute) for up to 2–3 min.
5. GDP (1 mM final concentration) was added to the chamber to inhibit oxygen consumption due to proton leak through UCP1.
6. Figure 3 demonstrates the effect of calcium on glycerol-3-phosphate fuelled oxygen consumption rate by brown adipose

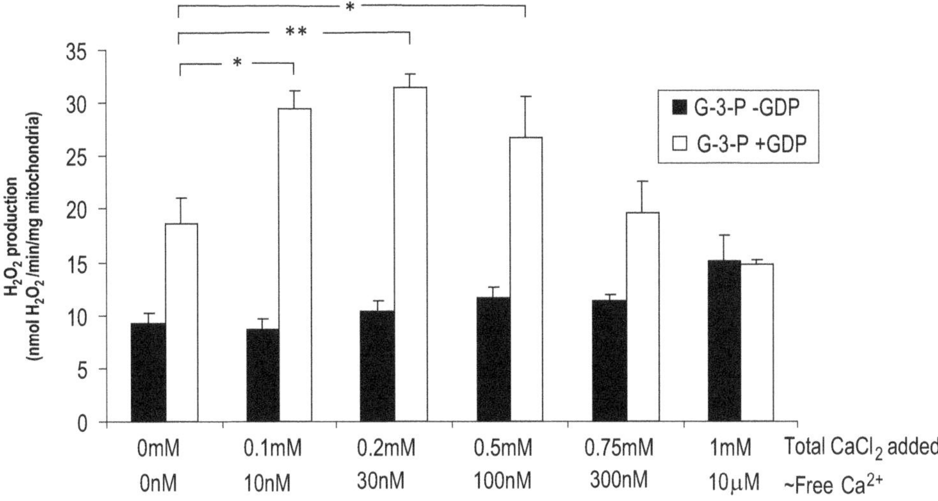

Fig. 2 GDP-sensitive H_2O_2 production by non-phosphorylating brown adipose tissue mitochondria respiring on glycerol-3-phosphate increases significantly upon addition of $CaCl_2$. H_2O_2 production rates by BAT mitochondria (100 µg/mL) were determined at 37 °C in the presence of 120 mM KCl, 5 mM Hepes-KOH pH 7.4, 1 mM EGTA, 1 µg/mL oligomycin, 5 µM atractyloside, 0.1% defatted BSA, 5 µM Amplex Red, 10 U/mL horseradish peroxidase, and 30 U/mL superoxide dismutase. Substrates were 10 mM glycerol-3-phosphate plus or minus 1 mM GDP. 1 mM GDP was added to inhibit UCP 1. Mitochondria were incubated with the desired concentration of $CaCl_2$ prior to addition of substrate. Fluorescence was detected by a Perkin Elmer LS 55 fluorometer with excitation set at 570 ± 8 nm and emission at 585 ± 4 nm. Fluorescence was calibrated using known amounts of H_2O_2 on each experimental day. Data is expressed as mean ± S.E.M. of at least three experiments, each experiment performed in triplicate (*$p < 0.05$; **$p < 0.01$). The data demonstrate no significant increase in H_2O_2 production with increasing calcium concentration by brown adipose tissue mitochondria respiring on glycerol-3-phosphate in the absence of GDP. However in the presence of GDP, H_2O_2 significantly increases (two-fold) with increasing calcium concentration (10–100 nM) but falls off with further increasing calcium concentrations. ROS production rates in this figure are consistent with data from the literature [2, 24]

tissue mitochondria in the presence and absence of GDP, an inhibitor of UCP1 function.

7. Figure 4 demonstrates the effect of calcium and ruthenium red on glycerol-3-phosphate fuelled oxygen consumption rate by brown adipose tissue mitochondria in the presence and absence of GDP.
8. Figure 5 demonstrates the effect of calcium succinate fuelled oxygen consumption rate by brown adipose tissue mitochondria in the presence and absence of GDP.

4 Notes

1. 2,6-dichloroindophenolate hydrate aka 2,6-dichlorophenolindophenol (DCPIP) sodium salt hydrate absorbs at 600 nm (blue) but when reduced is colorless.

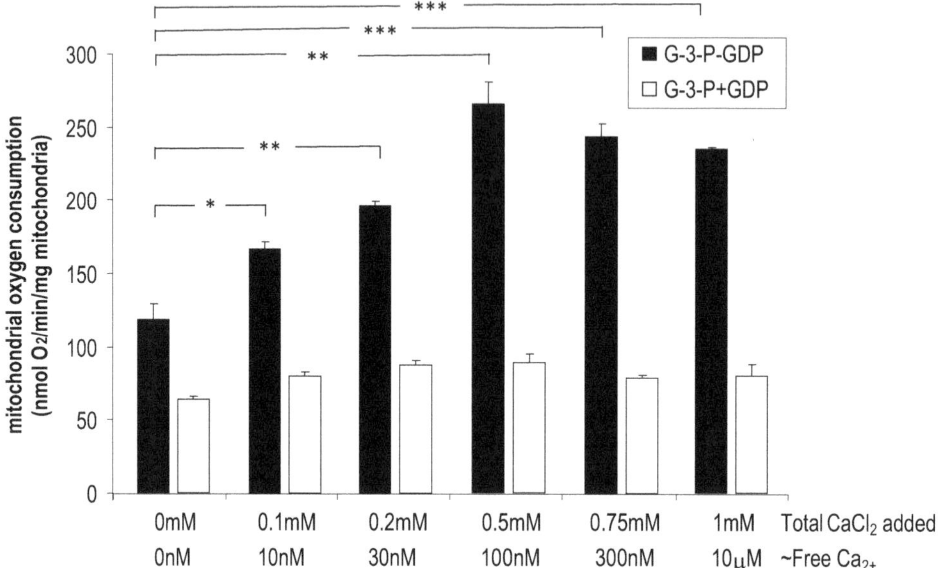

Fig. 3 Oxygen consumption by non-phosphorylating brown adipose tissue mitochondria respiring on glycerol-3-phosphate upon activation of mGPDH by $CaCl_2$. Oxygen consumption rates by BAT mitochondria (100 μg/mL) were determined at 37 °C in the presence of 120 mM KCl, 5 mM Hepes-KOH pH 7.4, 1 mM EGTA, 1 μg/mL oligomycin, 5 μM atractyloside, 0.1% defatted BSA, 10 mM glycerol-3-phosphate, and the appropriate concentration of $CaCl_2$ in a pre-calibrated Oroboros Oxygraph in the presence and absence of GDP. Steady-state oxygen consumption rates were then obtained. 1 mM GDP was added to inhibit UCP 1. Data is expressed as mean ± S.E.M. of at least three experiments, each experiment performed in triplicate (*$p < 0.05$; **$p < 0.01$; ***$p < 0.001$). The data demonstrate no significant increase in oxygen consumption with increasing calcium concentration by brown adipose tissue mitochondria respiring on glycerol-3-phosphate in the presence of GDP. However, the data show that oxygen consumption rates in the absence of GDP peak (~three-fold increase) in the 10–100 nM range (compared to zero calcium levels) with rates being maintained at higher calcium concentrations. Oxygen consumption rates in this figure are consistent with data from the literature [2, 24]

2. Mitochondria from whatever source are traditionally isolated in nonionic media and for good reason. The functionality of the resulting mitochondrial fraction is preserved in a preparation from a non-ionic medium when compared preparations with an ionic medium. In addition, the inclusion of the calcium chelator ethylene glycol-bis (2-aminoethylether)-N,N,N′, N′-tetraacetic acid (EGTA) results in a good quality mitochondrial preparation.
3. Defatted BSA is present to mop up free fatty acid which can decouple mitochondria directly and facilitate uncoupling through UCP1.
4. The tinfoil is to allow you to place the pestles in the ice without ice contacting the pestle. No ice/water should get onto the pestles or in the homogenizer.

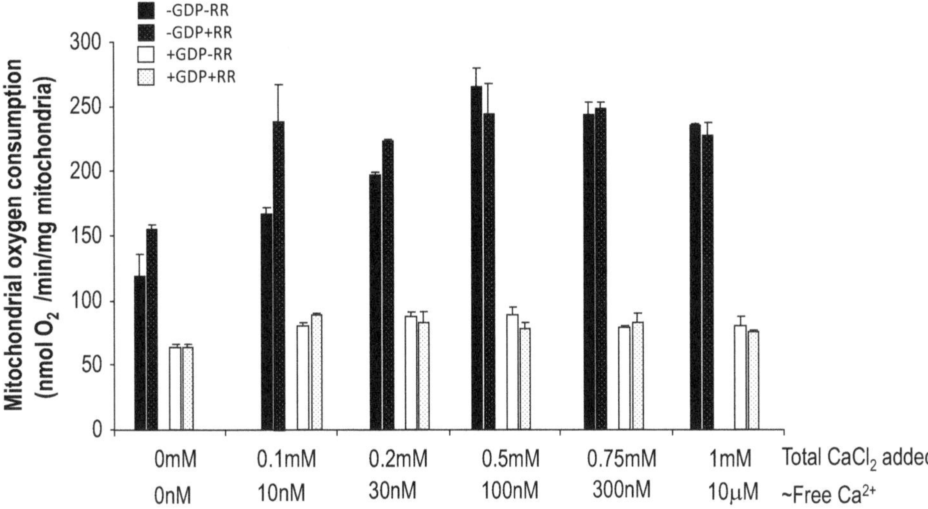

Fig. 4 Ruthenium red does not inhibit the $CaCl_2$-induced increase in oxygen consumption by brown adipose tissue mitochondria respiring on glycerol-3-phosphate. BAT mitochondria were isolated as described in Subheading 2.3 from female Wistar rats. Subsequent oxygen consumption rates by BAT mitochondria (100 μg/mL) were determined at 37 °C in the presence of 120 mM KCl, 5 mM Hepes-KOH pH 7.4, 1 mM EGTA, 1 μg/mL oligomycin, 5 μM atractyloside, 0.1% defatted BSA, 10 mM glycerol-3-phosphate, and various concentrations of $CaCl_2$ and plus or minus 5 μM ruthenium red in a pre-calibrated Oroboros Oxygraph. Steady-state oxygen consumption rates were then obtained. 1 mM GDP was added to inhibit UCP 1. Data is expressed as mean ± S.E.M. of at least three experiments, each experiment performed in triplicate. The data demonstrate a lack of effect of the calcium transporter inhibitor ruthenium red on oxygen consumption rates at varying calcium concentrations with glycerol-3-phosphate as substrate. It is concluded that calcium is manifesting its effect on mitochondrial oxygen consumption through allosteric activation of mGPDH in an extramitochondrial mechanism rather than through any intramitochondrial dehydrogenase activation

5. Young animals are selected as they have substantial brown adipose tissue when compared with older animals.
6. The Oroboros Respirometer is a very sensitive Clark-type oxygen electrode with 2 × 2 mL glass sealable chambers. It is the most sensitive oxygen electrode on the market, and further details can be obtained at the Oroboros website: http://www.oroboros.at/.
7. Amplex Red is converted to Resorufin in the presence of hydrogen peroxide by the enzyme horseradish peroxidase included in the assay medium. As electrons "escape" from the electron transport chain, they generate superoxide. Inclusion of superoxide dismutase in the assay medium converts superoxide to hydrogen peroxide and oxygen. So in short the Amplex Red assay indirectly measures superoxide production by mitochondria, by directly measuring hydrogen peroxide production rate, which in turn is detected by the increase in fluorescence due to the increase in abundance of Resorufin from Amplex Red. Steady-state hydrogen peroxide production

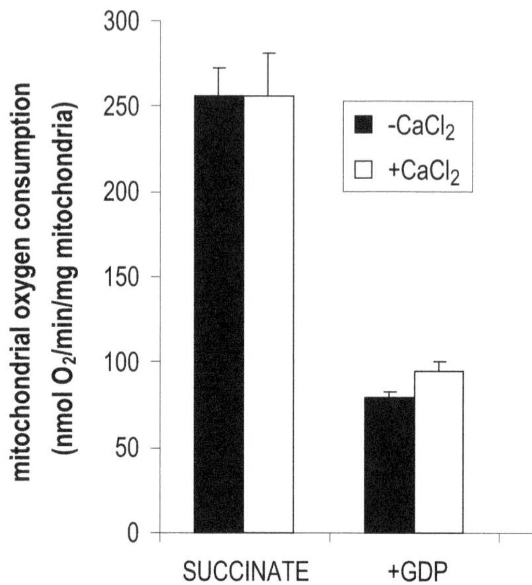

Fig. 5 $CaCl_2$ addition to succinate supported brown adipose tissue mitochondria does not affect oxygen consumption. Oxygen consumption rates by BAT mitochondria (100 μg/mL) were determined at 37 °C in the presence of 120 mM KCl, 5 mM Hepes-KOH pH 7.4, 1 mM EGTA, 1 μg/mL oligomycin, 5 μM atractyloside, 0.1% defatted BSA, 10 mM succinate, 1 μM rotenone, and 0.2 mM added $CaCl_2$ (~30 nM free Ca^{2+}) in a pre-calibrated Oroboros Oxygraph. Steady-state oxygen consumption rates were then obtained. 1 mM GDP was added to inhibit UCP 1. Data is expressed as mean ± S.E.M. of at least three experiments, each experiment performed in triplicate. The data demonstrate a lack of effect of calcium on mitochondrial oxygen consumption with succinate as substrate which again supports the observation in Fig. 3 that calcium is effecting mGPDH activity, but clearly not succinate dehydrogenase activity. Oxygen consumption rates in this figure are consistent with data from the literature [2, 24]

was measured using a Perkin Elmer LS 55 fluorometer set to detect Resorufin with excitation wavelength at 570 ± 8 nm and emission wavelength set at 585 ± 4 nm.

8. The algorithm for Ca-EGTA Calculator v1.3 using constants from Theo Schoenmakers' Chelator can be found at https://web.stanford.edu/~cpatton/CaEGTA-TS.htm.

9. Asphyxiation by carbon dioxide is an approved method for euthanasia of rats/mice. The method limits any damage or bleeding into the thymus when compared with other forms of euthanasia such as cervical dislocation.

10. In Scarpace et al. [27], mitochondria are isolated by differential centrifugation.

11. A second wash of the pellet in the absence of defatted BSA is so as not to contaminate the protein determination assay with other protein from the original sample.
12. We have used the bicinchoninic acid assay described by Smith et al. [28] as a means to quantify of protein concentration in the final mitochondrial suspension.
13. Interestingly, direct superoxide detection by ethidium bromide was not increased on ablation of UCP1 when measured in isolated mouse brown adipose tissue mitochondria [24]. Whereas increased hydrogen peroxide production was detectable in the presence of horseradish peroxidase and superoxide dismutase in isolated mouse brown adipose tissue mitochondria from UCP1 knockout mice when compared to wild-type mice when Amplex Red was used [24].
14. Atractyloside inhibits the adenine nucleotide carrier, and oligomycin inhibits the ATP synthase, insuring non-phosphorylating mitochondria.

References

1. Brown LJ, Koza RA, Everett C, Reitman ML, Marshall L, Fahien LA, Kozak LP, MacDonald MJ (2002) Normal thyroid thermogenesis but reduced viability and adiposity in mice lacking the mitochondrial glycerol phosphate dehydrogenase. J Biol Chem 277:32892–32898
2. Orr AL, Quinlan CL, Perevoshchikova IV, Brand MD (2012) A refined analysis of superoxide production by mitochondrial sn-glycerol-3-phosphate dehydrogenase. J Biol Chem 287:42921–42935
3. Mráček T, Drahota Z, Houštěk J (2013) The function and the role of the mitochondrial glycerol-3-phosphate dehydrogenase in mammalian tissues. Biochim Biophys Acta 1827:401–410
4. Drahota Z, Chowdhury SK, Floryk D, Mráček T, Wilhelm J, Rauchova H, Lenaz G, Houštěk J (2002) Glycerophosphate-dependent hydrogen peroxide production by brown adipose tissue mitochondria and its activation by ferricyanide. J Bioenerg Biomembr 34:105–113
5. Vrbacký M, Drahota Z, Mráček T, Vojtíšková A, Ješina P, Stopka P, Houštěk J (2007) Respiratory chain components involved in the glycerophosphate dehydrogenase-dependent ROS production by brown adipose tissue mitochondria. Biochim Biophys Acta 1767:989–997
6. Mráček T, Holzerová E, Drahota Z, Kovářová N, Vrbacký M, Ješina P, Houštěk J (2014) ROS generation and multiple forms of mammalian mitochondrial glycerol-3-phosphate dehydrogenase. Biochim Biophys Acta 1837:98–111
7. Brand MD (2016) Mitochondrial generation of superoxide and hydrogen peroxide as the source of mitochondrial redox signaling. Free Radic Biol Med 100:14–31
8. MacDonald MJ, Brown LJ (1996) Calcium activation of mitochondrial glycerol phosphate dehydrogenase restudied. Arch Biochem Biophys 326:79–84
9. Wohlrab H (1977) The divalent cation requirement of the mitochondrial glycerol-3-phosphate dehydrogenase. Biochim Biophys Acta 462:102–112
10. Beleznai Z, Szalay L, Jancsik V (1988) Ca^{2+} and Mg^{2+} as modulators of mitochondrial L-glycerol-3-phosphate dehydrogenase. Eur J Biochem 170:631–636
11. Denton RM (2009) Regulation of mitochondrial dehydrogenases by calcium ions. Biochim Biophys Acta 1787:1309–1316
12. Brown LJ, MacDonald MJ, Lehn DA, Moran SM (1994) Sequence of rat mitochondrial glycerol-3-phosphate dehydrogenase cDNA: evidence for EF-hand calcium-binding domains. J Biol Chem 269:14363–14366
13. Dummler K, Muller S, Seitz HJ (1996) Regulation of adenine nucleotide translocase and glycerol-3-phosphate dehydrogenase expression by thyroid hormones in different rat tissues. Biochem J 317:913–918
14. Koza RA, Kozak UC, Brown LJ, Leiter EH, MacDonald MJ, Kozak LP (1996) Sequence and tissue-dependent RNA expression of

mouse FAD-linked glycerol-3-phosphate dehydrogenase. Arch Biochem Biophys 336:97–104

15. MacDonald MJ (1981) High content of mitochondrial glycerol-3-phosphate dehydrogenase in pancreatic islets and its inhibition by diazoxide. J Biol Chem 256:8287–8290
16. Idahl LA, Lembert N (1995) Glycerol-3-phosphate-induced ATP production in intact mitochondria from pancreatic B-cells. Biochem J 312:287–292
17. Honzik T, Drahota Z, Bohm M, Ješina P, Mráček T, Paul J, Zeman J, Houštěk J (2006) Specific properties of heavy fraction of mitochondria from human-term placenta - glycerophosphate-dependent hydrogen peroxide production. Placenta 27:348–356
18. Nicholls DG, Locke RM (1984) Thermogenic mechanisms in brown fat. Physiol Rev 64:1–64
19. Cannon B, Nedergaard J (2004) Brown adipose tissue: function and physiological significance. Physiol Rev 84:277–359
20. Zhao J, Cannon B, Nedergaard J (1997) alpha1-Adrenergic stimulation potentiates the thermogenic action of beta3-adrenoreceptor-generated cAMP in brown fat cells. J Biol Chem 272:32847–32856
21. Koivisto A, Siemen D, Nedergaard J (2000) Norepinephrine-induced sustained inward current in brown fat cells: alpha(1)-mediated by nonselective cation channels. Am J Physiol Endocrinol Metab 279:E963–E977
22. Klingenberg M, Huang SG (1999) Structure and function of the uncoupling protein from brown adipose tissue. Biochim Biophys Acta 1415:271–296
23. Carroll AM, Porter RK, Morrice NA (2008) Identification of serine phosphorylation in mitochondrial uncoupling protein 1. Biochim Biophys Acta 1777:1060–1065
24. Dlasková A, Clarke KJ, Porter RK (2010) The role of UCP 1 in production of reactive oxygen species by mitochondria isolated from brown adipose tissue. Biochim Biophys Acta 1797:1470–1476
25. Dawson AP, Thorne CJR (1969) Preparation and some properties of L-3- glycerophosphate dehydrogenase from pig brain mitochondria. Biochem J 111:27–34
26. Schoenmakers TJ, Visser GJ, Flik G, Theuvenet AP (1992) CHELATOR: an improved method for computing metal ion concentrations in physiological solutions. BioTechniques 12:870–879
27. Scarpace PJ, Bender BS, Borst SE (1991) *Escherichia coli* peritonitis activates thermogenesis in brown adipose tissue: relationship to fever. Can J Physiol Pharmacol 69:761–766
28. Smith PK, Krohn RI, Hermanson GT, Mallia AK, Gartner FH, Provenzano MD, Fujimoto EK, Goeke NM, Olson BJ, Klenk DC (1985) Measurement of protein using bicinchoninic acid. Anal Biochem 150:76–85

Chapter 20

Isolation and Analysis of Mitochondrial Small RNAs from Rat Liver Tissue and HepG2 Cells

Julian Geiger and Louise T. Dalgaard

Abstract

The presence of noncoding RNAs, such as microRNAs (miRNAs), in mitochondria has been reported by several studies. The biological roles and functions of these mitochondrial miRNAs ("mitomiRs") have not been sufficiently characterized, but the mitochondrial localization of miRNAs has recently gained significance due to modified mitomiR-populations in certain states of diseases. Here, we describe the isolation and analysis of mitochondrial RNAs from rat liver tissue and HepG2 cells. The principle of the analysis is to prepare mitochondria by differential centrifugation. Cytosolic RNA contamination is eliminated by RNase A treatment followed by Percoll gradient purification and RNA extraction. Small RNA content is verified by capillary electrophoresis. Mitochondrial miRNAs are detected by qPCR following synthesis of cDNA. After qPCR-based mitomiR-profiling, the Normfinder algorithm is applied to identify the suitable reference miRNAs to use as normalizers for mitochondrial input and data analysis. The described procedure depicts a simple way of isolating and quantifying mitomiRs in tissue and cell culture samples.

Key words Mitochondria, RNA, MicroRNA, Mitochondrial purification, Percoll gradient, MitomiR, Normfinder, qPCR

1 Introduction

The mitochondrion constitutes a central organelle in cellular energy homeostasis, metabolism, and viability. Recently, studies have reported the presence of noncoding RNAs, such as microRNAs (miRNAs), in mitochondrial fractions of different model systems [1–5]. Although some noncoding RNAs originate from the mitochondrial genome [1, 5], the majority of mitochondrial miRNAs (mitomiRs) seem to be imported into mitochondria by currently unidentified mechanisms [6]. In general, mitochondrial RNA trafficking is an understudied area. Investigations on tRNA import suggest a high degree of species variation [7] which might also apply to other types of noncoding RNAs.

The biological roles of mitomiRs are still elusive, yet their mitochondrial localization is subject to dynamic regulation.

Carlos M. Palmeira and António J. Moreno (eds.), *Mitochondrial Bioenergetics: Methods and Protocols*, Methods in Molecular Biology, vol. 1782, https://doi.org/10.1007/978-1-4939-7831-1_20,

Occurrence of stress events, such as traumatic brain injury [8] or diabetes [9], are associated with a redistribution of mitochondrial miRNAs. These observations suggest the possibility that mitomiRs might be involved in modifying mitochondrial performance and could potentially contribute to the emergence of mitochondrial dysfunction in metabolic diseases such as diabetes. The liver, as one of the major target tissues of insulin action, has therefore gained great attention in diabetes research. Interestingly, hepatic mitochondria have also been shown to contain miRNAs [4, 10].

The following chapter describes experimental procedures to extract and analyze small RNAs from mitochondria of rat liver tissue and HepG2 cells. Cytosolic RNA contamination and RNA attached to the outer surface of mitochondria is removed from crude mitochondrial fractions by RNase A digestion. Mitochondria are further purified on a Percoll gradient, before RNA extraction. The purity of mitochondria is estimated by Western blotting for cytosolic (GAPDH) and mitochondrial (MnSOD) marker proteins. Quality of obtained RNA samples is assessed using an automated capillary electrophoresis system. MicroRNAs present in mitochondrial extracts are detected by reverse transcriptase-mediated quantitative polymerase chain reaction (RT-qPCR). Furthermore, we demonstrate how to identify potential reference genes from miRNA profiles with the use of the Normfinder algorithm [11].

2 Materials

All solutions are prepared using ultrapure water (18MΨ, 25 °C). Unless otherwise noted, all chemicals were of pro analytical grade and purchased from Sigma-Aldrich.

2.1 Isolation of Mitochondrial Fraction

2.1.1 Mitochondrial Isolation from Liver Tissue

1. Animal (rat, mouse).
2. Dounce homogenizer with Teflon pestle, glass Potter-Elvehjem homogenizer.
3. Power drill.
4. Scissors.
5. Tweezer, blunt.
6. Ice.
7. Serological pipets (5 mL or 10 mL).
8. Preparation buffer: 220 mM mannitol, 70 mM sucrose, 2 mM Hepes. Add 0.5 mg mL^{-1} BSA (fraction V, Calbiochem) fresh on the day of preparation. Adjust pH = 7.4 with KOH at 4 °C.
9. MSTPi buffer: 0.5 mM EDTA, 225 mM mannitol, 75 mM sucrose, 20 mM Tris, 10 mM KH_2PO_4. Adjust pH = 7.0 with KOH at room temperature.

10. RNase A (100 mg mL^{-1} dissolved in water, Sigma).
11. Percoll (15%, 23%, 40% vol/vol diluted with preparation buffer).
12. Centrifugation tubes, sterile (50 mL, 15 mL, 2 mL, 1.5 mL).
13. Glass beaker (100 mL, clean).
14. Table top centrifuge with cooling for 1.5 mL and 2 mL tubes (VWR).
15. Tube racks for tube sizes 50 mL, 15 mL, 2 mL, 1.5 mL.
16. Refrigerated centrifuge (Sorvall) with buckets for 50 mL and 15 mL tubes.
17. TriReagent (T9424, Sigma).

2.1.2 Mitochondrial Isolation from HepG2 Cells (Based on Ref. [12])

1. Culture of HepG2 cells (ATCC HB-8065): Culture medium: RPMI 1640 (Lonza) supplemented with 10% FBS (Hyclone) and 1% penicillin-streptomycin (Gibco). Cells are subcultivated 1:5 two times per week by trypsinization using trypsin-EDTA (Gibco).
2. Six 15 cm diameter plates with HepG2 cells (80% confluent).
3. STE buffer: 250 mM sucrose, 5 mM Tris, 2 mM EGTA. Adjust pH = 7.4 at 4 °C.
4. STE buffer supplemented with 0.5% (weight/vol) BSA.
5. RNase A (100 mg mL^{-1}, dissolved in water).
6. Percoll (25% (vol/vol) diluted with STE buffer, Sigma).
7. Ice-cold phosphate-buffered saline (PBS).
8. Cell scrapers (BD Falcon).
9. Glass-glass Dounce homogenizer (40 mL) with glass pestle.
10. 250 μM gauze filter.
11. Syringe (10 mL).
12. Needle (20G).
13. Centrifuge tubes, clean (50 mL, 15 mL, clear-sighted plastic, and adapters for centrifuge rotors).
14. Microcentrifuge tubes, sterile (2 mL, 1.5 mL).
15. Refrigerated centrifuge (Sorval) with buckets for 50 mL and 15 mL tubes.
16. Table top centrifuge with cooling for 1.5 mL and 2 mL tubes (VWR, pre-cooled).
17. TriReagent (T9424, Sigma).

2.2 Immunoblotting

1. 12% MOPS SDS-PAGE gel (NP0342BOX, NuPAGE, Thermo Fisher).
2. XCell SureLock Mini-Cell Electrophoresis System (Thermo Fisher).
3. NuPAGE MOPS SDS Running buffer (20×) (NP0001, Thermo Fisher).
4. BCA kit for protein determination (#23225, Pierce).
5. Nitrocellulose blotting membranes, precut (LC2001, Thermo Fisher).
6. NuPAGE transfer buffer (20×) (NP0006, Thermo Fisher).
7. Loading buffer: NuPAGE LDS Sample Buffer (4×) (NP0007, Thermo Fisher).
8. NuPAGE reducing agent (10×) (Thermo Fisher).
9. PageRuler protein ladder, 10–180 kDa (#26616, Thermo Fisher).
10. Ponceau S solution: 0.25% (weight/vol) Ponceau S dye powder, 1% (vol/vol) acetic acid, dissolved in water.
11. Tris-buffered saline (TBS): 50 mM Tris–HCl, 27 mM KCl, 138 mM NaCl
12. TBS-T: TBS, 0.1% Tween 20 (vol/vol).
13. Blocking buffer: TBS, 5% milk powder (weight/vol).
14. Antibodies: anti-GAPDH (ab8245, Abcam), anti-MnSOD (ADI-SOD-111, Enzo Life Sciences), HRP-coupled anti-mouse (Santa Cruz), HRP-coupled anti-rabbit (Dako).
15. Enhanced chemiluminescence Western blotting substrate (#32106, Pierce)
16. Thermoblock (Fisher Scientific).
17. G:Box CCD camera system (Syngene).

2.3 RNA Isolation

1. RNase-free water—diethylpyrocarbonate (DEPC) treated.
2. Ethanol (70%, vol/vol) freshly prepared with RNase-free water.
3. Chloroform.
4. Isopropanol.
5. RNase-free filter tips (1000 μL, 100 μL, 10 μL, ART).
6. RNase-free autoclaved microcentrifuge tubes (1.5 mL, 2 mL).

2.4 RNA Quality Control

1. Experion capillary gel electrophoresis instrument (BioRad).
2. RNA chip (standard sensitivity, #7007153, BioRad).
3. RNA analysis kit (standard sensitivity, #7007154, BioRad).

2.5 miRNA Expression Profiles

1. Thermocycler (Veriti thermal cycler, Thermo Fisher).
2. TaqMan MicroRNA Reverse Transcription Kit (#4366596, Thermo Fisher).
3. Rodent megaplex RT Primers (Rodent Pool Set v3.0, Thermo Fisher).
4. VIIA7 quantitative PCR instrument (Thermo Fisher).
5. TaqMan microRNA Array Cards (Thermo Fisher).
6. Centrifuge (Heraus Megafuge) with buckets and adaptors for TaqMan Array Cards (4442571, Thermo Fisher).
7. TaqMan Universal PCR Master Mix (4324018, Thermo Fisher).
8. RNase-free filter tips (1000 μL, 100 μL, 10 μL, ART).
9. QuantStudio Real-Time PCR system software (Version 1.2, Thermo Fisher).

2.6 Identification of Stable Reference MicroRNAs

1. R (https://www.r-project.org/).
2. R-Studio (https://www.rstudio.com/).
3. Normfinder (https://moma.dk/normfinder-software).

3 Methods

3.1 Isolation of Mitochondrial Fraction

3.1.1 Mitochondrial Isolation from Liver Tissue (Adapted from Ref. [13])

1. Dissect the liver out of a sacrificed animal. Place the tissue on ice in a beaker containing ca. 50 mL ice-cold preparation buffer (*see* **Note 1**).
2. Chop the liver into small pieces using scissors and tweezers. Rinse twice in preparation buffer.
3. Add ca. 50 mL ice-cold preparation buffer, and transfer the sample into a 50 cm^3 glass Potter-Elvehjem homogenizer.
4. Homogenize the liver using approximately ten up/down strokes of a Teflon pestle at medium speed.
5. After homogenization, the volume is distributed equally into two 50 mL plastic centrifugation tubes with screw on lids. Centrifuge at 600 × *g* for 15 min, at 4 °C.
6. Decant the supernatant carefully in two new 50 mL plastic centrifugation tubes with screw-on lids. Close the tubes after balancing and centrifuge at 8000 × *g* for 10 min, at 4 °C.
7. Carefully discard the supernatant, and wipe any residing lipid residues on the vessel walls with a non-fuzzing paper tissue.
8. Gently resuspend the pellet containing the crude mitochondrial fraction in 20 mL ice-cold preparation buffer using a 5 mL or 10 mL serological pipet.

9. Add 4 μL RNase A per 20 mL volume, mix by inversion, and incubate on ice for 10 min (*see* **Note 2**).
10. Centrifuge the tubes at 8000 × *g* for 10 min, at 4 °C. Discard the supernatant, and gently resuspend the pellet in 20 mL ice-cold preparation buffer. Centrifuge the tubes again at 8000 × *g* for 10 min, at 4 °C, and discard the supernatant to remove residual RNase A.
11. Resuspend each pellet in 3 mL Percoll solution (15%, vol/vol), and transfer the volume to the bottom of a clear plastic 15 mL centrifugation tube. Use a syringe with needle to apply 3 mL Percoll solution (23% vol/vol) to the bottom of the centrifugation tube. Apply the same technique to add a third layer of 3 mL Percoll solution (40%, vol/vol) to the bottom of the tube. After balancing the tubes with Percoll-solution, centrifuge for 30,700 × *g* for 10 min, at 4 °C. Disable any break at the end of the centrifugation cycle to preserve the layer structure of the sample.
12. Use a syringe with needle to harvest the lower of the two appearing layers, and transfer it into a 15 mL tube.
13. Wash by adding cold preparation buffer to a volume of 15 mL and centrifuging at 16,600 × *g* for 10 min, at 4 °C.
14. Carefully remove the supernatant, and resuspend the loose pellet in cold 10 mL MSTPi buffer. Centrifuge 6300 × *g* for 10 min, at 4 °C.
15. Discard the supernatant, and resuspend the pellet in 1 mL cold MSTPi-buffer. Transfer a 30 μL aliquot in a 1.5 mL tube for mitochondrial quality control. Centrifuge the rest in a 2 mL tube at 8000 × *g* for 10 min, 4 °C.
16. Remove the supernatant, and dissolve the mitochondrial pellet in 1 mL TriReagent using a pipet or a vortexer.
17. Store the sample at −80 °C until further analysis.

3.1.2 Mitochondrial Isolation from HepG2 Cells (Adapted from Ref. [12])

1. Prepare HepG2 cells from six 15 cm dishes for harvest by removing the medium and washing the cells twice with ice-cold PBS.
2. Add 1 mL cold STE buffer (supplemented with 0.5% BSA) to each plate, and use a cell scraper to detach the cells. Combine cells from all six plates in one 15 mL centrifugation tube.
3. Remove two 400 μL aliquots for later preparation of total cell lysates, and store them at −80 °C.
4. Centrifuge the rest of the cells at 500 × *g* for 5 min, at 4 °C.
5. Discard the supernatant, and resuspend the pellet in cold 20 mL STE buffer containing 0.5% BSA.

6. Homogenize the cells in a 40 mL glass Dounce homogenizer by applying 15 up/down strokes using a loose fitted glass pestle.
7. Centrifuge the homogenate at 1000 × *g* for 10 min, at 4 °C.
8. Transfer the supernatant into a new 50 mL centrifuge tube by filtering through 250 μM gauze.
9. Resuspend the pellet in 15 mL cold STE buffer containing 0.5% BSA. Homogenize again with 15 strokes as described. Centrifuge the homogenate at 1000 × *g* for 10 min, at 4 °C.
10. Transfer the supernatant obtained from the previous step into a new 50 mL centrifuge tube by filtering through 250 μM gauze.
11. Combine both filtered supernatants, and centrifuge at 10,400 × *g* for 10 min, at 4 °C to pellet the crude mitochondrial fraction. Discard the supernatant, and resuspend the pellet in 20 mL cold STE buffer.
12. Add 4 μL Rnase A solution per 20 mL volume, mix by inversion, and incubate on ice for 10 min.
13. Centrifuge the sample at 10,400 × *g* for 10 min, at 4 °C. For washing, remove the supernatant, and carefully resuspend the pellet in 20 mL cold STE buffer. Repeat the centrifugation step (10,400 × *g*, 10 min, at 4 °C), and remove the supernatant.
14. Resuspend the crude mitochondrial pellet carefully in 100 μL cold STE buffer. Transfer the sample into a 2 mL tube.
15. Centrifuge at 16,000 × *g* for 2 min, at 4 °C, and remove the supernatant.
16. Carefully resuspend the pellet in 200 μL STE. Slowly layer the suspension of top of 5 mL Percoll solution (25% vol/vol) in a 15 mL tube. After balancing the tubes with Percoll solution, centrifuge for 80,000 × *g* for 20 min, at 4 °C. Disable any break at the end of the centrifugation cycle to preserve the layer structure of the sample.
17. Collect the lower of the two appearing layers with a plastic pipet, and transfer it into a new 15 mL centrifuge tube.
18. Add cold STE buffer to a volume of 15 mL, and centrifuge at 10,000 × *g* for 10 min, at 4 °C.
19. Discard the supernatant, and dissolve the pellet in 300 μL STE buffer. Transfer a 50 μL aliquot in a 1.5 mL tube for mitochondrial quality control. Centrifuge the rest in a 2 mL tube at 16,000 × *g* for 2 min, at 4 °C.
20. Remove the supernatant, and resuspend the mitochondrial pellet in 1 mL TriReagent using a pipet and a vortexer.
21. Store the sample at −80 °C until further analysis.

3.2 Immunoblotting

1. For detection of marker proteins by Western blotting, mitochondrial aliquots can be used directly after one freezing/thawing cycle (*see* **Note 3**).
2. Determine protein concentration for each sample using the BCA kit.
3. 15–25 μg protein are mixed with the appropriate volume of NuPAGE LDS Sample Buffer (4×) and NuPAGE reducing agent (10×). Samples are denatured at 70 °C for 10 min and briefly spun down.
4. For gel electrophoresis, the Xcell SureLock system was used with NuPAGE reagents. The samples were loaded on a purchased 12% MOPS SDS-PAGE gel and separated for ca. 45 min at 180 V.
5. The separated proteins were transferred for 1 h at 30 V onto a nitrocellulose membrane via a wet blotting system (Xcell SureLock transfer module).
6. Check the successful transfer by Ponceau S staining. The membrane is stained in Ponceau S solution for a few minutes on a rocking device. Destain the membrane for a few minutes in water to visualize the protein bands.
7. Wash the membrane in TBS under gentle rocking for 5 min at room temperature.
8. Block the membrane in blocking buffer (5% milk in TBS) for 2 h at room temperature under gentle rocking.
9. Antibodies against desired marker proteins are diluted in blocking buffer (*see* **Note 4**). The membrane is incubated in the respective antibody solution under gentle rocking overnight, at 4 °C (*see* **Note 5**).
10. The next day, wash the membrane three times for 15 min in TBS-T under gentle rocking at room temperature.
11. A secondary antibody solution is diluted 1:1000 in blocking buffer and added on the membrane. Incubate for 2 h at room temperature using gentle rocking.
12. Wash for three times for 10 min in TBS-T using gentle rocking.
13. For ECL detection, prepare the Western blotting substrate solution as instructed by the manufacturer. Dry the membrane by tapping excess fluid on a paper towel. Then apply the ECL mix onto the membrane so that the whole area is covered with ECL reagent. Incubate for 5 min in the dark at room temperature before tapping off excessive detection. Capture the ECL signal by the G:Box camera system (Fig. 1).

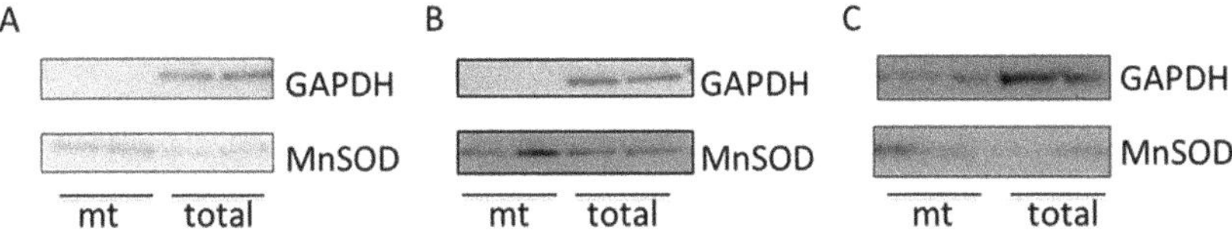

Fig. 1 Western blotting of the cytosolic marker protein GAPDH and the mitochondrial marker protein MnSOD in mitochondrial (mt) and total protein lysates (total). Samples were obtained from (**a**) rat liver tissue, (**b**) mouse liver tissue, and (**c**) HepG2 cells. GAPDH, glyceraldehyde-3-dehydrogenase; MnSOD, manganese-dependent superoxide dismutase

3.3 RNA Isolation

1. Thaw the frozen samples until they reach room temperature. Mix samples by inversion.
2. Incubate all samples at least 5 min at room temperature (*see* **Note 6**).
3. Add 200 μL chloroform, and vortex each tube for 15 s.
4. Incubate the samples for 15 min, at room temperature.
5. Centrifuge the tubes at 12,000 × *g* for 15 min, at 4 °C.
6. Transfer the RNA-containing upper phase to a new 1.5 mL plastic tube (*see* **Note 7**).
7. Add 500 μL isopropanol, mix by inversion, and incubate the samples for 10 min, at room temperature.
8. Centrifuge the tubes at 12,000 × *g* for 10 min, at 4 °C, and remove the supernatant.
9. Wash the white pellet by adding 1 mL ethanol (75% vol/vol), and briefly vortex and centrifuge at 7500 × *g* for 10 min, at 4 °C. Repeat the washing procedure once.
10. Open tubes and air-dry for ca. 10 min at room temperature.
11. Dissolve the RNA pellet in 30 μL RNase-free water or buffer.
12. Determine the RNA concentration using a nanospectrophotometer (*see* **Note 8**).

3.4 RNA Quality Control

1. Determine RNA integrity using the Experion bioanalyzer system from BioRad. Briefly, prepare 12 RNA samples as described by the manufacturer (*see* **Note 9**).
2. Prepare a standard RNA chip (*see* **Note 10**).
3. Load samples on the chip.
4. Run samples on the Experion bioanalyzer machine (Fig. 2).

3.5 miRNA Expression Analysis

For downstream analysis, the expression profiles of miRNAs can be detected using RT-qPCR-based array cards (*see* **Note 11**).

1. For detection of microRNA expression profiles, a Rodent microRNA TaqMan Array Card (also known as TaqMan Low Density Array, TLDA) is used. In short, 350 ng RNA is reverse

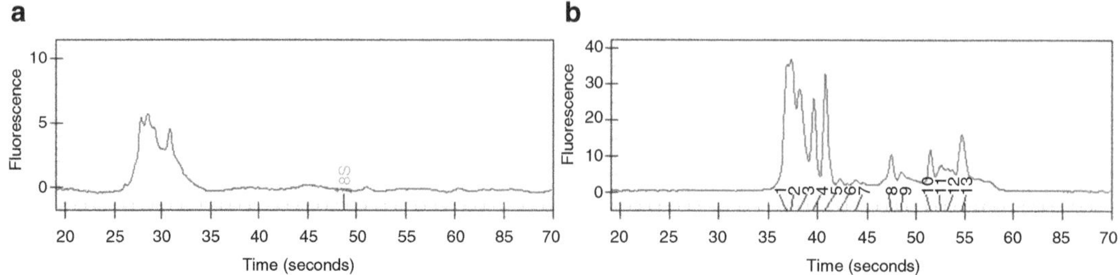

Fig. 2 Capillary gel electrophoresis of RNA isolated from (**a**) liver mitochondria and (**b**) liver tissue from BioRad Experion instrument

transcribed using the Reverse Transcription Kit with megaplex RT Primers (*see* **Note 12**).

2. Load the cDNA samples on the TLDA-card, and centrifuge the card in the centrifuge (Thermo Scientific Heraeus Megafuge 40, with card buckets) two times for 1 min, 1200 rpm (400 × *g*) at room temperature. Seal the card afterward as instructed by the manufacturer.
3. Quantitative PCR is run using the VIIA7 instrument with following thermal profile: The initial step consists of 50 °C for 2 min, followed by 95° for 10 min. Detection is performed for 40 cycles with data collection at the end of every cycle. Each cycle consists of 95 °C, for 15 s followed by 60 °C, for 1 min. All temperature changes are set to 1 °C/s.
4. PCR well definition is achieved by importing the setup file into the QuantStudio software from the CD delivered with the array cards.
5. The obtained raw data are analyzed by the QuantStudio software. Select the automatic baseline detection, and visually inspect all amplification curves for acceptable amplification. Figure 3a shows an example for acceptable amplification, i.e., miR-222, and poor amplification, i.e., miR-224.
6. Set a maximal Ct of 36 cycles to include only miRNAs with robust detection, and export the data set as an excel file or CSV file.

3.6 Identification of Potential Reference Genes

1. Prepare a data file with raw Ct values (*see* **Note 13**). Genes appear in rows, samples in columns (Fig. 3b). The header row contains sample names, while the first column contains the microRNA names.
2. Load Normfinder in R-Studio (*see* **Note 14**).
3. Run the analysis using the Normfinder command (Fig. 3c line 9) and the raw data file.
4. A ranked list of the variation of the different genes can be obtained from the result (Fig. 3c line 12, Fig. 3d). The lower

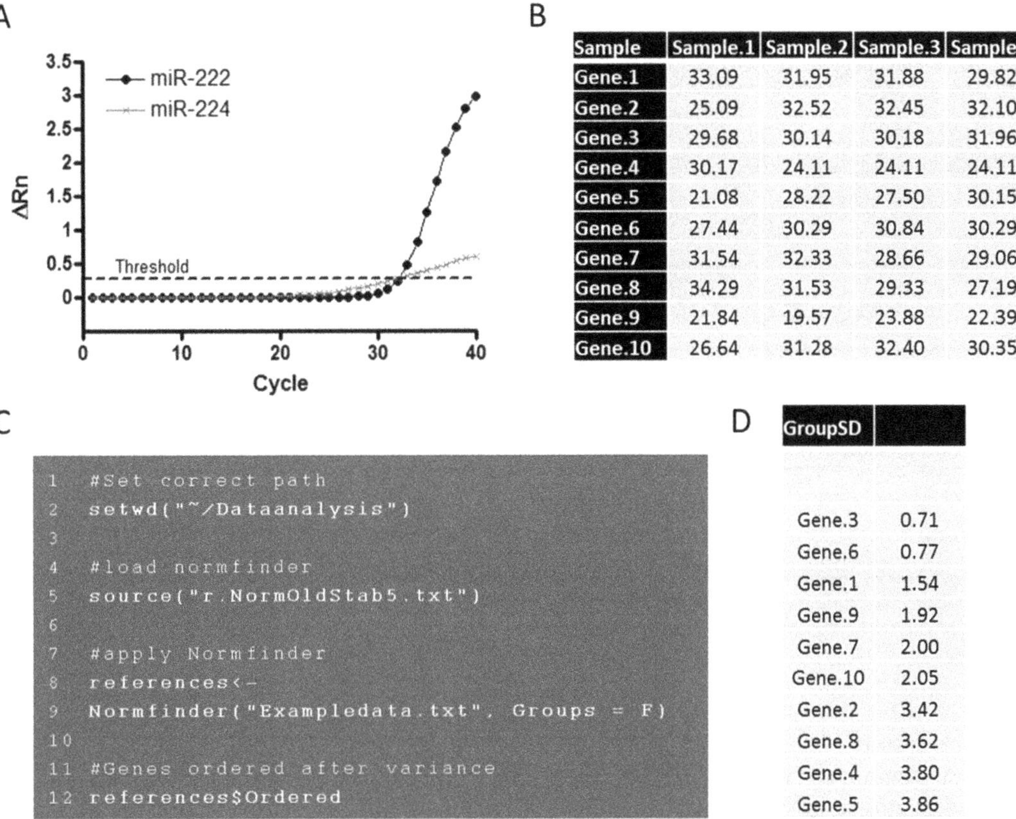

Sample	Sample.1	Sample.2	Sample.3	Sample.4
Gene.1	33.09	31.95	31.88	29.82
Gene.2	25.09	32.52	32.45	32.10
Gene.3	29.68	30.14	30.18	31.96
Gene.4	30.17	24.11	24.11	24.11
Gene.5	21.08	28.22	27.50	30.15
Gene.6	27.44	30.29	30.84	30.29
Gene.7	31.54	32.33	28.66	29.06
Gene.8	34.29	31.53	29.33	27.19
Gene.9	21.84	19.57	23.88	22.39
Gene.10	26.64	31.28	32.40	30.35

```
#Set correct path
setwd("~/Dataanalysis")

#load normfinder
source("r.NormOldStab5.txt")

#apply Normfinder
references<-
Normfinder("Exampledata.txt", Groups = F)

#Genes ordered after variance
references$Ordered
```

GroupSD	
Gene.3	0.71
Gene.6	0.77
Gene.1	1.54
Gene.9	1.92
Gene.7	2.00
Gene.10	2.05
Gene.2	3.42
Gene.8	3.62
Gene.4	3.80
Gene.5	3.86

Fig. 3 Analysis of qPCR raw data. (**a**) qPCR amplification plots of miR-222 and miR-224 in liver mitochondria. (**b**) Table with raw Ct values where each row represents one gene and each column one sample. (**c**) The R code to analyze the raw data Table. (**d**) Variation for the expression of the different genes calculated by Normfinder. The lower the stability index, the lower the variance across different samples

the index, the more stable the expression of a gene across all included samples.

5. Use microRNAs with low variation as references for the $2^{-\Delta\Delta C_T}$ calculations yielding relative miRNA quantities.
6. Calculate ΔC_T-values by subtracting the reference C_T from the C_T of the gene of interest (Fig. 4a, step 1). The ΔC_T-values are used to calculate fold changes relative to a control sample (Fig. 4a, steps 2 and 3) (*see* **Note 15**).

4 Notes

1. Set aside two small pieces of liver tissue in two 1.5 mL tubes, and freeze them in liquid nitrogen.
2. It is important to keep the stock of RNase A separate from all other reagents and out of touch with the RNA samples prepared later.

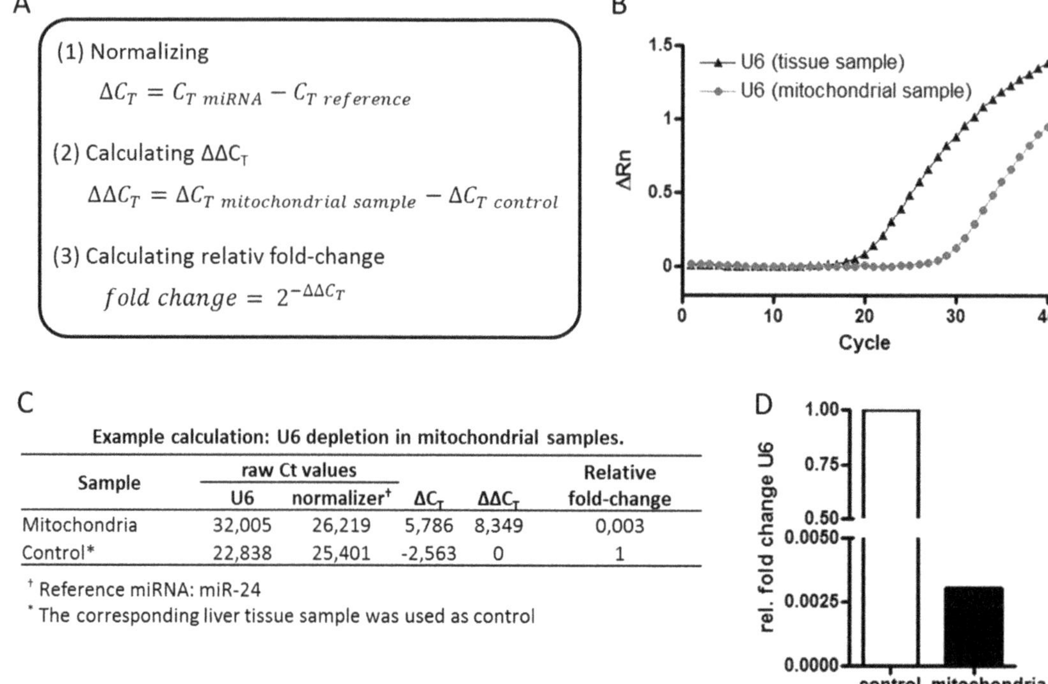

Example calculation: U6 depletion in mitochondrial samples.

Sample	raw Ct values				Relative
	U6	normalizer†	ΔC_T	$\Delta\Delta C_T$	fold-change
Mitochondria	32,005	26,219	5,786	8,349	0,003
Control*	22,838	25,401	-2,563	0	1

† Reference miRNA: miR-24

* The corresponding liver tissue sample was used as control

Fig. 4 Quantification of mitochondrial miRNAs. (**a**) Three-step process of calculating relative fold changes in miRNA expression using the $2^{-\Delta\Delta C_T}$-method. (**b**) U6 amplification in cDNA obtained from liver tissue samples or liver mitochondria. (**c**) Exemplary calculation of microRNA quantities using the formulas of "A". (**d**) Bar diagram showing relative U6 changes in liver mitochondria and control samples calculated in "C". RNA from liver tissue was used as control

3. Control lysates can be prepared by isolating proteins from a piece of liver tissue or whole cell aliquots.
4. In our lab, we use anti-GAPDH (Abcam, 1:10,000 dilution) and anti-MnSOD (Enzo Lifesciences, 1:1000 dilution).
5. To save antibody solution, the membrane may be incubated in a sealed plastic bag using a volume of 1–2 mL.
6. This is an important step. The proper incubation time is required to ensure sufficient lysis of the sample.
7. Avoid touching the interphase. It is better to leave a little bit of the upper phase in the tube than to risk getting contaminations from the middle phase. For valuable samples, Phase Lock Gel Heavy tubes (VWR) may be used to secure maximal sample recovery.
8. A ratio of A260/280 from 1.8–2.0 is indicative for a sufficient RNA purity. Obtained RNA concentrations can range from 20 to 1000 ng μL^{-1}.

9. As an alternative, RNA integrity can also be assessed using TBE gel electrophoresis followed by staining with ethidium bromide.
10. In case of low RNA yield, an RNA high sensitivity chip might be used.
11. Other methods of downstream analysis can be applied, such as RNA-seq, gene-specific RT-qPCR, Northern blot, or quantification of putative miRNA target genes.
12. For reverse transcription, use RNase-free filter tips to handle RNA samples.
13. Data tables can also be prepared in Excel, followed by export to a txt-file. Depending on the study design, it might be relevant to add a row containing grouping variables at the end of the data file.
14. The latest version of the Normfinder file "r.NormOldStab5.txt" can be downloaded from https://moma.dk/normfinder-software. The file needs to be copied in the folder that is set as current working directory in R-Studio. As an alternative, Normfinder is also available as Excel version.
15. As an example, the depletion of U6 in mitochondria from a liver samples (Fig. 4b) is quantified. Using raw C_T-data, the fold changes relative to the corresponding liver tissue are calculated (Fig. 4c) and visualized (Fig. 4d).

References

1. Ro S, Ma H-Y, Park C et al (2013) The mitochondrial genome encodes abundant small noncoding RNAs. Cell Res 23:759–774
2. Bandiera S, Rüberg S, Girard M et al (2011) Nuclear outsourcing of RNA interference components to human mitochondria. PLoS One 6: e20746
3. Geiger J, Dalgaard LT (2017) Interplay of mitochondrial metabolism and microRNAs. Cell Mol Life Sci 74(4):631–646
4. Kren B, Wong PY, Sarver A et al (2009) microRNAs identified in highly purified liver-derived mitochondria may play a role in apoptosis. RNA Biol 6(1):65–72
5. Rackham O, Shearwood A-MJ, Mercer TR et al (2011) Long noncoding RNAs are generated from the mitochondrial genome and regulated by nuclear-encoded proteins. RNA 17:2085–2093
6. Bandiera S, Matégot R, Girard M et al (2013) MitomiRs delineating the intracellular localization of microRNAs at mitochondria. Free Radic Biol Med 64:12–19
7. Tarassov I, Kamenski P, Kolesnikova O et al (2007) Import of nuclear DNA-encoded RNAs into mitochondria and mitochondrial translation. Cell Cycle 6:2473–2477
8. Wang W-X, Visavadiya NP, Pandya JD et al (2015) Mitochondria-associated microRNAs in rat hippocampus following traumatic brain injury. Exp Neurol 265:84–93
9. Jagannathan R, Thapa D, Nichols CE et al (2015) Translational regulation of the mitochondrial genome following redistribution of mitochondrial MicroRNA (MitomiR) in the diabetic heart. Circ Cardiovasc Genet 8 (6):785–802
10. Bian Z, Li L-M, Tang R et al (2010) Identification of mouse liver mitochondria-associated miRNAs and their potential biological functions. Cell Res 20:1076–1078
11. Andersen CL, Jensen JL, Ørntoft TF (2004) Normalization of real-time quantitative reverse transcription-PCR data: a model-based variance estimation approach to identify genes suited for normalization, applied to bladder

and colon cancer data sets. Cancer Res 64:5245–5250

12. Kappler L, Li J, Haring H-U et al (2016) Purity matters: a workflow for the valid high-resolution lipid profiling of mitochondria from cell culture samples. Sci Rep 6:21107

13. Jørgensen W, Jelnes P, Rud KA et al (2012) Progression of type 2 diabetes in GK rats affects muscle and liver mitochondria differently: pronounced reduction of complex II flux is observed in liver only. Am J Physiol Endocrinol Metab 303:E515–E523

Chapter 21

Imaging of Mitochondrial pH Using SNARF-1

Venkat K. Ramshesh and John J. Lemasters

Abstract

Laser scanning confocal microscopy provides the ability to image submicron sections in living cells and tissues. In conjunction with pH-indicating fluorescent probes, confocal microscopy can be used to visualize the distribution of pH inside living cells. Here we describe a confocal microscopic technique to image intracellular pH in living cells using carboxyseminaphthorhodafluor-1 (SNARF-1), a ratiometric pH-indicating fluorescent probe. SNARF-1 is ester-loaded into the cytosol and mitochondria of adult cardiac myocytes or other cell type. Using 568-nm excitation, emitted fluorescence longer and shorter than 595-nm is imaged and then ratioed after background subtraction. Ratio values for each pixel are converted to values of pH using a standard curve (lookup table). Images of the intracellular distribution of pH show cytosolic and nuclear areas to have a pH of ~7.1, but in regions corresponding to mitochondria, pH is 8.0, giving a mitochondrial ΔpH of 0.9. During hypoxia, mitochondrial pH decreases to cytosolic values, signifying the collapse of ΔpH. These results illustrate the ability of laser scanning confocal microscopy to image the intracellular distribution of pH in living cells and to determine mitochondrial ΔpH.

Key words Confocal microscopy, Cytosol, Mitochondria, Myocytes, pH, Ratio imaging, SNARF-1

1 Introduction

ATP is the source of energy for most biological reactions with mitochondria being the principal ATP generator in aerobic tissues like heart, brain, liver, and kidney. To synthesize ATP from ADP and phosphate via the mitochondrial F_1F_0-ATP synthase, mitochondria must generate a protonmotive force (Δp) across the inner membrane. Δp in millivolts equals $\Delta\Psi$-60ΔpH, where $\Delta\Psi$ is the mitochondrial membrane potential (negative inside) and ΔpH is the mitochondrial pH gradient (alkaline inside) [1]. Δp also supports other energy-requiring reactions, such as ion transport and the $NAD(P)^+$ transhydrogenase reaction. The $\Delta\Psi$ component

This work was supported, in part, by grants 1 R01 AA021191, 1 R01 CA184456, 2 R01 DE016572, 2 R01 DK073336, and 1 P20 GM103542 from the National Institutes of Health. Imaging facilities were supported, in part, by NIH Center Grant 5 P30 CA138313.

Carlos M. Palmeira and António J. Moreno (eds.), *Mitochondrial Bioenergetics: Methods and Protocols*, Methods in Molecular Biology, vol. 1782, https://doi.org/10.1007/978-1-4939-7831-1_21,

of Δp can be visualized by confocal microscopy by any of several membrane-permeant cationic fluorophores, such as rhodamine 123 and tetramethylrhodamine methylester that accumulate electrophoretically into polarized mitochondrial [2].

Several techniques have been developed to measure average mitochondrial ΔpH in isolated mitochondria, cell suspensions, and cell cultures [1]. However, the magnitude of ΔpH of individual mitochondria in single cells has been much more difficult to assess, since mitochondria are too small to measure ΔpH using microelectrodes. Here, we illustrate the use of laser scanning confocal microscopy to visualize pH of individual mitochondria in living cardiac myocytes under normal and hypoxic conditions by ratiometric imaging of carboxyseminaphthorhodafluor-1 (SNARF-1) [3]. This approach can be adapted to study other cell types, such as primary hepatocytes, and isolated arterially perfused rabbit papillary muscles [4, 5].

2 Materials

2.1 Buffer A

5 mM KCl, 110 mM NaCl, 1.2 mM NaH_2PO_4, 28 mM $NaHCO_3$, 30 mM glucose, 20 mM butanedione monoxime, 0.05 units/ml insulin, 250 μM adenosine, 1 mM creatine, 1 mM carnitine, 1 mM octanoic acid, 1 mM taurine, 10 units/ml penicillin, 10 μg/ml streptomycin, and 25 mM HEPES, pH 7.30 (*see* **Note 1**).

2.2 Buffer B

Joklik's medium and medium 199 (1:1 mixture) supplemented with 20 mM butanedione monoxime, 1 mM creatine, 1 mM taurine, 1 mM octanoic acid, 1 mM carnitine, 0.05 units/ml insulin, 10 units/ml penicillin, and 10 μg/ml streptomycin.

2.3 Culture Medium

Eagle's minimum essential medium supplemented with 5% newborn calf serum, 0.5 units/ml penicillin G potassium salt, 0.05 mg/ml streptomycin sulfate, and 0.5 μg/ml amphotericin B.

2.4 Krebs-Ringer-HEPES Buffer (KRH)

110 mM NaCl, 5 mM KCl, 1.25 mM $CaCl_2$, 1.0 mM Mg_2SO_4, 0.5 mM Na_2HPO_4, 0.5 mM KH_2PO_4, and 20 mM HEPES, pH 7.4.

3 Methods

3.1 Preparation of Cardiac Myocytes

Adult rabbit cardiac myocytes are isolated by enzymatic digestion, as described [3], and plated at a density of $15{,}000/cm^2$ on #1.5 glass coverslip-bottomed Petri dishes coated with laminin ($10\ \mu g/cm^2$). Experiments are conducted 1 day after initial plating. Cell lines or other primary cells (e.g., hepatocytes) may be substituted for myocytes.

3.2 Loading of SNARF-1

Intracellular pH is estimated with SNARF-1, a pH-sensitive fluorophore with a pK_a of about 7.5. To load SNARF-1, cultured myocytes are incubated with 5 μM SNARF-1 acetoxymethyl ester (SNARF-1 AM) for 45 min in culture medium at 37 °C. During incubation, intracellular esterases release and trap SNARF-1 free acid in the cytoplasm. The cells are washed twice with KRH and placed on the microscope stage in KRH or other physiological medium like Buffer A or B. Unlike other ester-loaded fluorescent indicators, SNARF-1 loads well into mitochondria, although such loading may be cell specific. To promote better mitochondrial uptake, cells can be loaded with SNARF-1 AM at a cooler temperature (4–12 °C) for a longer time [6, 7].

3.3 Confocal Imaging of pH of Cardiac Myocytes

Confocal imaging of cells is performed using 568-nm excitation of an argon-krypton laser, which is near the absorbance maximum for the dye [3]. At this excitation, SNARF fluorescence increases at >620 nm with increasing pH but remains unchanged at 585 nm (Fig. 1). Alternatively, excitation can be performed with the 561-nm line of a helium-neon laser. Emitted fluorescence is divided by a 595-nm long pass dichroic reflector with the shorter wavelengths directed through a 585-nm (10-nm band pass) barrier filter and longer wavelengths through a 620-nm long pass filter to separate detectors. Importantly, image oversaturation (pixels at highest gray level) and undersaturation (pixels with a zero gray level) should be kept to a minimum, and laser intensity should be kept at the lowest level possible consistent with an acceptable single-to-noise ratio (S/N). Because images are to be ratioed and background subtracted, S/N ratios higher than required for routine imaging are needed in both image channels. If necessary to improve S/N, binning or median filtering of pixels can be performed whereby each pixel is reassigned a value equal to the average

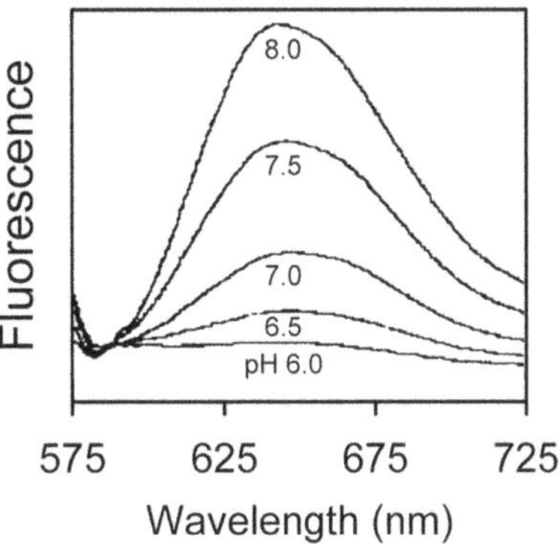

Fig. 1 Fluorescence emission spectra of SNARF-1. Fluorescence emission spectra of SNARF-1 in KRH buffer at different pH. Excitation wavelength is 568 nm

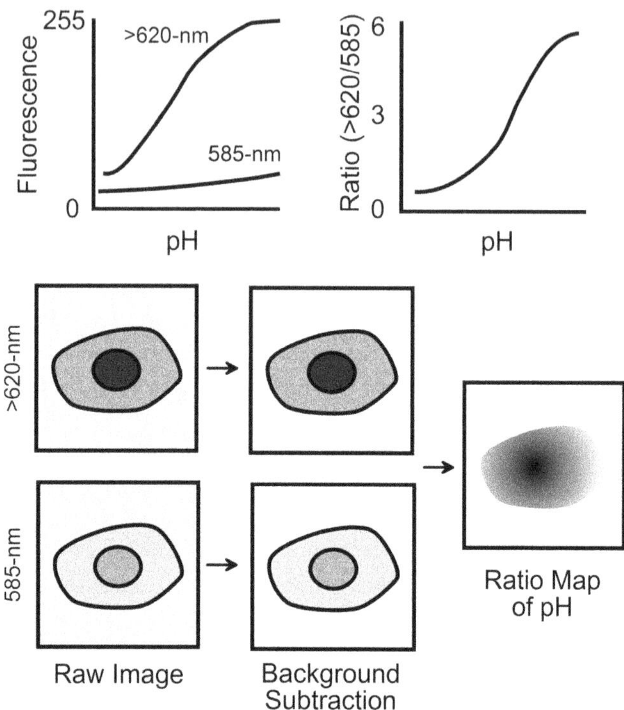

Fig. 2 Principle of background subtraction and ratio imaging. *See* text for details

of 2 × 2 or 3 × 3 groups of pixels or the median value of the pixel and its adjacent pixels. If instrumentation permits, images should be acquired using the multitrack option where each wavelength is acquired alternately on a line-by-line basis.

The intensity of fluorescence acquired at the two wavelengths must be corrected for background (background subtraction) (*see* **Note 2**) and then divided on a pixel-by-pixel basis (ratioing) (Fig. 2). The resulting ratios are converted to pH values based on an in situ pH calibration of SNARF-1 through the microscope optics (*see* **Note 3**). An example of measured pH in a myocyte before and after chemical hypoxia (*see* **Note 4**) is shown in Fig. 3. pH is estimated at 7.0–7.2 in cytosolic (e.g., subsarcolemmal areas) and nuclear regions and 8.0 in mitochondria, yielding a mitochondrial ΔpH of ~0.9. This gradient decreased to ~0.5 and 0 after 30 and 40 min of chemical hypoxia, respectively. After 42 min of hypoxia, the myocyte hypercontracted and died.

4 Notes

1. All solutions should be prepared in deionized distilled water that has a resistance of 18.2 MΩ.

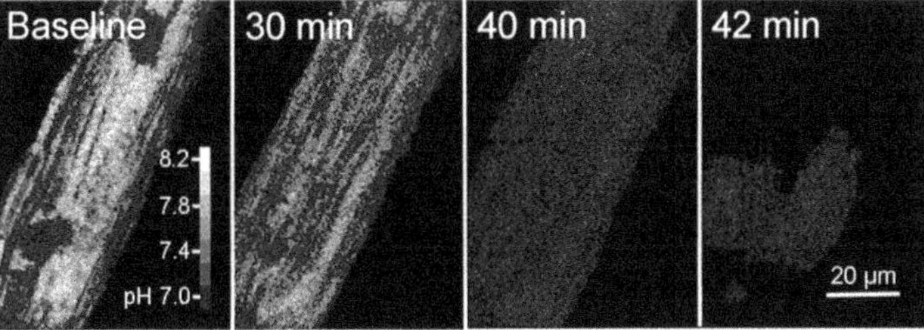

Fig. 3 Confocal SNARF-1 ratio images of intracellular pH. A 1-day cultured cardiac myocyte was loaded with SNARF-1-AM (5 μM) for 45 min in culture medium at 37 °C, and intracellular pH was measured by ratio imaging of SNARF-1 fluorescence before (baseline) and after 30, 40, and 42 min of chemical hypoxia

2. In confocal microscopy, detectors generate signals even in the absence of light. To quantify this dark signal, background images are collected by focusing the objective lens completely within the coverslip just underneath the cells using the same instrument settings as during acquisition of cell images. Average pixel intensity for each color channel of the background images is then determined and subtracted from each pixel of the fluorescence images of the cells at each of the two emission wavelengths. The resulting images are the background-subtracted fluorescence images (*see* Fig. 2).
3. For in situ calibration, SNARF-1-loaded myocytes are incubated with 5 μM valinomycin and 10 μM nigericin in modified KRH buffer in which KCl and NaCl are replaced by their corresponding gluconate salts to minimize swelling [8]. Images are then collected as extracellular pH is varied. Instrument settings should be the same. Alternatively, the fluorescence of SNARF-1 free acid (100–200 μM) in solution can be imaged through the microscope optics as pH is varied. After background subtraction, the >620-nm image channel is divided by the 585-nm channel on a pixel-by-pixel basis. Using thresholding to eliminate low pixel values of the extracellular space, a standard curve is created relating ratio values to pH. Lookup tables are then created assigning specific colors to different values of pH.
4. To simulate the ATP depletion and reductive stress of hypoxia, myocytes are exposed to 2.5 mM NaCN, an inhibitor of mitochondrial respiration, and 20 mM 2-deoxyglucose, an inhibitor of glycolysis. This treatment is termed chemical hypoxia [3].

References

1. Nicholls DG, Ferguson SJ (2013) Bioenergetics4. Elsevier, London
2. Lemasters JJ, Ramshesh VK (2007) Imaging of mitochondrial polarization and depolarization with cationic fluorophores. Methods Cell Biol 80:283–295
3. Chacon E, Reece JM, Nieminen AL, Zahrebelski G, Herman B, Lemasters JJ (1994) Distribution of electrical potential, pH, free Ca^{2+}, and volume inside cultured adult rabbit cardiac myocytes during chemical hypoxia: a multiparameter digitized confocal microscopic study. Biophys J 66:942–952
4. Muller-Borer BJ, Yang H, Marzouk SA, Lemasters JJ, Cascio WE (1998) pHi and pHo at different depths in perfused myocardium measured by confocal fluorescence microscopy 129. Am J Physiol 275:H1937–H1947
5. Lemasters JJ, Trollinger DR, Qian T, Cascio WE, Ohata H (1999) Confocal imaging of Ca^{2+}, pH, electrical potential, and membrane permeability in single living cells. Methods Enzymol 302:341–358
6. Nieminen AL, Saylor AK, Tesfai SA, Herman B, Lemasters JJ (1995) Contribution of the mitochondrial permeability transition to lethal injury after exposure of hepatocytes to t-butylhydroperoxide. Biochem J 307:99–106
7. Trollinger DR, Cascio WE, Lemasters JJ (1997) Selective loading of Rhod 2 into mitochondria shows mitochondrial Ca^{2+} transients during the contractile cycle in adult rabbit cardiac myocytes. Biochem Biophys Res Commun 236:738–742
8. Kawanishi T, Nieminen AL, Herman B, Lemasters JJ (1991) Suppression of Ca^{2+} oscillations in cultured rat hepatocytes by chemical hypoxia. J Biol Chem 266:20062–20069

Chapter 22

Relation Between Mitochondrial Membrane Potential and ROS Formation

Jan Suski, Magdalena Lebiedzinska, Massimo Bonora, Paolo Pinton, Jerzy Duszynski, and Mariusz R. Wieckowski

Abstract

Mitochondria are considered the main source of reactive oxygen species (ROS) in the cell. For this reason they have been recognized as a source of various pathological conditions as well as aging. Chronic increase in the rate of ROS production is responsible for the accumulation of ROS-associated damages in DNA, proteins, and lipids and may result in progressive cell dysfunctions and, in a consequence, apoptosis, increasing the overall probability of an organism's pathological conditions. The superoxide anion is the main undesired by-product of mitochondrial oxidative phosphorylation. Its production is triggered by a leak of electrons from the mitochondrial respiratory chain and the reaction of these electrons with O_2. Superoxide dismutase (MnSOD, SOD2) from the mitochondrial matrix, as well as superoxide dismutase (Cu/ZnSOD, SOD1) present in small amounts in the mitochondrial intramembrane space, converts superoxide anion to hydrogen peroxide, which can be then converted by catalase to harmless H_2O.

In the chapter we describe a relation between mitochondrial membrane potential and the rate of ROS formation. We present different methods applicable for isolated mitochondria or intact cells. We also present experiments demonstrating that a magnitude and a direction (increase or decrease) of a change in mitochondrial ROS production depend on the metabolic state of this organelle.

Key words ROS, MnSOD, Cu/ZnSOD, Brain mitochondria, Fibroblasts, Ehrlich ascites tumor cells

1 Introduction

It has been repeatedly demonstrated, on different experimental models, that a strong positive correlation exists between mitochondrial membrane potential ($\Delta\Psi$) and reactive oxygen species (ROS) production [1, 2]. At present, it is widely accepted that mitochondria produce more ROS at high membrane potential. It has been shown that ROS production dramatically increases above 140 mV [2]. Studies performed on mitochondria from *Drosophila*

Jan Suski and Magdalena Lebiedzinska have contributed equally to this work.

Carlos M. Palmeira and António J. Moreno (eds.), *Mitochondrial Bioenergetics: Methods and Protocols*, Methods in Molecular Biology, vol. 1782, https://doi.org/10.1007/978-1-4939-7831-1_22,

melanogaster showed that even a slight decrease in the ΔΨ (10 mV) can cause a significant decrease in ROS production (according to authors the decrease of ROS production diminished by approximately 70%) by complex I of the respiratory chain [3]. In contrast, an increase in the ΔΨ produced either by a closure of the mitochondrial permeability transition pore or an inhibition of ATP synthase [4] is associated with increased ROS production. Interestingly, in certain pathological conditions, opposite correlations between ΔΨ and ROS production can also be observed. In the case of ATP synthase dysfunction (mutation T8993G in the mitochondrial ATPase-6 gene), higher ΔΨ and increased ROS production are observed [5]. On the other hand, in the case of mitochondrial disorders associated with the dysfunctions of the respiratory chain components, lower ΔΨ and decreased activity of the respiratory chain are observed with a simultaneous increase in ROS production [6].

2 Materials

2.1 Isolation of Crude Mitochondria from Mouse Brain

1. Stirrer motor with electronic speed controller (Cole-Parmer).
2. Motor-driven tightly fitting glass/Teflon Potter-Elvehjem homogenizer.
3. Loose-fitting glass/Teflon Potter-Elvehjem homogenizer.
4. Sucrose (Merck, cat. no. 100892.9050) (*see* **Note 1**).
5. Trizma base (Sigma-Aldrich).
6. D-Mannitol (Sigma-Aldrich).
7. Ethylene-bis(oxyethylenenitrilo)tetraacetic acid (EGTA; Sigma-Aldrich).
   - Homogenization buffer: 320 *mM* sucrose, 10 *mM* Tris–HCl, 1 *mM* EDTA, pH 7.4. Store at 4 °C (*see* **Note 2**).
   - Mitochondria isolation buffer: 75 *mM* sucrose, 225 *mM* mannitol, 5 *mM* Tris–HCl, pH 7.4. Store at 4 °C (*see* **Note 2**).

2.2 Cell Culture

1. Dulbecco's Modified Eagle's Medium (DMEM) (Lonza) supplemented with 10% fetal bovine serum (heat-inactivated FBS, GIBCO), 2 mM L-glutamine (GIBCO), and 1.2% antibiotic: penicillin/streptomycin (penicillin-streptomycin solution, Sigma-Aldrich).
2. KRB saline: NaCl 135 mM, KCl 5 mM, KH_2PO_4 0.4 mM, $MgSO_4$ 1 mM, HEPES 20 mM, bring to pH 7.4 with NaOH. Glucose 1 g/l and 1 mM $CaCl_2$ should be added the day of experiment.
3. Fibroblasts:

- Primary culture of human skin fibroblasts grown from explants of skin biopsies of control individual and patient with mitochondrial disorder.
- NHDF (neonatal human dermal fibroblasts) (Lonza).

4. Ehrlich ascites tumor cells were cultivated in Swiss albino mice and harvested as described in [6].
5. 24-Well Cell Culture Cluster (Costar/Corning).

2.3 Measurement of Hydrogen Peroxide Production in Isolated Brain Mitochondria

1. Shimadzu Spectrofluorometer RF 5000.
2. Measurement medium: 75 mM sucrose, 225 mM mannitol, 5 mM Tris–HCl, pH 7.4. Store at 4 °C (*see* **Note 2**).
3. Succinate: 0.5 M, pH 7.4 (adjusted with KOH) (Sigma).
4. Antimycin A: 1 mM (ethanol solution) (Sigma).
5. Carbonyl cyanide m-chlorophenyl hydrazone (CCCP); 1 mM, (ethanol solution) (Sigma).
6. 5-(and-6)-chloromethyl-2′,7′-dichlorodihydrofluorescein diacetate, acetyl ester (CM-H_2DCFDA) (Invitrogen), prepared as a 5 mM stock solution in DMSO.

2.4 Measurement of Superoxide Production by Peroxidase/Amplex Red Assay in Isolated Mitochondria

1. Multiwell plate reader TECAN F200.
2. 24-Well Cell Culture Cluster (Costar/Corning).
3. Reaction buffer: 75 mM sucrose, 225 mM mannitol, 5 mM Tris–HCl, pH 7.4. Store at 4 °C (*see* **Note 2**).
4. Amplex Red: 5 μM (Invitrogen).
5. Peroxidase: 7 U/ml (Sigma).
6. Glutamate: 0.5 M, pH 7.4 (Sigma).
7. Malate: 0.5 M, pH 7.4 (Sigma).
8. Oligomycin: 1 mM (ethanol solution) (Sigma).
9. CCCP: 1 mM (ethanol solution) (Sigma).
10. Antimycin A: 1 mM (ethanol solution) (Sigma).

2.5 Measurement of the Mitochondrial Transmembrane Potential Using Safranin O in Isolated Mouse Brain Mitochondria

1. Shimadzu Spectrofluorometer RF 5000.
2. Measurement medium: 75 mM sucrose, 225 mM mannitol, 5 mM Tris–HCl, pH 7.4. Store at 4 °C (*see* **Note 2**).
3. Safranin O: 5 mM (Sigma).
4. Succinate: 0.5 M, pH 7.4 (adjusted with KOH) (Sigma).
5. Antimycin A: 1 mM (ethanol solution) (Sigma).
6. Oligomycin: 1 mM (ethanol solution) (Sigma).
7. CCCP: 1 mM, (ethanol solution) (Sigma).

2.6 Measurement of the Mitochondrial Transmembrane Potential in Human Fibroblasts Using JC-1

1. Multiwell plate reader TECAN F200.
2. 24-Well Cell Culture Cluster (Costar/Corning).
3. 5 mM JC1 (5,5′,6,6′-tetrachloro-1,1′,3,3′-tetraethylbenzimidazolylcarbocyanine iodide) (Invitrogen/Molecular Probes) in DMSO (Sigma). Store at −20 °C.
4. KRB saline: NaCl 135 mM, KCl 5 mM, KH_2PO_4 0.4 mM, $MgSO_4$ 1 mM, HEPES 20 mM, bring to pH 7.4 with NaOH. Glucose 1 g/l and 1 mM $CaCl_2$ should be added the day of experiment.
5. Antimycin A, 1 mM (ethanol solution) (Sigma).
6. Oligomycin, 1 mM (ethanol solution) (Sigma).

2.7 Measurement of "Mitochondrial Matrix" Superoxide Production in Human Fibroblasts Using MitoSOX Red

1. Multiwell plate reader TECAN F200.
2. 24-Well Cell Culture Cluster (Costar/Corning).
3. 5 mM MitoSOX Red (Invitrogen) in DMSO (Sigma). Store at −20 °C.
4. KRB saline: NaCl 135 mM, KCl 5 mM, KH_2PO_4 0.4 mM, $MgSO_4$ 1 mM, HEPES 20 mM, bring to pH 7.4 with NaOH. Glucose 1 g/l and 1 mM $CaCl_2$ should be added the day of experiment.
5. Antimycin A (Sigma), 1 *mM* (ethanol solution).
6. Oligomycin (Sigma), 1 *mM* (ethanol solution).

2.8 Measurement of the "Cytosolic" Superoxide Production in Human Fibroblasts Using DHE

1. Multiwell plate reader TECAN F200.
2. 24-Well Cell Culture Cluster (Costar/Corning).
3. 10 mM dihydroethidium (hydroethidine) (Invitrogen/Molecular Probes) in DMSO (Sigma). Store at −20 °C.
4. KRB saline: NaCl 135 mM, KCl 5 mM, KH_2PO_4 0.4 mM, $MgSO_4$ 1 mM, HEPES 20 mM, bring to pH 7.4 with NaOH. Glucose 1 g/l and 1 mM $CaCl_2$ should be added the day of experiment.
5. Antimycin A (Sigma), 1 mM (ethanol solution).
6. Oligomycin (Sigma), 1 mM (ethanol solution) .

2.9 Measurement of the Oxygen Consumption in Ehrlich Ascites Tumor Cells

1. Clark-type oxygen electrode (YSI, Yellow Springs, OH, USA) equipped with a unit calculating the equivalent to the rate of oxygen consumption (first derivative of the oxygen concentration trace).
2. KRB saline: NaCl 135 mM, KCl 5 mM, KH_2PO_4 0.4 mM, $MgSO_4$ 1 mM, HEPES 20 mM, bring to pH 7.4 with NaOH. Glucose 1 g/l and 1 mM $CaCl_2$ should be added the day of experiment.
3. Oligomycin (Sigma), 1 mM (ethanol solution).

4. Cyclosporin A (Sigma), 1 mM (ethanol solution).
5. Carbonylcyanide p-trifluoromethoxyphenylhydrazone (FCCP) (Sigma), 1 mM (ethanol solution).

2.10 Fluorometric Measurement of the Mitochondrial Membrane Potential Using TMRM in Ehrlich Ascites Tumor Cells

1. Shimadzu Spectrofluorometer RF 5000.
2. TMRM (Invitrogen), prepared as a 100 μM stock solution in H_2O.
3. KRB saline: NaCl 135 mM, KCl 5 mM, KH_2PO_4 0.4 mM, $MgSO_4$ 1 mM, HEPES 20 mM, bring to pH 7.4 with NaOH. Glucose 1 g/l and 1 mM $CaCl_2$ should be added the day of experiment.
4. Oligomycin (Sigma), 1 mM (ethanol solution).
5. Cyclosporin A (Sigma), 1 mM (ethanol solution).
6. FCCP (Sigma), 1 mM (ethanol solution).

2.11 Fluorometric Measurement of Hydrogen Peroxide Production in Ehrlich Ascites Tumor Cells

1. Shimadzu Spectrofluorometer RF 5000.
2. CM-H_2DCFDA (Invitrogen), prepared as a 5 mM stock solution in DMSO.
3. KRB saline: NaCl 135 mM, KCl 5 mM, KH_2PO_4 0.4 mM, $MgSO_4$ 1 mM, HEPES 20 mM, bring to pH 7.4 with NaOH. Glucose 1 g/l and 1 mM $CaCl_2$ should be added the day of experiment.
4. Oligomycin (Sigma), 1 mM (ethanol solution).
5. Cyclosporin A (Sigma), 1 mM (ethanol solution).
6. FCCP (Sigma), 1 mM (ethanol solution) .

2.12 Measurement of Mitochondrial Membrane Potential in HeLa Cells Using a Confocal Microscope

1. Nikon Swept Field Confocal Microscope or Zeiss LSM510 Confocal Microscope.
2. Microscope cover slips (24 mm diameter, 0.15 mm thickness) from VWR International.
3. KRB saline: NaCl 135 mM, KCl 5 mM, KH_2PO_4 0.4 mM, $MgSO_4$ 1 mM, HEPES 20 mM, bring to pH 7.4 with NaOH. Glucose 1 g/l and 1 mM $CaCl_2$ should be added the day of experiment.
4. TMRM (Invitrogen), prepared in a stock solution 10 μM in absolute ethanol.

2.13 Measurement of Hydrogen Peroxide Production in HeLa Cells Using a Confocal Microscope

1. Nikon Swept Field Confocal Microscope.
2. Microscope cover slips (24 mm diameter, 0.15 mm thickness) from VWR International.
3. KRB saline: NaCl 135 mM, KCl 5 mM, KH_2PO_4 0.4 mM, $MgSO_4$ 1 mM, HEPES 20 mM, bring to pH 7.4 with NaOH.

Glucose 1 g/l and 1 mM $CaCl_2$ should be added the day of experiment.

4. CM-H_2DCFDA (Invitrogen), prepared in a stock solution 5 mM in DMSO.

2.14 Measurement of Mitochondrial Superoxide Production in HeLa Cells Using a Confocal Microscope

1. Nikon Swept Field Confocal Microscope.
2. Microscope cover slips (24 mm diameter, 0.15 mm thickness) from VWR International.
3. KRB saline: NaCl 135 mM, KCl 5 mM, KH_2PO_4 0.4 mM, $MgSO_4$ 1 mM, HEPES 20 mM, bring to pH 7.4 with NaOH. Glucose 1 g/l and 1 mM $CaCl_2$ should be added on the day of experiment.
4. MitoSOX Red (Invitrogen), prepared in a stock solution 5 mM in DMSO.

2.15 Measurement of Cytosolic Calcium in HeLa Cells Using a Confocal Microscope

1. Zeiss LSM510 Confocal Microscope.
2. Microscope cover slips (24 mm diameter, 0.15 mm thickness) from VWR International.
3. KRB saline: NaCl 135 mM, KCl 5 mM, KH_2PO_4 0.4 mM, $MgSO_4$ 1 mM, HEPES 20 mM, bring to pH 7.4 with NaOH. Glucose 1 g/l and 1 mM $CaCl_2$ should be added the day of experiment.
4. Cell-permeant fluo-3 AM (Invitrogen), prepared in a stock solution 5 mM in DMSO.

2.16 Measurement of the Respiratory Chain Activity in Human Fibroblasts

1. Multiwell plate reader TECAN F200.
2. 24-Well Cell Culture Cluster (Costar/Corning).
3. Resazurin (7-Hydroxy-3H-phenoxazin-3-one-10-oxide sodium salt) (Sigma), 1 mM in H_2O. Store at −20 °C.
4. KRB saline: NaCl 135 mM, KCl 5 mM, KH_2PO_4 0.4 mM, $MgSO_4$ 1 mM, HEPES 20 mM, bring to pH 7.4 with NaOH. Glucose 1 g/l and 1 mM $CaCl_2$ should be added the day of experiment.
5. KCN (Sigma), 1 M in H_2O.

2.17 Fluorometric Measurement of Hydrogen Peroxide Production in Human Fibroblasts

1. Multiwell plate reader TECAN F200.
2. 24-Well Cell Culture Cluster (Costar/Corning).
3. CM-H_2DCFDA (Invitrogen), 5 mM in DMSO (Sigma). Store at −20 °C.
4. KRB saline: NaCl 135 mM, KCl 5 mM, KH_2PO_4 0.4 mM, $MgSO_4$ 1 mM, HEPES 20 mM, bring to pH 7.4 with NaOH. Glucose 1 g/l and 1 mM $CaCl_2$ should be added on the day of experiment.

2.18 Determination of Protein Concentration

1. Lysis buffer: 50 mM Tris–HCl pH 7.5; 150 mM NaCl; 1% TritonX; 0.1% SDS (sodium dodecyl sulfate); 1% sodium deoxycholate.
2. Bio-Rad Protein Assay (Bio-Rad Laboratories).
3. Spectrophotometer UV-1202 (Shimadzu).
4. Acryl Cuvettes 10 × 10 × 48 mm (SARSTEDT).

3 Methods

To study the relationship between mitochondrial ROS production and mitochondrial bioenergetic parameters (like membrane potential and respiration), a variety of methods can be used. Some of them, particularly those dedicated to human fibroblasts, could potentially be adapted as diagnostic procedures in case of suspected mitochondrial disorders [5].

3.1 Isolation of Crude Mouse Brain Mitochondria for Measurement of $\Delta\Psi$ and ROS Production

1. Kill the mouse by decapitation, remove the brain immediately, and cool it down at 4 °C in the homogenization medium (*see* **Note 3**).
2. Wash the brain with the homogenization medium (to remove blood). Add fresh homogenization medium in a proportion of 5 ml/g of brain.
3. Homogenize the brain in a glass Potter-Elvehjem homogenizer with a motor-driven Teflon pestle (*see* **Note 4**).
4. Centrifuge the homogenate for 3 min at 1330 × *g* at 4 °C.
5. Discard the pellet, and centrifuge the supernatant for 10 min at 21,000 × *g* at 4 °C.
6. Gently resuspend the resulting mitochondrial pellet in approx. 15 ml of the mitochondria isolation buffer, and centrifuge it again for 10 min at 21,000 × *g* at 4 °C.
7. Gently resuspend the final crude mitochondrial pellet in 1–2 ml (depending on the pellet volume) of the mitochondria isolation buffer using a loose Potter-Elvehjem homogenizer.
8. The material can now be used for further experiments. (Such isolated mitochondria contain synaptosomes. If necessary additional steps in the isolation procedure can be undertaken to isolate the pure mitochondrial fraction.)

3.2 Measurement of Hydrogen Peroxide Production with the Use of CM-H_2DCFDA in Isolated Mitochondria

1. Adjust the fluorometer: excitation, 513 nm; emission, 530 nm; slits (excitation and emission) ~3.
2. Fill the fluorometer cuvette with 3 ml of the measurement medium (75 *mM* sucrose, 225 *mM* mannitol, 5 *mM* Tris–HCl,

pH 7.4. Store at 4 °C (*see* **Note 2**)) containing 1 mg of mitochondrial protein.

3. Start measurement.
4. Add CM-H_2DCFDA to the final 5 μM concentration, and record changes in fluorescence.
5. Make traces in the presence of following additions:
   - First trace: 2 μM antimycin A.
   - Second trace: 2 μM FCCP.
6. An example of the results obtained is shown in Fig. 1a.

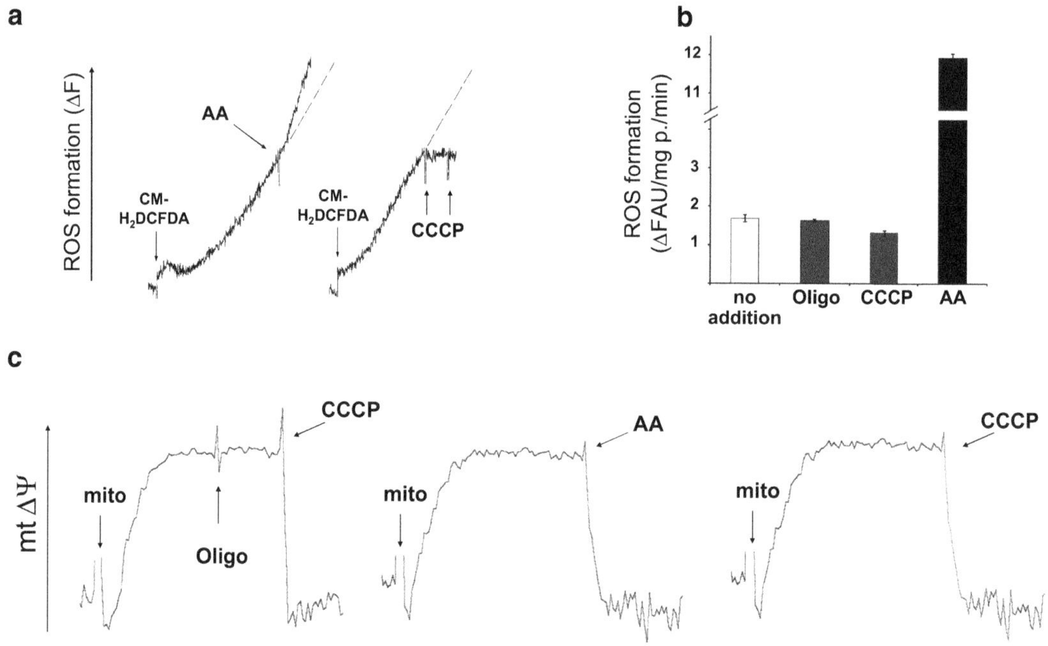

Fig. 1 Effect of antimycin A, CCCP, and oligomycin on mitochondrial membrane potential and ROS formation measured in isolated mitochondria. (**a**) Effect of antimycin A and CCCP on mitochondrial H_2O_2 production measured with the use of CM-H_2DCFDA (Subheading 3.2); (**b**) effect of oligomycin, CCCP, and antimycin A on superoxide production measured by peroxidase/Amplex Red assay (Subheading 3.3); (**c**) effect of oligomycin, antimycin A, and CCCP on $\Delta\Psi$ measured with the use of safranin O (Subheading 3.4). Addition of antimycin A or CCCP leads to the collapse of the $\Delta\Psi$ (panel **c**); however, each compound results in contradictory effects in terms of ROS formation in isolated mitochondria. These effects can be observed with the use of different probes, such as CM-H_2DCFDA (panel A) and peroxidase/Amplex Red assay (panel **b**). Antimycin A is an inhibitor of complex III of the respiratory chain and causes the accumulation of reduced intermediates leading to the chain's blockage and an increased leakage of electrons. This in turn results in an increase in ROS formation. CCCP as a mitochondrial uncoupler decreases the amount of reduced respiratory chain intermediates and decreases ROS formation (panels A and B). In "coupled" mitochondria, in the absence of ADP, no significant effect of oligomycin (an inhibitor of the mitochondrial ATP synthase) neither on the $\Delta\Psi$ nor on the ROS production is observed (panel B and C). A significant effect of oligomycin on both parameters can be seen in the intact cell model (*see* Fig. 2)

3.3 Mitochondrial Superoxide Measurement by Peroxidase/Amplex Red Assay

1. Prepare 13 ml of the reaction buffer (*see* **Note 5**).
2. Supplement the reaction buffer with 3 μl of Amplex Red, 13 μl of peroxidase, 130 μl of glutamate, and 130 μl of malate.
3. Aliquot 0.5 ml of the reaction solution into the wells.
4. Supplement individual wells with oligomycin, FCCP, and antimycin A.
5. Start the reaction with the addition of 100 μg of mitochondrial protein.
6. Place the plate in the microplate reader, and read the fluorescence at 510 nm excitation and 595 nm emission wavelengths. An example of the results obtained is shown in Fig. 1b.

3.4 Measurement of the Mitochondrial Membrane Potential with the Use of Safranin O

1. Adjust the fluorometer: excitation, 495 nm; emission, 586 nm; slits (excitation and emission) ~3.
2. Fill the fluorometer cuvette with 3 ml of the measurement medium (*see* **Note 5**).
3. Add 3 μl of safranin O and 30 μl of succinate.
4. Start measuring fluorescence.
5. Add the mitochondrial suspension corresponding to about 1 mg of protein, and observe changes in fluorescence.
6. Make traces with the following consecutive additions:
   - First trace: 5 μl oligomycin, 5 μl CCCP.
   - Second trace: 5 μl antimycin A.
   - Third trace: 5 μl CCCP.
7. An example of the results obtained is shown in Fig. 1c.

3.5 Measurement of Mitochondrial Membrane Potential in Human Fibroblasts with the Use of JC-1

1. Remove the culture medium, and wash the cells gently with warm KRB before preincubation with effectors. Dilute stock solutions of antimycin A and oligomycin to final concentration 2 μM in KRB. Add 0.5 ml of KRB with particular effector to selected wells and KRB alone to control wells. Preincubate the plate, in the incubator, for 15–30 min prior to the measurement (*see* **Note 6**).
2. Prepare 5 μM solution of JC-1 in KRB (*see* **Note 7**).
3. Remove the preincubation solution from wells, and add the KRB solution containing JC-1. As the effectors must be present till the end of measurement, add adequate amount of each to selected wells (*see* **Notes 5** and **8**).
4. Incubate the plate for 10 min in the incubator (*see* **Note 9**).
5. Gently wash the cells twice with KRB.
6. Add 0.5 ml of KRB to each well (*see* **Note 10**).

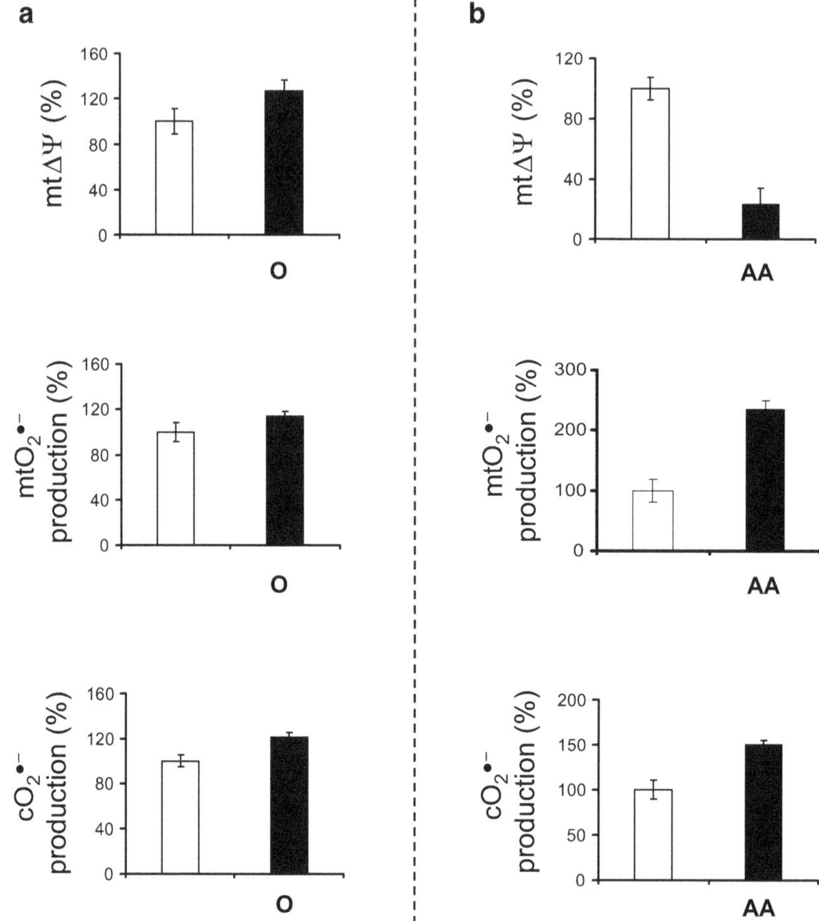

Fig. 2 Effect of oligomycin and antimycin A on mitochondrial membrane potential and superoxide production measured in intact human fibroblasts. Effect of (**a**) oligomycin (O) and (**b**) antimycin A (AA) on $\Delta\Psi$ measured with the use of JC-1 (Subheading 3.5), mitochondrial superoxide production measured with the use of MitoSOX Red (Subheading 3.6), and cytosolic superoxide production measured with the use of DHE (Subheading 3.7). All parameters were measured in a multiwell plate reader; O oligomycin; AA antimycin A; oligomycin by the inhibition of ATP synthase causes an increase of $\Delta\Psi$. It is a result of a drift from mitochondrial respiratory state III to state IV. Mitochondrial coupling is connected with the increase of superoxide production by mitochondria determined either in mitochondrial matrix ($mt0_2{\bullet}^-$) or in cytosol ($c0_2{\bullet}^-$). Similarly to isolated mitochondria (Fig. 1), in intact fibroblasts antimycin A increases superoxide formation with the simultaneous decrease of $\Delta\Psi$

7. Place the plate in reader and read the fluorescence. The read must be done twice, first at 485 nm excitation and 520 nm emission wavelengths for JC-1 green fluorescence detection and second at 535 nm excitation and 635 nm emission wavelengths for JC-1 red fluorescence detection. The value of mitochondrial potential is the ratio of red to green fluorescence (*see* **Note 11**). An example of the results obtained is shown in Figs. 2a, b and 8a.

3.6 Measurement of Mitochondrial Superoxide Production in Human Fibroblasts with the Use of MitoSOX Red

1. Remove the culture medium, and wash the cells gently with warm KRB before preincubation with effectors. Dilute stock solutions of antimycin A and oligomycin to final concentration 2 μM in KRB. Add 0.5 ml of KRB with particular effector to selected wells and KRB alone to control wells. Preincubate the plate, in the incubator, for 15–30 min prior to the measurement (*see* **Note 6**).
2. Prepare 5 μM solution of MitoSOX in KRB. Protect the solution from light (*see* **Notes 5** and **8**).
3. Remove the preincubation solution from wells, and add 0.5 ml of the MitoSOX Red solution per well. Add proper amounts of antimycin A or oligomycin to the selected wells (*see* **Notes 5, 8**, and **10**).
4. Incubate for 10 min in the incubator.
5. Gently wash the cells twice with warm KRB.
6. Add 0.5 ml KRB to each well (*see* **Note 10**).
7. Place the plate in the microplate reader, and read the fluorescence at 510 nm excitation and 595 nm emission wavelengths (*see* **Note 11**). An example of the results obtained is shown in Figs. 2a, b and 8d.

3.7 Measurement of Cytosolic Superoxide Production in Human Fibroblasts with the Use of DHE

1. Remove the culture medium, and wash the cells gently with warm KRB before preincubation with effectors. Dilute stock solutions of antimycin A and oligomycin to final 2 μM concentration in KRB. Add 0.5 ml of KRB with particular effector to selected wells and KRB alone to control wells. Preincubate the plate, in the incubator, for 15–30 min prior to the measurement (*see* **Note 6**).
2. Prepare 5 μM DHE solution in KRB (*see* **Notes 5, 8** and **10**).
3. Remove the preincubation solution from wells, and add 0.5 ml of DHE solution to each well of 24-well plate. As the effectors must be present till the end of measurement, add proper amounts of antimycin A or oligomycin to the selected wells.
4. Incubate for 20 min in the incubator.
5. Gently wash the cells twice with KRB.
6. Add KRB to each well (*see* **Note 10**).
7. Place the plate in the microplate reader, and read the fluorescence at 535 nm excitation and 635 nm emission wavelengths. This is a single read; no kinetic should be defined. An example of the results obtained is shown in Figs. 2a, b and 8c.

3.8 Measurement of Oxygen Consumption (Respiration) in Ehrlich Ascites Tumor Cells

1. Add cells (approx. 6 mg) to the chamber, fill the chamber with the measurement medium (NaCl 135 mM, KCl 5 mM, KH_2PO_4 0.4 mM, $MgSO_4$ 1 mM, HEPES 20 mM, bring to pH 7.4 with NaOH. Glucose 1 g/l and 1 mM $CaCl_2$), and close the chamber.

2. Start the oxygen consumption measurement (first derivative of the oxygen concentration trace).
3. Wait for the stable signal (usually it takes about 1–3 min).
4. Make traces with the following additions:
   - First trace: 1 μl of oligomycin → 1 μl of CsA → 1 μl of FCCP.
   - Second trace: 1 μl of CsA → 1 μl of oligomycin → 1 μl of FCCP.
5. An example of the results obtained is shown in Fig. 3a, b.

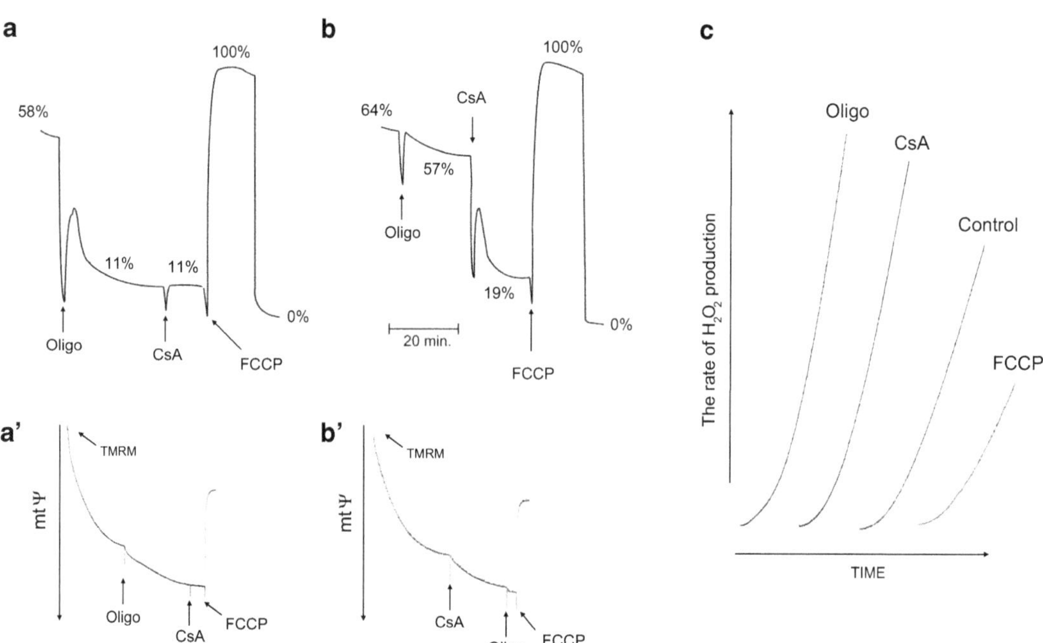

Fig. 3 Effect of oligomycin, cyclosporine A, and FCCP on oxygen consumption, mitochondrial membrane potential, and H_2O_2 production measured in Ehrlich ascites tumor cells. (**a, a′**) Oxygen consumption was measured with the use of a Clark-type oxygen electrode (Subheading 3.8); (**b, b′**) $\Delta\Psi$ was measured with the use of TMRM in a Shimadzu Spectrofluorometer RF 5000 (Subheaidng 3.9); (**c**) H_2O_2 production was measured with the use of CM-H_2DCFDA in Shimadzu Spectrofluorometer RF 5000 (Subheading 3.10); the addition of oligomycin (oligo) to the Ehrlich ascites tumor cells caused a decrease in oxygen consumption (manifested as a decrease of first derivative value of the oxygen concentration trace) (Panel **a**, **b**) and an increase in mt $\Delta\Psi$ (Panel **a′**, **b′**), corresponding to the resting-state (respiratory State 4) level. The effect of oligomycin on mitochondrial bioenergetic parameters is manifested in an increased rate of H_2O_2 production (Panel **c**). Further addition of FCCP resulted in the acceleration of oxygen consumption, a rapid collapse of the $\Delta\Psi$ and decreased rate of H_2O_2 production. Cyclosporine A (CsA), an inhibitor of the permeability transition pore, added before oligomycin also partially decreased oxygen consumption (Panel **b**), increased the mt $\Delta\Psi$ (Panel **b′**), and increased the rate of H_2O_2 production (Panel **c**). Incompatibility of the amplitude changes in TMRM fluorescence with the alterations in the rate of oxygen consumption after oligomycin and cyclosporine A addition was evoked by the effect of both of these compounds on the multidrug-resistant (MDR) proteins. Basing on the oxygen consumption data, it is recommended to use an inhibitor of MDR proteins as, e.g., sulfinpyrazone 100 μM

3.9 Fluorometric Measurement of Mitochondrial Membrane Potential in Ehrlich Ascites Tumor Cells with the Use of TMRM

1. Adjust the fluorometer: excitation, 556 nm; emission, 576 nm; slits (excitation and emission) ~3.
2. Fill the fluorometer cuvette with 3 ml of the measurement medium (NaCl 135 mM, KCl 5 mM, KH_2PO_4 0.4 mM, $MgSO_4$ 1 mM, HEPES 20 mM, bring to pH 7.4 with NaOH. Glucose 1 g/l and 1 mM $CaCl_2$) containing 1×10^7 of EAT cells.
3. Add 5 μl of 200 μM TMRM.
4. Start the fluorescence measurement.
5. Make traces with the following additions:
   - First trace: 5 μl oligomycin, 5 μl of CsA and 5 μl FCCP.
   - Second trace: 5 μl CsA, 5 μl of oligomycin and 5 μl FCCP.
6. An example of the results obtained is shown in Fig. 3a′, b′.

3.10 Fluorometric Measurement of H_2O_2 Production in Ehrlich Ascites Tumor Cells with the Use of CM-H_2DCFDA

1. Adjust the fluorometer: excitation, 513 nm; emission, 530 nm; slits (excitation and emission) ~3.
2. Fill the fluorometer cuvette with 3 ml of the measurement medium (NaCl 135 mM, KCl 5 mM, KH_2PO_4 0.4 mM, $MgSO_4$ 1 mM, HEPES 20 mM, bring to pH 7.4 with NaOH. Glucose 1 g/l and 1 mM $CaCl_2$) containing 5×10^6 of EAT cells.
3. Start measurement.
4. Add 10 μM of CM-H_2DCFDA and record changes in fluorescence.
5. Make similar traces in the presence of the following additions:
   - Second trace: 2 μM oligomycin.
   - Third trace: 2 μM antimycin A.
   - Fourth trace: 2 μM FCCP.
6. An example of the results obtained is shown in Fig. 3c.

3.11 Confocal Measurements of Mitochondrial Membrane Potential in HeLa Cells

1. Plate cells on 25 mm coverslips 2 days before the experiment, in a number determined for each cell type, in order to obtain not more than 90% confluent culture on the day of the experiment. Before plating cells the coverslips must be sterilized by UV exposure (30 joules) or by temperature (200 °C for 2 h).
2. Prepare a 10 nM solution of TMRM in KRB saline supplemented with glucose (1 g/l) and 1 mM $CaCl_2$ just before loading the cells. The total volume should be calculated considering 1 ml for each coverslip and 100 μl for each drug or chemical addition executed during experiment.
3. Prepare a 10× solution of each compound required for the experiment with an exceeding volume of the TMRM solution (*see* **Notes 12–14**).

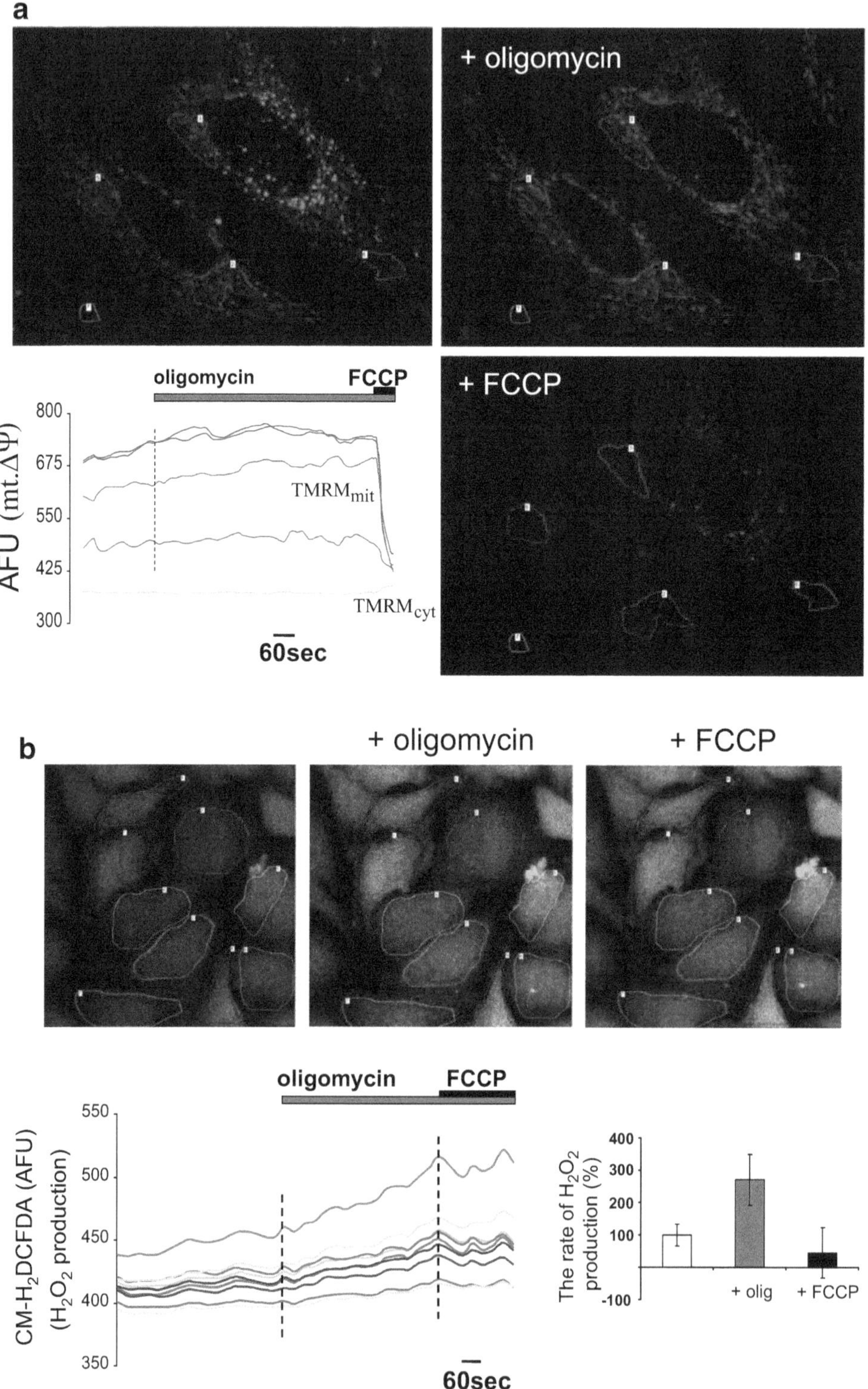

Fig. 4 Effect of oligomycin and FCCP on mitochondrial membrane potential and H_2O_2 production in HeLa cells. (**a**) $\Delta\Psi$ was measured with the use of TMRM in a confocal microscope (Subheading 3.11); (**b**) H_2O_2 production was measured with the use of CM-H_2DCFDA in a confocal microscope (Subheading 3.12); the addition of

4. Wash cells twice in order to remove dead cells and cell debris, add 1 ml of TMRM solution (room temperature) to the coverslip, and then incubate for 20–40 min at 37 °C. The correct loading time with TMRM may vary for different cell types.
5. After loading, the coverslip should be mounted in a metal cage, or a different appropriate support depending on the microscope model, and covered with 1 ml of the same TMRM solution as used for loading.
6. The coverslip and its support should be placed on the inverted confocal microscope equipped with a thermostated stage set at 37 °C. Correct visualization of mitochondria should be performed using a 40–100× oil immersion objective. Optimal illumination is obtained using a 543 nm HeNe gas laser or a 561 nm solid-state laser, while emission should be selected using a long-pass 580 filter (*see* **Note 15**).
7. We suggest time lapse with a delay of at least 10 s between each measurement step in order to avoid phototoxicity. To investigate the effect of the compound of interest (e.g., oligomycin), addition of 100 μl of the 10× concentrated solution is recommended. This is required in order to obtain a fast diffusion of the substance in the chamber. To obtain the basal fluorescence intensity level, terminate each experiment by adding 500 nM of FCCP.
8. After the experiment fluorescence intensity can be measured in selected regions drawn around mitochondria. An example of the results obtained is shown in Fig. 4a.

3.12 Confocal Measurement of Hydrogen Peroxide Production in HeLa Cells

1. Plate cells on 25 mm coverslips 2 days before the experiment, in a number determined for each cell type, in order to obtain not more than 90% confluent culture on the day of the experiment. Before plating cells the coverslips must be sterilized by UV exposure (30 joules) or by temperature (200 °C for 2 h).
2. Prepare a 5 μM solution of CM-H_2DCFDA in KRB saline supplemented with glucose (1 g/l) and 1 mM $CaCl_2$ just before loading the cells. The total volume should be calculated considering 1 ml for each coverslip (*see* **Note 16**).
3. Prepare a 10× solution of each compound used for the stimulation or inhibition of H_2O_2 production in complete KRB saline.

Fig. 4 (continued) oligomycin to the intact cells led to hyperpolarization of mitochondria (drift from mitochondrial respiratory state III to state IV) (Panel **a**) and an increase of the rate of H_2O_2 production (Panel **b**). Further addition of FCCP resulted in a rapid collapse of the $\Delta\Psi$ (represented as a decrease of the TMRM fluorescence) (Panel **a**) and a decrease rate of H_2O_2 production to a level lower than the initial threshold (Panel **b**)

4. Wash cells twice in order to remove dead cells and cell debris, add 1 ml of the H_2DCFDA solution (room temperature) to the coverslip, and incubate for 10 min at 37 °C.
5. After loading, the coverslip should be mounted in a metal cage, or a different appropriate support depending on the microscope model, and covered with 1 ml of the same H_2DCFDA solution used for loading.
6. The coverslip and its support should be placed on the inverted confocal microscope equipped with a thermostated stage set at 37 °C. Images should be recorded with a 40–100× oil immersion objective, illuminating with 488 Argon or solid-state laser. Emitted light will be preferably selected with a 505–550 band-pass filter (*see* **Note 17**).
7. We suggest time lapse with a delay of at least 15–30 s to avoid photoactivation not related to H_2O_2 production. Stimulation with chemical obtained by adding 100 μl of the 10× concentrated solution is recommended. This is required in order to obtain a fast diffusion of the substance in the chamber.
8. After the experiment fluorescence intensity will be measured drawing small region around the bright objects; the cytosolic area should be excluded to avoid artifacts (*see* **Note 18**). An example of the results obtained is shown in Fig. 4b.

3.13 Confocal Measurement of Superoxide Production in HeLa Cell

1. Plate cells on 25 mm coverslips 2 days before the experiment, in a number determined for each cell type, in order to obtain not more than 90% confluent culture on the day of the experiment. Before plating cells the coverslips must be sterilized by UV exposure (30 joules) or by temperature (200 °C for 2 h).
2. Apply the selected treatment to the sample before loading with MitoSOX Red. In the presented example, cells were incubated with 5 μM oligomycin for 15 min.
3. Prepare a 5 μM solution of MitoSOX Red in KRB saline supplemented with glucose (1 g/l) and 1 mM $CaCl_2$ just before loading the cells. The total volume should be calculated considering 1 ml for each coverslip and should be protected from light and high temperature.
4. Wash cells twice in order to remove dead cells and cell debris, add 1 ml of MitoSOX Red solution (room temperature) to the coverslip, and incubate for 15 min at 37 °C (*see* **Note 19**).
5. After loading, the coverslips should be washed three time with complete KRB saline and mounted in a metal cage or a different appropriated support depending on the microscope model.

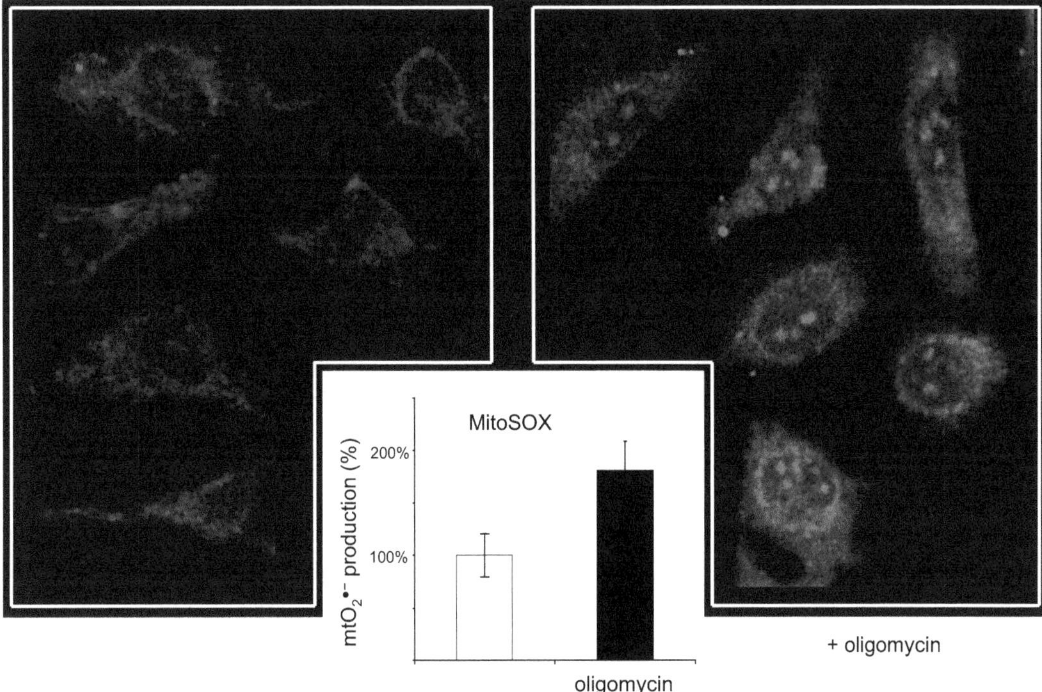

Fig. 5 Effect of oligomycin on mitochondrial superoxide production in HeLa cells. The mitochondrial superoxide production was measured with the use of MitoSOX Red in a confocal microscope (Subheading 3.13); addition of oligomycin (right panel) to the HeLa cells augmented mitochondrial superoxide production represented by increased fluorescence. These data are in line with the previous observations that in intact cells, the hyperpolarization of the inner mitochondrial membrane accelerates ROS formation

6. The coverslip and its support should be placed on the inverted confocal microscope. Recorded images with a 40–100× oil immersion objective, illuminating with 514 Argon or 488 solid-state laser. Emitted light should be selected with a 580 nm long-pass filter. A thermostated stage is optional (*see* **Note 20**).
7. In our experience MitSOX Red is not able to perform kinetics of superoxide production. After image collection, mean fluorescence intensity should be measured by drawing small regions around the bright objects for each cell. The cytosolic area should be excluded to avoid artifacts. An example of the results obtained is shown in Fig. 5.

3.14 Simultaneous Measurement of the Mitochondrial Membrane Potential and Cytosolic Calcium in HeLa Cells Using Confocal Microscope

1. Plate cells on 25 mm coverslips 2 days before the experiment, in a number determined for each cell type, in order to obtain not more than 90% confluent culture on the day of the experiment.
2. Prepare a KRB saline supplemented with 1 μM Fluo-3 AM, 10 nM TMRM, and glucose (1 g/l). The total volume should be calculated considering 1 ml for each coverslip and should be protected from light and high temperature.

3. Just before loading the cells, add $CaCl_2$ to the final 1 mM concentration to the KRB saline prepared in the previous step.
4. Wash cells twice in order to remove dead cells and cell debris; add 1 ml of room temperature KRB solution containing 1 μM Fluo-3 AM and 10 nM TMRM to the coverslip and incubate for 30 min, 37 °C.
5. After loading, the coverslip should be washed with KRB/Ca^{2+}, mounted in a metal cage, or a different appropriate support depending on the microscope model, and covered with 1 ml of the KRB saline supplemented with glucose (1 g/l) and 1 mM $CaCl_2$.
6. The coverslip and its support should be placed on the inverted Zeiss LSM510 Confocal Microscope equipped with a thermostated stage set at 37 °C.
7. Start recording the images sequentially with a 40–100× oil immersion objective, illuminating with 488 Argon or solid-state laser for Fluo-3 illumination and 543 HeNe or 561 solid-state laser for TMRM. Collect emitted light for the Fluo-3 in the range ≥505–≥535 nm and for the TMRM as total emission ≥570 nm.
8. Stimulate the cells by the addition of 100 μl of the 1 mM histamine solution.
9. After the experiment, analyze the changes in fluorescence intensity with Zeiss LSM510 software. An example of the results obtained is shown in Fig. 6.

3.15 Simultaneous Measurement of the Mitochondrial Membrane Potential and H_2O_2 Production in HeLa Cells Using Confocal Microscope

1. Plate cells on 25 mm coverslips 2 days before the experiment, in a number determined for each cell type, in order to obtain not more than 90% confluent culture on the day of the experiment.
2. Prepare the KRB saline supplemented with 5 μM CM-H_2DCFDA, 10 nM TMRM, glucose (1 g/l), and 1 mM $CaCl_2$. The total volume should be calculated considering 1 ml for each coverslip and should be protected from light and high temperature.
3. Wash cells twice in order to remove dead cells and cell debris, add 1 ml of the KRB solution (room temperature) containing 5 μM CM-H_2DCFDA and 10 nM TMRM to the coverslip, and incubate for 20 min, 37 °C.
4. After loading, the coverslip should be washed with KRB/Ca^{2+}, mounted in a metal cage, or a different appropriated support depending on the microscope model, and covered with 1 ml of the KRB saline supplemented with glucose (1 g/l) and 1 mM $CaCl_2$.

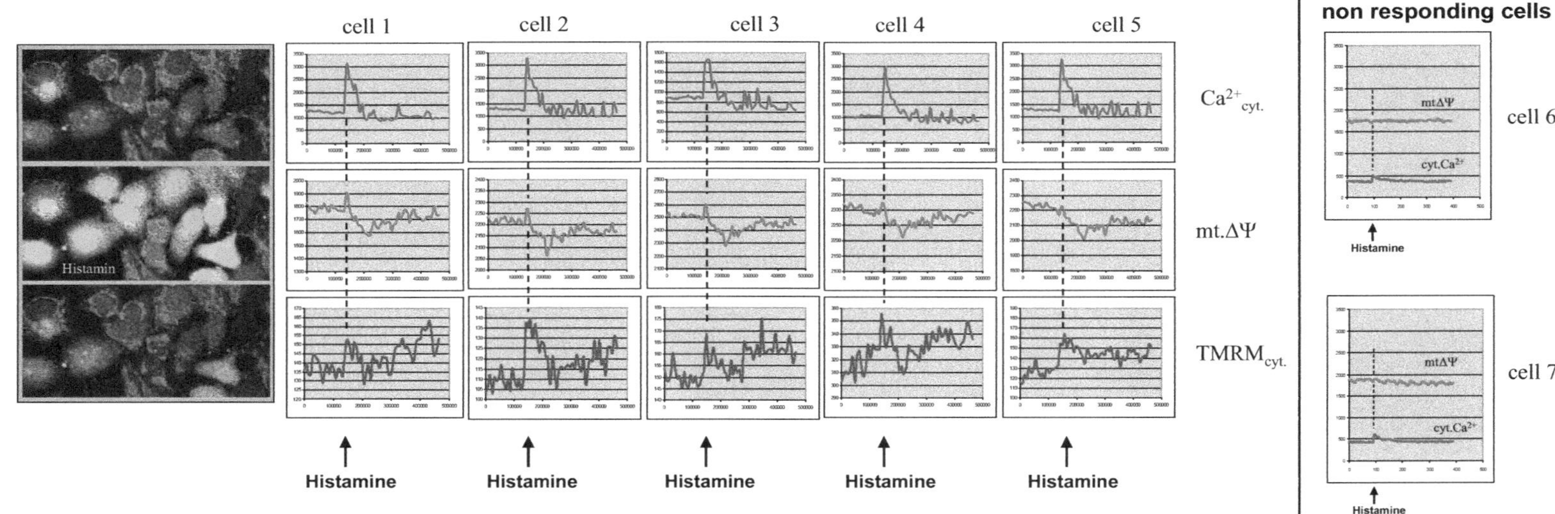

Fig. 6 Effect of histamine on cytosolic calcium and mitochondrial membrane potential in HeLa cells. Simultaneous measurements of cytosolic calcium and ΔΨ with the use of Fluo-3 and TMRM in confocal microscope (Subheading 3.14); the addition of histamine to the intact HeLa cells induced a transient cytosolic Ca^{2+} signal recorded as an increase of Fluo-3 fluorescence. Simultaneously, a transient partial mitochondrial depolarization (decrease of mitochondrial TMRM fluorescence and increased cytosolic TMRM signal) was observed. Decrease of ΔΨ occurs due to the calcium uptake by mitochondria. Cells, not responding to the histamine stimulation, have unchanged ΔΨ

5. The coverslip and its support should be placed on the inverted Zeiss LSM510 Confocal Microscope equipped with a thermostated stage set at 37 °C.
6. Start recording the images sequentially with a 40–100× oil immersion objective, illuminating with 488 Argon or solid-state laser for CM-H_2DCFDA illumination and 543 HeNe or 561 solid-state laser for TMRM. Collect the emitted light for the CM-H_2DCFDA in the range ≥505–≥535 nm and for the TMRM as total emission ≥570 nm.
7. Stimulate the cells by adding 100 μl of the 1 mM histamine solution.
8. After a few minutes, when the rate of free radical production is constant, add 50 μl of 12 μM FCCP.
9. After the experiment, analyze the changes in fluorescence intensity with Zeiss LSM510 software. An example of the results obtained is shown in Fig. 7.

3.16 Measurement of Mitochondrial Respiratory Chain Activity in Human Fibroblasts with the Use of Resazurin

1. Prepare a 6 μM resazurin solution in KRB.
2. Remove the culture medium, and wash the cells gently twice with warm KRB (*see* **Note 6**).
3. Add 0.5 ml of the resazurin solution to each well of the 24-well plate.
4. Immediately after addition, start the measurement in the kinetic mode at 510 nm excitation and 595 nm emission wavelengths (*see* **Note 21**). An example of the results obtained is shown in Fig. 8b.

3.17 Measurement of Hydrogen Peroxide Production in Human Fibroblasts with the Use of CM-H_2DCFDA

1. Prepare a 2 μM CM-H_2DCFDA solution in KRB.
2. Remove the culture medium, and wash the cells gently twice with warm KRB (*see* **Note 6**).
3. Add 0.5 ml of the CM-H_2DCFDA solution to each well of the 24-well plate. Immediately after addition, start the measurement in the kinetic mode at 495 nm excitation and 520 nm emission wavelengths. An example of the results obtained is shown in Fig. 8e.

3.18 Measurement of Protein Concentration in the Plate Wells After Measurement of ROS and Mitochondrial Respiratory Chain Activity in Human Fibroblasts

1. Usually cells are plated and grow in equal density at each well; however addition of different chemical compounds may induce cell death. Thus, protein concentration on each well should be determined after measurement to calculate appropriate values of bioenergetic parameters.
2. For spectrophotometric measurement of protein concentration, cells grown on multiwell plates must be suspended in the lysis buffer, about 500 μl per well. Protein assay is based on Bradford method. Add to the 3 ml spectrophotometer

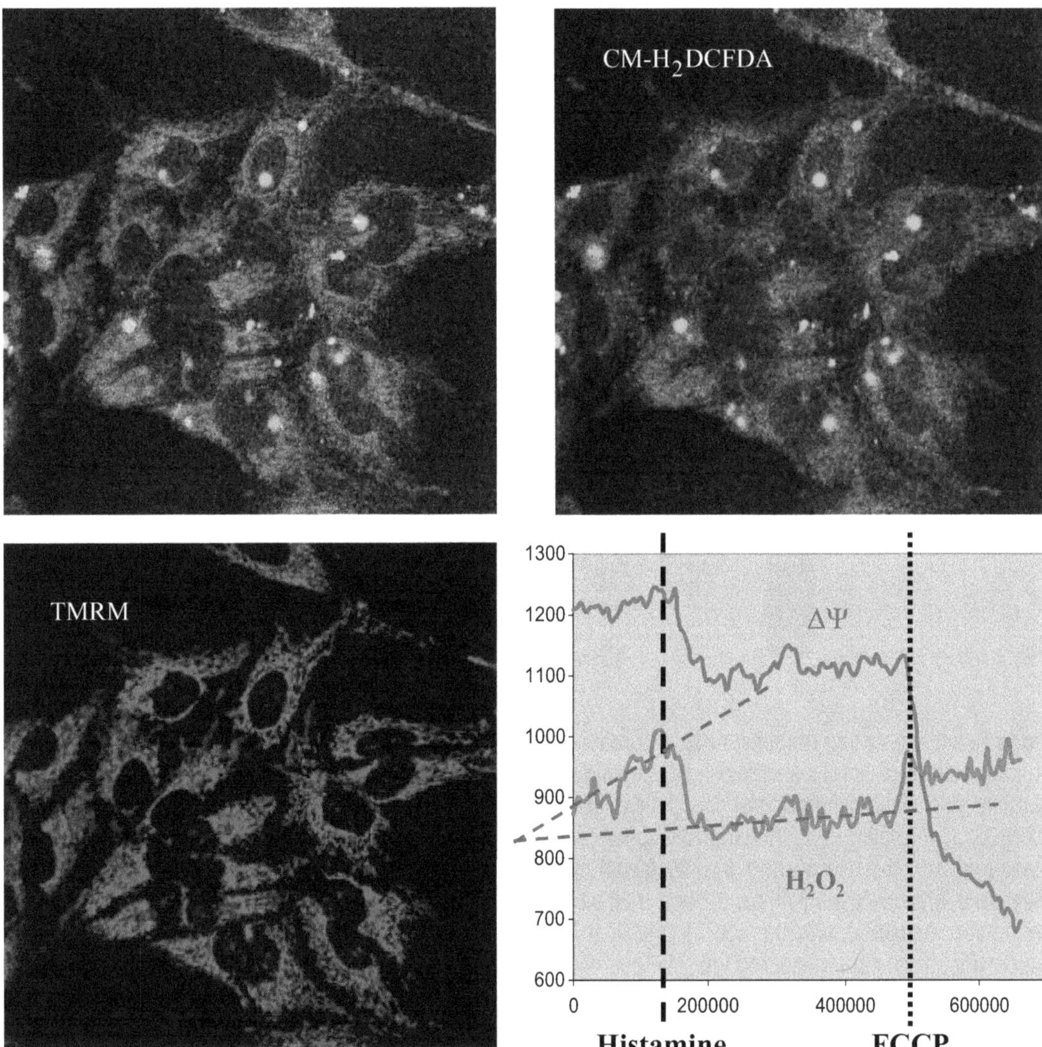

Fig. 7 Effect of histamine on mitochondrial membrane potential and H_2O_2 production in HeLa cells. Simultaneous measurement of $\Delta\Psi$ and H_2O_2 production with the use of TMRM and CM-H_2DCFDA in confocal microscope (Subheading 3.15); as was also presented in Fig. 6, addition of histamine to the intact HeLa cells induced a transient, partial mitochondrial depolarization (decrease of mitochondrial TMRM fluorescence) which is accompanied by the reduction of the rate of H_2O_2 production. These data are in line with the previous observations that in intact cells, partial depolarization of the inner mitochondrial membrane is manifested by a decreased ROS production

cuvette 2.4 ml of H_2O_2. Depending on the cell density, add the amount of sample (that the absorbance should be between 100 and 600 spectrophotometric units) from each well to separate cuvettes. Then add 600 μl of room temperature Bio-Rad Protein Assay, and shake the sample. Measure the absorbance at 595 nm.

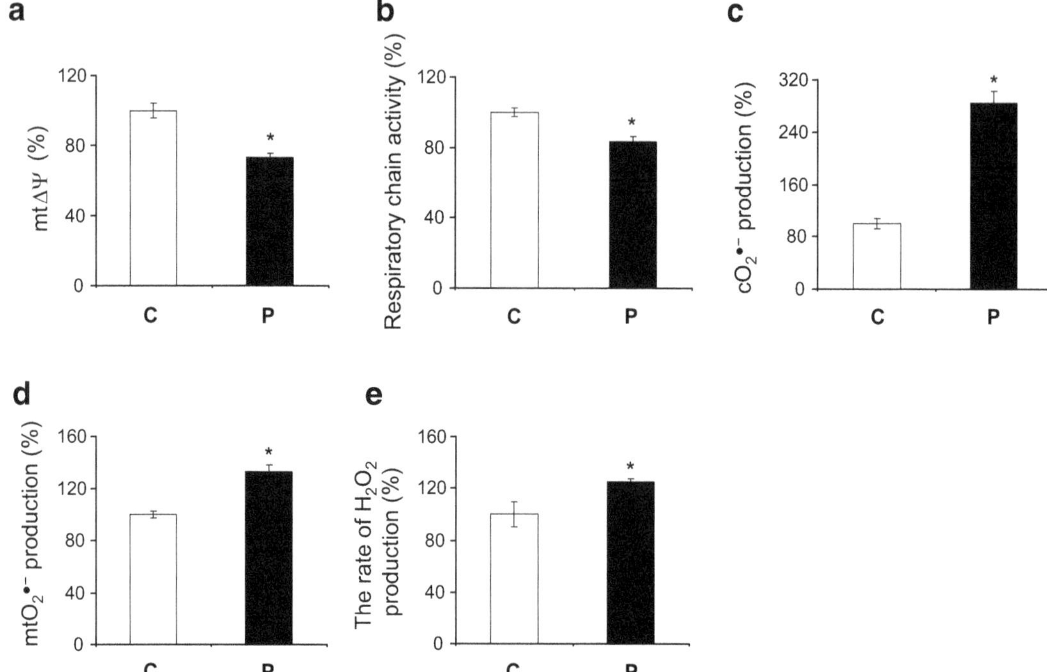

Fig. 8 Relation between mitochondrial membrane potential, respiratory chain activity, and ROS production in primary culture of fibroblasts from a healthy individual and a child with a mitochondrial disorder. C control fibroblasts, P patient fibroblasts; the patient used in these studies demonstrated a clinical phenotype of OXPHOS abnormality (mitochondrial encephalopathy) in muscle biopsies and in fibroblast culture. (**a**) $\Delta\Psi$ measured with the use of JC-1 in a multiwell plate reader (Subheading 3.5); (**b**) respiratory chain activity measured with the use of resazurin in a multiwell plate reader (Subheading 3.16); (**c**) cytosolic superoxide production measured with the use of DHE in a multiwell plate reader (Subheading 3.7); (**d**) mitochondrial superoxide production measured with the use of MitoSOX Red in a multiwell plate reader (Subheading 3.6); and (**e**) H_2O_2 production measured with the use of CM-H_2DCFDA in a multiwell plate reader (Subheading 3.17). Dysfunction of the respiratory chain in the patients' fibroblasts was represented as a decreased respiratory chain activity and lower mitochondrial potential compared to the healthy fibroblasts. A defect in the mitochondrial respiratory chain results in higher cytosolic (**c**) and mitochondrial (**d**) superoxide production. In such cells, the rate of H_2O_2 production is also increased (**e**)

4 Notes

1. To isolate intact mitochondria, it is necessary to use low-calcium sucrose.
2. The medium can be prepared in advance and stored in 4 °C for approx 2 weeks.
3. All solutions should be at 4 °C and all equipment precooled.
4. Extreme care should be taken to avoid contamination with the ice and tap water.

5. For the assay the reaction buffer should have room temperature.
6. Fibroblasts, plated with equal density, are grown on 24-well plates until they reach confluence, in conditions of 5% (v/v) CO_2 in air at 37 °C. The medium is changed every 2 days including the day before experiment.
7. Avoid higher JC-1 concentration as its precipitates are hard to wash out.
8. Avoid light exposure of the fluorescent probe solution and the loaded cells.
9. The time of incubation should not be longer than 10 min.
10. Keep in mind that antimycin A and oligomycin must be present during incubation with the fluorescent probe and during the measurement.
11. For the final calculation of the measured parameter, the background (basal fluorescence) should be subtracted.
12. TMRM solution may be kept for a few hours at 4 °C; however it is preferable not to add cold solution to the cells. Any changes of temperature during experiments should be avoided because it can cause alterations in TMRM distribution independently to $\Delta\Psi$.
13. In all experiments TMRM concentration must always be the same in order to avoid differences in the redistribution of the dye in the cell.
14. Correct TMRM loading can be confirmed by a short time acquisition (i.e., 10 min). If stable fluorescence intensity is recorded, the cells are loaded correctly. If the signal is increasing during acquisition, the loading time should be increased.
15. The laser power should be kept low in order to avoid photoactivation and bleaching of the dye. With a HeNe laser, it is recommended to set the transmission lower than 10%, while with the solid-state laser, the power should not be higher than 30%.
16. CM-H_2DCFDA solution should be protected from light and high temperature.
17. The laser power should be kept low in order to avoid photoactivation and bleaching of the dye. With a neon laser, it is recommended to set the transmission lower than 5%, while with the solid-state laser, the power should not be higher than 20%. A short recording can be executed before the actual experiment at basal conditions. If fast increase in fluorescence is observed, probably photoactivation of the CM-H_2DCFDA occurs. In this case it is necessary to reduce the laser power and/or the number of the readings of the same frame (typical

parameter for laser scanning confocal microscopes) and/or increase reading speed.

18. If a fast confocal system is available (i.e., spinning disk or swept field), Z stack acquisition is recommended. In conventional laser scanning, confocal microscopy Z stack acquisition will be too slow with the risk of recording artifacts.
19. No kinetic is recorded with MitoSOX Red dye, so loading time and washing must be carefully respected in order to avoid experimental artifacts.
20. Optimal excitation wavelength declared by the manufacturer is 510 nm, but we observed that even with the 488 nm solid-state laser, a good signal is recordable. In this case short exposure, time is essential to avoid phototoxic stress and photoactivation.
21. For the background fluorescence subtraction, selected wells on the plate can be incubated with 1 μl of 1 M KCN stock solution before (about 15 min) and during measurement.

Acknowledgments

This work was supported by the Polish Ministry of Science and Higher Education grants N301 092 32/3407 and N N407 075 137 for MRW, JD, ML, and JS. JS was also supported by a PhD fellowship from the Foundation for Polish Science (FNP), UE, European Regional Development Fund, and Operational Programme "Innovative Economy." PP and MB are supported by AIRC, Telethon (GGP09128); local funds from the University of Ferrara; the PRRIITT program of the Emilia-Romagna Region; the Italian Multiple Sclerosis Foundation (FISM Cod.2008/R/18); the Italian Ministry of Education, University and Research; and the Italian Ministry of Health.

Ethics

The studies with the use of human fibroblasts were carried out in accordance with the Declaration of Helsinki of the World Medical Association and were approved by the Committee on Bioethics at the Children's Memorial Health Institute. Informed consent was obtained from the parents before any biopsy or molecular analysis was performed.

References

1. Turrens JF (2003) Mitochondrial formation of reactive oxygen species. J Physiol 552:335–344
2. Korshunov SS, Skulachev VP, Starkov AA (1997) High protonic potential actuates a mechanism of production of reactive oxygen species in mitochondria. FEBS Lett 416:15–18
3. Miwa S, Brand MD (2003) Mitochondrial matrix reactive oxygen species production is very sensitive to mild uncoupling. Biochem Soc Trans 31(part 6):1300–1301
4. Wojtczak L, Teplova VV, Bogucka K, Czyz A, Makowska A, Wiechowski MR, Muszyński J,

Evtodienko YV (1999) Effect of glucose and deoxyglucose on the redistribution of calcium in ehrlich ascites tumour and Zajdela hepatoma cells and its consequences for mitochondrial energetics. Further arguments for the role of Ca(2+) in the mechanism of the crabtree effect. Eur J Biochem 263:495–501

5. Geromel V, Kadhom N, Cebalos-Picot I, Ouari O, Polidori A, Munnich A, Rotig A, Rustin P (2001) Superoxide-induced massive apoptosis in cultured skin fibroblasts harboring the neurogenic ataxia retinitis pigmentosa (narp) mutation in the atpase-6 gene of the mitochondrial DNA. Hum Mol Genet 10:1221–1228
6. Lebiedzinska M, Karkucinska-Wieckowska A, Giorgi C, Karczmarewicz E, Pronicka E, Pinton P, Duszyński J, Pronicki M, Wieckowski MR (2010) Oxidative stress-dependent p66Shc phosphorylation in skin fibroblasts of children with mitochondrial disorders. Biochim Biophys Acta 1797(6–7):952–960. *in press*

Chapter 23

Assessing Spatiotemporal and Functional Organization of Mitochondrial Networks

Felix T. Kurz, Miguel A. Aon, Brian O'Rourke, and Antonis A. Armoundas

Abstract

The functional and spatiotemporal organization of mitochondrial redox signaling networks can be studied in detail in cardiac myocytes and neurons by assessing the time-resolved signaling traits of their individual mitochondrial components. Perturbations of the mitochondrial network through oxidative stress can lead to coordinated, cluster-bound behavior in the form of synchronized limit-cycle oscillations of mitochondrial inner membrane potentials. These oscillations are facilitated by both structural coupling through changes in the local redox balance and signaling microdomains and functional coupling that is yet poorly understood. Thus, quantifiable measures of both coupling mechanisms, local dynamic mitochondrial coupling constants and functional clustering coefficients, are likely to offer valuable information on mitochondrial network organization. We provide step-by-step methodologies on how to acquire and assess these measures for inner membrane potential fluorescence fluctuations in laser-scanning two-photon microscope recordings of cardiac myocytes and neurons, that can be applied to other tissues as well.

Key words Mitochondrial oscillator, Mitochondrial network, Mitochondrial clustering, Cardiac myocyte, Wavelets, Mitochondrial coupling

1 Introduction

Mitochondria contribute to a wide range of cellular processes such as metabolic pathway control, ATP production, cellular homeostasis and apoptosis, redox oxidation and oxidative phosphorylation, β-oxidation of fatty acids, and calcium signaling. Their role as intracellular hubs that influence, control, and interfere with cellular functions implicates them in many diseases and disorders, e.g., cardiovascular and neurological pathologies [1–3].

There are numerous methodologies to measure mitochondrial activity, many of them linked to mitochondrial signals that are generated by changes in the local redox or pH environment [3–5]. These signals enable mitochondrial regulation of specific enzymes, transcription processes, and communication with other mitochondria [2, 6, 7]. In fact, mitochondria may produce changes

Carlos M. Palmeira and António J. Moreno (eds.), *Mitochondrial Bioenergetics: Methods and Protocols*, Methods in Molecular Biology, vol. 1782, https://doi.org/10.1007/978-1-4939-7831-1_23,

of their inner membrane shape [8] or potential [9] in response to external stimuli that alter the local redox status like oxidative stress, thiol-oxidizing substances (diamide), or substrate deprivation [10–13].

A well-studied example appears in mitochondrial inner membrane potential oscillations in cardiac myocytes where mitochondria form a lattice-like, cell-wide network of densely packed organelles to facilitate cardiac energy demand and supply [11]. These oscillations typically commence with a local perturbation of the mitochondrial redox environment and can be reliably reproduced by a focalized laser flash that depolarizes the inner membrane potential, $\Delta\Psi_m$, of a small fraction of mitochondria [14]. The stress-associated increase of locally confined reactive oxygen species (ROS) is sensed by mitochondrial inner membrane anion channels that may open to release ROS when ROS levels surpass a critical threshold [11, 14, 15].This autocatalytic or ROS-induced ROS release (RIRR) mechanism is controlled by ROS scavenging processes so that, in a confined environment, a cyclical activation of RIRR can lead to local mitochondrial $\Delta\Psi_m$ oscillations [14, 16, 17]. However, depending on the gradient and distribution of intracellular ROS densities, ROS-sensitive mitochondrial inner membrane potential channels, and availability of ROS scavenging molecules, $\Delta\Psi_m$ oscillations may not be restricted to the initial nucleus of mitochondria with a perturbed redox environment, but can expand to the whole cell by destabilizing $\Delta\Psi_m$ of neighboring mitochondria such that a wave of $\Delta\Psi_m$ depolarization transcends through the myocyte [18]. This process is also coordinated by electrochemical coupling at inter-mitochondrial junctions [8]. Typically, the experimentally observed myocyte-wide $\Delta\Psi_m$ oscillations are mainly sustained by a cluster of synchronously oscillating mitochondria distributed over the whole myocyte, which we call a spanning cluster [18]. Spanning clusters emerge when intracellular ROS values in a sufficiently large volume of the myocyte approach a critical threshold, a phenomenon that can be linked to percolation theory [12, 13, 18]. There is evidence, however, that not all mitochondria participate in synchronized $\Delta\Psi_m$ oscillations: some mitochondria oscillate with different frequencies, some do not oscillate at all, and others oscillate only temporarily, i.e., they may leave the cluster due to an exhaustion of ROS defense mechanisms [19, 20]. In addition, of those mitochondria that oscillate with different frequencies, some may lock to a common oscillatory mode to form a smaller synchronized cluster that coexists next to the spanning cluster. Since network mitochondria are structurally and functionally coupled through a plentitude of local and global mitochondrial ROS balance mechanisms, naturally, these processes are nonstationary and interdependent. Inter-mitochondrial coupling is further influenced by the interplay of mitochondria with other intracellular organelles (e.g., myofilaments or sarcoplasmic reticulum [21]), by mitochondrial functional heterogeneity [22, 23], and generally by any other intracellular, intercellular, and extracellular regulatory mechanism that exerts an influence on mitochondrial function. Mitochondrial structural heterogeneity adds to the complexity of the mitochondrial network [24, 25]. The observed mitochondrial $\Delta\Psi_m$ oscillations therefore reveal changing frequencies and amplitudes [19, 20], as does the size and growth of synchronized mitochondrial clusters; however, the mere presence of synchronized oscillations suggests a strong influence of local coupling in the mitochondrial network [26]. Furthermore, an experimentally observed inverse relation between size and frequency of synchronized clusters indicates that large clusters take longer to equilibrate their inter-mitochondrial coupling mechanisms and, thus, possess a lower common frequency than smaller clusters [20, 26]. A large cluster of synchronously oscillating mitochondria usually features a decrease in $\Delta\Psi_m$ amplitudes over time, which is accompanied by a decelerated cluster growth, indicating an exhaustion of ROS scavenging mechanisms and, therefore, is a sign of impending cell death. Measures of inter-mitochondrial coupling or functional network properties can thus provide important information about the network's response to perturbations and may be used to simulate large mitochondrial oscillator networks.

Recent modeling efforts of mitochondrial network behavior have shown that a reaction-diffusion model built on biochemical exchange rates for mitochondrial matrix constituents (calcium, NADH, ADP, Krebs cycle intermediates) and mitochondrial inner membrane potential physiology (including the RIRR mechanism) correctly predicts limit-cycle $\Delta\Psi_m$ oscillations [16, 27]. However, increasingly deterministic models of mitochondrial networks that include various forms of regulation and control on mitochondrial behavior, including functional and structural coupling of mitochondria within the network or with other (extra-)cellular organelles, biochemical pathways, as well as diffusion dynamics or fluxes of relevant biomolecules, need enormous computational resources. A more efficient approach is the introduction of stochastic noise in the description of individual mitochondrial signals [26]. The application of stochasticity in cardiac mitochondrial networks is justified for strong links between network components [28]. Based on an extension of the Kuramoto model of coupled oscillators [29, 30], stochastic differential equations for the mitochondrial $\Delta\Psi_m$ oscillation phase can be utilized to assign to every mitochondrion m a time-dependent coupling constant $K_m(t)$ (of local mean-field type, see below) to quantify its coupling to the nearest mitochondrial neighbors [26]. The model uses the observed drift of mitochondrial frequencies toward a local mean frequency described as stochastic Ornstein-Uhlenbeck frequencies [31]. It shows larger inter-mitochondrial coupling in smaller clusters at the early stages of myocyte-wide $\Delta\Psi_m$ oscillations, suggesting a strong effect of local coupling [26, 32]; also, see a discussion of coupling dynamics in mitochondrial frequency clusters [33].

The concept of a cluster of mitochondria with similar frequencies does not presume a spatially contiguous ensemble of mitochondria. In fact, some mitochondria may connect to the common oscillatory mode of a large cluster of synchronously oscillating mitochondria, although they are too distant from any cluster mitochondrion to experience local coupling. In analogy to functional networks of communicating neurons [34], such functional relations were recently examined for the network of mitochondrial oscillators [35]. They allow the quantification of a topological clustering coefficient that provides a measure of *functional connectedness* between mitochondrial oscillators. The analysis of functional clustering is not restricted to mitochondria from one major cluster of synchronously oscillating mitochondria, but involves all oscillating mitochondria. Briefly, functional clustering of mitochondrion *m* is a measure of the topological connections between all topological neighbors of *m*. Understanding of the dynamic changes in functional clustering provides valuable clues about the network's spatiotemporal organization in exploring the connectedness between mitochondria that cannot be explained by immediate (structural) coupling effects. It could be shown that functional clustering in mitochondrial networks is significantly higher than that in equivalent random networks [35]; this result supports the notion of a functional-structural unity in mitochondrial unity in mitochondrial networks [36].

Below we provide a detailed description of experimental and computational methods to assess mitochondrial spatiotemporal organization for the stress-induced cardiac myocyte mitochondrial inner membrane potential oscillations. Specifically, we present point-by-point procedural methodologies based on findings and results [19, 20, 26, 35] for the extraction of individual mitochondrial $\Delta\Psi_m$ signals and identification of nearest neighbors (Subheadings 3.1–3.4), the extraction of time-dependent frequency content for each mitochondrial $\Delta\Psi_m$ signal using wavelet analysis (Subheading 3.5), the identification of a major cluster of synchronously oscillating mitochondria and its frequency and cluster size analysis (Subheading 3.6), the analysis of local cluster coherence (Subheading 3.7), the analysis of propagation of mitochondrial signals following the onset of myocyte-wide oscillations (Subheading 3.8), the analysis of functional connectedness and functional mitochondrial clustering (Subheading 3.9), and the determination of local inter-mitochondrial coupling (Subheading 3.10).

2 Materials

2.1 Sets of Mitochondrial Signals

The assessment of the mitochondrial networks' spatiotemporal properties relies on time-resolved recordings of mitochondrial signals in an entire mitochondrial population. Naturally, in these recordings, mitochondrial signals must be associated with or arise from the single organelle level, i.e., they can be linked to mitochondrial metabolic pathway products, mitochondrial respiration, mitochondrial membrane channel signaling molecules, and/or mitochondrial membrane potential. The recordings also need to have a sufficiently high spatial resolution of ideally less than 500 nm to allow a differentiation of individual mitochondria. A further prerequisite is an adequate temporal resolution that captures dynamic changes in mitochondrial signals and signaling. For mitochondrial redox signaling or membrane potential fluctuations, the image sampling period is ideally less than 5 s [3, 20].

A well-studied example are mitochondrial inner membrane potentials in isolated cardiac myocytes: they can be marked with the fluorescence dyes tetramethylrhodamine-ethyl ester or methyl ester (TMRE or TMRM, hereafter referred to as TMRE) and measured with a laser-scanning two-photon microscope [13, 14]. Such mitochondrial signals can be recorded with sufficient temporal resolution to visualize $\Delta\Psi_m$ oscillations (*see* also Fig. 1). Other examples of time-resolved measurements of mitochondrial signaling within mitochondrial networks include the assessment of mitochondrial redox and mitochondrial matrix pH dynamics in mice intercostal axons and neuromuscular junctions [7] or cell cultures [37], stress-induced pulsing of mitochondrial membrane potentials in *Arabidopsis* [38], or in vivo mitochondrial superoxide signals for mouse skeletal muscle and sciatic nerve [39].

A transition of the mitochondrial network into myocyte-wide synchronized $\Delta\Psi_m$ oscillations could be measured ex vivo in freshly isolated adult guinea pig ventricular myocytes [14], but also in cellular structures of intact heart tissue of guinea pigs [40] and in vivo in rat glandular cells [41], demonstrating that mitochondrial $\Delta\Psi_m$ oscillations can prevail in a superordinate cellular network that contains additional control mechanisms, e.g., through intercellular gap junctions.

Every imaging methodology requires its own biological specimen, solutions, mitochondrial markers, and imaging protocols [3, 7, 14, 20, 26, 37–41]. However, since our group mostly worked with mitochondrial inner membrane potential signals from isolated cardiomyocyte experiments, we provide here a brief list of necessary materials to perform this specific experiment:

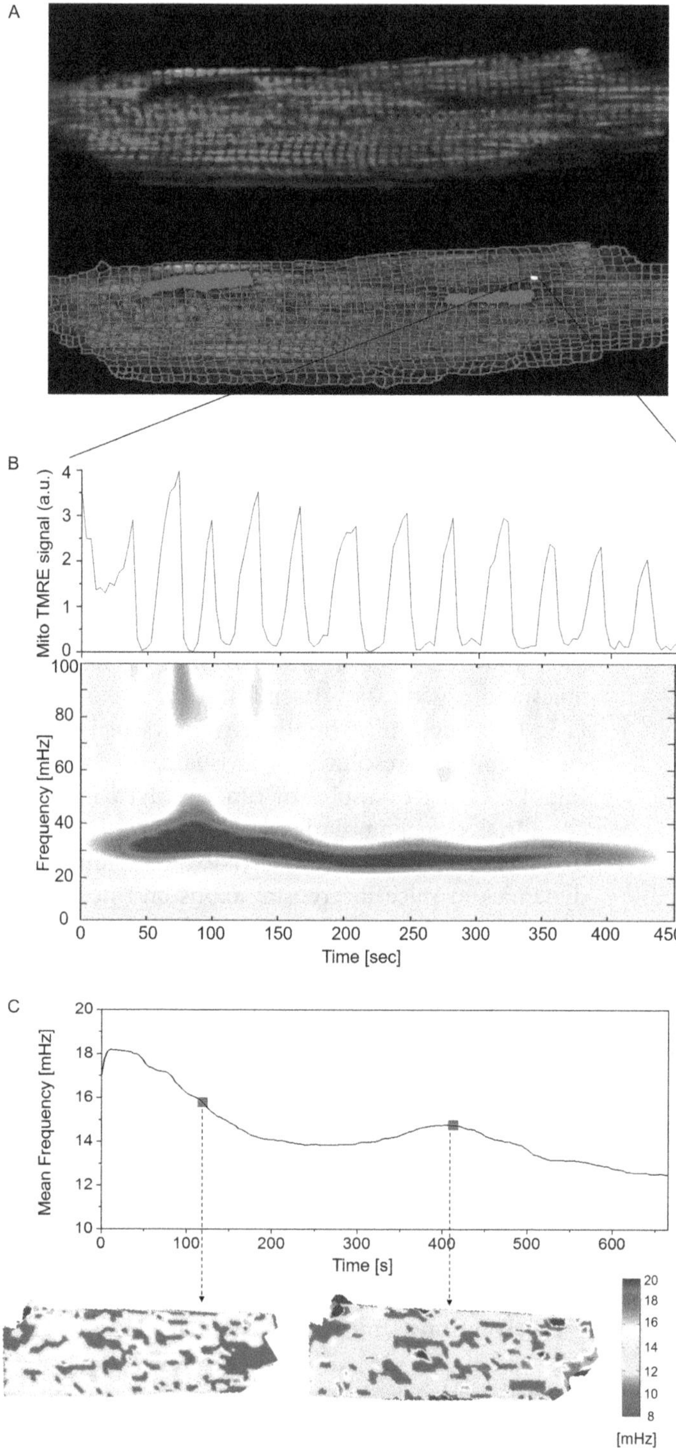

Fig. 1 Mitochondrial oscillations in a cardiac mitochondrial network. (**a**) Two-photon image of a TMRE-fluorescent guinea pig cardiac myocyte (upper panel) and with an overlaid grid that captures single mitochondria (lower panel), e.g., the yellow mitochondrion in the upper right corner of the cell. (**b**) TMRE

(a) *Materials for the Isolation of Cardiac Myocytes.*

- Enzymatically dispersed adult guinea pig ventricular myocytes (*see* also [9]).
- Dulbecco's modified Eagle's medium (10-013, Mediatech, Inc. Herndon, VA).
- Laminin-coated Petri dishes.
- 5% CO_2 incubator.
- Tyrode's solution containing 140 mM NaCL, 5 mM KCl, 1 mM $MgCl_2$, 10 mM HEPES, and 1 mM $CaCl_2$, with a pH = 7.5 (adjusted with NaOH), supplemented with 10 mM glucose (or other substrates such as pyruvate, lactate, or β-hydroxybutyrate; *see* [26]).
- Upright epifluorescence microscope (BX61W1; Olympus, Waltham, MA).
- 100 μM TMRE stock solution.

(b) *Materials for Image Acquisition.*

– Multiphoton laser scanning fluorescence system: Fluoview FV1000 MPE (Olympus) and a DeepSee ultrafast laser (Spectra Physics, Santa Clara, CA).

2.2 Software to Process the Mitochondrial Signals

Any imaging processing program, raster graphics editor, and technical computing software may be adequate for signal processing and network analysis; however, we used the following programs:

- ImageJ v1.38q or later, including the image stabilizer plug-in for ImageJ [42].
- Adobe Photoshop CS4 or later.
- Matlab v7.1.0.246 or later.
- OriginPro 8 SR0 v8.0724 or later.

3 Methods

We briefly describe the mitochondrial network imaging protocol, and we provide an example for the acquisition of mitochondrial

Fig. 1 (continued) intensity plot of the marked mitochondrion in (**a**) and the corresponding absolute squared wavelet transform (lower panel). The selected mitochondrion shows oscillations that range between 20 and 40 mHz. (**c**) Frequency distribution maps obtained from a different cardiac myocyte. The upper panel shows the mean mitochondrial oscillation frequency, averaged over all mitochondria, as a function of time. The lower panel shows maps of frequency distributions during two time-points (indicated in red in the upper panel). Missing pixels were interpolated. Most mitochondria oscillate with frequencies between 14 and 16 mHz while a small cluster in the right corner of the cell shows a strong increase in oscillation frequency during the two time-points between (Adapted from Fig. 1 in [26], with permission from ref. 26. Copyright 2015)

3.1 Studies in Isolated Cardiac Myocytes

signals in isolated cardiomyocytes; more detailed protocols can be found in [9, 14]:

1. Load the freshly isolated cardiac myocytes with 25 nM TMRE for 20–30 min in a thermostatically controlled flow chamber (at 37 °C) that is mounted on the stage of an upright epifluorescence microscope.
2. Washout the dye by perfusing the cardiac myocytes with Tyrode solution supplemented with 10 mM glucose (or other substrates) and 1 mM Ca^{2+} for 2–3 min. Visualize the myocytes with an objective 40×/1.0 W MP.
3. Record images using the multiphoton laser scanning fluorescence microscope (Fluoview FV 1000 MPE) with excitation at 740 nm and the red emission of TMRE collected at 605 nm using a 578–630 nm band-pass filter.
4. Trigger mitochondrial inner membrane oscillations with a localized ($5 \times 5\ \mu m^2$) laser flash.

3.2 Image Processing

1. For each stack of recorded images, use the ImageJ image stabilizer plug-in to minimize movement effects of the cardiac myocyte and the mitochondrial network.
2. For each stack of stabilized images, identify the onset of mitochondrial inner membrane potential oscillations (or an equivalent onset or relevant signal change for other types of experiments that involve mitochondrial signals/signaling), as the first image where the mean TMRE fluorescence intensity drops by more than 10% with respect to the averaged mean TMRE intensity of the previous images.
3. Identify the smallest period l of all TMRE oscillations that comprise a significant amount of the mitochondrial network (usually more than 30% of the network's mitochondria).
4. Arrange the images after the first significant TMRE intensity drop (*see* **step 2**) in intervals of l.
5. Average all images in each interval.

3.3 Extract Individual Mitochondrial Signals

1. Use a maximum-intensity averaged image from the set of images resulting from Subheading 3.2, **step 5** (above) as a template, and upload it into a raster graphics editor program to manually draw grid lines around each mitochondrion on a pixel-by-pixel basis for each separate image (layer) (*see* **Note 1**).
2. Save the grid image as a binary image.
3. Allocate a numerical identifier for each grid mesh, i.e., for each mitochondrion (e.g., using the function *bwlabel* in Matlab).
4. Average the TMRE intensity in each grid mesh across every time-point to obtain a mean intensity for each mitochondrion, at every time-point of the recording. This last step produces the

individual mitochondrial inner membrane potential time series signals.

3.4 Identification of Nearest Neighbors

1. Determine the "center of gravity" for each mitochondrial grid mesh.
2. For every mitochondrion *m*, determine all mitochondria other than *m* for which the connection between their centers of gravity by a straight line crosses exactly one grid line. Save their numerical identifiers as the nearest neighbors of *m*.

3.5 Wavelet Analysis of Individual Mitochondrial Signals

The wavelet transform of a time series signal provides a measure of the frequency content of the signal at a specific time-point [43, 44]. This is especially useful if one probes signals with time-varying (dynamic) frequency content, as is the case in mitochondrial oscillations during oxidative stress [19, 20, 26].

1. Normalize each mitochondrial time series signal by its standard deviation and pad the number of recorded images with zeros to the next higher power of 2. This will accelerate the ensuing computation and prevent a wraparound from the end of the time series to the beginning.
2. (a) Apply the wavelet transform to each mitochondrial signal: Matlab possesses a built-in wavelet toolbox; however, there are also wavelet software packages for other computational platforms [45].

   (b) Specific parameters for the mother wavelet should be adapted to the observed dynamic changes of the signal. For the continuous analysis of mitochondrial signals we recommend the Morlet wavelet (as opposed to other wavelet forms such as the Mexican hat wavelet or the Paul wavelet) for its higher frequency resolution.

   (c) Choose fixed wavelet scales to adequately sample all relevant frequencies that are present in the mitochondrial signal. It is convenient to choose the smallest resolvable scale as $s_0 = 4\ dt$ (dt being the sampling period), corresponding to at least four data points needed to adequately resolve one oscillation, and thus, to a maximum frequency of $f_{\max} = s_0^{-1}$. Larger scales s_k should be chosen as $s_k = s_0\, 2^{k\ dk}$, where $k = 0,1,\ldots K$, with $K = \log_2(T/s_0)/dk$, recording time T, and $dk = 0.1$–0.25, leading to at least four subscales within each scale. This oversampling is necessary to obtain adequate frequency information in between the scales. The largest scale can be constrained further by excluding very long periods that surpass 10% of the longest visible period L of a synchronized mitochondrial oscillation, thus setting a minimum frequency of $f_{\min} = (1.1\ \mathrm{L})^{-1}$. This procedure helps avoid time-consuming computation by focusing on the relevant frequencies.

3. Determine the wavelet power spectrum at each time-point by taking the squared absolute value of the wavelet transform.
4. Interpolate the wavelet power spectrum to a frequency resolution of 0.1 mHz.
5. Determine the frequency with the maximum wavelet power in the interpolated wavelet power spectrum from **step 4**, at each time-point (for each mitochondrion). This provides the time-resolved frequency for an individual mitochondrion.

3.6 Major Mitochondrial Frequency Clusters and Frequency Cluster Size Analysis

The major mitochondrial frequency cluster determines the cluster of (oscillating) mitochondria with similar frequency content at a specific time-point t (*see* also Fig. 2a in [20]).

1. Determine the distribution histogram (over a frequency resolution of $1/dt$) of mitochondrial frequencies at each time-point t.
2. Determine the maximum (peak) of each histogram, $P_{\text{max}}(t)$, and its associated frequency, $\nu_{\text{max}}(t)$, that corresponds to the major mitochondrial frequency.
3. Identify all mitochondria that contribute to $P_{\text{max}}(t)$.
4. Identify mitochondria in a peak that is directly adjacent to $P_{\text{max}}(t)$, $P_{\text{a}}(t)$, with a peak magnitude of at least 10% of the magnitude of $P_{\text{max}}(t)$.
5. Determine the mean (TMRE) signal $S_{\text{max}}(t)$ from mitochondria that belong to $P_{\text{max}}(t)$.
6. Determine the mean (TMRE) signal $S_{\text{a}}(t)$ from mitochondria that belong to $P_{\text{a}}(t)$.
7. Within a window of 1.1 $[\nu_{\text{max}}(t)]^{-1}$ around time-point t, find the correlation coefficient c_{a} of the signals S_{max} and S_{a}.
8. Add the mitochondria in $P_{\text{a}}(t)$ to $P_{\text{max}}(t)$ if $c_{\text{a}} \geq 0.95$ (corresponding to at least 95% correlation of the signals).
9. Repeat **steps 4–8** until you reach a $P_{\text{a}}(t)$ with $c_{\text{a}} < 0.95$ for frequencies both higher and lower than $\nu_{\text{max}}(t)$.
10. Determine all mitochondria that do not belong to $P_{\text{max}}(t)$.
11. For each mitochondrion m of the mitochondria in **step 10**, determine the correlation coefficient c_{m} of its (TMRE) signal with the mean (TMRE) signal of the mitochondria in $P_{\text{max}}(t)$ within a window of 1.1 $[\nu_{\text{max}}(t)]^{-1}$. If $c_{\text{m}} \geq 0.95$, include mitochondrion m in the cluster of mitochondria that form $P_{\text{max}}(t)$.
12. Assign a cluster frequency at time-point t being the mean frequency from all mitochondria in $P_{\text{max}}(t)$.
13. Determine the (normalized) space occupied by the cluster mitochondria at time-point t by adding the sizes of each mitochondrion in $P_{\text{max}}(t)$ and dividing by the size of the myocyte area.

14. In analogy to **step 13**, one may obtain a relative cluster size by dividing the number of mitochondria in $P_{max}(t)$ by the total number of mitochondria.
15. Determine the time-dependent mean cluster frequency as the averaged frequency from the wavelet frequencies at time-point t for all mitochondria belonging to the major cluster.
16. For the analysis of the relation between cluster frequency and size, produce a graph where you insert the values from **step 15** (x-axis) and **step 13** or **14** (y-axis) for all time-points, in analogy to Fig. 2c, d in [20].

3.7 Local Cluster Coherence Analysis

A complementary analysis to investigate the temporal and spatial properties of a major mitochondrial cluster can be achieved by determining the coherence of each of the cluster mitochondria with its nearest neighbors. A coherence analysis of two (oscillating) signals is sensitive to both a change in phase relationships and a change in power [46]. For two signals S_A and S_B, it results from the quotient of absolute squared cross-spectrum between S_A and S_B and the product of the auto-spectra of S_A and S_B. Coherence levels range between zero and one.

1. At each time-point t, create a window W around t of size 1.1 L (*see* Subheading 3.5, **step 2c**).
2. Choose a frequency band; a convenient frequency band for mitochondrial oscillations under oxidative stress ranges from 0 to 100 mHz.
3. Choose the number of sampling points for the discrete fast Fourier transform that is being used for the coherence analysis; a convenient choice is 2^{11} sampling points that, for the frequency band in **step 1**, lead to $(2^{11}/2) + 1$ segments or approximately 0.1 mHz per segment.
4. Choose a Hanning window of size $W/8$.
5. Choose an overlap of 50% between segments.
6. A convenient method to determine the coherence between two mitochondrial signals is then provided by Matlab's *mscohere* function with the parameters given in **steps 1–5**.
7. Determine the mean coherence at time-point t for each cluster mitochondrion m as the average coherence of m with all its nearest neighbors within the window W.
8. To compare the coherence of cluster mitochondria and non-cluster mitochondria, the procedure in **steps 1–7** can be repeated for all non-cluster mitochondria (*see* also Fig. 3 in [20]) (*see* **Note 2**).
9. To allow for comparison among different recordings, the time-resolved coherence values should be normalized in time through linear interpolation at a resolution that is equal or smaller than the smallest recorded sampling period.

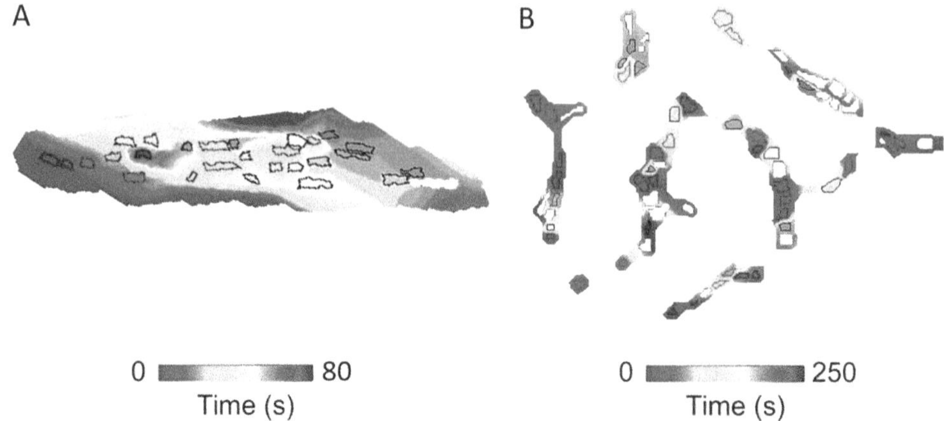

Fig. 2 Propagation of mitochondrial oxidation events in axons and neuromuscular junctions. (**a**) Mitochondrial redox event propagation in a triangularis sterni explant axon of a Thy1-mito-Grx1-roGFP2 mice after nerve crush injury. The signal propagates from left to right and indicates clustered oxidation (*see* also [3, 7]). (**b**) Propagation of mitochondrial matrix pH events in a triangularis sterni explant neuromuscular junction of mice injected with recombinant adeno-associated virus particles (rAAV-1/2-mito-SypHer; for details *see* [7]). Mitochondrial signal propagation is more fragmented and prolonged (Adapted from Figs. 3 and 5 in [7], with permission from ref. 7. Copyright 2016)

3.8 Propagation of Mitochondrial Signals

Properties of a propagating mitochondrial signal, in the myocyte or axon, can be measured with isochronal maps that are created with reference to the onset of the mitochondrial signal changes (*see* also Fig. 2 for an isochronal map of axonal mitochondria in Thy1-mito-Grx1-roGFP2 mice triangularis sterni explants [7]). Isochronal maps provide information not only about the signal's propagating velocity but also about the synchronization properties of the mitochondrial network; for example, propagation of the mitochondrial inner membrane potential depolarization may arise from a nucleus of already synchronized mitochondria [18, 20].

1. Determine the first signal change in a mitochondrion's signal by marking the onset of the signal change (*see* also Subheading 3.2, **step 2**), t_1, and the first signal maximum or minimum thereafter, t_2.
2. From all time-points in the interval $[t_1, t_2]$, exclude the 10% of time-points whose signal value is closest to that at t_1 and the 10% of time-points whose signal value is closest to the signal value at t_2.
3. Determine the earliest, t_I, and latest, t_E, of the remaining time-points.
4. Take the arithmetic mean of t_I and t_E as the mitochondrial signal change reference point, t_S, and repeat **steps 1–3** for all mitochondria.
5. Of all mitochondrial reference points t_S, use the earliest one as reference point and create a color-interpolated isochronal map over all t_S values.

3.9 Functional Connectedness and Functional Mitochondrial Clustering

The mitochondrial network's functional characteristics can be further studied using a recently proposed methodology [34] that is based on the correlation of mitochondrial signals [35]. This procedure acknowledges the fact that some mitochondria from the network may form a spanning cluster of synchronously oscillating mitochondria at different cellular positions that do not necessarily have close spatial proximity [13, 18, 20]. Mitochondria in these spatially non-contiguous clusters, however, may oscillate in synchrony, and there may be several of such spanning clusters hidden in the myocyte. This property constitutes the main difference with respect to the major mitochondrial frequency clusters (see above) that only consider one (major) cluster of similarly oscillating mitochondria.

1. At each time-point t, create a window W around t of size 1.1 L (*see* Subheading 3.5, **step 2c**).
2. For each pair of mitochondria, determine the cross-correlation coefficient c_p of their respective signals, in W. If coefficient $c_p \geq 0.9$, allocate a (undirected) link between both mitochondria. This correlation coefficient cutoff value is arbitrary; however, it may be chosen based on the behavior of the clustering coefficient (*see* below and Fig. 3a): the clustering coefficients reach a plateau for lower correlation cutoff values and drop to zero for correlation cutoff values that approach 1. One may therefore reasonably assume an intermediate correlation cutoff at the point where the curve of clustering coefficients versus correlation coefficients reaches its median derivative value (Fig. 3a). This procedure generates a dynamic functional topology of the mitochondrial network.
3. For each mitochondrion m, determine the set N of mitochondria that are topologically connected to m, according to 2.
4. Determine the number of mitochondria in N, m_N, and the number of topological connections between all mitochondria in N, D_N.
5. Determine the clustering coefficient for mitochondrion m at time t as $C_m(t) = 2\ D_N/(m_N(m_N - 1))$ (*see* also [35]).
6. Determine the mean clustering coefficient of the whole network at time-point t as $C(t) = (1/M)\ \Sigma_m\ C_m(t)$, where M represents the number of mitochondria in the network.
7. The functional network must be tested against a random network (*see* Fig. 3b). Construct the random network having the same number of network mitochondria, M, with the Erdös-Rényi model [47]: randomly assign $m_N(t)$ links to each mitochondrion m at time-point t, where $m_N(t)$ corresponds to the number of topologically connected mitochondria to m at t, obtained from the functional analysis in **steps 2–4**.

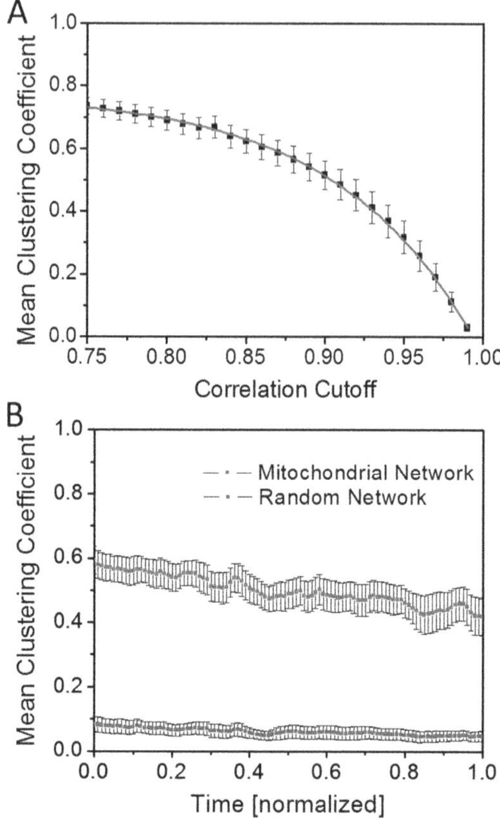

Fig. 3 Functional mitochondrial clustering. (**a**) Time- and myocyte-averaged mean clustering coefficient for different cutoffs of the correlation coefficient that defines the threshold for functional connectedness. The red curve was fitted with the function $f(x) = a{\cdot}\exp.(-x/b) + c$ ($a = -7.81{\cdot}10^{-6} \pm 1.33{\cdot}10^{-6}$, $b = -8.64{\cdot}10^{-2} \pm 0.13{\cdot}10^{-2}$, $c = 77.65{\cdot}10^{-2} \pm 0.35{\cdot}10^{-2}$; *see* also [35]). The 50% value of the mean clustering coefficient is approximately crossed at a correlation cutoff of 0.9, which approximately coincides with the median derivative value of the fitted curve. (**b**) Mean clustering coefficient for a correlation cutoff of 0.9, averaged over nine different glucose-perfused cardiac myocytes (red line; for details *see* [35]) and for equivalent random networks that are built on the same set of vertices and number of links (blue line). (Reprinted with permission from ref. 35. Copyright 2014)

3.10 Local Inter-Mitochondrial Coupling

In contrast to functional properties of the mitochondrial network that may connect distant mitochondria, inter-mitochondrial coupling between nearest neighbors provides a measure for local network properties. Local coupling of oscillating mitochondria can be probed using a stochastic phase model of coupled oscillators with drifting frequencies [26]. The model is based on the Kuramoto model for weakly coupled circadian oscillators [29]; however, instead of assigning one overall coupling constant for the whole network, the stochastic phase model assigns time-dependent coupling constants for each mitochondrion [31]. We provide a step-by-step algorithm to determine these coupling constants, as described in [26]. For a short description of the relevant model parameters and functions, *see* Table 1.

Table 1
Table of used symbols and parameters

Symbol	Description
$\Delta\Psi_m$	Inner mitochondrial membrane potential
dt	Sampling period of signal
T	Recording time of signal
s_0	Smallest wavelet scale for wavelet analysis
l	Smallest period of a synchronized mitochondrial oscillation during a signal recording
L	Longest period of a synchronized mitochondrial oscillation during a signal recording
$P_{max}(t)$	Maximum peak of mitochondrial frequency histogram at time t
c_m, c_a, c_n	Correlation coefficients
W	Running window for cluster coherence analysis and topological analysis
m_N	Number of mitochondria in the network N
D_N	Number of topological connections between all mitochondria in the network N
$C_m(t)$	Clustering coefficient of mitochondrion m at time t
$C(t)$	Mean network clustering coefficient
M	Number of network mitochondria
$R_m(t)$	Local order parameter in the stochastic phase model for mitochondrion m
$\psi_m(t)$	Local mean-field phase in the stochastic phase model for mitochondrion m
$\phi_m(t)$	Phase of mitochondrion m
$\omega_m(t)$	Intrinsic frequency of mitochondrion m
$\nu_m(t)$	Mean frequency for mitochondrion m, corresponds to the wavelet frequency
σ_m	Standard variation of $n_m(t)$
ζ	Random number drawn from a distribution with zero mean and variance s_m^2
γ	Decay rate parameter for the Ornstein-Uhlenbeck frequencies
λ	Optimization parameter for Tikhonov regularization
$K_m(t)$	Coupling constant for mitochondrion m
$s(t)$	Normalized intensity signal for the mitochondrial network

1. For each mitochondrion, determine the dynamic mitochondrial phase from the mitochondrial wavelet frequency, $\nu_m(t)$, as follows:
   - Determine the phase $\phi_m(t)$ of mitochondrion m from the wavelet transform at time t using the inverse tangent of the quotient of the imaginary and real part of the wavelet transform (*atan2* in Matlab).

- Determine the differential $d\phi_m(t)/dt$ as the difference of consecutive phase values divided by the sampling period.
- Apply a median filter of order 1.1 L (*medfilt1* in Matlab), to account for frequency changes at the turning point of phase cycles.
- Use linear interpolation of phase changes between the nearest positive values, to interpolate over negative values (*see* **Note 3**).

2. For each mitochondrion m at time-point t, sum the respective phases of the nearest neighbors of m, $\phi^{(m)}{}_n(t)$, as $\Sigma_n \exp(2\pi i\ \phi^{(m)}{}_n(t))$ and divide by the number of nearest neighbors (*see* also Eq. (2) in [26]).
3. Determine the local order parameter $R_m(\mathrm{t})$ as the absolute value of the result in **step 2**.
4. Determine the local phase angle $\psi_m(t)$ of the result in **step 2**.
5. Determine the phase sum $P_m(t) = R_m(t)\sin(2\pi[\psi_m(t) - \phi_m(t)])$, which is part of the coupling term in the phase differential equation, Eq. (3) in [26].
6. Determine the parameter σ_m as the standard deviation of the dynamic wavelet frequency of mitochondrion m.
7. For each mitochondrion m, create a Gaussian distribution with zero mean and variance $\sigma_m{}^2$.
8. Use an initial random phase value $\omega_m(0)$ from a distribution with mean $\nu_m(dt)$ and variance $\sigma_m{}^2$.
9. Update the intrinsic phase $\omega_m(t)$ with the equation

$$\omega_m(t + dt) = \omega_m(t)\exp(-\gamma \cdot dt) + \nu_m(t)[1 - \exp(-\gamma \cdot dt)] \\ + \mathrm{Sqrt}(1 - \exp(-2\gamma \cdot dt))\zeta,$$

where, for cardiac myocytes with substrate perfusion, $\gamma \approx 0.07$ [26], and ζ a random number drawn from the distribution in **step 7** (*see* also Eq. (1) of the supplementary information in [26]) (*see* **Note 4**).

10. Determine the time-points t_z for which the phase sum $P_m(t)$ (**step 5**) has a sign shift (from negative to positive or vice versa).
11. At each time-point t, determine the coupling constant

$$[\nu_m(t) - \omega_m(t)]/P_m(t)$$

from the update equation in **step 9** (*see* also Eq. (3) in the supplementary information in [26]). If this quotient is smaller than zero, repeat the analysis with a new random number from **step 9** until the quotient becomes positive.

Perform at least 10 × 10 runs (better 10 × 100, as used in [26]) for ten different random initial phase values in **step 8** and for ten different runs through the update phases in **step 9** (with a different random number ζ at each time-point). Average the resulting coupling constants, $K_m(t)$, over all runs and starting points.

- if $P_m(t)$ is close to zero, determine the parameter λ for Tikhonov regularization from the minimization of the function $f(\lambda) = \surd(1 + 1/\lambda)\ \surd(r_1{}^T(\lambda)\ r_0(\lambda))$ on the interval $[M/100, 100\ M]$ with $M = \|\ P_m\ \|\ /\ T$, and (here we use vector notation for vector elements corresponding to time-points and $\mathbf{1}_T$ to the unit vector of size T), *see* also [26, 48]:

$$x_0 = \left[P_m{}^2 + \lambda\ \mathbf{1}_T\right]^{-1} P_m{}^T[\nu_m - \omega_m],$$

$$r_0(\lambda) = [\nu_m(t) - \omega_m(t)] - P_m\ x_0,$$

$$x_1 = \left[P_m{}^2 + \lambda\ 1_T\right]^{-1} P_m{}^T\ r_0 + x_0,$$

$$r_1(\lambda) = [\nu_m(t) - \omega_m(t)] - P_m\ x_1.$$

As a starting value one may choose $\lambda_i = \|\ P_m\ \|\ /\ \surd(T)$. A convenient optimization function is provided by *fminsearch* in Matlab. If the optimization does not converge, set λ to the lower bound $\lambda = M/100$.

- For the optimized parameter λ, apply a median filter (*medfilt1* in Matlab with order 1.1 P) on the coupling constants $[P_m{}^2 + \lambda\ \mathbf{1}_T]^{-1}\ P_m{}^T\ [\nu_m - \omega_m]$.
- Average over all runs and starting points as above to determine $K_m(t)$.

12. The forward model can be tested by determining the update phases $\phi_m(t)$ directly from the equation

$$\mathrm{d}\phi_m(t)/\mathrm{d}t = \omega_m(t) + K_m(t)\ \ P_m(t),$$

with updated intrinsic frequencies $\omega_m(t)$ as above. The (normalized) mitochondrial network signal $s(t)$ is then constructed as

$$s(t) = (1/M)\Sigma_m \cos\left(\phi_m(t)\right),$$

where the sum runs over all mitochondria divided by the total number of mitochondria, M.

4 Notes

1. When drawing the grid to separate individual mitochondria in Subheading 3.3, **step 1**, the overlay image of the averaged images can sometimes be very blurry due to insufficient

movement stabilization with the ImageJ image stabilizer plug-in in Subheading 3.2, **step 1**. It is then best to only focus on the averaged image closest to the onset of myocyte-wide synchronized mitochondrial depolarizations.

2. The "non-cluster" coherence in Subheading 3.7, **step 8** can be higher than the coherence of the major frequency cluster. This may be due to a minority of spatially close mitochondria that oscillate in a highly coherent fashion but with a frequency or phase difference that is largely different to those mitochondria in the major cluster.
3. It often occurs that, for some time-points, a few mitochondria do not show any signal changes apart from noise; however, they are still assigned a wavelet frequency which is usually either the upper or lower cutoff frequency. To eliminate these mitochondria from the model calculations in Subheading 3.10, one may exclude, at every time-point, all mitochondria with values at the cutoff frequencies.
4. The optimal decay rate parameter γ in Subheading 3.10, **step 9**, if not known, should be determined with an error estimation in a forward model that matches the averaged signal of all network mitochondria (Subheading 3.10, **step 12**) over many γ values (see also Fig. S6 in the supplementary information in [26]). A convenient choice is to take increasing γ values as multiples of 0.01 s^{-1} in the interval [0 s^{-1},1 s^{-1}]. Typically, increasing γ values approach an asymptotic (constant) error value (that accounts for the deviation between the forward model and the averaged mitochondrial network signal); the optimal decay rate parameter may then be chosen as the first γ whose error value is within 1% of the asymptotic error value.

Acknowledgments

The work was supported by a grant-in-aid (#15GRNT23070001) from the American Heart Association (AHA), the Ricbac Foundation, and NIH grant 1 R01 HL135335–01.

References

1. Nunnari J, Suomalainen A (2012) Mitochondria: in sickness and in health. Cell 148 (6):1145–1159
2. Chouchani ET, Methner C, Nadtochiy SM et al (2013) Cardioprotection by S-nitrosation of a cysteine switch on mitochondrial complex I. Nat Med 19(6):753–759
3. Breckwoldt MO, Pfister FMJ, Bradley PM et al (2014) Multiparametric optical analysis of mitochondrial redox signals during neuronal physiology and pathology in vivo. Nat Med 20(5):555–560
4. Hou T, Wang X, Ma Q et al (2014) Mitochondrial flashes: new insights into mitochondrial ROS signalling and beyond. J Physiol 592 (17):3703–3713
5. Cortassa S, Aon MA (2013) Dynamics of mitochondrial redox and energy networks: insights

from an experimental-computational synergy. In: Aon MA (ed.) et al Systems biology of metabolic and signaling networks. Energy, mass and information transfer, 1, Springer-Verlag, Berlin

6. Dröge W (2002) Free radicals in the physiological control of cell function. Physiol Rev 82 (1):47–95
7. Breckwoldt MO, Armoundas AA, Aon MA et al (2016) Mitochondrial redox and pH signaling occurs in axonal and synaptic organelle clusters. Sci Rep 6:23251
8. Picard M, McManus MJ, Csordás G et al (2015) Trans-mitochondrial coordination of cristae at regulated membrane junctions. Nat Commun 6:6259
9. O'Rourke B, Ramza B, Marban E (1994) Oscillations of membrane current and excitability driven by metabolic oscillations in heart cells. Science 265:962–966
10. Aon MA, Cortassa S, Maack C et al (2007) Sequential opening of mitochondrial ion channels as a function of glutathione redox thiol status. J Biol Chem 282(30):21889–21900
11. Aon MA, Cortassa S, O'Rourke B (2008) Mitochondrial oscillations in physiology and pathophysiology. Adv Exp Med Biol 641:98–117
12. Aon MA, Cortassa S, Akar FG et al (2006) Mitochondrial criticality: a new concept at the turning point of life or death. Biochim Biophys Acta 1762(2):232–240
13. Aon MA, Cortassa S, O'Rourke B (2006) The fundamental organization of cardiac mitochondria as a network of coupled oscillators. Biophys J 91(11):4317–4327
14. Aon MA, Cortassa S, Marban E et al (2003) Synchronized whole cell oscillations in mitochondrial metabolism triggered by a local release of reactive oxygen species in cardiac myocytes. J Biol Chem 278(45):44735–44744
15. Aon MA, Cortassa S, Akar FG et al (2009) From mitochondrial dynamics to arrhythmias. Int J Biochem Cell Biol 41(10):1940–1948
16. Zhou L, Aon MA, Almas T et al (2010) A reaction-diffusion model of ROS-induced ROS release in a mitochondrial network. PLoS Comput Biol 6(1):e1000657
17. Zorov DB, Filburn CR, Klotz LO et al (2000) Reactive oxygen species (ROS)-induced ROS release: a new phenomenon accompanying induction of the mitochondrial permeability transition in cardiac myocytes. J Exp Med 192:1001–1014
18. Aon MA, Cortassa S, O'Rourke B (2004) Percolation and criticality in a mitochondrial network. Proc Natl Acad Sci U S A 101 (13):4447–4452
19. Kurz FT, Aon MA, O'Rourke B et al (2010) Wavelet analysis reveals heterogeneous time-dependent oscillations of individual mitochondria. Am J Physiol Heart Circ Physiol 299(5): H1736–H1740
20. Kurz FT, Aon MA, O'Rourke B et al (2010) Spatio-temporal oscillations of individual mitochondria in cardiac myocytes reveal modulation of synchronized mitochondrial clusters. Proc Natl Acad Sci U S A 107 (32):14315–14320
21. Ruiz-Meana M, Fernandez-Sanz C, Garcia-Dorado D (2010) The SR-mitochondria interaction: a new player in cardiac pathophysiology. Cardiovasc Res 88:30–39
22. Kuznetsov AV, Usson Y, Leverve X et al (2004) Subcellular heterogeneity of mitochondrial function and dysfunction: evidence obtained by confocal imaging. Mol Cell Biochem 256–257:359–365
23. Manneschi L, Federico A (1995) Polarographic analyses of subsarcolemmal and intermyofibrillar mitochondria from rat skeletal and cardiac muscle. J Neurol Sci 128(2):151–156
24. Kuznetsov AV, Mayboroda O, Kunz D et al (1998) Functional imaging of mitochondria in saponin-permeabilized mice muscle fibers. J Cell Biol 140(5):1091–1099
25. Lesnefsky EJ, Tandler B, Ye J et al (1997) Myocardial ischemia decreases oxidative phosphorylation through cytochrome oxidase in subsarcolemmal mitochondria. Am J Phys 273:H1544–H1554
26. Kurz FT, Derungs T, Aon MA et al (2015) Mitochondrial networks in cardiac myocytes reveal dynamic coupling behavior. Biophys J 108(8):1922–1933
27. Yang L, Paavo K, Weiss JN et al (2010) Mitochondrial oscillations and waves in cardiac myocytes: insights from computational models. Biophys J 98:1428–1438
28. Rosenfeld S, Kapetanovic I (2008) Systems biology and cancer prevention: all options on the table. Gene Regul Syst Bio 2:307–319
29. Kuramoto Y (1984) Chemical oscillations, waves, and turbulence. Springer-Verlag, Berlin
30. Acebrón JA, Bonilla LL, Vicente P et al (2005) The Kuramoto model: a simple paradigm for synchronization phenomena. Revies Mod Phys 77:137–185
31. Rougemont J, Naef F (2007) Dynamical signatures of cellular fluctuations and oscillator stability in peripheral circadian clocks. Mol Syst Biol 3:93

32. Kurz FT, Kembro JM, Flesia AG et al (2017) Network dynamics: quantitative analysis of complex behavior in metabolism, organelles, and cells, from experiments to models and back. Wiley Interdiscip Rev Syst Biol Med 9 (1):e1352
33. Kurz FT, Aon MA, O'Rourke B et al (2017) Functional implications of cardiac mitochondria clustering. Adv Exp Med Biol 982:1–24
34. Eguíluz VM, Chialvo DR, Cecchi GA et al (2005) Scale-free brain functional networks. Phys Rev Lett 94(1):18102
35. Kurz FT, Aon MA, O'Rourke B et al (2014) Cardiac mitochondria exhibit dynamic functional clustering. Front Physiol 5:329
36. Viola HM, Arthur PG, Hool LC (2009) Evidence for regulation of mitochondrial function by the L-type Ca2+ channel in ventricular myocytes. J Mol Cell Cardiol 46(6):1016–1026
37. Santo-Domingo J, Giacomello M, Poburko D et al (2013) OPA1 promotes pH flashes that spread between contiguous mitochondria without matrix protein exchange. EMBO J 32 (13):1927–1940
38. Schwarzländer M, Finkemeier I (2013) Mitochondrial energy and redox signaling in plants. Antioxid Redox Signal 18(16):2122–2144
39. Fang H, Chen M, Ding Y et al (2011) Imaging superoxide flash and metabolism-coupled mitochondrial permeability transition in living animals. Cell Res 21(9):1295–1304
40. Slodzinski MK, Aon MA, O'Rourke B (2008) Glutathione oxidation as a trigger of mitochondrial depolarization and oscillation in intact hearts. J Mol Cell Cardiol 45(5):650–660
41. Porat-Shliom N, Chen Y, Tora M et al (2014) In vivo tissue-wide synchronization of mitochondrial metabolic oscillations. Cell Rep 9 (2):514–521
42. Li K (2008) The image stabilizer plugin for ImageJ. www.cs.cmu.edu/~kangli/code/Image_Stabilizer.html
43. Grossmann A, Morlet J (1984) Decomposition of hardy functions into square Integrable wavelets of constant shape. SIAM J Math Anal 15 (4):723–736
44. Grossmann A, Morlet J, Paul T (1985) Transforms associated to square integrable group representations. I. General results. J Math Phys 26(10):2473–2479
45. Torrence C, Compo GP (1998) A practical guide to wavelet analysis. Bull Amer Meteor Soc 79:61–78
46. Guevara MA, Corsi-Cabrera M (1996) EEG coherence or EEG correlation? Int J Psychophysiol 23(3):145–153
47. Erdős P, Rényi A (1960) On the evolution of random graphs. Publ Math Inst Hung Acad Sci 5:17–61
48. O'Leary DP (2001) Near-optimal parameters for Tikhonov and other regularization methods. SIAM J Sci Comput 23:1161–1171

Chapter 24

Measurement of Mitochondrial ROS Formation

Soni Deshwal, Salvatore Antonucci, Nina Kaludercic, and Fabio Di Lisa

Abstract

Reactive oxygen species (ROS) are involved in both physiological and pathological processes. This widely accepted concept is based more on the effects of antioxidant interventions than on reliable assessments of rates and sites of intracellular ROS formation. This argument applies also to mitochondria that are generally considered the major site for ROS formation, especially in skeletal and cardiac myocytes.

Detection of oxidative modifications of intracellular or circulating molecules is frequently used as a marker of ROS formation. However, this approach provides limited information on spatiotemporal aspects of ROS formation that have to be defined in order to elucidate the role of ROS in a given pathophysiological condition. This information can be obtained by means of fluorescent probes that allow monitoring ROS formation in cell-free extracts and isolated cells. Thus, this approach can be used to characterize ROS formation in both isolated mitochondria and mitochondria within intact cells. This chapter describes three major examples of the use of fluorescent probes for monitoring mitochondrial ROS formation. Detailed methods description is accompanied by a critical analysis of the limitations of each technique, highlighting the possible sources of errors in performing the assay and results interpretation.

Key words Mitochondria, Reactive oxygen species, Fluorescence, Amplex Red, HyPer, MitoTracker Red

1 Introduction

Although the involvement of reactive oxygen species (ROS) in physiological and pathological processes is widely accepted [1, 2], information on ROS production in biological samples derives from methods that are far from being optimal. An ideal method should provide reliable estimates of the species generated, the cellular site, the kinetics, and the amounts. Also, factors that favor ROS formation or are modified by these oxidant species should be identified to elucidate causal relationships. This is especially the case when mitochondrial ROS formation is investigated in isolated cells or intact organs. Indeed, ROS accumulations can perturb mitochondrial

Soni Deshwal and Salvatore Antonucci contributed equally to this work.

Carlos M. Palmeira and António J. Moreno (eds.), *Mitochondrial Bioenergetics: Methods and Protocols*, Methods in Molecular Biology, vol. 1782, https://doi.org/10.1007/978-1-4939-7831-1_24,

function, yet in most cases mitochondrial dysfunction results in an increased ROS generation further exacerbating the initial mitochondrial derangement. The primary cause of this vicious cycle can hardly be established by means of loss-of-function approaches. Indeed, in most cases the inhibition of mitochondrial pathways for ROS generation, such as respiratory chain complexes, hampers inevitably vital processes of energy conservation [3].

The impact of ROS on mitochondrial and/or cellular homeostasis can be investigated by detecting the oxidation of biomolecules. However, this approach cannot provide a real-time monitoring of ROS formation in living cells. To this aim, imaging techniques have been developed using fluorescent probes or genetically encoded fluorescent proteins [4, 5]. The former compounds are small aromatic molecules that generate fluorescent products upon oxidation. This heterogeneous group of compounds includes lipophilic cations that localize preferentially to energized mitochondria. The degree of specificity and sensitivity of small molecule based fluorescent probes is significantly lower than that displayed by genetically encoded fluorescent proteins. In addition, these proteins can be targeted to specific cellular sites providing nonambiguous evidence of variations in ROS formation in the various cellular compartments [6]. The obvious disadvantage with fluorescent proteins is the procedure necessary for inducing their cellular expression. Indeed, transfection procedures can hardly be applied to primary cells, such as freshly isolated cardiac myocytes.

Advantages and limitations of the various compounds available for monitoring ROS formation should be carefully considered in relation to the experimental protocol. For instance, as detailed in the following sections, Amplex Red is quite useful in studies involving cell-free extracts, while it cannot be used in intact cells. On the other hand, the limited sensitivity and specificity of small molecule fluorescent probes is frequently tolerated in studies employing primary cells where the use of more efficacious genetically encoded proteins is hardly feasible.

This review is aimed at detailing three different approaches for assessing mitochondrial ROS formation in isolated mitochondria (exemplifying cell-free extracts) and isolated neonatal rat ventricular myocytes (NRVMs) in culture, respectively.

2 Methods

2.1 Measurement of ROS in Isolated Mitochondria

The Amplex Red assay is a widely used procedure for assessing ROS formation in cell-free extracts, including mitochondria [6–8]. This technique is devoid of limitations affecting other assays that are frequently used. Among these the fluorescence of 2′,7′-dichlorofluorescein (DCFH) is hampered by low specificity for H_2O_2 [6, 7]. Amplex Red is a substrate of horseradish peroxidase (HRP), which in presence of H_2O_2 oxidizes Amplex Red resulting in the production of a red fluorescent compound resorufin (excitation/emission: 571/585 nm) [9]. Amplex Red reacts with H_2O_2 in a 1:1 stoichiometry to produce resorufin. The major shortcoming is the significant light sensitivity of Amplex Red that can lead to the formation of resorufin even in the absence of HRP and H_2O_2 [10]. Thus, necessary precautions should be taken to prevent photooxidation of this probe. Moreover, the fact that Amplex Red is cell impermeable limits its use to the permeabilized cells [9].

Amplex Red assay is commonly used to measure ROS produced by the mitochondrial respiratory chain in the presence of substrates, such as succinate or glutamate/malate [9]. Moreover, it is also possible to measure the activity of several enzymes that produce H_2O_2, including monoamine oxidases (MAO). These mitochondrial flavoenzymes catalyze the oxidative deamination of catecholamines and biogenic amines, resulting in the production of H_2O_2, aldehydes and ammonia [11–13].

A method for detection of MAO-generated H_2O_2 formation in isolated heart mitochondria by Amplex Red probe is described below.

2.1.1 Materials

1. *Isolated mitochondria:* Heart mitochondria were isolated from C57BL/6 male as described in ref. 14.
2. *Probes:* Amplex Red (ThermoFisher Scientific, A12222) and HRP (Sigma, P6782).
3. *Buffer:* 137 mM KCl, 2 mM KH_2PO_4, 20 mM Hepes, and 20 μM EGTA.
4. H_2O_2.
5. *Fluorimeter:* Fluorimeter equipped with proper excitation/emission filters at 571 and 585 nm, respectively.
6. Black 96-well plate.

2.1.2 Methods

(a) Generation of hydrogen peroxide calibration curve.

1. Set the fluorimeter at 571 nm excitation and 585 nm emission wavelengths. Set the temperature at 37°C.
2. Prepare a buffer aliquot containing 5 μM Amplex Red and 4 μg/mL HRP. Pipette 200 μL of this master mix in each well and put the plate in the fluorimeter.
3. Record the fluorescence changes at baseline.
4. Add increasing concentrations of H_2O_2 starting from 0 to 1000 pmol, mix and record the fluorescence intensity until a flat line is observed as shown in Fig. 1a.
5. Calculate the change in fluorescence intensity by subtracting the basal fluorescence level (measured at **step 3**) from

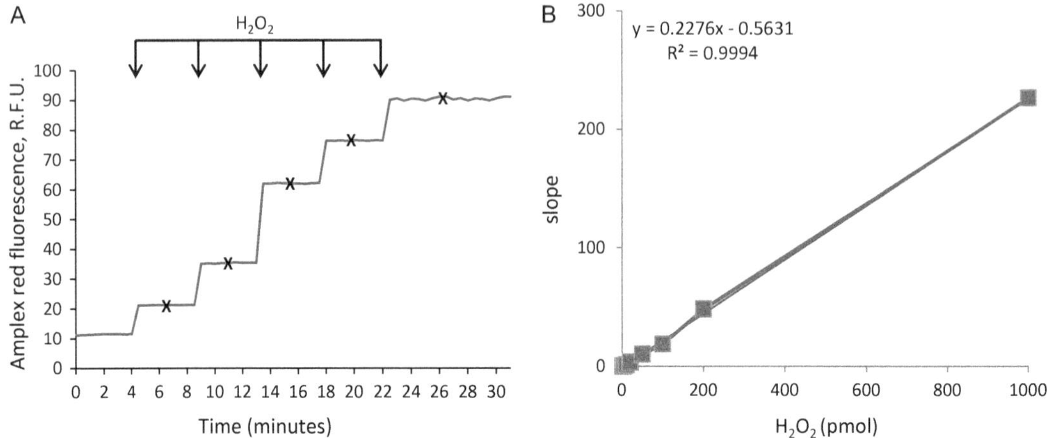

Fig. 1 H_2O_2 calibration curve using Amplex Red. (**a**) Increasing H_2O_2 concentrations, from 0 to 1000 pmol, were added to each well containing the Amplex Red and HRP master-mix and fluorescence intensity was recorded. (**b**) Final fluorescence intensities (fluorescence intensity after addition of H_2O_2—basal fluorescence intensity) were calculated for each point and plotted against the amount of H_2O_2 added. The slope and coefficient of determination (R^2) were calculated

fluorescence intensity recorded after each addition of H_2O_2 (as marked by X in Fig. 1a).

6. Plot the graph with Δ fluorescence intensity on y-axis vs the amount of H_2O_2 added on x-axis as shown in Fig. 1b and calculate the slope of the line.
   Note: The coefficient of multiple determination for multiple regression (R^2) should always be higher than 0.95.

(b) Measurement of MAO-dependent H_2O_2 formation in isolated mitochondria

1. Set the fluorimeter as described in **step 1**. Add 5 μM Amplex Red, 4 μg/mL HRP, and 0.1 μg/μL mitochondria in the buffer. Pipette 200 μL of this master-mix in each well, and put the plate in the fluorimeter.
2. Record the fluorescence changes at basal level for ~5 min.
3. Add MAO substrate tyramine in each well and read the fluorescence up to 30 min. As shown in Fig. 2a, addition of tyramine results in an increase in H_2O_2 production as reflected by an increase in fluorescence intensity.

   Note: MAO catabolizes tyramine and leads to the production of H_2O_2, aldehydes, and ammonia [11] (Fig. 2c). Thus, increasing tyramine concentration results in an increase in H_2O_2 production.
4. To confirm that this increase in H_2O_2 production is indeed due to MAO activity, preincubate mitochondria with 1 μM pargyline, an irreversible inhibitor of these flavoenzymes, and repeat **steps 1–3** (Fig. 2a).

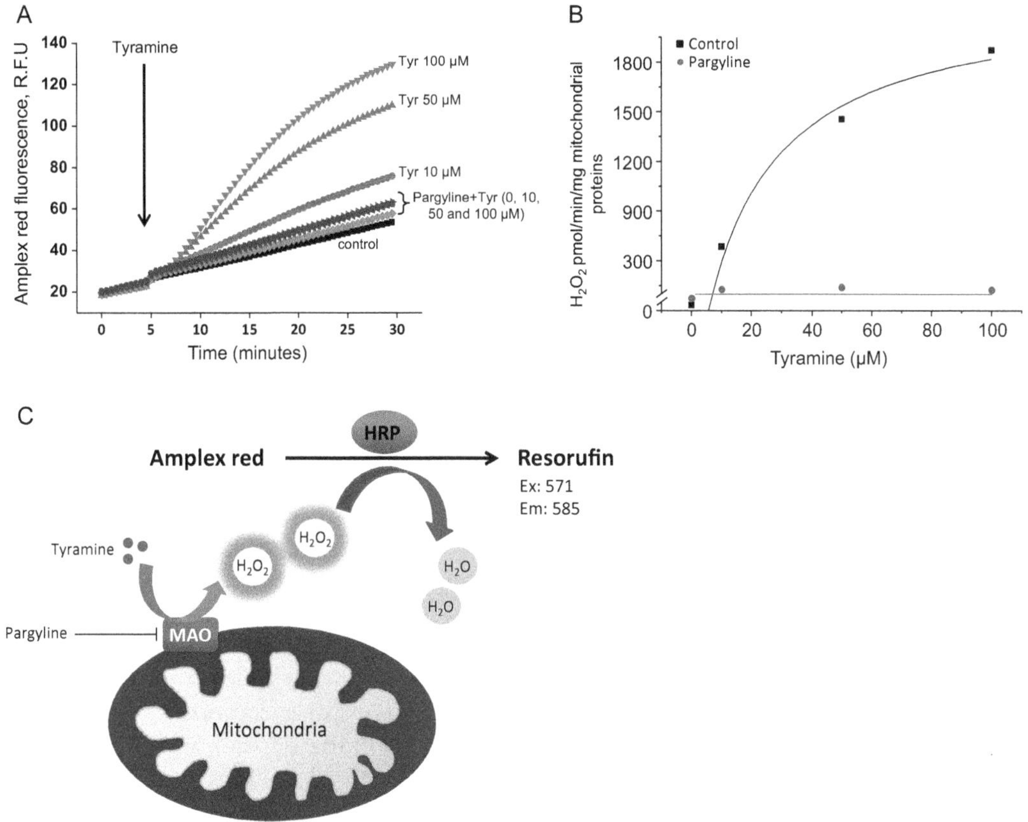

Fig. 2 MAO-generated H_2O_2 measurement in isolated heart mitochondria with Amplex Red. (**a**) Heart mitochondria were added to the mixture of Amplex Red and HRP, and basal H_2O_2 formation was measured for 5 min. To assess MAO-dependent mitochondrial ROS formation, tyramine, a MAO substrate, was added to each well at different concentrations in the presence or absence of the MAO inhibitor pargyline (1 μM). The fluorescence intensity was measured at excitation 571 and emission 585 nm. (**b**) Rate of H_2O_2 formation expressed as pmol/min/0.2 mg of mitochondrial proteins was calculated and plotted vs the concentration of tyramine. (**c**) Scheme showing MAO as an outer mitochondrial flavoenzyme that catabolizes the oxidative deamination of catecholamines and biogenic amines, such as serotonin, tyramine, and dopamine, and leads to the production of H_2O_2, ammonia, and aldehydes. Pargyline is an inhibitor of MAO that irreversibly blocks the activity of these flavoenzymes. In the presence of H_2O_2, HRP oxidizes Amplex Red and results in the production of a red fluorescent compound resorufin, which is excited and emits at 571 and 585 nm, respectively

5. To quantify mitochondrial H_2O_2 formation, calculate the slope of the fluorescence traces obtained in each condition and extrapolate the amount of H_2O_2 generated using the slope and intercept values obtained in A. This value is then normalized to the amount of mitochondrial protein used in the assay and expressed as the rate of H_2O_2 emission in pmol/min/mg of mitochondrial proteins (Fig. 2b).

2.2 Measurement of ROS in Isolated Cardiomyocytes

2.2.1 Fluorescent Redox-Sensitive Probes

In the past decade, several probes have been developed to measure intracellular and compartmentalized ROS formation inside the cell [6, 7, 15]. Small molecule fluorescent dyes, such as MitoSOX and reduced MitoTracker dyes, are commonly used for detection of mitochondrial ROS formation in intact cells.

MitoSOX Red is a derivative of hydroethidine and is widely used for the measurement of O^{2-} formation in the active mitochondria. This dye is specifically targeted to mitochondria because it contains the lipophilic cation triphenyl phosphonium substituent [6, 16]. MitoSOX Red is oxidized by O^{2-} to form a red fluorescent product 2-hydroxyethidium, which is excited at 510 nm and emits at 580 nm [16]. Even though MitoSOX is used to specifically detect O^{2-} formation, it has been reported that this probe can also be oxidized by other oxidants to form ethidium, which overlaps with the fluorescence peak of 2-hydroxyethidium [17, 18]. Moreover, at high concentrations, MitoSOX displays some non-mitochondrial staining, for instance, in the nucleus [19].

MitoTracker Orange CM-H_2TMRos and MitoTracker Red CM-H_2XRos are derivatives of dihydrotetramethyl rosamine and dihydro-X-rosamine, respectively [6]. Reduced MitoTracker dyes do not fluoresce until entering viable cells, where they get oxidized and become positively charged. The cationic fluorescent compound then accumulates in the mitochondria depending on ROS levels and mitochondrial membrane potential and forms fluorescent conjugate with thiol groups [20]. The excitation/emission wavelenghts for MitoTracker Orange CM-H_2TMRos and MitoTracker Red CM-H_2XRos are 554/576 and 579/599 nm, respectively [6]. Unlike MitoSOX, these reduced MitoTracker dyes are not specific for single oxidant species, and thus detect general mitochondrial ROS [6]. Moreover, the fact that their mitochondrial localization and accumulation depend on the mitochondrial membrane potential is a crucial aspect to bear in mind for the correct interpretation of the fluorescence intensity levels [17, 19]. Indeed, it is necessary to measure mitochondrial membrane potential in order to avoid erroneous interpretations.

A method to detect mitochondrial ROS formation in NRVMs using reduced MitoTracker Red CMH_2XRos dye is described below.

1. Materials
   - (a) *Cells:* NRVMs were isolated from 0–3-day-old Wistar rats as described in [12].
   - (b) *Fluorescent probes:* MitoTracker™ Red CM-H_2XRos (ThermoFisher Scientific, M7153).
   - (c) *Culture medium:*

     Medium A: Minimum essential medium (MEM, ThermoFisher Scientific, 2175022) supplemented with 1%

nonessential amino acids, 1% penicillin/streptomycin, 10% fetal bovine serum (FBS), and BrdU (0.1 mM).

Medium B: MEM supplemented with 1% nonessential amino acids, 1% penicillin/streptomycin, and 1% FBS/insulin-transferrin-selenium (ITS).

(d) *Buffer:* Hanks' balanced salt solution (HBSS, 136.9 mM NaCl, 5.36 mM KCl, 0.4 mM $MgSO_4$, 0.5 mM $MgCl_2$, 0.4 mM KH_2PO_4, 0.4 mM Na_2HPO_4, 4 mM $NaHCO_3$, 5 mM glucose, and 2 mM $CaCl_2$).

(e) *Microscope:* Fluorescence microscope equipped with appropriate filters to detect the fluorescence at 579/599 nm. We use an inverted Leica DMI6000 B fluorescence microscope equipped with DFC365FX camera.

(f) *Incubator:* To culture cells use a CO_2 incubator set at 37°C with 5% CO_2 and 96% relative humidity.

2. Methods

(a) After isolation of NRVMs, plate cells in MEM *medium A* in a 6-well plate containing gelatin-coated glass slides. Place NRVMs in the incubator for 24 h.

Note: It is always suggested to pre-warm the medium at 37°C prior to use for cell culture. To coat the plastic plates with gelatin, dissolve 0.1% gelatin in H_2O and autoclave. Place sterile glass slides into each well and add 2 mL gelatin solution. Incubate the plates at 37°C for 30 min to create a layer of gelatin over the glass slides. After incubation, wash plates with PBS to remove excess gelatin. Plates are then ready to use.

(b) After 24 h of culture, aspirate the medium and wash cells with HBSS to remove any dead cells. Replace the medium with MEM medium B and place the cells in the incubator.

(c) Treat NRVMs in different conditions as required. In Fig. 3, NRVMs are treated either with vehicle or 5 μM doxorubicin for 1 h.

Note: Doxorubicin, a well-known anticancer drug, is a redox cycler that localizes to mitochondria [21]. Doxorubicin-induced ROS production is known to trigger cardiac dilation, contractile dysfunction, and ultimately heart failure [22–25]. Here, we used doxorubicin as a ROS-inducing agent.

(d) While the cells are in treatment, dissolve one vial of MitoTracker Red CMH_2XRos in DMSO to make a 100 μM stock solution. Further dilute MitoTracker Red to 25 nM in HBSS and cover it with aluminum foil to protect it from the light.

Note: Always prepare fresh mixture of MitoTracker Red and HBSS. The optimal MitoTracker Red concentration for different cell types should be determined empirically. Using too high concentration of the dye can lead to non-specific staining and higher fluorescence background.

(e) After the treatment, aspirate the medium and incubate cells with 2 mL of HBSS + MitoTracker Red solution at 37°C for 30 min. MitoTracker Red will enter in the living cells and accumulate inside the respiring mitochondria.

(f) Aspirate the medium after 30 min and wash three to four times with HBSS to remove any excess dye. Leave the cells in HBSS medium for the rest of the experiment.

(g) Pick the glass slide with forceps and place it carefully in the holder (cells should be on the side facing up). Tighten the holder and add 1 mL HBSS on the top of the cells.

Note: Tightening the holder is a sensitive step, since tightening too much can break the glass slide, while leaving it loose can result in leaking of the medium.

(h) Fix the holder containing cells on the microscope stage, and set the temperature at 37 °C. Focus the cells and start capturing images of the different field of views (~15 fields per slide). Remember to use appropriate filters (excitation at 579 nm and emission at 599 nm). Cells loaded with MitoTracker Red will show mitochondria network in red color (Fig. 3a). Increase in ROS production will lead to an increase in fluorescence intensity, since the probe will be oxidized more, as shown in doxorubicin-treated cardiomyocytes compared to control cells (Fig. 3).

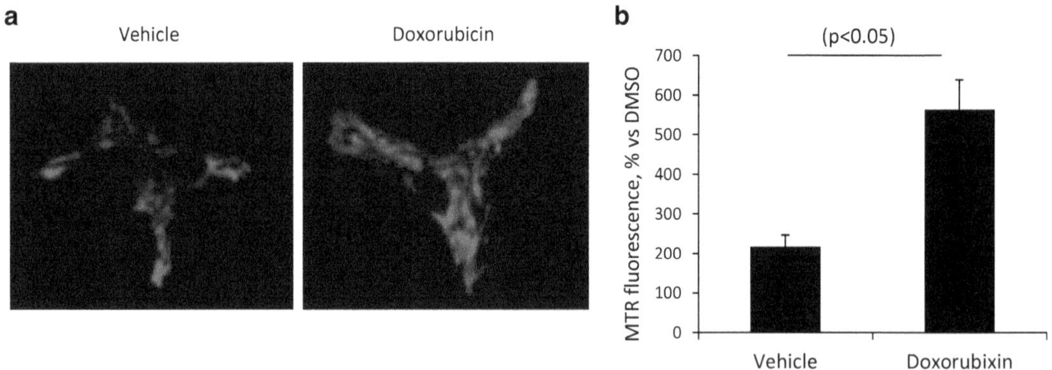

Fig. 3 Mitochondrial ROS measurement in neonatal cardiomyocytes. (**a**) NRVMs were treated either with vehicle or 5 μM doxorubicin for 1 h, and cells were loaded with fluorescent probe MitoTracker Red CM-H_2XRos for 30 min. Images were captured with inverted fluorescence microscope using 579 nm excitation and 599 nm emission, wavelengths. (**b**) Quantification of ROS levels in cardiomyocytes following doxorubicin treatment

Note: It is crucial to keep the exposure time, gain, lamp intensity, and all other settings the same for all the slides to be able to make the comparisons between the groups.

(i) To quantify the fluorescence intensity, Java-based image processing program ImageJ (NIH) can be used.
Fluorescence intensity analysis using ImageJ:

- Open the images in ImageJ.
- Click on the freehand selection tool, and select the area within the cell where fluorescence intensity is to be measured.
- Open ROI manager (Analyze>Tools>ROI manager), and click on "Add" to add the area in the list. Select several regions of interest in the image, and add all the areas in the list.

  Note: In each image, remember to select a background region where there are no cells.
- Press "measure", ROI manager will calculate the fluorescence intensity for all the selected regions. Remember to subtract the background fluorescence from each region of interest.

  Note: In ImageJ several parameters can be measured, for instance area, centroid, skewness, and many more. To quantify the fluorescence intensity, "mean gray value" should be measured. To set this parameter, click analyze>set measurements>check mean gray value box and press ok.
- Repeat the same steps for all the acquired images, and calculate the mean and standard deviation.

  Note: Doxorubicin has an intrinsic fluorescence. Therefore, to correctly interpret the data, doxorubicin autofluorescence intensity was subtracted from the MitoTracker Red fluorescence levels in doxorubicin-treated cells.

2.2.2 Genetically Encoded Fluorescent Probe-HyPer

As described above, fluorescent redox probes such as MitoSOX and MitoTracker dyes present limitations in terms of selectivity, sensitivity, and also localization. Therefore, genetically encoded biosensors have been developed to measure H_2O_2 or other species fluctuations in different compartments of living cells, such as roGFP [26], Orp1-roGFP2 [27], and HyPer ([26], Fig. 4a).

HyPer contains an H_2O_2-sensitive regulatory domain of *E. coli* transcription factor OxyR [28, 29], bound to a yellow fluorescent protein (cpYFP). OxyR can be oxidized by H_2O_2, leading to the formation of a disulfide bond between Cys-199 and Cys-208; this oxidative modification can be reversed by the activity of

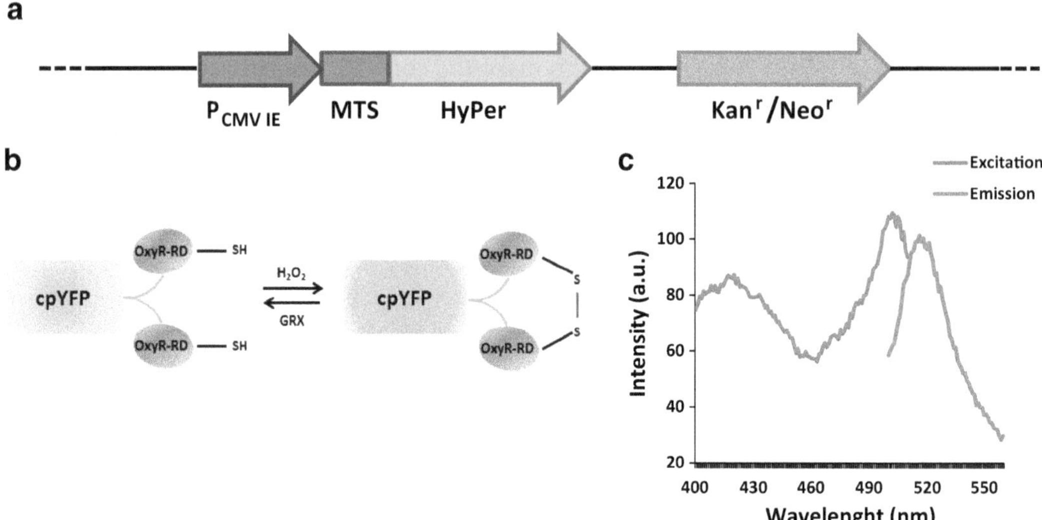

Fig. 4 pHyPer-dMito: structure and main properties. (**a**) Mammalian expression vector encoding mitochondria-targeted HyPer. Duplicated mitochondria targeting sequence (MTS) derived from the subunit VIII of human cytochrome C oxidase is fused with HyPer N-terminus. The vector backbone contains immediate early cytomegalovirus promoter ($P_{CMV\ IE}$) for protein expression. The vector has a neomycin resistance gene (Neo^r), to select stably transfected eukaryotic cells using G418 and kanamycin resistance gene expression (Kan^r) to select in *E. coli*. (**b**) Scheme of HyPer oxidation and reduction reactions. In presence of H_20_2, the two cysteines are oxidized, inducing an increase in the cpYFP fluorescence. Glutaredoxin (GRX) activity reverses the process. (**c**) HyPer excitation (blue line) and emission (green line) spectra. In the absence of H_20_2, HyPer has two excitation peaks with maxima at 420 nm and 500 nm and one emission peak with maximum at 516 nm. Upon exposure to H_20_2, the excitation peak at 420 nm decreases proportionally to the increase in the peak at 500 nm, allowing ratiometric measurement of H_20_2

endogenous glutaredoxins. The conformational change induced by the oxidation is transmitted to the cpYFP located between amino acids 205 and 206 [29] (Fig. 4b).

The major advantages of this genetically encoded redox sensor are that it's ratiometric and reversible and can be targeted to specific compartments of the cell. Several variants of HyPer are commercially available, such as cytosolic (pHyPer-cyto), mitochondrial (pHyPer-dMito), and nuclear (pHyPer-nuc) constructs. In addition, other versions of the plasmid targeted to endoplasmic reticulum, lysosomes, and different compartments within the mitochondria have been reported in the literature. Thus, HyPer is a useful tool to measure compartmentalized H_2O_2 formation [30].

HyPer has two excitation maxima (420/500 nm) and a single emission peak maximum (516 nm). Upon oxidation, the intensity of the 420 nm peak decreases proportionally to the increase of the intensity of the 500 nm peak, thus making HyPer a ratiometric sensor. An increase in H_2O_2 levels is directly proportional to the increase in fluorescence ratio F500/F420 [31] (Figs. 4c and 5b).

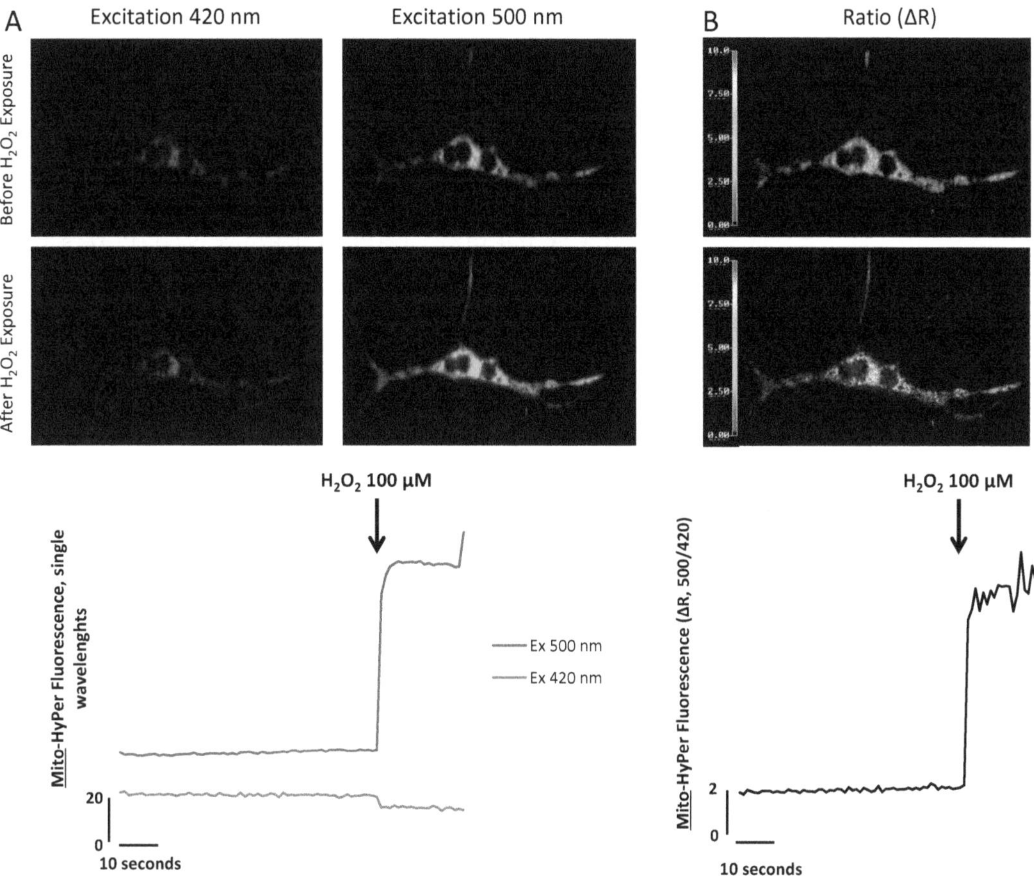

Fig. 5 Measurement of mitochondrial H_2O_2 in neonatal cardiomyocytes. NRVMs transfected with mitochondria-targeted HyPer were placed under a microscope, and images were captured using an external filter wheel composed by excitation filters CFP_{ex} (BP427/10) and YFP_{ex} (BP504/12) and a 535/30m T515lp emission filter. Baseline kinetics was followed for 60 s, and then 100 μM H_2O_2 was added to the cells. (**a**) The fluorescence in response to 420 nm excitation decreases, while the fluorescence in response to 500 nm excitation increases. (**b**) The ratio of the emission of the two wavelengths (F500/F420) is shown

Although HyPer is widely used to detect H_2O_2, its fluorescence levels can be influenced by pH, potentially leading to erroneous result interpretation. Hence, it is important to monitor pH in the same compartment and experimental conditions in which HyPer is being used. This can be accomplished using SypHer, a form of HyPer bearing a mutation in one of the two H_2O_2-sensing cysteines of the OxyR domain, making it H_2O_2 insensitive but pH sensitive sensor [6].

1. Materials
   (a) *Cells:* NRVMs were isolated as described in Subheading 2.2.1, **step 1**. Coat plates as described in Subheading 2.2.1, **step 2**.

(b) *Plasmid:* pHyPer-dMito (Evrogen, FP942).

(c) *Medium:* Use the same media described in Subheading 2.2.1, **step 1**.

(d) *Buffers:*

- 1× Hanks' balanced salt solution (HBSS, 136.9 mM NaCl, 5.36 mM KCl, 0.4 mM $MgSO_4$, 0.5 mM $MgCl_2$, 0.4 mM KH_2PO_4, 0.4 mM Na_2HPO_4, 4 mM $NaHCO_3$, 5 mM glucose, and 2 mM $CaCl_2$, pH 7.4).
- 2× HBS (274 mM NaCl, 10 mM KCl, 1.4 mM Na_2HPO_4, and 42 mM HEPES, pH 7.1).
- 1× PBS (134 mM NaCl, 2.7 mM KCl, 10 mM Na_2HPO_4, 1.8 mM KH_2PO_4, pH 7.4).
  pH of all buffers is adjusted with NaOH.

(e) *Microscope:* We use an inverted fluorescence microscope with the same features described in Subheading 2.2.1, **step 1**. We use an external filter wheel containing excitation filters for CFP_{ex} (BP427/10) and YFP_{ex} (BP504/12), and a 535/30m T515lp emission filter to detect the emission fluorescence.

(f) *Incubator:* As described in Subheading 2.2.1, **step 1**.

2. Methods

(a) Plate NRVMs at a density of 3×10^5 cells/well in MEM medium A in a 6-well plate containing gelatin-coated glass slides. Place NRVMs in the incubator for 24 h.

(b) After 24 h of culture, aspirate the medium from the cells and wash them with HBSS to remove any dead cells. Replace the medium with MEM medium B, place the cells in the incubator, and start preparing for the HyPer transfection.

(c) NRVMs are transfected using the calcium phosphate method. For each transfection, dilute pHyPer-dmito to 2 μg with sterile water, mix with ice-cold 0.25 M $CaCl_2$, add drop by drop to the 2× HBS buffer, and leave for 4 min to precipitate. Add this mixture to cells and incubate for 4 h. Rinse cells with PBS and add fresh MEM medium B.

Note: In order to obtain a good transfection, 2.5 M $CaCl_2$ must be stored at −20 °C, and the ratio between $CaCl_2$/DNA/HBS must be 0.1:0.9:1 (i.e., 10 μL $CaCl_2$:90 μL Water/DNA:100 μL HBS). The mixture $CaCl_2$/DNA must be added to HBS drop by drop while vortexing HBS. An important point is to wash the cells with PBS for 6–10 times in order to remove all the deposits of calcium that would otherwise be toxic for the cells.

(d) After 48 h from transfection, wash three to four times with HBSS to remove any dead cells. Pick the glass slide with forceps and place it in the holder carefully. Tighten the holder and add 1 mL HBSS on the top of cells.

(e) *Treatment:*

- *End Point Experiment:* Pretreat the cells in the conditions you would like to investigate. Place the slide under the microscope and set the temperature at 37°C. Focus the cells and capture images of the different fields of view. In order to obtain consistent data, capture at least five fields of view per slide.
- *Kinetics Experiment:* Place the slide under the microscope, set the temperature at 37°C, and focus the cells in the field of view of interest. Record continuously for at least 30 frames to obtain a baseline (i.e., 1 frame/1 s as shown in Fig. 5), and then add the acute treatment. Remember to use the correct filters based on the HyPer excitation/emission spectrum. Transfected cells will show mitochondria network in green color (Fig. 5a). In conditions in which H_2O_2 levels will increase, the fluorescence at 420 nm will decrease, while 500 nm will increase, as shown in Fig. 5a. In this experiment addition of 100 μM H_2O_2 led to an increase the ratio of F500/F420 (Fig. 5b).

  Note: The expression efficiency of HyPer may vary in different cells due to different expression levels of the sensor. This difference or other artifacts (i.e., mitochondrial movements or cell contractions) will not influence the result since HyPer is a ratiometric sensor.

(f) To quantify the fluorescence intensity, Java-based image processing program ImageJ (NIH) can be used.
Analysis using ImageJ program:

- Open the images in ImageJ.
- Click on the freehand selection tool and select the area within the cell where fluorescence intensity is to be measured.
- Open ROI manager (Analyze>Tools>ROI manager) and click on "Add" to add the area in the list (ROI, Region of Interest). Select several regions of interest in the image and add all the areas in the list.

  Note: In each image, remember to select a background region where there are no cells. You will have two sets of images, one deriving from the excitation at 420 nm and the other one deriving from the excitation at 500 nm. Do the same procedure for both sets and calculate the ratio. In order to select the same area, once the ROI list is done, click on ROI manager>More>Save to save the list. Open the second set of images and click on ROI manager>More>Open to load the list.

- Press "measure", ROI manager will calculate the fluorescence intensity for all the selected regions. Remember to subtract the background fluorescence from each region of interest.

  Note: In ImageJ several parameters could be measured, for instance, area, centroid, skewness, and many more. To quantify the fluorescence intensity, "mean gray value" should be measured. To set this parameter, click analyze>set measurements>check mean gray value box and press ok.

- Repeat the same steps for all the acquired images, and calculate the mean and standard deviation.

  Note: In kinetics experiments, click File>Import>Image Sequence to open all the files. In ROI manager, click on more>Multi Measure to calculate the fluorescence intensity for all the selected regions in all the images.

References

1. Egea J, Fabregat I, Frapart YM, Ghezzi P, Gorlach A, Kietzmann T, Kubaichuk K, Knaus UG, Lopez MG et al (2017) European contribution to the study of ROS: a summary of the findings and prospects for the future from the COST action BM1203 (EU-ROS). Redox Biol 13:94–162. https://doi.org/10.1016/j.redox.2017.05.007
2. Casas AI, Dao VT, Daiber A, Maghzal GJ, Di Lisa F, Kaludercic N, Leach S, Cuadrado A, Jaquet V, Seredenina T, Krause KH, Lopez MG, Stocker R, Ghezzi P, Schmidt HH (2015) Reactive oxygen-related diseases: therapeutic targets and emerging clinical indications. Antioxid Redox Signal 23 (14):1171–1185. https://doi.org/10.1089/ars.2015.6433
3. Murphy E, Ardehali H, Balaban RS, DiLisa F, Dorn GW 2nd, Kitsis RN, Otsu K, Ping P, Rizzuto R, Sack MN, Wallace D, Youle RJ (2016) Mitochondrial function, biology, and role in disease: a scientific statement from the American Heart Association. Circ Res 118 (12):1960–1991. https://doi.org/10.1161/RES.0000000000000104
4. Lukyanov KA, Belousov VV (2014) Genetically encoded fluorescent redox sensors. Biochim Biophys Acta 1840(2):745–756. https://doi.org/10.1016/j.bbagen.2013.05.030
5. Winterbourn CC (2014) The challenges of using fluorescent probes to detect and quantify specific reactive oxygen species in living cells. Biochim Biophys Acta 1840(2):730–738. https://doi.org/10.1016/j.bbagen.2013.05.004
6. Kaludercic N, Deshwal S, Di Lisa F (2014) Reactive oxygen species and redox compartmentalization. Front Physiol 5:285. https://doi.org/10.3389/fphys.2014.00285
7. Dikalov SI, Harrison DG (2014) Methods for detection of mitochondrial and cellular reactive oxygen species. Antioxid Redox Signal 20 (2):372–382. https://doi.org/10.1089/ars.2012.4886
8. Mailloux RJ (2015) Teaching the fundamentals of electron transfer reactions in mitochondria and the production and detection of reactive oxygen species. Redox Biol 4:381–398. https://doi.org/10.1016/j.redox.2015.02.001
9. Starkov AA (2010) Measurement of mitochondrial ROS production. Methods Mol Biol

648:245–255. https://doi.org/10.1007/978-1-60761-756-3_16

10. Zhao B, Summers FA, Mason RP (2012) Photooxidation of Amplex Red to resorufin: implications of exposing the Amplex Red assay to light. Free Radic Biol Med 53(5):1080–1087. https://doi.org/10.1016/j.freeradbiomed.2012.06.034
11. Deshwal S, Di Sante M, Di Lisa F, Kaludercic N (2017) Emerging role of monoamine oxidase as a therapeutic target for cardiovascular disease. Curr Opin Pharmacol 33:64–69. https://doi.org/10.1016/j.coph.2017.04.003
12. Kaludercic N, Takimoto E, Nagayama T, Feng N, Lai EW, Bedja D, Chen K, Gabrielson KL, Blakely RD, Shih JC, Pacak K, Kass DA, Di Lisa F, Paolocci N (2010) Monoamine oxidase A-mediated enhanced catabolism of norepinephrine contributes to adverse remodeling and pump failure in hearts with pressure overload. Circ Res 106(1):193–202. https://doi.org/10.1161/CIRCRESAHA.109.198366
13. Kaludercic N, Carpi A, Nagayama T, Sivakumaran V, Zhu G, Lai EW, Bedja D, De Mario A, Chen K, Gabrielson KL, Lindsey ML, Pacak K, Takimoto E, Shih JC, Kass DA, Di Lisa F, Paolocci N (2014) Monoamine oxidase B prompts mitochondrial and cardiac dysfunction in pressure overloaded hearts. Antioxid Redox Signal 20(2):267–280. https://doi.org/10.1089/ars.2012.4616
14. Di Lisa F, Menabo R, Canton M, Barile M, Bernardi P (2001) Opening of the mitochondrial permeability transition pore causes depletion of mitochondrial and cytosolic NAD+ and is a causative event in the death of myocytes in postischemic reperfusion of the heart. J Biol Chem 276(4):2571–2575. https://doi.org/10.1074/jbc.M006825200
15. Zhang H, Goodman HM, Jansson S (1997) Antisense inhibition of the photosystem I antenna protein Lhca4 in Arabidopsis thaliana. Plant Physiol 115(4):1525–1531
16. Cottet-Rousselle C, Ronot X, Leverve X, Mayol JF (2011) Cytometric assessment of mitochondria using fluorescent probes. Cytometry A 79(6):405–425. https://doi.org/10.1002/cyto.a.21061
17. Zielonka J, Kalyanaraman B (2010) Hydroethidine- and MitoSOX-derived red fluorescence is not a reliable indicator of intracellular superoxide formation: another inconvenient truth. Free Radic Biol Med 48(8):983–1001. https://doi.org/10.1016/j.freeradbiomed.2010.01.028
18. Kalyanaraman B, Darley-Usmar V, Davies KJ, Dennery PA, Forman HJ, Grisham MB, Mann GE, Moore K, Roberts LJ 2nd, Ischiropoulos H (2012) Measuring reactive oxygen and nitrogen species with fluorescent probes: challenges and limitations. Free Radic Biol Med 52(1):1–6. https://doi.org/10.1016/j.freeradbiomed.2011.09.030
19. Robinson KM, Janes MS, Pehar M, Monette JS, Ross MF, Hagen TM, Murphy MP, Beckman JS (2006) Selective fluorescent imaging of superoxide in vivo using ethidium-based probes. Proc Natl Acad Sci U S A 103(41):15038–15043. https://doi.org/10.1073/pnas.0601945103
20. Hsieh CW, Chu CH, Lee HM, Yuan Yang W (2015) Triggering mitophagy with far-red fluorescent photosensitizers. Sci Rep 5:10376. https://doi.org/10.1038/srep10376
21. Kuznetsov AV, Margreiter R, Amberger A, Saks V, Grimm M (2011) Changes in mitochondrial redox state, membrane potential and calcium precede mitochondrial dysfunction in doxorubicin-induced cell death. Biochim Biophys Acta 1813(6):1144–1152. https://doi.org/10.1016/j.bbamcr.2011.03.002
22. Tocchetti CG, Ragone G, Coppola C, Rea D, Piscopo G, Scala S, De Lorenzo C, Iaffaioli RV, Arra C, Maurea N (2012) Detection, monitoring, and management of trastuzumab-induced left ventricular dysfunction: an actual challenge. Eur J Heart Fail 14(2):130–137. https://doi.org/10.1093/eurjhf/hfr165
23. Tocchetti CG, Carpi A, Coppola C, Quintavalle C, Rea D, Campesan M, Arcari A, Piscopo G, Cipresso C, Monti MG, De Lorenzo C, Arra C, Condorelli G, Di Lisa F, Maurea N (2014) Ranolazine protects from doxorubicin-induced oxidative stress and cardiac dysfunction. Eur J Heart Fail 16(4):358–366. https://doi.org/10.1002/ejhf.50
24. Kim SY, Kim SJ, Kim BJ, Rah SY, Chung SM, Im MJ, Kim UH (2006) Doxorubicin-induced reactive oxygen species generation and intracellular Ca2+ increase are reciprocally modulated in rat cardiomyocytes. Exp Mol Med 38(5):535–545. https://doi.org/10.1038/emm.2006.63
25. Khouri MG, Douglas PS, Mackey JR, Martin M, Scott JM, Scherrer-Crosbie M, Jones LW (2012) Cancer therapy-induced cardiac toxicity in early breast cancer: addressing the unresolved issues. Circulation 126(23):2749–2763. https://doi.org/10.1161/CIRCULATIONAHA.112.100560

26. Dooley CT, Dore TM, Hanson GT, Jackson WC, Remington SJ, Tsien RY (2004) Imaging dynamic redox changes in mammalian cells with green fluorescent protein indicators. J Biol Chem 279(21):22284–22293. https://doi.org/10.1074/jbc.M312847200
27. Albrecht SC, Barata AG, Grosshans J, Teleman AA, Dick TP (2011) In vivo mapping of hydrogen peroxide and oxidized glutathione reveals chemical and regional specificity of redox homeostasis. Cell Metab 14(6):819–829. https://doi.org/10.1016/j.cmet.2011.10.010
28. Belousov VV, Fradkov AF, Lukyanov KA, Staroverov DB, Shakhbazov KS, Terskikh AV, Lukyanov S (2006) Genetically encoded fluorescent indicator for intracellular hydrogen peroxide. Nat Methods 3(4):281–286. https://doi.org/10.1038/nmeth866
29. Choi H, Kim S, Mukhopadhyay P, Cho S, Woo J, Storz G, Ryu SE (2001) Structural basis of the redox switch in the OxyR transcription factor. Cell 105(1):103–113
30. Malinouski M, Zhou Y, Belousov VV, Hatfield DL, Gladyshev VN (2011) Hydrogen peroxide probes directed to different cellular compartments. PLoS One 6(1):e14564. https://doi.org/10.1371/journal.pone.0014564
31. Bilan DS, Belousov VV (2016) HyPer family probes: state of the art. Antioxid Redox Signal 24(13):731–751. https://doi.org/10.1089/ars.2015.6586

INDEX

A

B

C

D

E

Carlos M. Palmeira and António J. Moreno (eds.), *Mitochondrial Bioenergetics: Methods and Protocols*, Methods in Molecular Biology, vol. 1782, https://doi.org/10.1007/978-1-4939-7831-1,

F

G

H

I

J

K

L

M

CPSIA information can be obtained
at www.ICGtesting.com
Printed in the USA
LVHW05s1524030618
579414LV00003B/33/P

9 781493 97830